Lessons In Industrial Instrumentation 3/3

A catalogue record for this book is available from the Hong Kong Public Libraries.

Published in Hong Kong by Samurai Media Limited.

Email: info@samuraimedia.org

ISBN 978-988-8407-10-1

Contents

Preface   1

1 Calculus   5
   1.1 Introduction to calculus   6
   1.2 The concept of differentiation   9
   1.3 The concept of integration   14
   1.4 How derivatives and integrals relate to one another   23
   1.5 Symbolic versus numerical calculus   27
   1.6 Numerical differentiation   31
   1.7 Numerical integration   41

2 Physics   55
   2.1 Terms and Definitions   56
   2.2 Metric prefixes   57
   2.3 Areas and volumes   57
        2.3.1 Common geometric shapes   58
   2.4 Unit conversions and physical constants   60
        2.4.1   Unity fractions   61
        2.4.2   Conversion formulae for temperature   64
        2.4.3   Conversion factors for distance   65
        2.4.4   Conversion factors for volume   65
        2.4.5   Conversion factors for velocity   65
        2.4.6   Conversion factors for mass   65
        2.4.7   Conversion factors for force   65
        2.4.8   Conversion factors for area   65
        2.4.9   Conversion factors for pressure (either all gauge or all absolute) . . .   66
        2.4.10 Conversion factors for pressure (absolute pressure units only)   66
        2.4.11 Conversion factors for energy or work   66
        2.4.12 Conversion factors for power   66
        2.4.13 Terrestrial constants   67
        2.4.14 Properties of water   67
        2.4.15 Miscellaneous physical constants   68
        2.4.16 Weight densities of common materials   69
   2.5 Dimensional analysis   71

2.6 The International System of Units 72

2.7 Conservation Laws 73

2.8 Classical mechanics 73

    2.8.1 Newton's Laws of Motion 74

    2.8.2 Work, energy, and power 75

    2.8.3 Mechanical springs 91

    2.8.4 Rotational motion 93

2.9 Simple machines 99

    2.9.1 Levers 99

    2.9.2 Pulleys 101

    2.9.3 Inclined planes 104

    2.9.4 Gears 106

    2.9.5 Belt drives 117

    2.9.6 Chain drives 120

2.10 Elementary thermodynamics 121

    2.10.1 Heat versus Temperature 122

    2.10.2 Temperature 123

    2.10.3 Heat 125

    2.10.4 Heat transfer 126

    2.10.5 Specific heat and enthalpy 137

    2.10.6 Phase changes 144

    2.10.7 Phase diagrams and critical points 152

    2.10.8 Saturated steam table 155

    2.10.9 Thermodynamic degrees of freedom 158

    2.10.10 Applications of phase changes 159

2.11 Fluid mechanics 170

    2.11.1 Pressure 170

    2.11.2 Pascal's Principle and hydrostatic pressure 175

    2.11.3 Fluid density expressions 180

    2.11.4 Manometers 182

    2.11.5 Systems of pressure measurement 186

    2.11.6 Negative pressure 189

    2.11.7 Buoyancy 191

    2.11.8 Gas Laws 197

    2.11.9 Fluid viscosity 199

    2.11.10 Reynolds number 201

    2.11.11 Law of Continuity 205

    2.11.12 Viscous flow 207

    2.11.13 Bernoulli's equation 208

    2.11.14 Torricelli's equation 218

    2.11.15 Flow through a venturi tube 219

3 Chemistry   **225**
  3.1   Terms and Definitions 228
  3.2   Atomic theory and chemical symbols 230
  3.3   Periodic table of the elements 236
  3.4   Electronic structure 241
  3.5   Spectroscopy 249
      3.5.1   Emission spectroscopy 253
      3.5.2   Absorption spectroscopy 256
  3.6   Formulae for common chemical compounds 258
  3.7   Molecular quantities 262
  3.8   Stoichiometry 265
      3.8.1   Balancing chemical equations by trial-and-error 266
      3.8.2   Balancing chemical equations using algebra 268
      3.8.3   Stoichiometric ratios 271
  3.9   Energy in chemical reactions 273
      3.9.1   Heats of reaction and activation energy 274
      3.9.2   Heats of formation and Hess's Law 278
  3.10 Periodic table of the ions 281
  3.11 Ions in liquid solutions 282
  3.12 pH 284

4 DC electricity   **291**
  4.1   Electrical voltage 292
  4.2   Electrical current 298
      4.2.1   Electron versus conventional flow 301
  4.3   Electrical sources and loads 306
  4.4   Electrical power 312
  4.5   Electrical resistance and Ohm's Law 313
  4.6   Series versus parallel circuits 316
  4.7   Kirchhoff's Laws 320
  4.8   Circuit fault analysis 324
  4.9   Bridge circuits 327
      4.9.1   Component measurement 328
      4.9.2   Sensor signal conditioning 330
  4.10 Null-balance voltage measurement 336
  4.11 Electromagnetism 340
  4.12 Capacitors 346
  4.13 Inductors 350

5 AC electricity   **355**
  5.1   RMS quantities 356
  5.2   Resistance, Reactance, and Impedance 360
  5.3   Series and parallel circuits 361
  5.4   Transformers 361
      5.4.1   Basic principles 362
      5.4.2   Loading effects 365

    5.4.3   Step ratios 367
  5.5   Phasors 369
    5.5.1   Circles, sine waves, and cosine waves 370
    5.5.2   Phasor expressions of phase shifts 375
    5.5.3   Phasor expressions of impedance 384
    5.5.4   Phasor arithmetic 388
    5.5.5   Phasors and circuit measurements 392
  5.6   The s variable 400
    5.6.1   Meaning of the s variable 401
    5.6.2   Impedance expressed using the s variable 404
  5.7   Transfer function analysis 409
    5.7.1   Example: LR low-pass filter circuit 410
    5.7.2   Example: RC high-pass filter circuit 416
    5.7.3   Example: LC "tank" circuit 421
    5.7.4   Example: RLC band-pass filter circuit 426
    5.7.5   Summary of transfer function analysis 435
  5.8   Polyphase AC power 437
    5.8.1   Delta and Wye configurations 445
    5.8.2   Power in three-phase circuits 449
    5.8.3   Grounded three-phase circuits 452
    5.8.4   Symmetrical components 455
  5.9   Phasor analysis of transformer circuits 464
    5.9.1   Phasors in single-phase transformer circuits 466
    5.9.2   Phasors in three-phase transformer circuits 472
  5.10 Transmission lines 473
    5.10.1 Open-ended transmission lines 475
    5.10.2 Shorted transmission lines 477
    5.10.3 Properly terminated transmission lines 479
    5.10.4 Discontinuities 481
    5.10.5 Velocity factor 481
    5.10.6 Cable losses 481
  5.11 Antennas 482
    5.11.1 Maxwell and Hertz 486
    5.11.2 Antenna size 488
    5.11.3 Antenna orientation and directionality 489

6 Introduction to industrial instrumentation   495
  6.1   Example: boiler water level control system 498
  6.2   Example: wastewater disinfection 503
  6.3   Example: chemical reactor temperature control 505
  6.4   Other types of instruments 508
    6.4.1   Indicators 509
    6.4.2   Recorders 512
    6.4.3   Process switches and alarms 515
  6.5   Summary 524
  6.6   Review of fundamental principles 525

7 Instrumentation documents   **527**

7.1   Process Flow Diagrams 529

7.2   Process and Instrument Diagrams 531

7.3   Loop diagrams 533

7.4   Functional diagrams 536

7.5   Instrument and process equipment symbols 539

     7.5.1   Line types 540

     7.5.2   Process/Instrument line connections 541

     7.5.3   Instrument bubbles 541

     7.5.4   Process valve types 542

     7.5.5   Valve actuator types 543

     7.5.6   Valve failure mode 544

     7.5.7   Liquid level measurement devices 545

     7.5.8   Flow measurement devices (flowing left-to-right) 546

     7.5.9   Process equipment 548

     7.5.10   Functional diagram symbols 549

     7.5.11   Single-line electrical diagram symbols 550

     7.5.12   Fluid power diagram symbols 552

7.6   Instrument identification tags 554

8 Instrument connections   **559**

8.1   Pipe and pipe fittings 559

     8.1.1   Flanged pipe fittings 560

     8.1.2   Tapered thread pipe fittings 567

     8.1.3   Parallel thread pipe fittings 570

     8.1.4   Sanitary pipe fittings 571

8.2   Tube and tube fittings 575

     8.2.1   Compression tube fittings 576

     8.2.2   Common tube fitting types and names 581

     8.2.3   Bending instrument tubing 584

     8.2.4   Special tubing tools 586

8.3   Electrical signal and control wiring 588

     8.3.1   Connections and wire terminations 589

     8.3.2   DIN rail 598

     8.3.3   Cable routing 601

     8.3.4   Signal coupling and cable separation 609

     8.3.5   Electric field (capacitive) de-coupling 613

     8.3.6   Magnetic field (inductive) de-coupling 619

     8.3.7   High-frequency signal cables 622

8.4   Fiber optics 623

     8.4.1   Fiber optic data communication 624

     8.4.2   Fiber optic sensing applications 627

     8.4.3   Fiber optic cable construction 632

     8.4.4   Multi-mode and single-mode optical fibers 635

     8.4.5   Fiber optic cable connectors, routing, and safety 637

     8.4.6   Fiber optic cable testing 640

8.5   Review of fundamental principles 646

9   Discrete process measurement   649
9.1   "Normal" status of a switch 650
9.2   Hand switches 655
9.3   Limit switches 657
9.4   Proximity switches 659
9.5   Pressure switches 664
9.6   Level switches 669
    9.6.1   Float-type level switches 670
    9.6.2   Tuning fork level switches 672
    9.6.3   Paddle-wheel level switches 673
    9.6.4   Ultrasonic level switches 674
    9.6.5   Capacitive level switches 675
    9.6.6   Conductive level switches 676
9.7   Temperature switches 678
9.8   Flow switches 682
9.9   Review of fundamental principles 684

10 Discrete control elements   685
10.1 On/off valves 686
10.2 Fluid power systems 688
10.3 Solenoid valves 698
    10.3.1 2-way solenoid valves 699
    10.3.2 3-way solenoid valves 702
    10.3.3 4-way solenoid valves 706
    10.3.4 Normal energization states 711
10.4 On/off electric motor control circuits 714
    10.4.1 AC induction motors 715
    10.4.2 Motor contactors 724
    10.4.3 Motor protection 726
    10.4.4 Motor control circuit wiring 734
10.5 Review of fundamental principles 741

11 Relay control systems   743
11.1 Control relays 745
11.2 Relay circuits 749
11.3 Interposing relays 756
11.4 Review of fundamental principles 759

12 Programmable Logic Controllers   761
12.1 PLC examples 762
12.2 Input/Output (I/O) capabilities 771
    12.2.1 Discrete I/O 773
    12.2.2 Analog I/O 780
    12.2.3 Network I/O 782

12.3 Logic programming 783
    12.3.1 Relating I/O status to virtual elements 784
    12.3.2 Memory maps and I/O addressing 792
12.4 Ladder Diagram (LD) programming 798
    12.4.1 Contacts and coils 800
    12.4.2 Counters 818
    12.4.3 Timers 823
    12.4.4 Data comparison instructions 828
    12.4.5 Math instructions 831
    12.4.6 Sequencers 834
12.5 Structured Text (ST) programming 844
12.6 Instruction List (IL) programming 844
12.7 Function Block Diagram (FBD) programming 844
12.8 Sequential Function Chart (SFC) programming 844
12.9 Human-Machine Interfaces 845
12.10 How to teach yourself PLC programming 851
12.11 Review of fundamental principles 854

13 Analog electronic instrumentation   **857**
13.1 4 to 20 mA analog current signals 857
13.2 Relating 4 to 20 mA signals to instrument variables 861
    13.2.1 Example calculation: controller output to valve 864
    13.2.2 Example calculation: flow transmitter 865
    13.2.3 Example calculation: temperature transmitter 867
    13.2.4 Example calculation: pH transmitter 870
    13.2.5 Example calculation: reverse-acting I/P transducer signal 872
    13.2.6 Example calculation: PLC analog input scaling 874
    13.2.7 Graphical interpretation of signal ranges 877
    13.2.8 Thinking in terms of per unit quantities 879
13.3 Controller output current loops 882
13.4 4-wire ("self-powered") transmitter current loops 885
13.5 2-wire ("loop-powered") transmitter current loops 887
13.6 4-wire "passive" versus "active" output transmitters 889
13.7 Troubleshooting current loops 890
    13.7.1 Using a standard milliammeter to measure loop current 892
    13.7.2 Using a clamp-on milliammeter to measure loop current 894
    13.7.3 Using "test" diodes to measure loop current 895
    13.7.4 Using shunt resistors to measure loop current 897
    13.7.5 Troubleshooting current loops with voltage measurements 898
    13.7.6 Using loop calibrators 902
    13.7.7 NAMUR signal levels 909
13.8 Review of fundamental principles 910

14 Pneumatic instrumentation   **911**
  14.1 Pneumatic sensing elements 917
  14.2 Self-balancing pneumatic instrument principles 919
  14.3 Pilot valves and pneumatic amplifying relays 924
  14.4 Analogy to opamp circuits 934
  14.5 Analysis of practical pneumatic instruments 946
      14.5.1 Foxboro model 13A differential pressure transmitter 947
      14.5.2 Foxboro model E69 "I/P" electro-pneumatic transducer 952
      14.5.3 Fisher model 546 "I/P" electro-pneumatic transducer 957
      14.5.4 Fisher-Rosemount model 846 "I/P" electro-pneumatic transducer 962
  14.6 Proper care and feeding of pneumatic instruments 965
  14.7 Advantages and disadvantages of pneumatic instruments 966
  14.8 Review of fundamental principles 967

15 Digital data acquisition and networks   **969**
  15.1 Digital representation of numerical data 973
      15.1.1 Integer number formats 974
      15.1.2 Fixed-point number formats 976
      15.1.3 Floating-point number formats 977
      15.1.4 Example of industrial number formats 979
  15.2 Digital representation of text 982
      15.2.1 Morse and Baudot codes 983
      15.2.2 EBCDIC and ASCII 984
      15.2.3 Unicode 986
  15.3 Analog-digital conversion 986
      15.3.1 Converter resolution 987
      15.3.2 Converter sampling rate and aliasing 991
  15.4 Analog signal conditioning and referencing 994
      15.4.1 Instrumentation amplifiers 995
      15.4.2 Analog input references and connections 1004
  15.5 Digital data communication theory 1017
      15.5.1 Serial communication principles 1019
      15.5.2 Physical encoding of bits 1022
      15.5.3 Communication speed 1025
      15.5.4 Data frames 1027
      15.5.5 Channel arbitration 1035
      15.5.6 The OSI Reference Model 1041
  15.6 EIA/TIA-232, 422, and 485 networks 1044
      15.6.1 EIA/TIA-232 1045
      15.6.2 EIA/TIA-422 and EIA/TIA-485 1049
  15.7 Ethernet networks 1056
      15.7.1 Repeaters (hubs) 1057
      15.7.2 Ethernet cabling 1060
      15.7.3 Switching hubs 1064
  15.8 Internet Protocol (IP) 1066
      15.8.1 IP addresses 1068

15.8.2 Subnetworks and subnet masks 1072

15.8.3 Routing tables 1076

15.8.4 IP version 6 1077

15.8.5 ARP 1078

15.8.6 DNS 1079

15.8.7 Command-line diagnostic utilities 1079

15.9 Transmission Control Protocol (TCP) and User Datagram Protocol (UDP) 1083

15.10 The HART digital/analog hybrid standard 1086

15.10.1 Basic concept of HART 1087

15.10.2 HART physical layer 1092

15.10.3 HART multidrop mode 1099

15.10.4 HART multi-variable transmitters and burst mode 1100

15.11 Modbus 1101

15.11.1 Modbus overview 1102

15.11.2 Modbus data frames 1106

15.11.3 Modbus function codes and addresses 1108

15.11.4 Modbus relative addressing 1110

15.11.5 Modbus function command formats 1113

15.11.6 Floating-point values in Modbus 1121

15.12 Review of fundamental principles 1124

16 FOUNDATION Fieldbus instrumentation **1129**

16.1 FF design philosophy 1130

16.2 H1 FF Physical layer 1136

16.2.1 Segment topology 1137

16.2.2 Coupling devices 1141

16.2.3 Electrical parameters 1144

16.2.4 Cable types 1146

16.2.5 Segment design 1149

16.3 H1 FF Data Link layer 1151

16.3.1 Device addressing 1153

16.3.2 Communication management 1154

16.3.3 Device capability 1163

16.4 FF function blocks 1163

16.4.1 Analog function blocks versus digital function blocks 1164

16.4.2 Function block location 1165

16.4.3 Standard function blocks 1170

16.4.4 Device-specific function blocks 1172

16.4.5 FF signal status 1173

16.4.6 Function block modes 1175

16.5 H1 FF device configuration and commissioning 1176

16.5.1 Configuration files 1177

16.5.2 Device commissioning 1179

16.5.3 Calibration and ranging 1188

16.6 H1 FF segment troubleshooting 1196

16.6.1 Cable resistance 1196

     16.6.2 Signal strength 1197

     16.6.3 Electrical noise 1197

     16.6.4 Using an oscilloscope on H1 segments 1198

     16.6.5 Message re-transmissions 1200

  16.7 Review of fundamental principles 1200

17 Wireless instrumentation   **1203**

  17.1 Radio systems 1203

     17.1.1 Antennas 1204

     17.1.2 Decibels 1206

     17.1.3 Antenna radiation patterns 1213

     17.1.4 Antenna gain calculations 1216

     17.1.5 Effective radiated power 1218

     17.1.6 RF link budget 1220

     17.1.7 Fresnel zones 1226

  17.2 *Wireless*HART 1228

     17.2.1 Introduction to *Wireless*HART 1229

     17.2.2 *Wireless*HART network protocol 1232

     17.2.3 *Wireless*HART network gateway device 1236

     17.2.4 WirelessHART device commissioning and configuration 1241

  17.3 Review of fundamental principles 1245

18 Instrument calibration   **1247**

  18.1 Calibration versus re-ranging 1247

  18.2 Zero and span adjustments (analog instruments) 1248

  18.3 Calibration errors and testing 1250

     18.3.1 Typical calibration errors 1251

     18.3.2 As-found and as-left documentation 1256

     18.3.3 Up-tests and Down-tests 1257

     18.3.4 Automated calibration 1258

  18.4 Damping adjustments 1261

  18.5 LRV and URV settings, digital trim (digital transmitters) 1265

  18.6 An analogy for calibration versus ranging 1271

  18.7 Calibration procedures 1272

     18.7.1 Linear instruments 1274

     18.7.2 Nonlinear instruments 1275

     18.7.3 Discrete instruments 1276

  18.8 Instrument turndown 1277

  18.9 NIST traceability 1278

  18.10 Practical calibration standards 1281

     18.10.1 Electrical standards 1283

     18.10.2 Temperature standards 1285

     18.10.3 Pressure standards 1291

     18.10.4 Flow standards 1296

     18.10.5 Analytical standards 1297

  18.11 Review of fundamental principles 1301

19 Continuous pressure measurement   **1303**

  19.1 Manometers 1304

  19.2 Mechanical pressure elements 1310

  19.3 Electrical pressure elements 1316

      19.3.1 Piezoresistive (strain gauge) sensors 1317

      19.3.2 Differential capacitance sensors 1322

      19.3.3 Resonant element sensors 1330

      19.3.4 Mechanical adaptations 1333

  19.4 Force-balance pressure transmitters 1334

  19.5 Differential pressure transmitters 1337

      19.5.1 DP transmitter construction and behavior 1338

      19.5.2 DP transmitter applications 1345

      19.5.3 Inferential measurement applications 1351

  19.6 Pressure sensor accessories 1354

      19.6.1 Valve manifolds 1355

      19.6.2 Bleed (vent) fittings 1361

      19.6.3 Pressure pulsation damping 1364

      19.6.4 Remote and chemical seals 1367

      19.6.5 Filled impulse lines 1377

      19.6.6 Purged impulse lines 1379

      19.6.7 Heat-traced impulse lines 1383

      19.6.8 Water traps and pigtail siphons 1386

      19.6.9 Mounting brackets 1388

      19.6.10 Heated enclosures 1389

  19.7 Process/instrument suitability 1391

  19.8 Review of fundamental principles 1393

20 Continuous level measurement   **1395**

  20.1 Level gauges (sightglasses) 1395

      20.1.1 Basic concepts of sightglasses 1396

      20.1.2 Interface problems 1399

      20.1.3 Temperature problems 1401

  20.2 Float 1402

  20.3 Hydrostatic pressure 1408

      20.3.1 Pressure of a fluid column 1409

      20.3.2 Bubbler systems 1414

      20.3.3 Transmitter suppression and elevation 1418

      20.3.4 Compensated leg systems 1422

      20.3.5 Tank expert systems 1430

      20.3.6 Hydrostatic interface level measurement 1435

  20.4 Displacement 1441

      20.4.1 Buoyant-force instruments 1442

      20.4.2 Torque tubes 1447

      20.4.3 Displacement interface level measurement 1454

  20.5 Echo 1458

      20.5.1 Ultrasonic level measurement 1460

20.5.2   Radar level measurement 1465

20.5.3   Laser level measurement 1477

20.5.4   Magnetostrictive level measurement 1478

20.6   Weight 1482

20.7   Capacitive 1487

20.8   Radiation 1489

20.9   Level sensor accessories 1494

20.10 Review of fundamental principles 1498

21 Continuous temperature measurement   1501

21.1   Bi-metal temperature sensors 1503

21.2   Filled-bulb temperature sensors 1505

21.3   Thermistors and Resistance Temperature Detectors (RTDs) 1509

21.3.1   Temperature coefficient of resistance (α) 1510

21.3.2   Two-wire RTD circuits 1512

21.3.3   Four-wire RTD circuits 1513

21.3.4   Three-wire RTD circuits 1514

21.3.5   Proper RTD sensor connections 1517

21.3.6   Self-heating error 1520

21.4   Thermocouples 1521

21.4.1   Dissimilar metal junctions 1522

21.4.2   Thermocouple types 1524

21.4.3   Connector and tip styles 1525

21.4.4   Manually interpreting thermocouple voltages 1529

21.4.5   Reference junction compensation 1531

21.4.6   Law of Intermediate Metals 1535

21.4.7   Software compensation 1539

21.4.8   Extension wire 1541

21.4.9   Side-effects of reference junction compensation 1546

21.4.10 Burnout detection 1552

21.5   Non-contact temperature sensors 1553

21.5.1   Concentrating pyrometers 1554

21.5.2   Distance considerations 1557

21.5.3   Emissivity 1562

21.5.4   Thermal imaging 1562

21.6   Temperature sensor accessories 1566

21.7   Process/instrument suitability 1570

21.8   Review of fundamental principles 1571

22 Continuous fluid flow measurement   1573

22.1   Pressure-based flowmeters 1574

22.1.1   Venturi tubes and basic principles 1580

22.1.2   Volumetric flow calculations 1586

22.1.3   Mass flow calculations 1589

22.1.4   Square-root characterization 1592

22.1.5   Orifice plates 1601

22.1.6 Other differential producers 1617

22.1.7 Proper installation 1625

22.1.8 High-accuracy flow measurement 1630

22.1.9 Equation summary 1637

22.2 Laminar flowmeters 1640

22.3 Variable-area flowmeters 1641

22.3.1 Rotameters 1642

22.3.2 Weirs and flumes 1645

22.4 Velocity-based flowmeters 1653

22.4.1 Turbine flowmeters 1654

22.4.2 Vortex flowmeters 1663

22.4.3 Magnetic flowmeters 1667

22.4.4 Ultrasonic flowmeters 1679

22.4.5 Optical flowmeters 1687

22.5 Positive displacement flowmeters 1691

22.6 Standardized volumetric flow 1694

22.7 True mass flowmeters 1700

22.7.1 Coriolis flowmeters 1703

22.7.2 Thermal flowmeters 1718

22.8 Weighfeeders 1722

22.9 Change-of-quantity flow measurement 1725

22.10 Insertion flowmeters 1728

22.11 Process/instrument suitability 1734

22.12 Review of fundamental principles 1735

23 Continuous analytical measurement **1741**

23.1 Conductivity measurement 1743

23.1.1 Dissociation and ionization in aqueous solutions 1744

23.1.2 Two-electrode conductivity probes 1745

23.1.3 Four-electrode conductivity probes 1748

23.1.4 Electrodeless conductivity probes 1751

23.2 pH measurement 1753

23.2.1 Colorimetric pH measurement 1753

23.2.2 Potentiometric pH measurement 1755

23.3 Chromatography 1776

23.3.1 Manual chromatography methods 1777

23.3.2 Automated chromatographs 1778

23.3.3 Species identification 1780

23.3.4 Chromatograph detectors 1781

23.3.5 Measuring species concentration 1783

23.3.6 Industrial applications of chromatographs 1786

23.3.7 Chromatograph sample valves 1789

23.3.8 Improving chromatograph analysis time 1792

23.4 Introduction to optical analyses 1797

23.5 Dispersive spectroscopy 1804

23.6 Non-dispersive Luft detector spectroscopy 1807

23.6.1 Single-beam analyzer 1810
23.6.2 Dual-beam analyzer 1811
23.6.3 Luft detectors 1814
23.6.4 Filter cells 1818
23.7 Gas Filter Correlation (GFC) spectroscopy 1821
23.8 Fluorescence 1826
23.9 Chemiluminescence 1835
23.10 Analyzer sample systems 1839
23.11 Safety gas analyzers 1844
23.11.1 Oxygen gas 1848
23.11.2 Lower explosive limit (LEL) 1849
23.11.3 Hydrogen sulfide gas 1850
23.11.4 Carbon monoxide gas 1851
23.11.5 Chlorine gas 1852
23.12 Review of fundamental principles 1854

24 Machine vibration measurement   **1859**
24.1 Vibration physics 1859
24.1.1 Sinusoidal vibrations 1860
24.1.2 Non-sinusoidal vibrations 1865
24.2 Vibration sensors 1871
24.3 Monitoring hardware 1876
24.4 Mechanical vibration switches 1879
24.5 Review of fundamental principles 1880

25 Electric power measurement and control   **1883**
25.1 Introduction to power system automation 1884
25.2 Electrical power grids 1896
25.3 Interconnected generators 1899
25.4 Single-line electrical diagrams 1903
25.5 Circuit breakers and disconnects 1909
25.5.1 Low-voltage circuit breakers 1911
25.5.2 Medium-voltage circuit breakers 1914
25.5.3 High-voltage circuit breakers 1921
25.5.4 Reclosers 1929
25.6 Electrical sensors 1931
25.6.1 Potential transformers 1932
25.6.2 Current transformers 1936
25.6.3 Transformer polarity 1943
25.6.4 Instrument transformer safety 1950
25.6.5 Instrument transformer test switches 1952
25.6.6 Instrument transformer burden and accuracy 1960
25.7 Introduction to protective relaying 1972
25.8 ANSI/IEEE function number codes 1979
25.9 Instantaneous and time-overcurrent (50/51) protection 1983
25.10 Differential (87) current protection 1991

25.11 Directional overcurrent (67) protection 2010

25.12 Distance (21) protection 2012

    25.12.1 Zone overreach and underreach 2013

    25.12.2 Line impedance characteristics 2016

    25.12.3 Using impedance diagrams to characterize faults 2019

    25.12.4 Distance relay characteristics 2025

25.13 Auxiliary and lockout (86) relays 2036

25.14 Review of fundamental principles 2039

26 Signal characterization **2043**

  26.1 Flow measurement from differential pressure 2044

  26.2 Flow measurement in open channels 2051

  26.3 Material volume measurement 2053

  26.4 Radiative temperature measurement 2061

  26.5 Analytical measurements 2062

  26.6 Review of fundamental principles 2065

27 Control valves **2067**

  27.1 Sliding-stem valves 2068

    27.1.1 Globe valves 2069

    27.1.2 Gate valves 2078

    27.1.3 Diaphragm valves 2078

  27.2 Rotary-stem valves 2080

    27.2.1 Ball valves 2081

    27.2.2 Butterfly valves 2081

    27.2.3 Disk valves 2082

  27.3 Dampers and louvres 2083

  27.4 Valve packing 2086

  27.5 Valve seat leakage 2095

  27.6 Control valve actuators 2096

    27.6.1 Pneumatic actuators 2097

    27.6.2 Hydraulic actuators 2105

    27.6.3 Self-operated valves 2106

    27.6.4 Electric actuators 2110

    27.6.5 Hand (manual) actuators 2114

  27.7 Valve failure mode 2115

    27.7.1 Direct/reverse actions 2116

    27.7.2 Available failure modes 2118

    27.7.3 Selecting the proper failure mode 2119

  27.8 Actuator bench-set 2121

  27.9 Pneumatic actuator response 2126

  27.10 Valve positioners 2130

    27.10.1 Force-balance pneumatic positioners 2135

    27.10.2 Motion-balance pneumatic positioners 2139

    27.10.3 Electronic positioners 2142

  27.11 Split-ranging 2147

27.11.1 Complementary valve sequencing 2148
27.11.2 Exclusive valve sequencing 2151
27.11.3 Progressive valve sequencing 2153
27.11.4 Valve sequencing implementations 2156
27.12 Control valve sizing 2163
27.12.1 Physics of energy dissipation in a turbulent fluid stream 2164
27.12.2 Importance of proper valve sizing 2169
27.12.3 Gas valve sizing 2174
27.12.4 Relative flow capacity 2175
27.13 Control valve characterization 2176
27.13.1 Inherent versus installed characteristics 2177
27.13.2 Control valve performance with constant pressure 2179
27.13.3 Control valve performance with varying pressure 2182
27.13.4 Characterized valve trim 2185
27.14 Control valve problems 2191
27.14.1 Mechanical friction 2192
27.14.2 Flashing 2197
27.14.3 Cavitation 2201
27.14.4 Choked flow 2208
27.14.5 Valve noise 2210
27.14.6 Erosion 2212
27.14.7 Chemical attack 2217
27.15 Review of fundamental principles 2218

28 Variable-speed motor controls   **2223**
28.1 DC motor speed control 2225
28.2 AC motor speed control 2233
28.3 AC motor braking 2238
28.3.1 DC injection braking 2239
28.3.2 Dynamic braking 2240
28.3.3 Regenerative braking 2242
28.3.4 Plugging 2244
28.4 Motor drive features 2245
28.5 Use of line reactors 2246
28.6 Metering pumps 2250
28.7 Review of fundamental principles 2252

29 Closed-loop control   **2253**
29.1 Basic feedback control principles 2254
29.2 Diagnosing feedback control problems 2261
29.3 On/off control 2263
29.4 Proportional-only control 2265
29.5 Proportional-only offset 2272
29.6 Integral (reset) control 2278
29.7 Derivative (rate) control 2283
29.8 Summary of PID control terms 2285

29.8.1 Proportional control mode (P) 2285

29.8.2 Integral control mode (I) 2286

29.8.3 Derivative control mode (D) 2287

29.9 P, I, and D responses graphed 2287

29.9.1 Responses to a single step-change 2288

29.9.2 Responses to a momentary step-and-return 2289

29.9.3 Responses to two momentary steps-and-returns 2291

29.9.4 Responses to a ramp-and-hold 2292

29.9.5 Responses to an up-and-down ramp 2293

29.9.6 Responses to a multi-slope ramp 2294

29.9.7 Responses to a multiple ramps and steps 2295

29.9.8 Responses to a sine wavelet 2296

29.9.9 Note to students regarding quantitative graphing 2298

29.10 Different PID equations 2302

29.10.1 Parallel PID equation 2303

29.10.2 Ideal PID equation 2304

29.10.3 Series PID equation 2305

29.11 Pneumatic PID controllers 2306

29.11.1 Proportional control action 2307

29.11.2 Automatic and manual modes 2311

29.11.3 Derivative control action 2313

29.11.4 Integral control action 2314

29.11.5 Fisher MultiTrol 2317

29.11.6 Foxboro model 43AP 2320

29.11.7 Foxboro model 130 2322

29.11.8 External reset (integral) feedback 2325

29.12 Analog electronic PID controllers 2327

29.12.1 Proportional control action 2328

29.12.2 Derivative and integral control actions 2330

29.12.3 Full-PID circuit design 2334

29.12.4 Single-loop analog controllers 2337

29.12.5 Multi-loop analog control systems 2340

29.13 Digital PID controllers 2343

29.13.1 Stand-alone digital controllers 2344

29.13.2 Direct digital control (DDC) 2349

29.13.3 SCADA and telemetry systems 2356

29.13.4 Distributed Control Systems (DCS) 2361

29.13.5 Fieldbus control 2367

29.14 Practical PID controller features 2370

29.14.1 Manual and automatic modes 2371

29.14.2 Output and setpoint tracking 2372

29.14.3 Alarm capabilities 2374

29.14.4 Output and setpoint limiting 2374

29.14.5 Security 2375

29.15 Digital PID algorithms 2376

29.15.1 Introduction to pseudocode 2376

29.15.2 Position versus velocity algorithms 2382
29.16 Note to students 2387
    29.16.1 Proportional-only control action 2388
    29.16.2 Integral-only control action 2389
    29.16.3 Proportional plus integral control action 2390
    29.16.4 Proportional plus derivative control action 2391
    29.16.5 Full PID control action 2392
29.17 Review of fundamental principles 2393

30 Process dynamics and PID controller tuning   **2395**
30.1 Process characteristics 2396
    30.1.1 Self-regulating processes 2397
    30.1.2 Integrating processes 2400
    30.1.3 Runaway processes 2407
    30.1.4 Steady-state process gain 2412
    30.1.5 Lag time 2417
    30.1.6 Multiple lags (orders) 2423
    30.1.7 Dead time 2430
    30.1.8 Hysteresis 2435
30.2 Before you tune 2439
    30.2.1 Identifying operational needs 2440
    30.2.2 Identifying process and system hazards 2442
    30.2.3 Identifying the problem(s) 2443
    30.2.4 Final precautions 2444
30.3 Quantitative PID tuning procedures 2445
    30.3.1 Ziegler-Nichols closed-loop ("Ultimate Gain") 2446
    30.3.2 Ziegler-Nichols open-loop 2451
30.4 Heuristic PID tuning procedures 2454
    30.4.1 Features of P, I, and D actions 2456
    30.4.2 Tuning recommendations based on process dynamics 2457
    30.4.3 Recognizing an over-tuned controller by phase shift 2458
    30.4.4 Recognizing a "porpoising" controller 2462
30.5 Tuning techniques compared 2463
    30.5.1 Tuning a "generic" process 2464
    30.5.2 Tuning a liquid level process 2469
    30.5.3 Tuning a temperature process 2473
30.6 Note to students 2477
    30.6.1 Electrically simulating a process 2478
    30.6.2 Building a "Desktop Process" unit 2479
    30.6.3 Simulating a process by computer 2483
30.7 Review of fundamental principles 2484

31 Basic process control strategies **2487**
31.1 Supervisory control 2488
31.2 Cascade control 2490
31.3 Ratio control 2498
31.4 Relation control 2506
31.5 Feedforward control 2508
   31.5.1 Load Compensation 2509
   31.5.2 Proportioning feedforward action 2520
31.6 Feedforward with dynamic compensation 2525
   31.6.1 Dead time compensation 2526
   31.6.2 Lag time compensation 2532
   31.6.3 Lead/Lag and dead time function blocks 2538
31.7 Limit, Selector, and Override controls 2549
   31.7.1 Limit controls 2552
   31.7.2 Selector controls 2557
   31.7.3 Override controls 2566
31.8 Techniques for analyzing control strategies 2571
   31.8.1 Explicitly denoting controller actions 2571
   31.8.2 Determining the design purpose of override controls 2579
31.9 Review of fundamental principles 2581

32 Process safety and instrumentation **2583**
32.1 Classified areas and electrical safety measures 2584
   32.1.1 Classified area taxonomy 2585
   32.1.2 Explosive limits 2587
   32.1.3 Protective measures 2590
32.2 Concepts of probability 2596
   32.2.1 Mathematical probability 2598
   32.2.2 Laws of probability 2600
   32.2.3 Applying probability laws to real systems 2611
32.3 Practical measures of reliability 2614
   32.3.1 Failure rate and MTBF 2614
   32.3.2 The "bathtub" curve 2620
   32.3.3 Reliability 2622
   32.3.4 Probability of failure on demand (PFD) 2625
32.4 High-reliability systems 2626
   32.4.1 Design and selection for reliability 2627
   32.4.2 Preventive maintenance 2628
   32.4.3 Component de-rating 2630
   32.4.4 Redundant components 2631
   32.4.5 Proof tests and self-diagnostics 2636
32.5 Overpressure protection devices 2641
   32.5.1 Rupture disks 2643
   32.5.2 Direct-actuated safety and relief valves 2644
   32.5.3 Pilot-operated safety and relief valves 2654
32.6 Safety Instrumented Functions and Systems 2655

          32.6.1 SIS sensors 2660
          32.6.2 SIS controllers (logic solvers) 2666
          32.6.3 SIS final control elements 2668
          32.6.4 Safety Integrity Levels 2671
          32.6.5 SIS example: burner management systems 2673
          32.6.6 SIS example: water treatment oxygen purge system 2680
          32.6.7 SIS example: nuclear reactor scram controls 2684
     32.7 Review of fundamental principles 2688

33 Instrumentation cyber-security   **2691**
     33.1 Stuxnet 2692
          33.1.1 A primer on uranium enrichment 2693
          33.1.2 Gas centrifuge vulnerabilities 2699
          33.1.3 The Natanz uranium enrichment facility 2701
          33.1.4 Stuxnet version 0.5 2702
          33.1.5 Stuxnet version 1.x 2703
     33.2 Motives 2703
          33.2.1 Technical challenge 2704
          33.2.2 Profit 2704
          33.2.3 Espionage 2705
          33.2.4 Sabotage 2706
          33.2.5 Terrorism 2707
     33.3 Lexicon of cyber-security terms 2708
     33.4 Fortifying strategies 2711
          33.4.1 Policy-based strategies 2711
          33.4.2 Design-based strategies 2719
     33.5 Review of fundamental principles 2725

34 Problem-solving and diagnostic strategies   **2729**
     34.1 Learn principles, not procedures 2730
     34.2 Active reading 2730
          34.2.1 Don't limit yourself to one text! 2731
          34.2.2 Marking versus outlining a text 2732
     34.3 General problem-solving techniques 2737
          34.3.1 Identifying and classifying all "known" conditions 2737
          34.3.2 Re-cast the problem in a different format 2737
          34.3.3 Working backwards from a known solution 2738
          34.3.4 Using thought experiments 2741
          34.3.5 Explicitly annotating your thoughts 2742
     34.4 Mathematical problem-solving techniques 2746
          34.4.1 Manipulating algebraic equations 2747
          34.4.2 Linking formulae to solve mathematical problems 2750
          34.4.3 Double-checking calculations 2764
          34.4.4 Maintaining relevance of intermediate calculations 2769
     34.5 Problem-solving by simplification 2771
          34.5.1 Limiting cases 2772

34.6 Scientific system diagnosis 2777
    34.6.1 Scientific method 2778
    34.6.2 Occam's Razor 2778
    34.6.3 Diagnosing intermittent problems 2779
    34.6.4 Strategy: tracing data paths 2780
34.7 Common diagnostic mistakes 2782
    34.7.1 Failing to gather data 2783
    34.7.2 Failing to use relevant documentation 2784
    34.7.3 Fixating on the first hypothesis 2785
    34.7.4 Failing to build and test a new system in stages 2787
34.8 Helpful "tricks" using a digital multimeter (DMM) 2789
    34.8.1 Recording unattended measurements 2789
    34.8.2 Avoiding "phantom" voltage readings 2790
    34.8.3 Non-contact AC voltage detection 2793
    34.8.4 Detecting AC power harmonics 2794
    34.8.5 Identifying noise in DC signal paths 2795
    34.8.6 Generating test voltages 2796
    34.8.7 Using the meter as a temporary jumper 2797

A Flip-book animations   **2799**
  A.1 Polyphase light bulbs animated 2799
  A.2 Polyphase induction motor animated 2826
  A.3 Rotating phasor animated 2851
  A.4 Differentiation and integration animated 2892
  A.5 Guided-wave radar level measurement 3029
  A.6 Basic chromatograph operation 3082

B *Doctor Strangeflow*, or how I learned to relax and love Reynolds numbers   **3143**

C Disassembly of a sliding-stem control valve   **3153**

D How to use this book – some advice for teachers   **3163**
  D.1 Teaching technical theory 3164
    D.1.1 The problem with lecture 3166
    D.1.2 A more accurate model of learning 3168
    D.1.3 The ultimate goal of education 3169
  D.2 Teaching technical practices (labwork) 3172
  D.3 Teaching diagnostic principles and practices 3180
    D.3.1 Deductive diagnostic exercises 3181
    D.3.2 Inductive diagnostic exercises 3189
  D.4 Practical topic coverage 3196
  D.5 Principles, not procedures 3197
  D.6 Assessing student learning 3199
  D.7 Summary 3201

E Contributors   **3203**
   E.1 Error corrections 3204
   E.2 New content 3208

F Creative Commons Attribution License   **3209**
   F.1 A simple explanation of your rights 3210
   F.2 Legal code 3211

Chapter 29

Closed-loop control

Instrumentation is the science of automated measurement and control. Applications of this science abound in modern research, industry, and everyday living. From automobile engine control systems to home thermostats to aircraft autopilots to the manufacture of pharmaceutical drugs, automation surrounds us. This chapter explains some of the fundamental principles of automatic process control.

29.1 Basic feedback control principles

Before we begin our discussion on process control, we must define a few key terms. First, we have what is known as the *process*: the physical system we wish to monitor and control. For the sake of illustration, consider a heat exchanger that uses high-temperature steam to transfer heat to a lower-temperature liquid. Heat exchangers are used frequently in the chemical industries to maintain the necessary temperature of a chemical solution, so the desired blending, separation, or reactions can occur. A very common design of heat exchanger is the "shell-and-tube" style, where a metal shell serves as a conduit for the chemical solution to flow through, while a network of smaller tubes runs through the interior of the shell, carrying steam or some other heat-transfer fluid. The hotter steam flowing through the tubes transfers heat energy to the cooler process fluid surrounding the tubes, inside the shell of the heat exchanger:

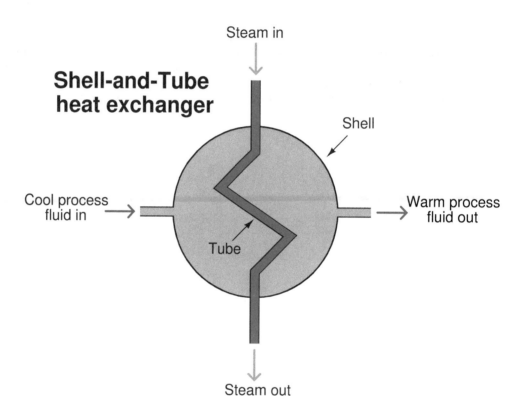

In this case, the *process* is the entire heating system, consisting of the fluid we wish to heat, the heat exchanger, and the steam delivering the required heat energy. In order to maintain steady control of the process fluid's exiting temperature, we must find a way to measure it and represent that measurement in signal form so it may be interpreted by other instruments taking some form of control action. In instrumentation terms, the measuring device is known as a *transmitter*, because it *transmits* the process measurement in the form of a signal.

Transmitters are represented in process diagrams by small circles with identifying letters inside, in this case, "TT," which stands for **T**emperature **T**ransmitter:

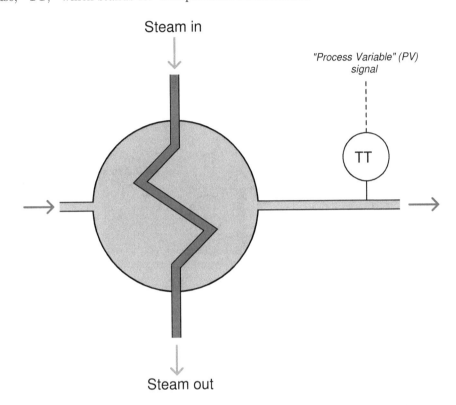

The signal output by the transmitter (represented by the "PV" dashed line), representing the heated fluid's exiting temperature, is called the *process variable*. Like a variable in a mathematical equation that represents some story-problem quantity, this signal represents the measured quantity we wish to control in the process.

In order to exert control over the process variable, we must have some way of altering fluid flow through the heat exchanger, either of the process fluid, the steam, or both. Generally, it makes more sense to alter the flow of the heating medium (the steam), and let the process fluid flow rate be dictated by the demands of the larger process. If this heat exchanger were part of an oil refinery unit, for example, it would be far better to throttle steam flow to control oil temperature rather than to throttle the oil flow itself, since altering the oil's flow will also affect other process variables upstream and downstream of the exchanger. Ideally, the heat exchanger temperature control system would provide consistent temperature of the exiting oil, for any given incoming oil temperature and flow-rate of oil through it.

One convenient way to throttle steam flow into the heat exchanger is to use a control valve (labeled "TV" because it is a **T**emperature **V**alve). In general terms, a control valve is known as a *final control element*. Other types of final control elements exist (servo motors, variable-flow pumps, and other mechanical devices used to vary some physical quantity at will), but valves are the most common, and probably the simplest to understand. With a final control element in place, the steam flow becomes known as the *manipulated variable*, because it is the quantity we will manipulate in order to gain control over the process variable:

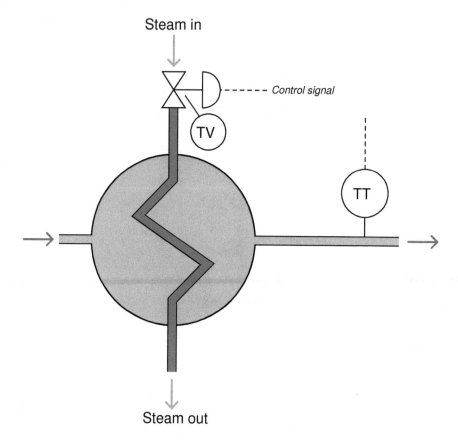

Valves come in a wide variety of sizes and styles. Some valves are hand-operated: that is, they have a "wheel" or other form of manual control that may be moved to "pinch off" or "open up" the flow passage through the pipe. Other valves come equipped with signal receivers and positioner devices, which move the valve mechanism to various positions at the command of a signal (usually an electrical signal, like the type output by transmitter instruments). This feature allows for remote control, so a human operator or computer device may exert control over the manipulated variable from a distance. In the previous illustration, the steam control valve is equipped with such an electrical signal input, represented by the "control signal" dashed line.

This brings us to the final component of the heat exchanger temperature control system: the *controller*. This is a device designed to interpret the transmitter's process variable signal and decide how far open the control valve needs to be in order to maintain that process variable at the desired value.

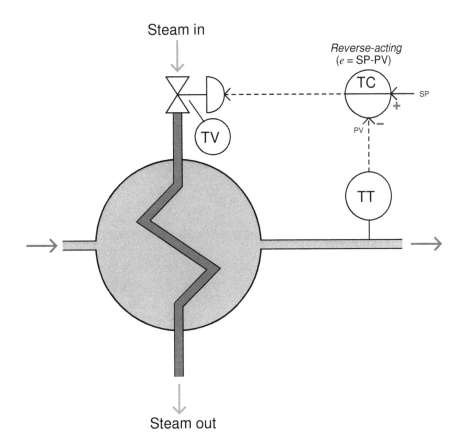

Here, the circle with the letters "TC" in the center represents the controller. Those letters stand for **T**emperature **C**ontroller, since the process variable being controlled is the process fluid's *temperature*. Usually, the controller consists of a computer making automatic decisions to open and close the valve as necessary to stabilize the process variable at some predetermined *setpoint*.

Note that the controller's circle has a solid line going through the center of it, while the transmitter and control valve circles are open. An open circle represents a field-mounted device according to the ISA standard for instrumentation symbols, and a single solid line through the middle of a circle tells us the device is located on the front of a control panel in a main control room location. So, even though the diagram might appear as though these three instruments are located close to one another, they in fact may be quite far apart. Both the transmitter and the valve must be located near the heat exchanger (out in the "field" area rather than inside a building), but the controller may be located a long distance away where human operators can adjust the setpoint from inside a safe and secure control room.

These elements comprise the essentials of a *feedback control system*: the *process* (the system to be controlled), the *process variable* (the specific quantity to be measured and controlled), the *transmitter* (the device used to measure the process variable and output a corresponding signal), the *controller* (the device that decides what to do to bring the process variable as close to setpoint as possible), the *final control element* (the device that directly exerts control over the process), and the *manipulated variable* (the quantity to be directly altered to effect control over the process variable).

Feedback control may be viewed as a sort of information "loop," from the transmitter (measuring the process variable), to the controller, to the final control element, and through the process itself, back to the transmitter. Ideally, a process control "loop" not only holds the process variable at a steady level (the setpoint), but also maintains control over the process variable given changes in setpoint, and even changes in other variables of the process:

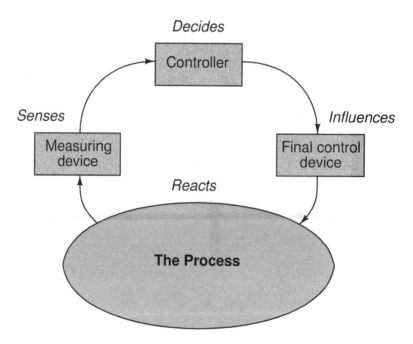

Specifically, the type of feedback we are employing here to control the process is *negative* or *degenerative* feedback. The term "negative" refers to the direction of action the control system takes in response to any measured change in the process variable. If something happens to drive the process variable up, the control system will automatically respond in such a way as to bring the process variable back down where it belongs. If the process variable happens to sag below setpoint, the control system will automatically act to drive the process variable back up to setpoint. Whatever the process variable does in relation to setpoint, the control system takes the opposite (inverse, or negative) action in an attempt to stabilize it at setpoint.

For example, if the unheated process fluid flow rate were to suddenly increase, the heat exchanger outlet temperature would fall due to the physics of heat transfer, but once this drop was detected by the transmitter and reported to the controller, the controller would automatically call for additional steam flow to compensate for the temperature drop, thus bringing the process variable back in agreement with the setpoint. Ideally, a well designed and well-tuned control loop will sense and compensate for *any* change in the process or in the setpoint, the end result being a process variable value that always holds steady at the setpoint value.

The unheated fluid flow rate is an example of an uncontrolled, or *wild*, variable because our control system here has no ability to influence it. This flow is also referred to as a *load* because it "loads" or affects the process variable we are trying to stabilize. Loads are present in nearly every controlled system, and indeed are the primary factor necessitating a control system at all. Referring back to our heat exchanger process again, we could adequately control the operating temperature of it with just a manually-set steam control valve if only none of the other factors (steam temperature, fluid flow rate, incoming fluid temperature, etc.) ever changed!

Many types of processes lend themselves to feedback control. Consider an aircraft autopilot system, keeping an airplane on a steady course heading despite the effects of loads such as side-winds: reading the plane's heading (process variable) from an electronic compass and using the rudder as a final control element to change the plane's "yaw." An automobile's "cruise control" is another example of a feedback control system, with the process variable being the car's velocity, and the final control element being the engine's throttle. The purpose of a cruise control is to maintain constant driving speed despite the influence of loads such as hills, head-winds, tail-winds, and road roughness. Steam boilers with automatic pressure controls, electrical generators with automatic voltage and frequency controls, and water pumping systems with automatic flow controls are further examples of how feedback may be used to maintain control over certain process variables.

Modern technology makes it possible to control nearly anything that may be measured in an industrial process. This extends beyond the pale of simple pressure, level, temperature, and flow variables to include even certain chemical properties.

In municipal water and wastewater treatment systems, for example, numerous chemical quantities must be measured and controlled automatically to ensure maximum health and minimum environmental impact. Take for instance the chlorination of treated wastewater, before it leaves the wastewater treatment facility into a large body of water such as a river, bay, or ocean. Chlorine is added to the water to kill any residual bacteria so they do not consume oxygen in the body of water they are released to. Too little chlorine added, and not enough bacteria are killed, resulting in a high *biological oxygen demand* or *BOD* in the water which will asphyxiate the fish swimming in it. Too much chlorine added, and the chlorine itself poses a hazard to marine life. Thus, the chlorine content must be carefully controlled at a particular setpoint, and the control system must take aggressive action if the dissolved chlorine concentration strays too low or too high:

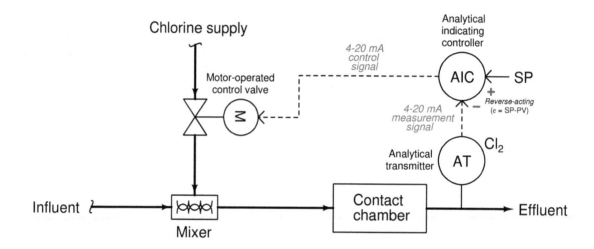

Now that we have seen the basic elements of a feedback control system, we will concentrate on the *algorithms* used in the controller to maintain a process variable at setpoint. For the scope of this topic, an "algorithm" is a mathematical relationship between the process variable and setpoint inputs of a controller, and the output (manipulated variable). Control algorithms determine *how* the manipulated variable quantity is deduced from PV and SP inputs, and range from the elementary to the very complex. In the most common form of control algorithm, the so-called "PID" algorithm, calculus is used to determine the proper final control element action for any combination of input signals.

29.2 Diagnosing feedback control problems

Negative feedback systems, in general, tend to cause much confusion for those first learning their fundamental principles and behaviors. The closed-cycle "loop" formed by the interaction of sensing element, controller, final control element, and process means essentially that *everything affects everything else*. This is especially problematic when the feedback control system in question contains a fault and must be diagnosed. For example, if an operator happens to notice that the process variable (as indicated by a manual measurement or by some trusted indicating instrument) is not holding to setpoint, it could be the result of a fault in *any* portion of the system (sensor, controller, FCE, or even the process itself).

Recall that every feedback control loop consists of four basic elements: an element that *senses* the process variable (e.g. primary sensing element, transmitter), an element that *decides* what how to regulate this process variable (e.g. a PID controller), an element that *influences* the process variable (e.g. a control valve, motor drive, or some other final control device), and finally the process itself which *reacts* to the final control device's actions:

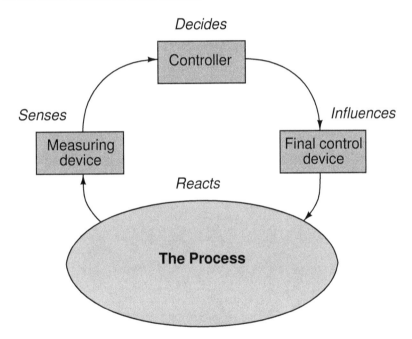

One of the basic diagnostic strategies for any instrumentation system is to assess whether the *input value(s)* and *output value(s)* correspond for each instrument. We may apply this same strategy to each of the four elements of a feedback control "loop" to identify where the problem might exist. If you encounter one of these four system portions whose output does not correspond with its input, you know that portion of the system is faulted.

You can check each element of your feedback control loop by comparing its input with its output to see if each element is doing what it should. I recommend beginning with the controller (the decision-making element) because typically those values are the most easily monitored:

- **Decision-making:** Carefully examine the controller faceplate, looking at the values of PV, SP, and Output. Is the controller taking appropriate action to force PV equal to SP? In other words, is the Output signal at a value you would expect if the controller were functioning properly to regulate the process variable at setpoint? If so, then the controller's action and tuning are most likely not at fault. If not, then the problem definitely lies with the controller.

- **Sensing:** Compare the controller's displayed value for PV with the actual process variable value as indicated by local gauges, by feel, or by any other means of detection. If there is good correspondence between the controller's PV display and the real process variable, then there probably isn't anything wrong with the measurement portion of the control loop (e.g. transmitter, impulse lines, PV signal wiring, analog input of controller, etc.). If the displayed PV disagrees with the actual process variable value, then something is definitely wrong here.

- **Influencing:** Compare the controller's displayed value for Output with the actual status of the final control element. If there is good correspondence between the controller's Output display and the FCE's status, then there probably isn't anything wrong with the output portion of the control loop (e.g. FCE, output signal wiring, analog output of controller, etc.). If the controller Output value differs from the FCE's state, then something is definitely wrong here.

- **Reacting:** Compare the process variable value with the final control element's state. Is the process doing what you would expect it to? If so, the problem is most likely not within the process (e.g. manual valves, relief valves, pumps, compressors, motors, and other process equipment). If, however, the process is not reacting the way you would expect it to given the final control element's state, then something is definitely awry with the process itself.

29.3 On/off control

Once while working as an instrument technician in an aluminum foundry, a mechanic asked me what it was that I did. I began to explain my job, which was essentially to calibrate, maintain, troubleshoot, document, and modify (as needed) all automatic control systems in the facility. The mechanic seemed puzzled as I explained the task of "tuning" loop controllers, especially those controllers used to maintain the temperature of large, gas-fired industrial furnaces holding many tons of molten metal. "Why does a controller have to be 'tuned'?" he asked. "All a controller does is turn the burner on when the metal's too cold, and turn it off when it becomes too hot!"

In its most basic form, the mechanic's assessment of the control system was correct: to turn the burner on when the process variable (molten metal temperature) drops below setpoint, and turn it off when it rises above setpoint. However, the actual algorithm is much more complex than that, finely adjusting the burner intensity according to the amount of *error* between PV and SP, the amount of time the error has accumulated, and the rate-of-change of the error over time. In his casual observation of the furnace controllers, though, he had noticed nothing more than the full-on/full-off action of the controller.

The technical term for a control algorithm that merely checks for the process variable exceeding or falling below setpoint is *on/off control*. In colloquial terms, it is known as *bang-bang* control, since the manipulated variable output of the controller rapidly switches between fully "on" and fully "off" with no intermediate state. Control systems this crude usually provide very imprecise control of the process variable. Consider our example of the shell-and-tube heat exchanger, if we were to implement simple on/off control[1]:

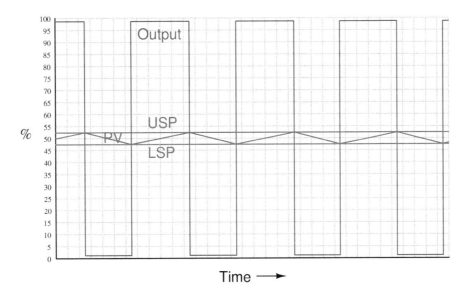

As you can see, the degree of control is rather poor. The process variable "cycles" between the upper and lower setpoints (USP and LSP) without ever stabilizing at the setpoint, because that

---

[1]To be precise, this form of on/off control is known as *differential gap* because there are two setpoints with a gap in between. While on/off control is possible with a single setpoint (FCE on when below setpoint and off when above), it is usually not practical due to the frequent cycling of the final control element.

would require the steam valve to be position somewhere *between* fully closed and fully open.

This simple control algorithm may be adequate for temperature control in a house, but not for a sensitive chemical process! Can you imagine what it would be like if an automobile's cruise control system relied on this algorithm? Not only is the lack of precision a problem, but the frequent cycling of the final control element may contribute to premature failure due to mechanical wear. In the heat exchanger scenario, thermal cycling (hot-cold-hot-cold) will cause metal fatigue in the tubes, resulting in a shortened service life. Furthermore, every excursion of the process variable above setpoint is wasted energy, because the process fluid is being heated to a greater temperature than what is necessary.

Clearly, the only practical answer to this dilemma is a control algorithm able to *proportion* the final control element rather than just operate it at zero or full effect (the control valve fully closed or fully open). This, in its simplest form, is called *proportional control*.

29.4 Proportional-only control

Imagine a liquid-level control system for a vessel, where the position of a level-sensing float directly sets the stem position of a control valve. As the liquid level rises, the valve opens up proportionally:

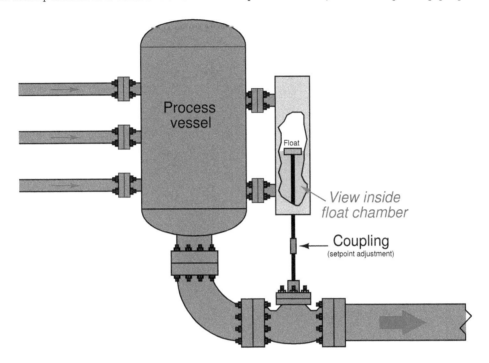

Despite its crude mechanical nature, this *proportional* control system would in fact help regulate the level of liquid inside the process vessel. If an operator wished to change the "setpoint" value of this level control system, he or she would have to adjust the coupling between the float and valve stems for more or less distance between the two. Increasing this distance (lengthening the connection) would effectively raise the level setpoint, while decreasing this distance (shortening the connection) would lower the setpoint.

We may generalize the proportional action of this mechanism to describe *any* form of controller where the output is a direct function of process variable (PV) and setpoint (SP):

$$m = K_p e + b$$

Where,
  m = Controller output
  e = Error (difference between PV and SP)
  K_p = Proportional gain
  b = Bias

A new term introduced with this formula is e, the "error" or difference between process variable and setpoint. Error may be calculated as SP−PV or as PV−SP, depending on whether or not the controller must produce an *increasing* output signal in response to an increase in the process variable ("direct" acting), or output a *decreasing* signal in response to an increase in the process variable ("reverse" acting):

$$m = K_p(\text{PV} - \text{SP}) + b \qquad \text{(Direct-acting proportional controller)}$$

$$m = K_p(\text{SP} - \text{PV}) + b \qquad \text{(Reverse-acting proportional controller)}$$

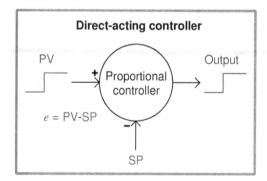 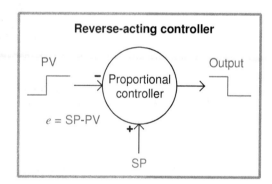

The optional "+" and "−" symbols clarify the effect each input has on the controller output: a "−" symbol representing an *inverting* effect and a "+" symbol representing a *noninverting* effect. When we say that a controller is "direct-acting" or "reverse-acting" we are referring to it reaction to the PV signal, therefore the output signal from a "direct-acting" controller goes in the same direction as the PV signal and the output from a "reverse-acting" controller goes in the opposite direction of its PV signal. It is important to note, however, that the response to a change in setpoint (SP) will yield the *opposite* response as does a change in process variable (PV): a rising SP will drive the output of a direct-acting controller *down* while a rising SP drives the output of a reverse-acting controller *up*. "+" and "−" symbols explicitly show the effect both inputs have on the controller output, helping to avoid confusion when analyzing the effects of PV changes versus the effects of SP changes.

The direction of action required of the controller is determined by the nature of the process, transmitter, and final control element. In the case of the crude mechanical level controller, the action needs to be *direct* so that a greater liquid level will result in a further-open control valve to drain the vessel faster. In the case of the automated heat exchanger shown earlier, we are assuming that an increasing output signal sent to the control valve results in increased steam flow, and consequently higher temperature, so our controller will need to be reverse-acting (i.e. an increase in measured temperature results in a decrease in output signal; error calculated as SP−PV):

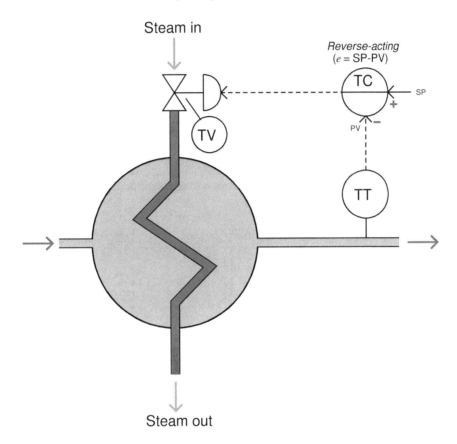

After the error has been calculated, the controller then multiplies the error signal by a constant value called the *gain*, which is programmed into the controller. The resulting figure, plus a "bias" quantity, becomes the output signal sent to the valve to proportion it. The "gain" value is exactly what it seems to be for anyone familiar with electronic amplifier circuits: a ratio of output to input. In this case, the gain of a proportional controller is the ratio of output signal change to input signal change, or how *aggressive* the controller reacts to changes in input (PV or SP).

To give a numerical example, a loop controller set to have a gain of 4 will change its output signal by 40% if it sees an input change of 10%: the ratio of output change to input change will be 4:1. Whether the input change comes in the form of a setpoint adjustment, a drift in the process variable, or some combination of the two does not matter to the magnitude of the output change.

The bias value of a proportional controller is simply the value of its output whenever process variable happens to be equal to setpoint (i.e. a condition of zero *error*). Without a bias term in the proportional control formula, the valve would always return to a fully shut (0%) condition if ever the process variable reached the setpoint value. The bias term allows the final control element to achieve a non-zero state at setpoint.

If the controller could be configured for infinite gain, its response would duplicate on/off control. That is, *any* amount of error will result in the output signal becoming "saturated" at either 0% or 100%, and the final control element will simply turn on fully when the process variable drops below setpoint and turn off fully when the process variable rises above setpoint. Conversely, if the controller is set for zero gain, it will become completely unresponsive to changes in either process variable *or* setpoint: the valve will hold its position at the bias point no matter what happens to the process.

Obviously, then, we must set the gain somewhere between infinity and zero in order for this algorithm to function any better than on/off control. Just how much gain a controller needs to have depends on the process and all the other instruments in the control loop.

If the gain is set too high, there will be oscillations as the PV converges on a new setpoint value:

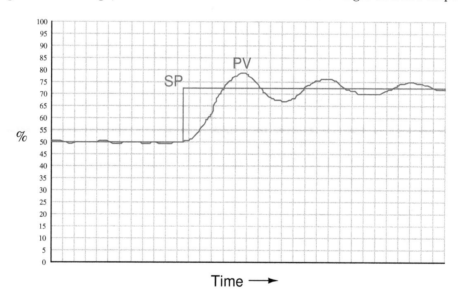

If the gain is set too low, the process response will be stable under steady-state conditions but relatively slow to respond to changes in setpoint, as shown in the following trend recording:

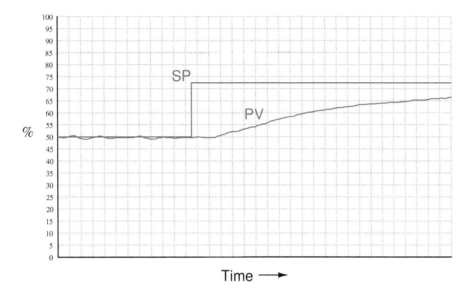

A characteristic deficiency of proportional control action, exacerbated with low controller gain values, is a phenomenon known as *proportional-only offset* where the PV never fully reaches SP. A full explanation of proportional-only offset is too lengthy for this discussion and will be presented in a subsequent section of the book, but may be summarized here simply by drawing attention to the proportional controller equation which tells us the output always returns to the bias value when PV reaches SP (i.e. $m = b$ when PV = SP). If anything changes in the process to require a different output value than the bias (b) to stabilize the PV, an error between PV and SP *must* develop to drive the controller output to that necessary output value. This means it is only by chance that the PV will settle precisely at the SP value – most of the time, the PV will deviate from SP in order to generate an output value sufficient to stabilize the PV and prevent it from drifting. This persistent error, or offset, worsens as the controller gain is reduced. Increasing controller gain causes this offset to decrease, but at the expense of oscillations.

With proportional-only control, the choice of gain values is really a compromise between excessive oscillations and excessive offset. A well-tuned proportional controller response is shown here:

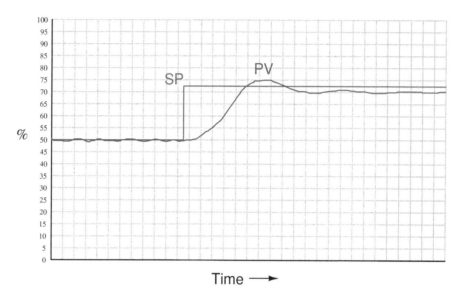

An unnecessarily confusing aspect of proportional control is the existence of two completely different ways to express controller proportionality. In the proportional-only equation shown earlier, the degree of proportional action was specified by the constant K_p, called *gain*. However, there is another way to express the sensitivity of proportional action, and that is to state the percentage of error change necessary to make the output (m) change by 100%. Mathematically, this is the inverse of gain, and it is called *proportional band* (PB):

$$K_p = \frac{1}{\text{PB}} \qquad \text{PB} = \frac{1}{K_p}$$

Gain is always specified as a unitless value[2], whereas proportional band is always specified as a percentage. For example, a gain value of 2.5 is equivalent to a proportional band value of 40%, because the error input to this controller must change by 40% in order to make the output change a full 100%.

---

[2]In electronics, the unit of *decibels* is commonly used to express gains. Thankfully, the world of process control was spared the introduction of decibels as a unit of measurement for controller gain. The last thing we need is a *third* way to express the degree of proportional action in a controller!

Due to the existence of these two completely opposite conventions for specifying proportional action, you may see the proportional term of the control equation written differently depending on whether the author assumes the use of gain or the use of proportional band:

$$K_p = \text{gain} \qquad \text{PB} = \text{proportional band}$$

$$K_p e \qquad\qquad \frac{1}{\text{PB}} e$$

Many modern digital electronic controllers allow the user to conveniently select the unit they wish to use for proportional action. However, even with this ability, anyone tasked with adjusting a controller's "tuning" values may be required to translate between gain and proportional band, especially if certain values are documented in a way that does not match the unit configured for the controller.

When you communicate the proportional action setting of a process controller, you should always be careful to specify either "gain" or "proportional band" to avoid ambiguity. *Never* simply say something like, "The proportional setting is twenty," for this could mean either:

- Proportional band = 20%; Gain = 5   . . . *or* . . .

- Gain = 20; Proportional band = 5%

As you can see here, the real-life difference in controller response to an input disturbance (wave) depending on whether it has a proportional band of 20% or a gain of 20 is quite dramatic:

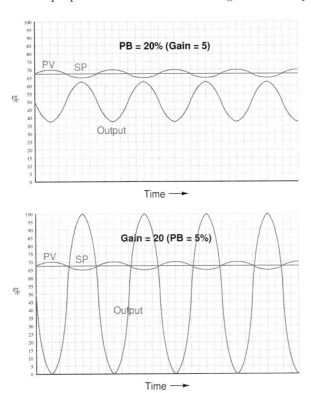

29.5 Proportional-only offset

A fundamental limitation of proportional control has to do with its response to changes in setpoint and changes in process *load*. A "load" in a controlled process is any variable not controlled by the loop controller which nevertheless affects the process variable the controller is trying to regulate. In other words, a "load" is any factor the loop controller must compensate for while maintaining the process variable at setpoint.

In our hypothetical heat exchanger system, the temperature of the incoming process fluid is an example of a load:

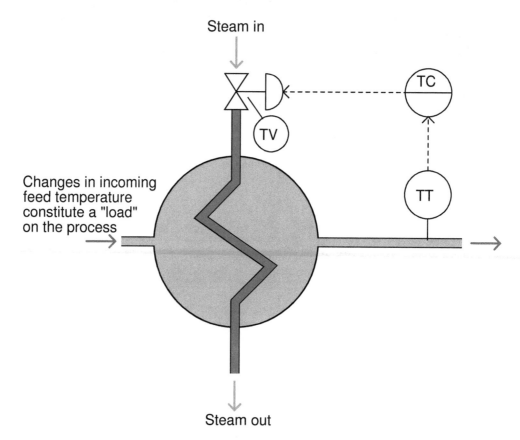

If the incoming fluid temperature were to suddenly decrease, the immediate effect this would have on the process would be to decrease the outlet temperature (which is the temperature we are trying to maintain at a steady value). It should make intuitive sense that a colder incoming fluid will require more heat input to raise it to the same outlet temperature as before. If the heat input remains the same (at least in the immediate future), this colder incoming flow must make the outlet flow colder than it was before. Thus, incoming feed temperature has an impact on the outlet temperature whether we like it or not, and the control system must compensate for these unforeseen and uncontrolled changes. This is precisely the definition of a "load": a burden[3] on the control

---

[3]One could argue that the presence of loads actually *justifies* a control system, for if there were no loads, there

system.

Of course, it is the job of the controller to counteract any tendency for the outlet temperature to stray from setpoint, but as we shall soon see this cannot be perfectly achieved with proportional control alone.

Let us perform a "thought experiment" to demonstrate this phenomenon of proportional-only offset. Imagine the controller has been controlling outlet temperature exactly at setpoint (PV = SP), and then suddenly the inlet feed temperature drops and remains colder than before. Recall that the equation for a reverse-acting proportional controller is as follows:

$$m = K_p(\text{SP} - \text{PV}) + b$$

Where,
m = Controller output
K_p = Proportional gain
SP = Setpoint
PV = Process variable
b = Bias

The introduction of colder feed fluid to the heat exchanger makes the outlet temperature (PV) begin to fall. As the PV falls, the controller calculates a positive error (SP − PV). This positive error, when multiplied by the controller's gain value, drives the output to a greater value. This opens up the steam valve, adding more heat to the exchanger.

As more heat is added, the rate of temperature drop slows down. The further the PV drops, the more the steam valve opens, until enough additional heat is being added to the heat exchanger to maintain a constant outlet temperature. However, this new stable PV value will be less than it was prior to the introduction of colder feed (i.e. less than the SP). In fact, the controller's automatic action can *never* return the PV to its original (SP) value so long as the feed remains colder than before. The reason for this is that a greater flow of steam is necessary to balance a colder feed coming in, and the only way a proportional controller is ever going to automatically drive the steam valve to this greater-flow position is if an error develops between PV and SP. Thus, an *offset* inevitably develops between PV and SP due to the load (colder feed).

We may prove the inevitability of this offset another way: imagine somehow that the PV did actually return to the SP value despite the colder feed fluid (remaining colder). If this happened, the steam valve would also return to its former throttling position where it was before the feed temperature dropped. However, we know that this former position will not allow enough steam through to the exchanger to overcome the colder feed – if it did, the PV never would have decreased to begin with! A further-open valve is precisely what we need to stabilize the PV given this colder feed, yet the only way the proportional-only controller can achieve this is if the PV actually falls below SP.

To summarize: the only way a proportional-only controller can automatically generate a new output value (m) is if the PV deviates from SP. Therefore, load changes (requiring new output values to compensate) force the PV to deviate from SP.

---

would be nothing to compensate for, and therefore no need for an automatic control system at all! In the total absence of loads, a manually-set final control element would be enough to hold most process variables at setpoint.

Another "thought experiment" may be helpful to illustrate the phenomenon of proportional-only offset. Imagine building your own cruise control system for your automobile based on the proportional-only equation: the engine's throttle position is a function of the difference between PV (road speed) and SP (the desired "target" speed). Let us further suppose that you carefully adjust the bias value of your cruise control system to achieve PV = SP on level ground at a speed of 70 miles per hour (70% on a 0 to 100 MPH speedometer scale), with the throttle at a position of 40%, and a gain (K_p) of 2:

$$m = K_p(\text{SP} - \text{PV}) + b$$

$$40\% = 2(70 - 70) + 40\%$$

Imagine now that after cruising precisely at setpoint (70% = 70 MPH), the road begins to incline uphill for several miles. This, obviously, is a load on the cruise control system. With the cruise control disengaged, the automobile would slow down because the same throttle position (40%) sufficient to maintain setpoint (70 MPH) on level ground is not enough power to maintain that same setpoint on an incline.

With the cruise control engaged, the engine throttle will automatically open further as speed drops. At a speed of 69 MPH, the throttle opens up to 42%. At a speed of 68 MPH, the throttle opens up to 44%. Every drop in speed of 1 MPH results in a 2% further-open throttle to send more power to the wheels.

Suppose the demands of this particular inclined road require a 50% throttle position for this automobile to maintain a constant speed. In order for your proportional-only cruise control system to deliver this necessary 50% throttle position, the speed will have to "droop" by 5 MPH below setpoint:

$$m = K_p(\text{SP} - \text{PV}) + b$$

$$50\% = 2(70 - 65) + 40\%$$

There is simply no other way for your proportional-only controller to automatically achieve the requisite 50% throttle position aside from letting the speed sag below setpoint by 5% (5 MPH). Given this fact, the only way the proportional-only cruise control will ever return the speed to setpoint (70 MPH) is if and when the load conditions change to allow for a lesser throttle position of 40%. So long as the load demands a different throttle position than the bias value, the speed *must* deviate from the setpoint value of 70 MPH.

This necessary error developing between PV and SP is called *proportional-only offset*, sometimes called *droop*. The amount of droop depends on how severe the load change is, and how aggressive the controller responds (i.e. how much gain it has). The term "droop" is very misleading, as it is possible for the error to develop the other way (i.e. the PV might rise above SP due to a load change!). Imagine the opposite load-change scenario in our steam heat exchanger process, where the incoming feed temperature suddenly *rises* instead of falls. If the controller was controlling exactly at setpoint before this upset, the final result will be an outlet temperature that settles at some point *above* setpoint, enough so the controller is able to pinch the steam valve far enough closed to stop any further rise in temperature.

Proportional-only offset also occurs as a result of setpoint changes. We could easily imagine the same sort of effect following an operator's increase of setpoint for the temperature controller on the heat exchanger. After increasing the setpoint, the controller immediately increases the output signal, sending more steam to the heat exchanger. As temperature rises, though, the proportional algorithm causes the output signal to decrease. When the rate of heat energy input by the steam equals the rate of heat energy carried away from the heat exchanger by the heated fluid (a condition of *energy balance*), the temperature stops rising. This new equilibrium temperature will not be at setpoint, assuming the temperature was holding at setpoint prior to the human operator's setpoint increase. The new equilibrium temperature indeed *cannot* ever achieve any setpoint value higher than the one it did in the past, for if the error ever returned to zero (PV = SP), the steam valve would return to its old position, which we know would be insufficient to raise the temperature of the heated fluid to a new value.

An example of proportional-only control in the context of electronic power supply circuits is the following opamp voltage regulator, used to stabilize voltage to a load with power supplied by an unregulated voltage source:

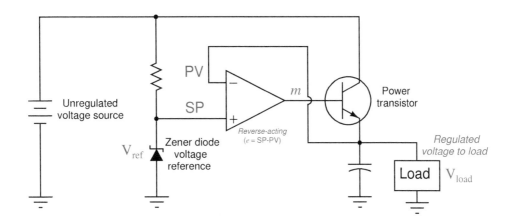

In this circuit, a zener diode establishes a "reference" voltage (which may be thought of as a "setpoint" for the controlling opamp to follow). The operational amplifier acts as the proportional-only controller, sensing voltage at the load (PV), and sending a driving output voltage to the base of the power transistor to keep load voltage constant despite changes in the supply voltage or changes in load current (both "loads" in the process-control sense of the word, since they tend to influence voltage at the load circuit without being under the control of the opamp).

If everything functions properly in this voltage regulator circuit, the load's voltage will be stable over a wide range of supply voltages and load currents. However, the load voltage cannot ever *precisely* equal the reference voltage established by the zener diode, even if the operational amplifier (the "controller") is without defect. The reason for this incapacity to perfectly maintain "setpoint" is the simple fact that in order for the opamp to generate any output signal at all, there *absolutely must be* a differential voltage between the two input terminals for the amplifier to amplify. Operational amplifiers (ideally) generate an output voltage equal to the enormously high gain value (A_V) multiplied by the difference in input voltages (in this case, $V_{ref} - V_{load}$). If V_{load} (the "process variable") were to ever achieve equality with V_{ref} (the "setpoint"), the operational amplifier would experience absolutely no differential input voltage to amplify, and its output signal driving the power transistor would fall to zero. Therefore, there must always exist some *offset* between V_{load} and V_{ref} (between process variable and setpoint) in order to give the amplifier some input voltage to amplify.

The amount of offset is ridiculously small in such a circuit, owing to the enormous gain of the operational amplifier. If we take the opamp's transfer function to be $V_{out} = A_V(V_{(+)} - V_{(-)})$, then we may set up an equation predicting the load voltage as a function of reference voltage (assuming a constant 0.7 volt drop between the base and emitter terminals of the transistor):

$$V_{out} = A_V(V_{(+)} - V_{(-)})$$

$$V_{out} = A_V(V_{ref} - V_{load})$$

$$V_{load} + 0.7 = A_V(V_{ref} - V_{load})$$

$$V_{load} + 0.7 = A_V V_{ref} - A_V V_{load}$$

$$V_{load} + A_V V_{load} = A_V V_{ref} - 0.7$$

$$(A_V + 1)V_{load} = A_V V_{ref} - 0.7$$

$$V_{load} = \frac{A_V V_{ref} - 0.7}{A_V + 1}$$

If, for example, our zener diode produced a reference voltage of 5.00000 volts and the operational amplifier had an open-loop voltage gain of 250000, the load voltage would settle at a theoretical value of 4.9999772 volts: just barely below the reference voltage value. If the opamp's open-loop voltage gain were much less – say only 100 – the load voltage would only be 4.94356 volts. This still is quite close to the reference voltage, but definitely not as close as it would be with a greater opamp gain!

Clearly, then, we can minimize proportional-only offset by increasing the gain of the process controller gain (i.e. decreasing its proportional band). This makes the controller more "aggressive" so it will move the control valve further for any given change in PV or SP. Thus, not as much error needs to develop between PV and SP to move the valve to any new position it needs to go. However, too much controller gain makes the control system unstable: at best it will exhibit residual oscillations after setpoint and load changes, and at worst it will oscillate out of control altogether. Extremely high gains work well to minimize offset in operational amplifier circuits, only because time delays are negligible between output and input. In applications where large physical processes are being controlled (e.g. furnace temperatures, tank levels, gas pressures, etc.) rather than voltages across small electronic loads, such high controller gains would be met with debilitating oscillations.

If we are limited in how much gain we can program in to the controller, how do we minimize this offset? One way is for a human operator to periodically place the controller in manual mode and move the control valve just a little bit more so the PV once again reaches SP, then place the controller back into automatic mode. In essence this technique adjusts the "Bias" term of the controller equation. The disadvantage of this technique is rather obvious: it requires human intervention. What is the point of having an automation system requiring periodic human intervention to maintain setpoint?

A more sophisticated method for eliminating proportional-only offset is to add a different control action to the controller: one that takes action based on the amount of error between PV and SP and the amount of time that error has existed. We call this control mode *integral*, or *reset*.

29.6   Integral (reset) control

Imagine a liquid-level control system for a vessel, where the position of a level-sensing float sets the position of a potentiometer, which then sets the *speed* of a motor-actuated control valve. If the liquid level is above setpoint, the valve continually opens up; if below setpoint, the valve continually closes off:

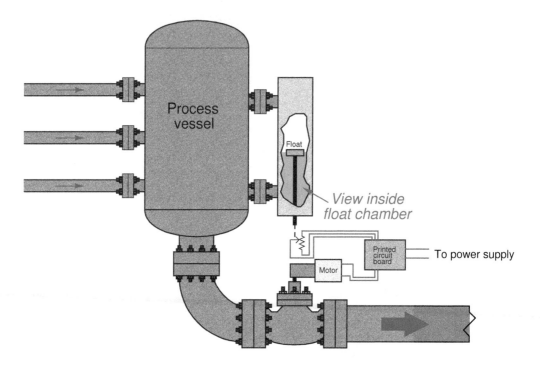

Unlike the *proportional* control system where valve position was a direct function of float position, this control system sets the *speed* of the motor-driven valve according to the float position. The further away from setpoint the liquid level is, the *faster* the valve moves open or closed. In fact, the only time the valve will ever halt its motion is when the liquid level is precisely at setpoint; otherwise, the control valve will be in constant motion.

This control system does its job in a very different manner than the all-mechanical float-based proportional control system illustrated previously. Both systems are capable of regulating liquid level inside the vessel, but they take very different approaches to doing so. One of the most significant differences in control behavior is how the proportional system would inevitably suffer from *offset* (a persistent error between PV and SP), whereas this control system actively works at all times to eliminate offset. The motor-driven control valve literally does not rest until all error has been eliminated!

Instead of characterizing this control system as *proportional*, we call it *integral*[4] in honor of the calculus principle ("integration") whereby small quantities are accumulated over some span to form a total. Don't let the word "calculus" scare you! You are probably already familiar with the concept of numerical integration even though you may have never heard of the term before.

Calculus is a form of mathematics dealing with *changing* variables, and how rates of change relate between different variables. When we "integrate" a variable with respect to time, what we are doing is *accumulating* that variable's value as time progresses. Perhaps the simplest example of this is a vehicle odometer, accumulating the total distance traveled by the vehicle over a certain time period. This stands in contrast to a speedometer, indicating the rate of distance traveled *per* unit of time.

Imagine a car moving along at exactly 30 miles per hour. How far will this vehicle travel after 1 hour of driving this speed? Obviously, it will travel 30 miles. Now, how far will this vehicle travel if it continues for another 2 hours at the exact same speed? Obviously, it will travel 60 more miles, for a total distance of 90 miles since it began moving. If the car's speed is a constant, calculating total distance traveled is a simple matter of multiplying that speed by the travel time.

The odometer mechanism that keeps track of the mileage traveled by the car may be thought of as *integrating* the speed of the car with respect to time. In essence, it is multiplying speed times time continuously to keep a running total of how far the car has gone. When the car is traveling at a high speed, the odometer "integrates" at a faster rate. When the car is traveling slowly, the odometer "integrates" slowly.

If the car travels in reverse, the odometer will decrement (count down) rather than increment (count up) because it sees a negative quantity for speed[5]. The rate at which the odometer decrements depends on how fast the car travels in reverse. When the car is stopped (zero speed), the odometer holds its reading and neither increments nor decrements.

Now let us return to the context of an automated process to see how this calculus principle works inside a process controller. Integration is provided either by a pneumatic mechanism, an electronic opamp circuit, or by a microprocessor executing a digital integration algorithm. The variable being integrated is *error* (the difference between PV and SP) over time. Thus the integral mode of the controller ramps the output either up or down over time in response to the amount of error existing between PV and SP, and the sign of that error. We saw this "ramping" action in the behavior of the liquid level control system using a motor-driven control valve commanded by a float-positioned potentiometer: the valve stem continuously moves so long as the liquid level deviates from setpoint. The reason for this ramping action is to increase or decrease the output *as far as it is necessary* in order to completely eliminate any error and force the process variable to precisely equal setpoint. Unlike proportional action, which simply moves the output an amount proportional to any change in PV or SP, integral control action never stops moving the output until all error is eliminated.

---

[4]An older term for this mode of control is *floating*, which I happen to think is particularly descriptive. With a "floating" controller, the final control element continually "floats" to whatever value it must in order to completely eliminate offset.

[5]At least the old-fashioned mechanical odometers would. Modern cars use a pulse detector on the driveshaft which cannot tell the difference between forward and reverse, and therefore their odometers always increment. Shades of the movie *Ferris Bueller's Day Off*.

If proportional action is defined by the error telling the output how *far* to move, integral action is defined by the error telling the output how *fast* to move. One might think of integral as being how "impatient" the controller is, with integral action constantly ramping the output as far as it needs to go in order to eliminate error. Once the error is zero (PV = SP), of course, the integral action stops ramping, leaving the controller output (valve position) at its last value just like a stopped car's odometer holds a constant value.

If we add an integral term to the controller equation, we get something that looks like this[6]:

$$m = K_p e + \frac{1}{\tau_i} \int e \, dt + b$$

Where,

m = Controller output

e = Error (difference between PV and SP)

K_p = Proportional gain

τ_i = Integral time constant (minutes)

t = Time

b = Bias

The most confusing portion of this equation for those new to calculus is the part that says "$\int e \, dt$". The integration symbol (looks like an elongated letter "S") tells us the controller will accumulate ("sum") multiple products of error (e) over tiny slices of time (dt). Quite literally, the controller multiplies error by time (for very short segments of time, dt) and continuously adds up all those products to contribute to the output signal which then drives the control valve (or other final control element). The integral time constant (τ_i) is a value set by the technician or engineer configuring the controller, proportioning this cumulative action to make it more or less aggressive over time.

To see how this works in a practical sense, let's imagine how a proportional + integral controller would respond to the scenario of a heat exchanger whose inlet temperature suddenly dropped. As we saw with proportional-only control, an inevitable offset occurs between PV and SP with changes in load, because an error *must* develop if the controller is to generate the different output signal value necessary to halt further change in PV. We called this effect *proportional-only offset*.

Once this error develops, though, integral action begins to work. Over time, a larger and larger quantity accumulates in the integral mechanism (or register) of the controller due to the persistent error between PV and SP. That accumulated value adds to the controller's output, driving the steam control valve further and further open. This, of course, adds heat at a faster rate to the heat exchanger, which causes the outlet temperature to rise. As the temperature re-approaches setpoint, the error becomes smaller and thus the integral action proceeds at a slower rate (like a car's odometer incrementing at a slower rate as the car's speed decreases). So long as the PV is below SP (the outlet temperature is still too cool), the controller will continue to integrate upwards, driving the control valve further and further open. Only when the PV rises to exactly meet SP does

---

[6]The equation for a proportional + integral controller is often written without the bias term (b), because the presence of integral action makes it unnecessary. In fact, if we let the integral term completely replace the bias term, we may consider the integral term to be a self-*resetting* bias. This, in fact, is the meaning of the word "reset" in the context of PID controller action: the "reset" term of the controller acts to eliminate offset by continuously adjusting (resetting) the bias as necessary.

integral action finally rest, holding the valve at a steady position. Integral action tirelessly works to eliminate any offset between PV and SP, thus neatly eliminating the offset problem experienced with proportional-only control action.

As with proportional action, there are (unfortunately) two completely opposite ways to specify the degree of integral action offered by a controller. One way is to specify integral action in terms of *minutes* or *minutes per repeat*. A large value of "minutes" for a controller's integral action means a less aggressive integral action over time, just as a large value for proportional band means a less aggressive proportional action. The other way to specify integral action is the inverse: how many *repeats per minute*, equivalent to specifying proportional action in terms of gain (large value means aggressive action). For this reason, you will sometimes see the integral term of a PID equation written differently:

$$\tau_i = \text{minutes per repeat} \qquad K_i = \text{repeats per minute}$$

$$\frac{1}{\tau_i} \int e \, dt \qquad\qquad K_i \int e \, dt$$

Many modern digital electronic controllers allow the user to select the unit they wish to use for integral action, just as they allow a choice between specifying proportional action as gain or as proportional band.

Integral is a highly effective mode of process control. In fact, some processes respond so well to integral controller action that it is possible to operate the control loop on integral action alone, without proportional. Typically, though, process controllers implement some form of proportional plus integral ("PI") control.

Just as too much proportional gain will cause a process control system to oscillate, too much integral action (i.e. an integral time constant that is too short) will also cause oscillation. If the integration happens at too fast a rate, the controller's output will "saturate" either high or low before the process variable can make it back to setpoint. Once this happens, the only condition that will "unwind" the accumulated integral quantity is for an error to develop of the opposite sign, and remain that way long enough for a canceling quantity to accumulate. Thus, the PV must cross over the SP, guaranteeing at least another half-cycle of oscillation.

A similar problem called *reset windup* (or *integral windup*) happens when external conditions make it impossible for the controller to achieve setpoint. Imagine what would happen in the heat exchanger system if the steam boiler suddenly stopped producing steam. As outlet temperature dropped, the controller's proportional action would open up the control valve in a futile effort to raise temperature. If and when steam service is restored, proportional action would just move the valve back to its original position as the process variable returned to its original value (before the boiler died). This is how a proportional-only controller would respond to a steam "outage": nice and predictably. If the controller had integral action, however, a much worse condition would result. All the time spent with the outlet temperature below setpoint causes the controller's integral term to "wind up" in a futile attempt to admit more steam to the heat exchanger. This accumulated quantity can only be un-done by the process variable rising above setpoint for an equal error-time product[7], which means when the steam supply resumes, the temperature will rise well above setpoint

---

[7]Since integration is fundamentally a process of multiplication followed by addition, the units of measurement are always the product (multiplication) of the function's variables. In the case of reset (integral) control, we are multiplying

until the integral action finally "unwinds" and brings the control valve back to a same position again.

Various techniques exist to manage integral windup. Controllers may be built with limits to restrict how far the integral term can accumulate under adverse conditions. In some controllers, integral action may be turned off completely if the error exceeds a certain value. The surest fix for integral windup is human operator intervention, by placing the controller in manual mode. This typically resets the integral accumulator to a value of zero and loads a new value into the bias term of the equation to set the valve position wherever the operator decides. Operators usually wait until the process variable has returned at or near setpoint before releasing the controller into automatic mode again.

While it might appear that operator intervention is again a problem to be avoided (as it was in the case of having to correct for proportional-only offset), it is noteworthy to consider that the conditions leading to integral windup usually occur only during shut-down conditions. It is customary for human operators to run the process manually anyway during a shutdown, and so the switch to manual mode is something they would do anyway and the potential problem of windup often never manifests itself.

Integral control action has the unfortunate tendency to create loop oscillations ("cycling") if the final control element exhibits hysteresis, such as the case with a "sticky" control valve. Imagine for a moment our steam-heated heat exchanger system where the steam control valve possesses excessive packing friction and therefore refuses to move until the applied air pressure changes far enough to overcome that friction, at which point the valve "jumps" to a new position and then "sticks" in that new position. If the valve happens to stick at a stem position resulting in the product temperature settling slightly below setpoint, the controller's integral action will continually increase the output signal going to the valve in an effort to correct this error (as it should). However, when that output signal has risen far enough to overcome valve friction and move the stem further open, it is very likely the stem will once again "stick" but this time do so at a position making the product temperature settle *above* setpoint. The controller's integral action will then ramp downward in an effort to correct this new error, but due to the valve's friction making precise positioning impossible, the controller can never achieve setpoint and therefore it cyclically "hunts" above and below setpoint.

The best solution to this "reset cycling" phenomenon, of course, is to correct the hysteresis in the final control element. Eliminating friction in the control valve will permit precise positioning and allow the controller's integral action to achieve setpoint as designed. Since it is practically impossible to eliminate *all* friction from a control valve, however, other solutions to this problem exist. One of them is to program the controller to stop integrating whenever the error is less than some pre-configured value (sometimes referred to as the "integral deadband" or "reset deadband" of the controller). By activating reset control action only for significant error values, the controller ignores small errors rather than "compulsively" trying to correct for any detected error no matter how small.

---

controller error (the difference between PV and SP, usually expressed in percent) by time (usually expressed in minutes or seconds). Therefore the result will be an "error-time" product. In order for an integral controller to self-recover following windup, the error must switch signs and the error-time product accumulate to a sufficient value to cancel out the error time product accumulated during the windup period.

29.7 Derivative (rate) control

The final element of PID control is the "D" term, which stands for *derivative*. This is a calculus concept like integral, except most people consider it easier to understand. Simply put, derivative is the expression of a variable's *rate-of-change* with respect to another variable. Finding the derivative of a function (differentiation) is the inverse operation of integration. With integration, we calculated accumulated value of some variable's product with time. With derivative, we calculate the ratio of a variable's change per unit of time. Whereas integration is fundamentally a multiplicative operation (products), differentiation always involves division (ratios).

A controller with derivative (or *rate*) action looks at how fast the process variable changes per unit of time, and takes action proportional to that rate of change. In contrast to integral (reset) action which represents the "impatience" of the controller, derivative (rate) action represents the "caution" of the controller.

If the process variable starts to change at a high rate of speed, the job of derivative action is to move the final control element in such a direction as to counteract this rapid change, and thereby moderate the speed at which the process variable changes. In simple terms, derivative action works to limit how fast the error can change.

What this will do is make the controller "cautious" with regard to rapid changes in process variable. If the process variable is headed toward the setpoint value at a rapid rate, the derivative term of the equation will diminish the output signal, thus tempering the controller's response and slowing the process variable's approach toward setpoint. This is analogous to a truck driver preemptively applying the brakes to slow the approach to an intersection, knowing that the heavy truck doesn't "stop on a dime." The heavier the truck's load, the sooner a cautious driver will apply the brakes, to avoid "overshoot" beyond the stop sign and into the intersection. For this reason, derivative control action is also called *pre-act* in addition to being called *rate*, because it acts "ahead of time" to avoid overshoot.

If we modify the controller equation to incorporate differentiation, it will look something like this:

$$m = K_p e + \frac{1}{\tau_i} \int e \, dt + \tau_d \frac{de}{dt} + b$$

Where,
   m = Controller output
   e = Error (difference between PV and SP)
   K_p = Proportional gain
   τ_i = Integral time constant (minutes)
   τ_d = Derivative time constant (minutes)
   t = Time
   b = Bias

The $\frac{de}{dt}$ term of the equation expresses the rate of change of error (e) over time (t). The lower-case letter "d" symbols represent the calculus concept of *differentials* which may be thought of in this context as very tiny increments of the following variables. In other words, $\frac{de}{dt}$ refers to the ratio of a very small change in error (de) over a very small increment of time (dt). On a graph, this is interpreted as the slope of a curve at a specific point (slope being defined as *rise over run*).

It is also possible to build a controller with proportional and derivative actions, but lacking integral action. These are most commonly used in applications prone to wind-up[8], and where the elimination of offset is not critical:

$$m = K_p e + \tau_d \frac{de}{dt} + b$$

Many PID controllers offer the option of calculating derivative response based on rates of change for the process variable (PV) only, rather than the error (PV − SP or SP − PV). This avoids huge "spikes" in the output of the controller if ever a human operator makes a sudden change in setpoint[9]. The mathematical expression for such a controller would look like this:

$$m = K_p e + \frac{1}{\tau_i} \int e \, dt + \tau_d \frac{d\text{PV}}{dt} + b$$

Even when derivative control action is calculated on PV alone (rather than on error), it is still useful for controlling processes dominated by large lag times. The presence of derivative control action in a PID controller generally means the proportional (P) and integral (I) terms may be adjusted more aggressively than before, since derivative (D) will act to limit overshoot. In other words, the judicious presence of derivative action in a PID controller lets us "get away" with using a bit more P and I action than we ordinarily could, resulting in faster approach to setpoint with minimal overshoot.

It should be mentioned that derivative mode should be used with caution. Since it acts on rates of change, derivative action will "go crazy" if it sees substantial noise in the PV signal. Even small amounts of noise possess extremely large rates of change (defined as percent PV change per minute of time) owing to the relatively high frequency of noise compared to the timescale of physical process changes.

Ziegler and Nichols, the engineers who wrote the ground-breaking paper entitled "Optimum Settings for Automatic Controllers" had these words to say regarding "pre-act" control (page 762 of the November 1942 *Transactions of the A.S.M.E.*):

> The latest control effect made its appearance under the trade name "Pre-Act." On some control applications, the addition of pre-act response made such a remarkable improvement that it appeared to be in embodiment of mythical "anticipatory" controllers. On other applications it appeared to be worse than useless. Only the difficulty of predicting the usefulness and adjustment of this response has kept it from being more widely used.

---

[8]An example of such an application is where the output of a loop controller may be "de-selected" or otherwise "over-ridden" by some other control function. This sort of control strategy is often used in energy-conserving controls, where multiple controllers monitoring different process variables selectively command a single FCE.

[9]It should not be assumed that such spikes are always undesirable. In processes characterized by long lag times, such a response may be quite helpful in overcoming that lag for the purpose of rapidly achieving new setpoint values. Slave (secondary) controllers in cascaded systems – where the controller receives its setpoint signal from the output of another (primary, or master) controller – may similarly benefit from derivative action calculated on error instead of just PV. As usual, the specific needs of the application dictate the ideal controller configuration.

29.8 Summary of PID control terms

PID control can be a confusing concept to understand. Here, a brief summary of each term within PID (P. I, and D) is presented for your learning benefit.

29.8.1 Proportional control mode (P)

Proportional – sometimes called *gain* or *sensitivity* – is a control action reproducing changes in input as changes in output. Proportional controller action responds to present changes in input by generating immediate and commensurate changes in output. When you think of "proportional action" (P), think *prompt*: this control action works immediately (never too soon or too late) to match changes in the input signal.

Mathematically defined, proportional action is the ratio of output change to input change. This may be expressed as a quotient of differences, or as a derivative (a rate of change, using calculus notation):

$$\text{Gain value} = \frac{\Delta \text{Output}}{\Delta \text{Input}}$$

$$\text{Gain value} = \frac{d\text{Output}}{d\text{Input}} = \frac{dm}{de}$$

For example, if the PV input of a proportional-only process controller with a gain of 2 suddenly changes ("steps") by 5 percent, and the output will immediately jump by 10 percent (ΔOutput = Gain $\times$ ΔInput). The direction of this output jump in relation to the direction of the input jump depends on whether the controller is configured for direct or reverse action.

A legacy term used to express this same concept is *proportional band*: the mathematical reciprocal of gain. "Proportional band" is defined as the amount of input change necessary to evoke full-scale (100%) output change in a proportional controller. Incidentally, it is always expressed as a percentage, never as fraction or as a per unit value:

$$\text{Proportional Band value} = \frac{\Delta \text{Input}}{\Delta \text{Output}}$$

$$\text{Proportional Band value} = \frac{d\text{Input}}{d\text{Output}} = \frac{de}{dm}$$

Using the same example of a proportional controller exhibiting an output "step" of 10% in response to a PV "step" of 5%, the proportional band would be 50%: the reciprocal of its gain ($\frac{1}{2} = 50\%$). Another way of saying this is that a 50% input "step" would be required to change the output of this controller by a full 100%, since its gain is set to a value of 2.

29.8.2   Integral control mode (I)

Integral – sometimes called *reset* or *floating control* – is a control action causing the output signal to change over time at a rate proportional to the amount of error (the difference between PV and SP values). Integral controller action responds to error accumulated over time, ramping the output signal are far as it needs to go to completely eliminate error. If proportional (P) action tells the output how *far* to move when an error appears, integral (I) action tells the output how *fast* to move when an error appears. If proportional (P) action acts on the *present*, integral (I) action acts on the *past*. Thus, how far the output signal gets driven by integral action depends on the *history* of the error over time: how much error existed, and for how long. When you think of "integral action" (I), think *impatience*: this control action drives the output further and further the longer PV fails to match SP.

Mathematically defined, integral action is the ratio of output *velocity* to input error:

$$\text{Integral value (repeats per minute)} = \frac{\text{Output velocity}}{\text{Input error}}$$

$$\text{Integral value (repeats per minute)} = \frac{\frac{dm}{dt}}{e}$$

An alternate way to express integral action is to use the reciprocal unit of "minutes per repeat." If we define integral action in these terms, the defining equations must be reciprocated:

$$\text{Integral time constant (minutes per repeat)} = \tau_i = \frac{\text{Input error}}{\text{Output velocity}}$$

$$\text{Integral time constant (minutes per repeat)} = \tau_i = \frac{e}{\frac{dm}{dt}}$$

For example, if an error of 5% appears between PV and SP on an integral-only process controller with an integral value of 3 repeats per minute (i.e. an integral time constant of 0.333 minutes per repeat), the output will begin ramping at a rate of 15% per minute ($\frac{dm}{dt} = \text{Integral\_value} \times e$, or $\frac{dm}{dt} = \frac{e}{\tau_i}$). In most PI and PID controllers, integral response is also multiplied by proportional gain, so the same conditions applied to a PI controller that happened to also have a gain of 2 would result in an output ramping rate of 30% per minute ($\frac{dm}{dt} = \text{Gain\_value} \times \text{Integral\_value} \times e$, or $\frac{dm}{dt} = \text{Gain\_value} \times \frac{e}{\tau_i}$). The direction of this ramping in relation to the direction (sign) of the error depends on whether the controller is configured for direct or reverse action.

29.8.3 Derivative control mode (D)

Derivative – sometimes called *rate* or *pre-act* – is a control action causing the output signal to be offset by an amount proportional to the rate at which the input is changing. Derivative controller action responds to how quickly the input changes over time, biasing the output signal commensurate with that rate of input change. If proportional (P) action tells the output how *far* to move when an error appears, derivative (D) action tells the output how far to move when the input *ramps*. If proportional (P) action acts on the *present* and integral (I) action acts on the *past*, derivative (D) action acts on the *future*: it effectively "anticipates" overshoot by tempering the output response according to how fast the process variable is rising or falling. When you think of "derivative action" (D), think *discretion*: this control action is cautious and prudent, working against change.

Mathematically defined, derivative action is the ratio of output offset to input *velocity*:

$$\text{Derivative time constant (minutes)} = \tau_d = \frac{\text{Output offset}}{\text{Input velocity}}$$

$$\text{Derivative time constant (minutes)} = \tau_d = \frac{\Delta \text{Output}}{\frac{de}{dt}}$$

For example, if the PV signal begins to ramp at a rate of 5% per minute on a process controller with a derivative time constant of 4 minutes, the output will immediately become offset by 20% (ΔOutput = Derivative_value $\times \frac{de}{dt}$). In most PD and PID controllers, derivative response is also multiplied by proportional gain, so the same conditions applied to a PD controller that happened to also have a gain of 2 would result in an immediate offset of 40% (ΔOutput = Gain_value $\times$ Derivative_value $\times \frac{de}{dt}$). The direction (sign) of this offset in relation to the direction of the input ramping depends on whether the controller is configured for direct or reverse action.

29.9 P, I, and D responses graphed

A very helpful method for understanding the operation of proportional, integral, and derivative control terms is to analyze their respective responses to the same input conditions over time. This section is divided into subsections showing P, I, and D responses for several different input conditions, in the form of graphs. In each graph, the controller is assumed to be *direct-acting* (i.e. an increase in process variable results in an increase in output).

It should be noted that these graphic illustrations are all qualitative, not quantitative. There is too little information given in each case to plot exact responses. The illustrations of P, I, and D actions focus only on the *shapes* of the responses, not their exact numerical values.

In order to *quantitatively* predict PID controller responses, one would have to know the values of all PID settings, as well as the original starting value of the output before an input change occurred and a time index of when the change(s) occurred.

29.9.1   Responses to a single step-change

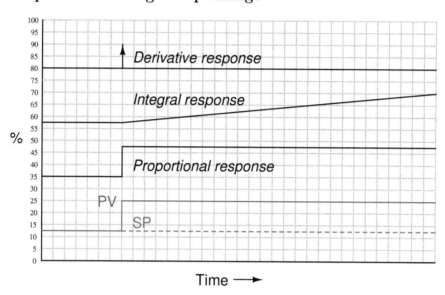

Proportional action directly mimics the shape of the input change (a step). Integral action ramps at a rate proportional to the magnitude of the input step. Since the input step holds a constant value, the integral action ramps at a constant rate (a constant *slope*). Derivative action interprets the step as an *infinite* rate of change, and so generates a "spike[10]" driving the output to saturation.

When combined into one PID output, the three actions produce this response:

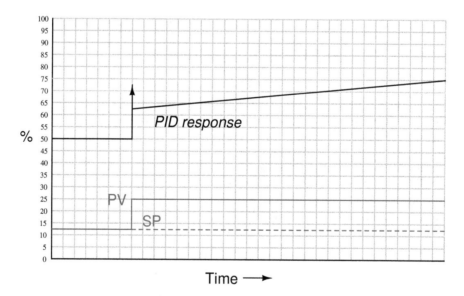

---

[10]This is the meaning of the vertical-pointing arrowheads shown on the trend graph: momentary saturation of the output all the way up to 100%.

29.9.2 Responses to a momentary step-and-return

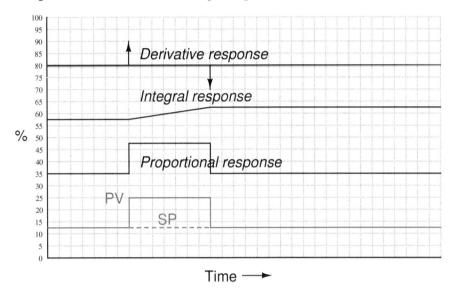

Proportional action directly mimics the shape of the input change (an up-and-down step). Integral action ramps at a rate proportional to the magnitude of the input step, for as long as the PV is unequal to the SP. Once PV = SP again, integral action stops ramping and simply holds the last value[11]. Derivative action interprets both steps as *infinite* rates of change, and so generates "spikes[12]" at the leading and at the trailing edges of the step. Note how the leading (rising) edge causes derivative action to saturate high, while the trailing (falling) edge causes it to saturate low.

---

[11]This is a good example of how integral controller action represents the *history* of the PV − SP error. The continued offset of integral action from its starting point "remembers" the area accumulated under the rectangular "step" between PV and SP. This offset will go away only if a *negative* error appears having the same percent-minute product (area) as the positive error step.

[12]This is the meaning of the vertical-pointing arrowheads shown on the trend graph: momentary saturation of the output all the way up to 100% (or down to 0%).

When combined into one PID output, the three actions produce this response:

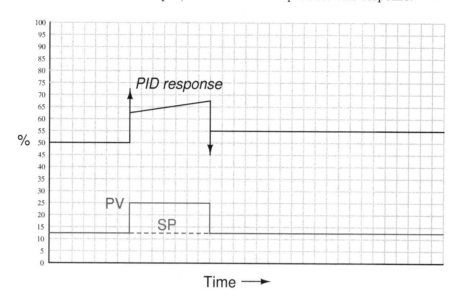

29.9.3   Responses to two momentary steps-and-returns

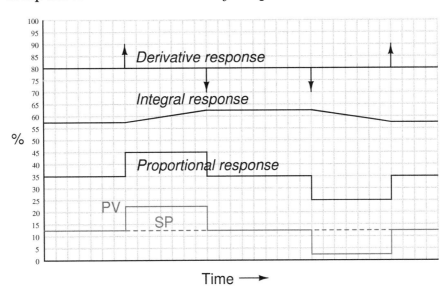

Proportional action directly mimics the shape of all input changes. Integral action ramps at a rate proportional to the magnitude of the input step, for as long as the PV is unequal to the SP. Once PV = SP again, integral action stops ramping and simply holds the last value. Derivative action interprets each step as an *infinite* rate of change, and so generates a "spike" at the leading and at the trailing edges of each step. Note how a leading (rising) edge causes derivative action to saturate high, while a trailing (falling) edge causes it to saturate low.

When combined into one PID output, the three actions produce this response:

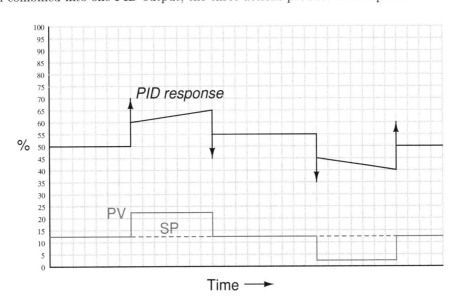

29.9.4   Responses to a ramp-and-hold

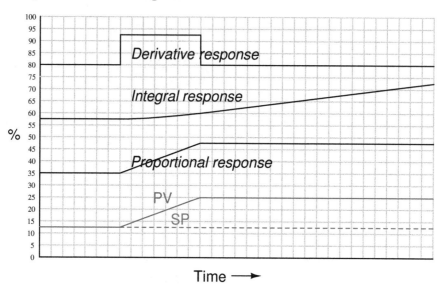

Proportional action directly mimics the ramp-and-hold shape of the input. Integral action ramps slowly at first (when the error is small) but increases ramping rate as error increases. When error stabilizes, integral rate likewise stabilizes. Derivative action offsets the output according to the input's ramping rate.

When combined into one PID output, the three actions produce this response:

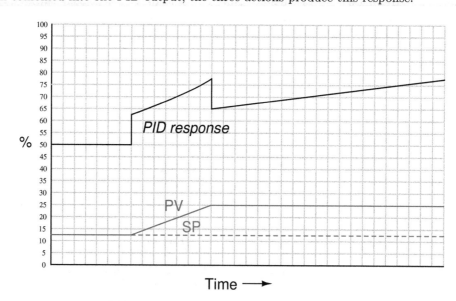

29.9.5 Responses to an up-and-down ramp

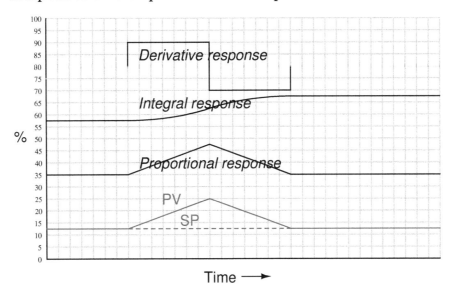

Proportional action directly mimics the up-and-down ramp shape of the input. Integral action ramps slowly at first (when the error is small) but increases ramping rate as error increases, then ramps slower as error decreases back to zero. Once PV = SP again, integral action stops ramping and simply holds the last value. Derivative action offsets the output according to the input's ramping rate: first positive then negative.

When combined into one PID output, the three actions produce this response:

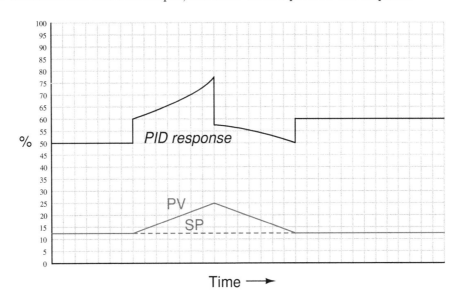

29.9.6   Responses to a multi-slope ramp

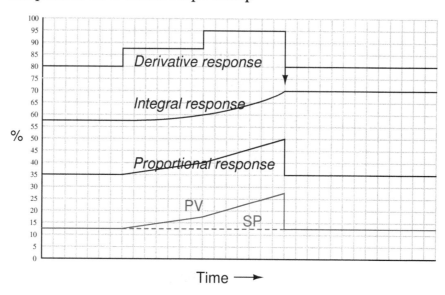

Proportional action directly mimics the ramp shape of the input. Integral action ramps slowly at first (when the error is small) but increases ramping rate as error increases, then accelerates its increase as the PV ramps even steeper. Once PV = SP again, integral action stops ramping and simply holds the last value. Derivative action offsets the output according to the input's ramping rate: first positive, then more positive, then it spikes negative when the PV suddenly returns to SP.

When combined into one PID output, the three actions produce this response:

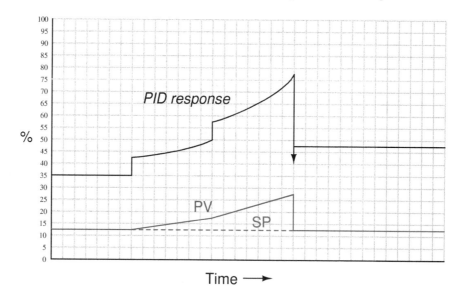

29.9.7 Responses to a multiple ramps and steps

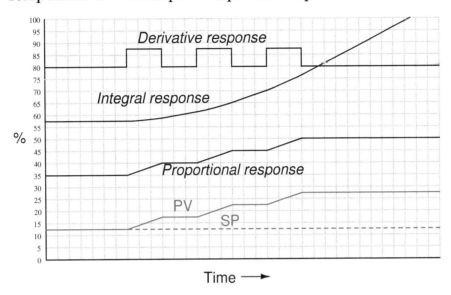

Proportional action directly mimics the ramp-and-step shape of the input. Integral action ramps slowly at first (when the error is small) but increases ramping rate as error increases. Which each higher ramp-and-step in PV, integral action winds up at an ever-increasing rate. Since PV never equals SP again, integral action never stops ramping upward. Derivative action steps with each ramp of the PV.

When combined into one PID output, the three actions produce this response:

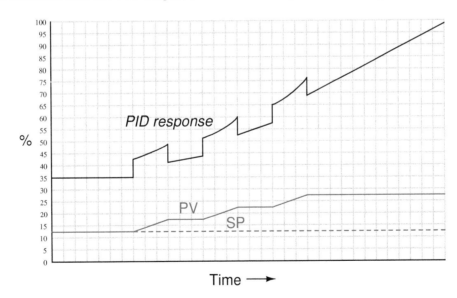

29.9.8   Responses to a sine wavelet

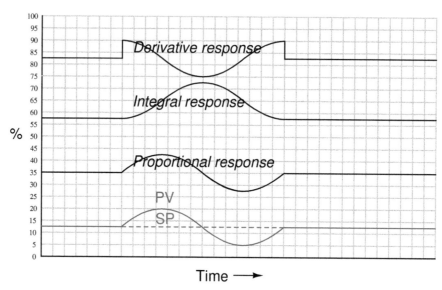

As always, proportional action directly mimics the shape of the input. The 90° phase shift seen in the integral and derivative responses, compared to the PV wavelet, is no accident or coincidence. The derivative of a sinusoidal function is *always* a cosine function, which is mathematically identical to a sine function with the angle advanced by 90°:

$$\frac{d}{dx}(\sin x) = \cos x = \sin(x + 90^o)$$

Conversely, the integral of a sine function is *always* a negative cosine function[13], which is mathematically identical to a sine function with the angle retarded by 90°:

$$\int \sin x \, dx = -\cos x = \sin(x - 90^o)$$

In summary, the derivative operation always adds a positive (leading) phase shift to a sinusoidal input waveform, while the integral operation always adds a negative (lagging) phase shift to a sinusoidal input waveform.

---

[13]In this example, I have omitted the constant of integration (C) to keep things simple. The actual integral is as such: $\int \sin x \, dx = -\cos x + C = \sin(x - 90^o) + C$. This constant value is essential to explaining why the integral response does not immediately "step" like the derivative response does at the beginning of the PV sine wavelet.

When combined into one PID output, these particular integral and derivative actions mostly cancel, since they happen to be sinusoidal wavelets of equal amplitude and opposite phase. Thus, the only way that the final (PID) output differs from proportional-only action in this particular case is the "steps" caused by derivative action responding to the input's sudden rise at the beginning and end of the wavelet:

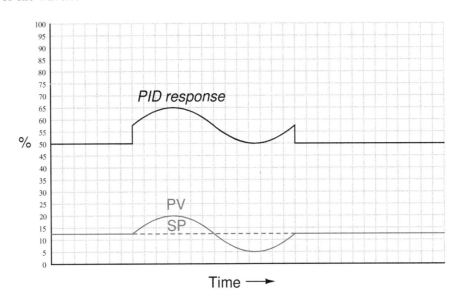

If the I and D tuning parameters were such that the integral and derivative responses were *not* equal in amplitude, their effects would not completely cancel. Rather, the resultant of P, I, and D actions would be a sine wavelet having a phase shift somewhere between -90° and $+90^o$ exclusive, depending on the relative strengths of the P, I, and D actions.

The 90 degree phase shifts associated with the integral and derivative operations are useful to understand when tuning PID controllers. If one is familiar with these phase shift relationships, it is relatively easy to analyze the response of a PID controller to a sinusoidal input (such as when a process oscillates following a sudden load or setpoint change) to determine if the controller's response is dominated by any one of the three actions. This may be helpful in "de-tuning" an over-tuned (overly aggressive) PID controller, if an excess of P, I, or D action may be identified from a phase comparison of PV and output waveforms.

29.9.9   Note to students regarding quantitative graphing

A common exercise for students learning the function of PID controllers is to practice graphing a controller's output given input (PV and SP) conditions, either qualitatively or quantitatively. This can be a frustrating experience for some students, as they struggle to accurately combine the effects of P, I, and/or D responses into a single output trend. Here, I will present a way to ease the pain.

Suppose for example you were tasked with graphing the response of a PD (proportional + derivative) controller to the following PV and SP inputs over time. You are told the controller has a gain of 1, a derivative time constant of 0.3 minutes, and is reverse-acting:

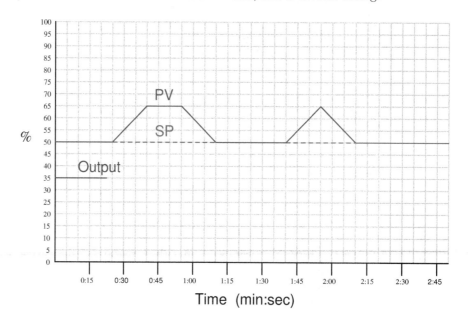

My first recommendation is to *qualitatively* sketch the individual P and D responses. Simply draw two different trends, each one right above or below the given PV/SP trends, showing the shapes of each response over time. You might even find it easier to do if you re-draw the original PV and SP trends on a piece of non-graph paper with the qualitative P and D trends also sketched on the same piece of non-graph paper. The purpose of the qualitative sketches is to separate the task of determining shapes from the task of determining numerical values, in order to simplify the process.

After sketching the separate P and D trends, label each one of the "features" (changes either up or down) in these qualitative trends. This will allow you to more easily combine the effects into one output trend later:

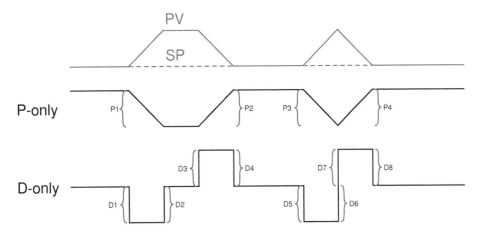

Now, you may qualitatively sketch an output trend combining each of these "features" into one graph. Be sure to label each ramp or step originating with the separate P or D trends, so you know where each "feature" of the combined output graph originates from:

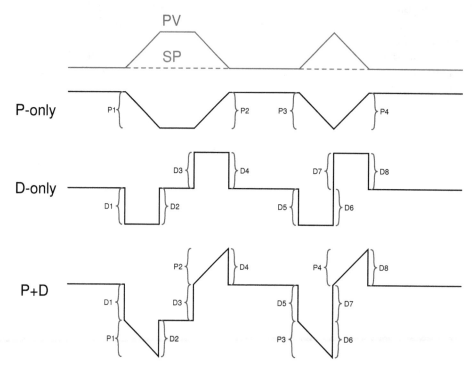

Once the general shape of the output has been qualitatively determined, you may go back to the separate P and D trends to calculate numerical values for each of the labeled "features."

Note that each of the PV ramps is 15% in height, over a time of 15 seconds (one-quarter of a minute). With a controller gain of 1, the proportional response to each of these ramps will also be a ramp that is 15% in height.

Taking our given derivative time constant of 0.3 minutes and multiplying that by the PV's rate-of-change ($\frac{d\text{PV}}{dt}$) during each of its ramping periods (15% per one-quarter minute, or 60% per minute) yields a derivative response of 18% during each of the ramping periods. Thus, each derivative response "step" will be 18% in height.

Going back to the qualitative sketches of P and D actions, and to the combined (qualitative) output sketch, we may apply the calculated values of 15% for each proportional ramp and 18% for each derivative step to the labeled "features." We may also label the starting value of the output trend as given in the original problem (35%), to calculate actual output values at different points in time. Calculating output values at specific points in the graph becomes as easy as cumulatively adding and subtracting the P and D "feature" values to the starting output value:

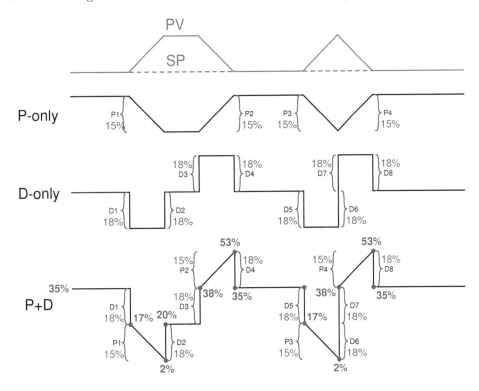

Now that we know the output values at all the critical points, we may quantitatively sketch the output trend on the original graph:

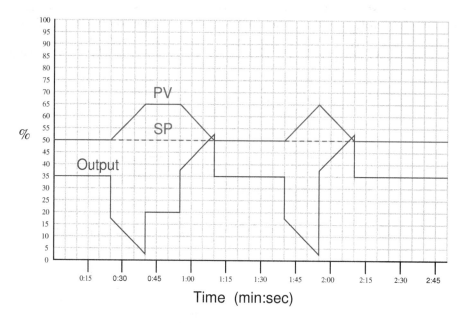

29.10   Different PID equations

For better or worse, there are no fewer than *three* different forms of PID equations implemented in modern PID controllers: the *parallel*, *ideal*, and *series*. Some controllers offer the choice of more than one equation, while others implement just one. It should be noted that more variations of PID equation exist than these three, but that these are the three major variations.

29.10.1 Parallel PID equation

The equation used to describe PID control so far in this chapter is the simplest form, sometimes called the *parallel* equation, because each action (P, I, and D) occurs in separate terms of the equation, with the combined effect being a simple sum:

$$m = K_p e + \frac{1}{\tau_i} \int e\, dt + \tau_d \frac{de}{dt} + b \qquad \textbf{Parallel PID equation}$$

In the parallel equation, each action parameter (K_p, τ_i, τ_d) is independent of the others. At first, this may seem to be an advantage, for it means each adjustment made to the controller should only affect one aspect of its action. However, there are times when it is better to have the gain parameter affect all three control actions (P, I, and D)[14].

We may show the independence of the three actions mathematically, by breaking the equation up into three different parts, each one describing its contribution to the output (Δm):

$$\Delta m = K_p \Delta e \qquad \text{Proportional action}$$

$$\Delta m = \frac{1}{\tau_i} \int e\, dt \qquad \text{Integral action}$$

$$\Delta m = \tau_d \frac{de}{dt} \qquad \text{Derivative action}$$

As you can see, the three portions of this PID equation are completely separate, with each tuning parameter (K_p, τ_i, and τ_d) acting independently within its own term of the equation.

---

[14]An example of a case where it is better for gain (K_p) to influence all three control modes is when a technician re-ranges a transmitter to have a larger or smaller span than before, and must re-tune the controller to maintain the same loop gain as before. If the controller's PID equation takes the parallel form, the technician must adjust the P, I, *and* D tuning parameters proportionately. If the controller's PID equation uses K_p as a factor in all three modes, the technician need only adjust K_p to re-stabilize the loop.

29.10.2   Ideal PID equation

An alternate version of the PID equation designed such that the gain (K_p) affects all three actions is called the *Ideal* or *ISA* equation:

$$m = K_p \left(e + \frac{1}{\tau_i} \int e\, dt + \tau_d \frac{de}{dt} \right) + b \qquad \textbf{Ideal or ISA PID equation}$$

Here, the gain constant (K_p) is distributed to all terms within the parentheses, equally affecting all three control actions. Increasing K_p in this style of PID controller makes the P, the I, *and* the D actions equally more aggressive.

We may show this mathematically, by breaking the "ideal" equation up into three different parts, each one describing its contribution to the output (Δm):

$$\Delta m = K_p \Delta e \qquad \text{Proportional action}$$

$$\Delta m = \frac{K_p}{\tau_i} \int e\, dt \qquad \text{Integral action}$$

$$\Delta m = K_p \tau_d \frac{de}{dt} \qquad \text{Derivative action}$$

As you can see, all three portions of this PID equation are influenced by the gain (K_p) owing to algebraic distribution, but the integral and derivative tuning parameters (τ_i and τ_d) act independently within their own terms of the equation.

29.10.3 Series PID equation

A third version, with origins in the peculiarities of pneumatic controller mechanisms and analog electronic circuits, is called the *Series* or *Interacting* equation:

$$m = K_p \left[\left(\frac{\tau_d}{\tau_i} + 1 \right) e + \frac{1}{\tau_i} \int e \, dt + \tau_d \frac{de}{dt} \right] + b \qquad \textbf{Series or Interacting PID equation}$$

Here, the gain constant (K_p) affects all three actions (P, I, and D) just as with the "ideal" equation. The difference, though, is the fact that both the integral and derivative constants have an effect on proportional action as well! That is to say, adjusting either τ_i or τ_d does not merely adjust those actions, but also influences the aggressiveness of proportional action[15].

We may show this mathematically, by breaking the "series" equation up into three different parts, each one describing its contribution to the output (Δm):

$$\Delta m = K_p \left(\frac{\tau_d}{\tau_i} + 1 \right) \Delta e \qquad \text{Proportional action}$$

$$\Delta m = \frac{K_p}{\tau_i} \int e \, dt \qquad \text{Integral action}$$

$$\Delta m = K_p \tau_d \frac{de}{dt} \qquad \text{Derivative action}$$

As you can see, all three portions of this PID equation are influenced by the gain (K_p) owing to algebraic distribution. However, the proportional term is also affected by the values of the integral and derivative tuning parameters (τ_i and τ_d). Therefore, adjusting τ_i affects both the I and P actions, adjusting τ_d affects both the D and P actions, and adjusting K_p affects all three actions.

This "interacting" equation is an artifact of certain pneumatic and electronic controller designs. Back when these were the dominant technologies, and PID controllers were modularly designed such that integral and derivative actions were separate hardware modules included in a controller at additional cost beyond proportional-only action, the easiest way to implement the integral and derivative actions was in a way that just happened to have an interactive effect on controller gain. In other words, this odd equation form was a sort of compromise made for the purpose of simplifying the physical design of the controller.

Interestingly enough, many digital PID controllers are programmed to implement the "interacting" PID equation even though it is no longer an artifact of controller hardware. The rationale for this programming is to have the digital controller behave identically to the legacy analog electronic or pneumatic controller it is replacing. This way, the proven tuning parameters of the old controller may be plugged into the new digital controller, yielding the same results. In essence, this is a form of "backward compatibility" between digital PID control and analog (electronic or pneumatic) PID control.

---

[15]This becomes especially apparent when using derivative action with low values of τ_i (aggressive integral action). The error-multiplying term $\frac{\tau_d}{\tau_i} + 1$ may become quite large if τ_i is small, even with modest τ_d values.

29.11   Pneumatic PID controllers

A *pneumatic* controller receives a process variable (PV) signal as a variable air pressure, compares that signal against a desired setpoint (SP) value, and then mechanically generates another air pressure signal as the output, driving a final control element.

Throughout this section I will make reference to a pneumatic controller mechanism of my own design. This mechanism does not directly correspond to any particular manufacturer or model of pneumatic controller, but shares characteristics common to many. This design is shown here for the purpose of illustrating the development of P, I, and D control actions in as simple a context as possible.

29.11.1 Proportional control action

Many pneumatic PID controllers use the *force-balance* principle. One or more input signals (in the form of pneumatic pressures) exert a force on a beam by acting through diaphragms, bellows, and/or bourdon tubes, which is then counter acted by the force exerted on the same beam by an output air pressure acting through a diaphragm, bellows, or bourdon tube. The self-balancing mechanical system "tries" to keep the beam motionless through an exact balancing of forces, the beam's position precisely detected by a nozzle/baffle mechanism:

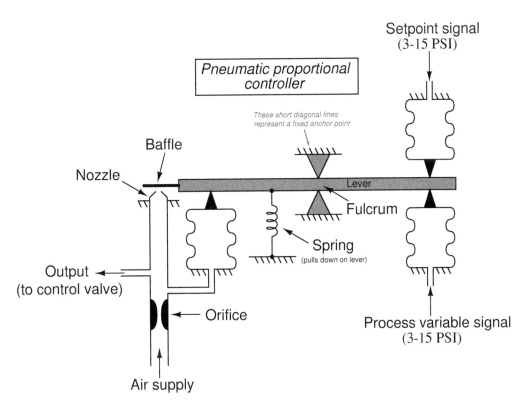

The action of this particular controller is *direct*, since an increase in process variable signal (pressure) results in an increase in output signal (pressure). Increasing process variable (PV) pressure attempts to push the right-hand end of the beam up, causing the baffle to approach the nozzle. This blockage of the nozzle causes the nozzle's pneumatic backpressure to increase, thus increasing the amount of force applied by the output feedback bellows on the left-hand end of the beam and returning the flapper (very nearly) to its original position. If we wished to reverse the controller's action, all we would need to do is swap the pneumatic signal connections between the input bellows, so that the PV pressure was applied to the upper bellows and the SP pressure to the lower bellows.

Any factor influencing the ratio of input pressure(s) to output pressure may be exploited as a gain (proportional band) adjustment in this mechanism. Changing bellows area (either both the PV and SP bellows equally, or the output bellows by itself) would influence this ratio, as would a change in output bellows position (such that it pressed against the beam at some difference distance from the fulcrum point). Moving the fulcrum left or right is also an option for gain control, and in fact is usually the most convenient to engineer.

In this illustration the fulcrum is shown moved to two different positions, to effect a change in gain:

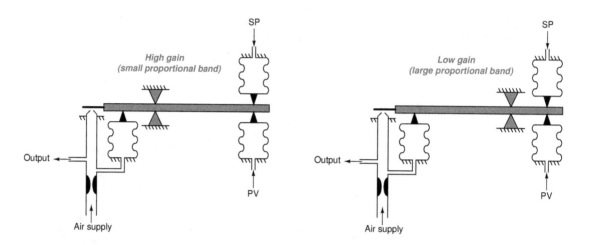

Moving the fulcrum closer to the output bellows places that bellows at a mechanical disadvantage for generating torque (leverage) on the beam. This means any given change in input (PV or SP) force is more difficult for the output bellows to counterbalance. The output pressure, therefore, must change to a greater degree in order for this force-balance mechanism to achieve balance. A greater change in output pressure for a given change in input pressure is the definition of a gain *increase*.

Conversely, moving the fulcrum farther away from the output bellows increases that bellows' mechanical advantage. This additional leverage makes it easier for the output bellows to counteract changes in input force, resulting in less output pressure change required to balance any given input pressure change. A lesser change in output pressure for a given change in input pressure is characteristic of a gain *decrease*.

Some pneumatic controllers employ the *motion-balance* principle instead of the force-balance principle in their operation. In contrast to a force-balance system where opposing forces cancel each other to restrain motion of the mechanism, a motion-balance system freely moves as the signal pressures traverse their working ranges. A simple motion-balance proportional controller design appears here:

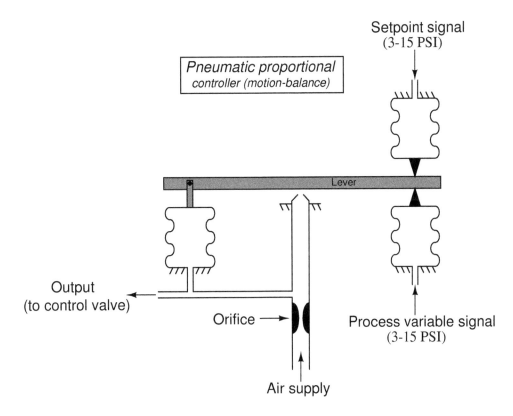

As the process variable signal increases, the right-hand end of the lever is forced up. This motion draws the lever away from the nozzle, resulting in decreased nozzle backpressure. The decreased backpressure causes the output bellows to collapse[16], moving the left-hand end of the lever down and returning the nozzle/lever gap to (approximately) where it was before the PV signal change. This behavior identifies this controller as *reverse-acting*. If direct action were desired, all we would need to do is swap the process variable and setpoint input pressure connections.

Unlike the force-balance controller mechanism where the lever is maintained in an essentially stationary position by equal and opposite forces, the lever in this motion-balance system is free to tilt. In fact, tilting is precisely how a (nearly) constant nozzle gap is maintained: as one end of the lever moves (either up or down), the other end moves in the opposite direction to keep the nozzle/lever gap constant in the middle.

---

[16]Being a motion-balance mechanism, these bellows must act as spring elements in order to produce consistent pressure/motion behavior. Some pneumatic controllers employ coil springs inside the brass bellows assembly to provide the necessary "stiffness" and repeatability.

The gain of such a mechanism may be changed by moving the position of the nozzle along the lever's length. However, it must be understood that this position change will have the opposite effect on gain compared with the fulcrum position change described for the force-balance mechanism. Here in the motion-balance system, it is the relative *travel* of each bellows that matters for gain, not the relative *leverage* (torque):

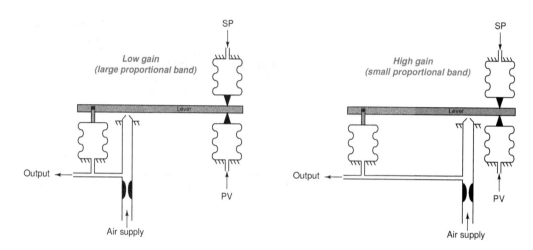

With the nozzle positioned closer to the output bellows, that bellows need not stretch or collapse as much in order to maintain the nozzle gap constant even with a large motion at the input (right-hand) end of the lever. The output pressure in this case will change only slightly for large changes in PV or SP pressures: characteristic of a low gain.

Moving the nozzle closer to the input (PV and SP) bellows gives those bellows more influence over the nozzle/lever gap. The output bellows must expand and contract quite a bit more than the input bellows in order to maintain a constant nozzle gap for any motion at the input side. This requires a greater change in output pressure for a given change in input pressure: the definition of increased gain.

29.11.2 Automatic and manual modes

A more practical pneumatic proportional controller mechanism is shown in the next illustration, complete with setpoint and bias adjustments, and a manual control mode:

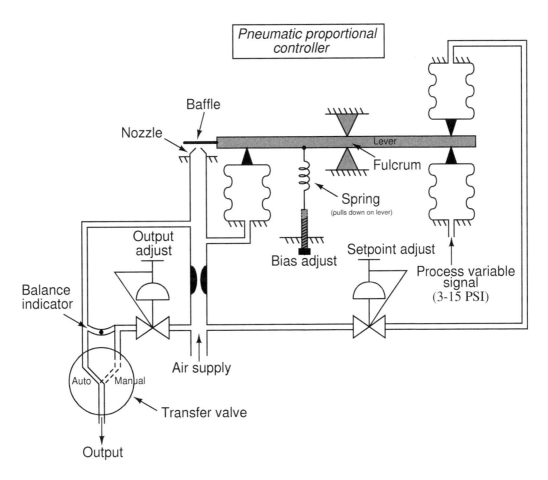

"Bumpless" transfer between automatic and manual modes is a very important feature for any loop controller because it allows human operators to change the mode of the controller without introducing an unnecessary disturbance to the process being controlled. Without provision for bumpless transfer, the output signal of the controller may suddenly change whenever the mode is switched between automatic and manual. This sudden signal change will cause the final control element to suddenly "step" to some new level of effect on the process.

In this particular pneumatic controller, bumpless auto/manual transfer is accomplished by the operator paying attention to the *balance indicator* revealing any air pressure difference between the output bellows and the output adjust pressure regulator. When in automatic mode, a switch to manual mode involves adjusting the output regulator until the balance indicator registers zero pressure difference, then switching the transfer valve to the "manual" position. The controller output is then at the direct command of the output adjust pressure regulator, and will not respond to changes in either PV or SP. "Bumplessly" switching back to automatic mode requires that the setpoint pressure regulator be adjusted until the balance indicator once again registers zero pressure difference, then switching the transfer valve to the "auto" position. The controller output will once again respond to changes in PV and SP.

A photograph showing a Foxboro model 43AP pneumatic controller manual/auto transfer switch and balance indicator appears here:

The metal ball within the curved plastic tube indicates equal pressures between automatic and manual modes when centered in the tube. To achieve bumpless transfer between automatic and manual modes, one must never switch the auto/manual valve unless that ball is centered. To center the ball while in automatic mode, the manual output pressure must be adjusted to achieve balance with the automatic-mode output pressure. To center the ball while in manual mode, the automatic-mode output pressure must be adjusted to achieve balance with the manual-mode output pressure – a condition attained by adjusting the *setpoint* knob.

29.11.3 Derivative control action

Derivative (rate) control action is relatively easy to add to this pneumatic controller mechanism. All we need to do is place a restrictor valve between the nozzle tube and the output feedback bellows, causing the bellows to delay filling or emptying its air pressure over time:

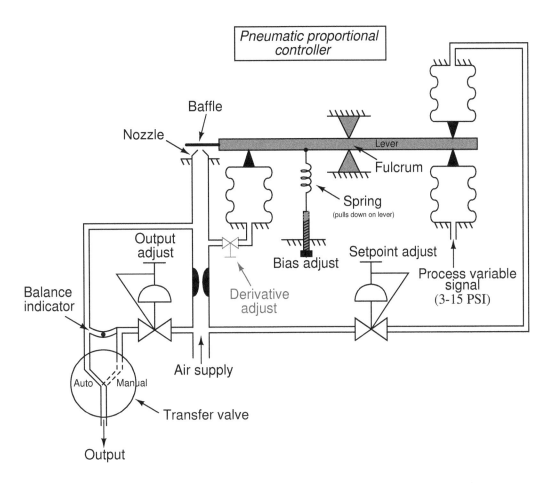

If any sudden change occurs in PV or SP, the output pressure will saturate before the output bellows has the opportunity to equalize in pressure with the output signal tube. Thus, the output pressure "spikes" with any sudden "step change" in input: exactly what we would expect with derivative control action.

If either the PV or the SP ramps over time, the output signal will ramp in direct proportion (proportional action), but there will *also* be an added offset of pressure at the output signal in order to keep air flowing either in or out of the output bellows at a constant rate to generate the force necessary to balance the changing input signal. Thus, derivative action causes the output pressure to shift either up or down (depending on the direction of input change) more than it would with just proportional action alone in response to a ramping input: exactly what we would expect from a controller with both proportional and derivative control actions.

29.11.4 Integral control action

Adding integral action to our hypothetical pneumatic controller mechanism requires the placement of a second bellows (a "reset" bellows) opposite the output feedback bellows, and another restrictor valve to the mechanism[17]:

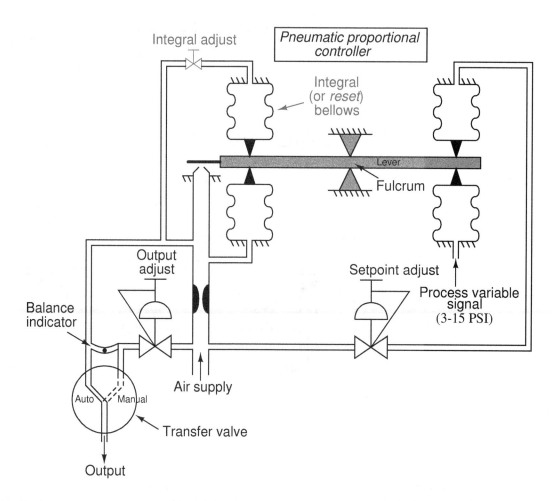

This second bellows takes air pressure from the output line and translates it into force that opposes the original feedback bellows. At first, this may seem counter-productive, for it nullifies the ability of this mechanism to continuously balance the force generated by the PV and SP bellows. Indeed, it would render the force-balance system completely ineffectual if this new "reset" bellows were allowed to inflate and deflate with no time lag. However, with a time lag provided by the

---

[17]Practical integral action also requires the elimination of the bias spring and adjustment, which formerly provided a constant downward force on the left-hand side of the beam to give the output signal the positive offset necessary to avoid saturation at 0 PSI. Not only is a bias adjustment completely unnecessary with the addition of integral action, but it would actually cause problems by making the integral action "think" an error existed between PV and SP when there was none.

restriction of the integral adjustment valve and the volume of the bellows (a sort of pneumatic "RC time constant"), the nullifying force of this bellows becomes delayed over time. As this bellows slowly fills (or empties) with pressurized air from the nozzle, the change in force on the beam causes the regular output bellows to have to "stay ahead" of the reset bellows action by constantly filling (or emptying) at some rate over time.

To better understand this integrating action, let us perform a "thought experiment" on a simplified version of the controller. The following mechanism has been stripped of all unnecessary complexity so that we may focus on just the proportional and integral actions. Here, we envision the PV and SP air pressure signals differing by 3 PSI, causing the force-balance mechanism to instantly respond with a 3 PSI output pressure to the feedback bellows (assuming a central fulcrum location, giving a controller gain of 1). The reset (integral) valve has been completely shut off at the start of this thought experiment:

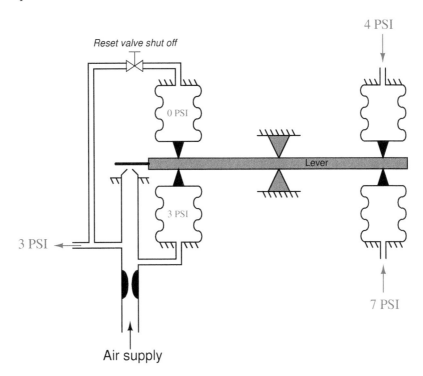

With 0 PSI of air pressure in the reset bellows, it is as though the reset bellows does not exist at all. The mechanism is a simple proportional-only pneumatic controller.

Now, imagine opening up the reset valve just a little bit, so that the output air pressure of 3 PSI begins to slowly fill the reset bellows. As the reset bellows fills with pressurized air, it begins to push down on the left-hand end of the force beam. This forces the baffle closer to the nozzle, causing the output pressure to rise. The regular output bellows has no restrictor valve to impede its filling, and so it *immediately* applies more upward force on the beam with the rising output pressure. With this greater output pressure, the reset bellows has an even greater "final" pressure to achieve, and so its rate of filling continues.

The result of these two bellows' opposing forces (one instantaneous, one time-delayed) is that the lower bellows' pressure must always *lead 3 PSI ahead of the upper bellows' pressure* in order to maintain a pressure difference of 3 PSI necessary to balance force with the PV and SP bellows (whose pressures differ by 3 PSI). This creates a constant 3 PSI differential pressure across the reset restriction valve, resulting in a constant flow of air into the reset bellows at a rate determined by that pressure drop and the opening of the restrictor valve. Eventually this will cause the output pressure to saturate at maximum, but until then the practical importance of this rising pressure action is that the mechanism now exhibits *integral control response* to the constant error between PV and SP:

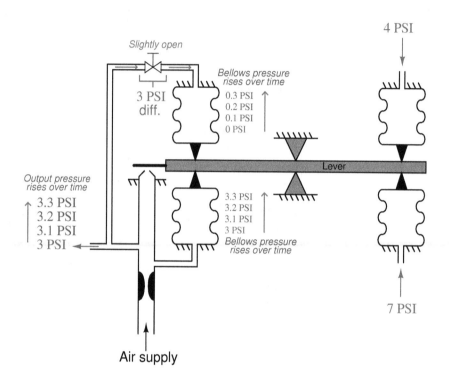

The greater the difference in pressures between PV and SP (i.e. the greater the *error*), the more pressure drop will develop across the reset restriction valve, causing the reset bellows to fill (or empty, depending on the sign of the error) with compressed air at a faster rate[18], causing the output pressure to change at a faster rate. Thus, we see in this mechanism the defining nature of integral control action: that the magnitude of the error determines the *velocity* of the output signal (its rate of change over time, or $\frac{dm}{dt}$). The rate of integration may be finely adjusted by changing the opening of the restrictor valve, or adjusted in large steps by connecting *capacity tanks* to the reset bellows to greatly increase its effective volume.

---

[18]These restrictor valves are designed to encourage laminar air flow, making the relationship between volumetric flow rate and differential pressure drop *linear* rather than quadratic as it is for large control valves. Thus, a doubling of pressure drop across the restrictor valve results in a doubling of flow rate into (or out of) the reset bellows, and a consequent doubling of integration rate. This is precisely what we desire and expect from a controller with integral action.

29.11.5 Fisher MultiTrol

Front (left) and rear (right) photographs of a real pneumatic controller (a Fisher "MultiTrol" unit) appear here:

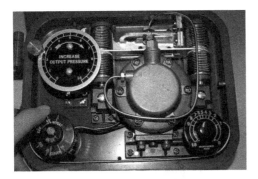 

The mechanism is remarkably similar to the one used throughout the explanatory discussion, with the important distinction of being *motion-balance* instead of force balance. Proportional and integral control modes are implemented through the actions of four brass bellows pushing as opposing pairs at either end of a beam:

The nozzle may be seen facing down at the middle of the beam, with the center of the beam acting as a baffle. Setpoint control is achieved by moving the position of the nozzle up and down with respect to the beam. A setpoint dial (labeled "Increase Output Pressure") turns a cam which moves the nozzle closer to or farther away from the beam. This being a motion-balance system, an offset in nozzle position equates to a biasing of the output signal, causing the controller to seek a new process variable value.

Instead of altering the position of a fulcrum to alter the gain (proportional band) of this controller, gain control is effected through the use of a "pressure divider" valve proportioning the amount of output air pressure sent to the feedback bellows. Integral rate control is implemented exactly the same way as in the hypothetical controller mechanism illustrated in the discussion: by adjusting a valve restricting air flow to and from the reset bellows. Both valves are actuated by rotary knobs with calibrated scales. The reset knob is actually calibrated in units of minutes per repeat, while the proportional band knob is labeled with a scale of arbitrary numbers:

Selection of direct versus reverse action is accomplished in the same way as selection between proportional and snap-action (on-off) control: by movable manifolds re-directing air pressure signals to different bellows in the mechanism. The direct/reverse manifold appears in the left-hand photograph (the letter "D" stands for *direct* action) while the proportional/snap manifold appears in the right-hand photograph (the letter "P" stands for *proportional* control):

Either setting is made by removing the screw holding the manifold plate to the controller body, rotating the plate one-quarter turn, and re-attaching. The following photograph shows one of the manifold plates removed and turned upside-down for inspection of the air passages:

The two quarter-circumference slots seen in the manifold plate connect adjacent air ports together. Rotating the plate 90 degrees connects the four air ports together as two different pairs.

29.11.6 Foxboro model 43AP

The Fisher MultiTrol pneumatic controller is a very simple device, intended for field-mounting near the pneumatic transmitter and control valve to form a control loop for non-precision applications. A more sophisticated field-mounted pneumatic controller is the Foxboro model 43AP, sporting actual PV and SP indicating pointers, plus more precise tuning controls. The following photographs show one of these controllers, with the access door closed (left) and open (right):

At the heart of this controller is a motion-balance "pneumatic control unit" mechanism. A dial for setting proportional band (and direct/reverse action) appears on the front of the mechanism:

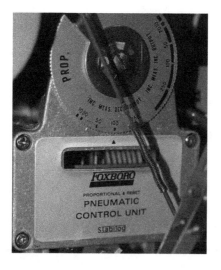 

Note the simple way in which direct and reverse actions are described on this dial: either *increasing measurement, decreasing output* (reverse action) or *increasing measurement, increasing output* (direct action).

29.11.7   Foxboro model 130

Foxboro also manufactured panel-mounted pneumatic controllers, the model 130 series, for larger-scale applications where multiple controllers needed to be located in one compact space. A bank of four Foxboro model 130 pneumatic controllers appears in the next photograph:

Each controller may be partially removed (slid out) from its slot in the rack, the P, I, and D settings adjustable on the left side panel with a screwdriver:

 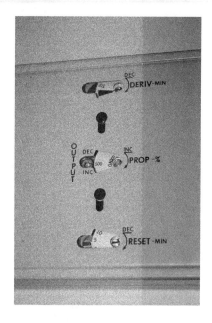

With the side panel removed, the entire mechanism is open to viewing:

The heart of the model 130 controller is a four-bellows force-balance mechanism, identical in principle to the hypothetical force-balance PID controller mechanism used throughout the explanatory discussion. Instead of the four bellows acting against a straight beam, however, these bellows push against a circular disk:

A nozzle (shown in the next photograph) detects if the disk is out of position (unbalanced), sending a back-pressure signal to an amplifying relay which then drives the feedback bellows:

The disk rocks along an axis established by a movable bar. As this bar is rotated at different angles relative to the face of the disk, the fulcrum shifts with respect to the four bellows, providing a simple and effective gain adjustment:

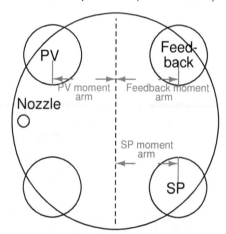
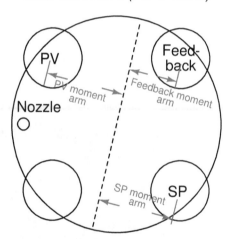

If the moment arms (lever lengths) between the input (PV and SP) bellows and the feedback bellows are equal, both sets of bellows will have equal leverage, and the gain will be one (a proportional band setting of 100%). However, if the fulcrum bar is rotated to give the input bellows more leverage and the feedback bellows less leverage, the feedback bellows will have to "work harder" (exert more force) to counteract any imbalance of force created by the input (PV and SP) bellows, thus creating a greater gain: more output pressure for the same amount of input pressure.

The fourth (lower-left) bellows acting on the disk provides an optional reset (integral) function. Its moment arm (lever length) of course is always equal to that of the feedback bellows, just as the PV and SP bellows' moment arm lengths are always equal, being positioned opposite the fulcrum line.

Selection between direct and reverse action works on the exact same principle as in the Fisher MultiTrol controller – by connecting four air ports in one of two paired configurations. A selector (movable with a hex wrench) turns an air signal port "switch" on the bottom of the four-bellows unit, effectively switching the PV and SP bellows:

An interesting characteristic of most pneumatic controllers is modularity of function: it is possible to order a pneumatic controller that is proportional-only (P), proportional plus integral (P+I), or full PID. Since each control mode requires additional components to implement, a P-only pneumatic controller costs less than a P+I pneumatic controller, which in turn costs less than a full PID pneumatic controller. This explains the relative scarcity of full PID pneumatic controllers in industry: why pay for additional functionality if less will suffice for the task at hand?

29.11.8 External reset (integral) feedback

Some pneumatic controllers come equipped with an option for *external reset*: a feature useful in control systems to avoid integral windup if and when the process stops responding to changes in controller output. Instead of receiving a pneumatic signal directly from the output line of the controller, the reset bellows receives its signal through another pneumatic line, connected to a location in the control system where the final *effect* of the output signal (m) is seen. If for some reason the final control element cannot achieve the state called for by the controller, the controller will sense this through the external reset signal, and will cease integration to avoid "wind-up."

In the following illustration[19], the external reset signal comes from a pneumatic *position transmitter* (ZT) mounted to the sliding stem of the control valve, sending back a 3-15 PSI signal representing valve stem position:

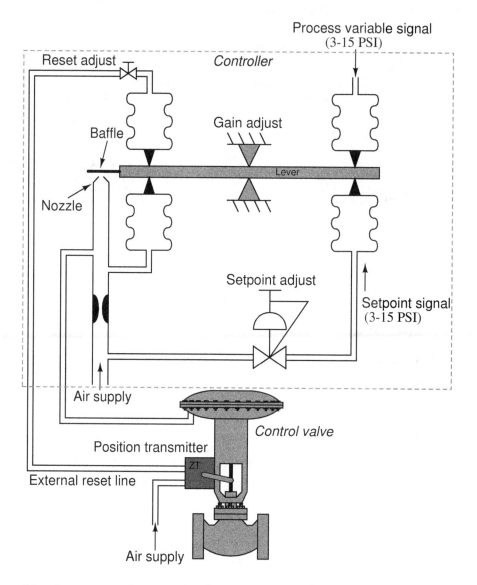

If something happens to the control valve causing it to freeze position when the controller commands it to move – suppose the stem encounters a mechanical "stop" limiting travel, or a piece of solid material jams the valve trim so it cannot close further – the pneumatic pressure signal sent from the position transmitter to the controller's reset bellows will similarly freeze. After the

---

[19]In case you are wondering, this controller happens to be *reverse-acting* instead of direct. This is of no consequence to the feature of external reset.

pneumatic lag caused by the reset restrictor valve and bellows passes, the reset bellows force will remain fixed. This halts the controller's integral action, which was formerly based on a "race" between the output feedback bellows and the reset bellows, causing the feedback bellows to "lead" the reset bellows pressure by an amount proportional to the error between PV and SP. This "race" caused the output pressure to wind either up or down depending on the sign of the error. Now that the reset bellows pressure is frozen due to the control valve stem position being frozen, however, the "race" comes to an end and the controller exhibits only proportional action. Thus, the dreaded effect of integral windup – where the integral action of a controller continues to act even though the change in output is of no effect on the process – is averted.

29.12 Analog electronic PID controllers

Although analog electronic process controllers are considered a newer technology than pneumatic process controllers, they are actually "more obsolete" than pneumatic controllers. Panel-mounted (inside a control room environment) analog electronic controllers were a great improvement over panel-mounted pneumatic controllers when they were first introduced to industry, but they were superseded by digital controller technology later on. Field-mounted pneumatic controllers were either replaced by panel-mounted electronic controllers (either analog or digital) or left alone. Applications still exist for field-mounted pneumatic controllers, even now at the beginning of the 21st century, but very few applications exist for analog electronic controllers in any location.

Analog electronic controllers enjoy two inherent advantages over digital electronic controllers: greater reliability[20] and faster response. However, these advantages have been diminishing as digital control technology has advanced. Today's digital electronic technology is far more reliable than the digital technology available during the heyday of analog electronic controllers. Now that digital controls have achieved very high levels of reliability, the first advantage of analog control is largely academic[21], leaving only the second advantage for practical consideration. The advantage of faster speed may be fruitful in applications such as motion control, but for most industrial processes even the slowest digital controller is fast enough[22]. Furthermore, the numerous advantages offered by digital technology (data recording, networking capability, self-diagnostics, flexible configuration, function blocks for implementing different control strategies) severely weaken the relative importance of reliability and speed.

Most analog electronic PID controllers utilize *operational amplifiers* in their designs. It is relatively easy to construct circuits performing amplification (gain), integration, differentiation, summation, and other useful control functions with just a few opamps, resistors, and capacitors.

---

[20]The reason for this is the low component count compared to a comparable digital control circuit. For any given technology, a simpler device will tend to be more reliable than a complex device if only due to there being fewer components to fail. This also suggests a third advantage of analog controllers over digital controllers, and that is the possibility of easily designing and constructing your own for some custom application such as a hobby project. A digital controller is not outside the reach of a serious hobbyist to design and build, but it is definitely more challenging due to the requirement of programming expertise in addition to electronic hardware expertise.

[21]It is noteworthy that analog control systems are completely immune from "cyber-attacks" (malicious attempts to foil the integrity of a control system by remote access), due to the simple fact that their algorithms are fixed by physical laws and properties of electronic components rather than by code which may be edited. This new threat constitutes an inherent weakness of digital technology, and has spurred some thinkers in the field to reconsider analog controls for the most critical applications.

[22]The real problem with digital controller speed is that the time delay between successive "scans" translates into dead time for the control loop. Dead time is the single greatest impediment to feedback control.

29.12.1   Proportional control action

The basic proportional-only control algorithm follows this formula:

$$m = K_p e + b$$

Where,

m = Controller output

e = Error (difference between PV and SP)

K_p = Proportional gain

b = Bias

The "error" variable (e) is the mathematical difference between process variable and setpoint. If the controller is direct-acting, $e = \mathrm{PV} - \mathrm{SP}$. If the controller is reverse-acting, $e = \mathrm{SP} - \mathrm{PV}$. Thus,

$$m = K_p(\mathrm{PV} - \mathrm{SP}) + b \qquad \text{Direct-acting}$$

$$m = K_p(\mathrm{SP} - \mathrm{PV}) + b \qquad \text{Reverse-acting}$$

Mathematical operations such as subtraction, multiplication by a constant, and addition are quite easy to perform using analog electronic (operational amplifier) circuitry. Prior to the advent of reliable digital electronics for industrial applications, it was natural to use analog electronic circuitry to perform proportional control for process control loops.

For example, the subtraction function necessary to calculate error (e) from process variable and setpoint signals may be performed with a three-amplifier "subtractor" circuit:

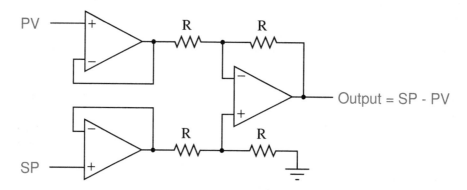

This particular subtractor circuit calculates error for a reverse-acting controller. As the PV signal increases, the error signal decreases (becomes more negative). It could be modified for direct action simply by swapping the two inputs: SP on top and PV on bottom such that the Output becomes PV − SP.

Gain is really nothing more than multiplication by a constant, in this case the constant being K_p. A very simple one-amplifier analog circuit for performing this multiplication is the *inverting*[23] *amplifier* circuit:

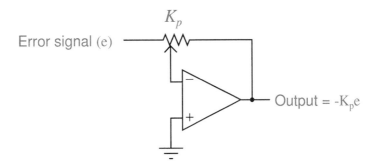

With the potentiometer's wiper in mid-position, the voltage gain of this circuit will be 1 (with an inverted polarity which we shall ignore for now). Moving the wiper toward the left-hand side of the potentiometer increases the circuit's gain past unity, while moving the wiper toward the right-hand side of the potentiometer decreases the gain toward zero.

In order to add the bias (*b*) term in the proportional control equation, we need an analog circuit capable of summing two voltage signals. This need is nicely met in the *inverting summer* circuit, shown here:

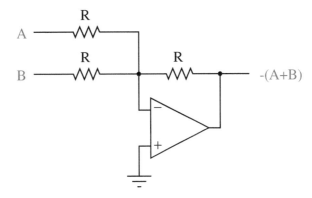

---

[23]This circuit configuration is called "inverting" because the mathematical sign of the output is always opposite that of the input. This sign inversion is not an intentional circuit feature, but rather a consequence of the input signal facing the opamp's inverting input. Non-inverting multiplier circuits also exist, but are more complicated when built to achieve multiplication factors less than one.

Combining all these analog functions together into one circuit, and adding a few extra features such as direct/reverse action selection, bias adjustment, and manual control with a null voltmeter to facilitate bumpless mode transfer, gives us this complete analog electronic proportional controller:

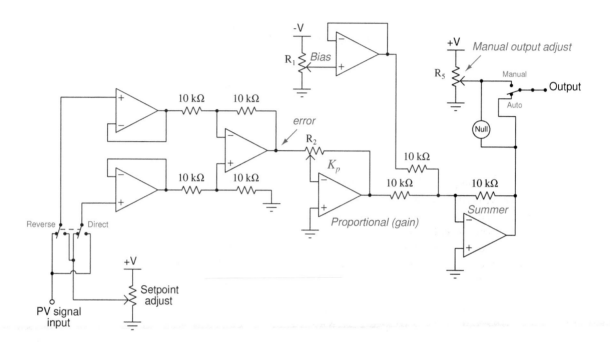

29.12.2   Derivative and integral control actions

Differentiating and integrating live voltage signals with respect to time is quite simple using operational amplifier circuits. Instead of using all resistors in the negative feedback network, we may implement these calculus functions by using a combination of *capacitors* and resistors, exploiting the capacitor's natural derivative relationship between voltage and current:

$$I = C\frac{dV}{dt}$$

Where,

I = Current through the capacitor (amperes)

C = Capacitance of capacitor (farads)

V = Voltage across the capacitor (volts)

$\frac{dV}{dt}$ = Rate-of-change of voltage across the capacitor (volts per second)

If we build an operational amplifier with a resistor providing negative feedback current through a capacitor, we create a *differentiator* circuit where the output voltage is proportional to the rate-of-change of the input voltage:

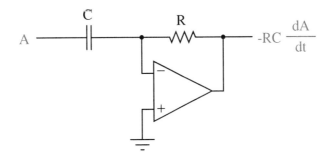

Since the inverting input of the operational amplifier is held to ground potential by feedback (a "virtual ground"), the capacitor experiences the full input voltage of signal A. So, as A varies over time, the current through that capacitor will directly represent the signal A's rate of change over time ($I = C\frac{dA}{dt}$). This current passes through the feedback resistor, creating a voltage drop at the output of the amplifier directly proportional to signal A's rate of change over time. Thus, the output voltage of this circuit reflects the input voltage's instantaneous rate of change, albeit with an inverted polarity. The mathematical term RC is the *time constant* of this circuit. For a differentiator circuit such as this, we typically symbolize its time constant as τ_d (the "derivative" time constant).

For example, if the input voltage to this differentiator circuit were to ramp at a constant rate of $+4.3$ volts per second (rising) with a resistor value of 10 kΩ and a capacitor value of 33 μF (i.e. τ_d = 0.33 seconds), the output voltage would be a constant -1.419 volts:

$$V_{out} = -RC\frac{dV_{in}}{dt}$$

$$V_{out} = -(10000\ \Omega)(33 \times 10^{-6}\ \text{F})\left(\frac{4.3\ \text{V}}{\text{s}}\right)$$

$$V_{out} = -(0.33\ \text{s})\left(\frac{4.3\ \text{V}}{\text{s}}\right)$$

$$V_{out} = -1.419\ \text{V}$$

Recall that the purpose of derivative action in a PID controller is to react to sudden changes in either the error (e) or the process variable (PV). This circuit fulfills that function, by generating an output proportional to the input voltage's rate of change.

If we simply swap[24] the locations of the resistor and capacitor in the feedback network of this operational amplifier circuit, we create an *integrator* circuit where the output voltage rate-of-change is proportional to the input voltage:

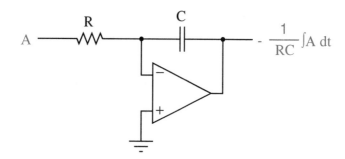

This integrator circuit provides the exact inverse function of the differentiator. Rather than a changing input signal generating an output signal proportional to the input's rate of change, an input signal in this circuit controls the rate at which the output signal changes.

The way it works is by acting as a *current source*, pumping current into the capacitor at a value determined by the input voltage and the resistor value. Just as in the previous (differentiator) circuit where the inverting terminal of the amplifier was a "virtual ground" point, the input voltage in this circuit is impressed across the resistor R. This creates a current which must go through capacitor C on its way either to or from the amplifier's output terminal. As we have seen in the capacitor's equation ($I = C\frac{dV}{dt}$), a current forced through a capacitor causes the capacitor's voltage to change over time. This changing voltage becomes the output signal of the integrator circuit. As in the case of the differentiator circuit, the mathematical term RC is the *time constant* of this circuit as well. Being an integrator, we customarily represent this "integral" time constant as τ_i.

Any amount of change in output voltage (ΔV_{out}) occurring between some initial time (t_0) and a finishing time (t_f) may be calculated by the following integral:

$$\Delta V_{out} = -\frac{1}{RC} \int_{t_0}^{t_f} V_{in}\, dt$$

If we wish to know the absolute output voltage at the end of that time interval, all we need to do is add the circuit's initial output voltage (V_0, i.e. the voltage stored in the capacitor at the initial time t_0) to the calculated change:

$$V_{out} = -\frac{1}{RC} \int_{t_0}^{t_f} V_{in}\, dt + V_0$$

---

[24]This inversion of function caused by the swapping of input and feedback components in an operational amplifier circuit points to a fundamental principle of negative feedback networks: namely, that placing a mathematical element within the feedback loop causes the amplifier to exhibit the inverse of that element's intrinsic function. This is why voltage dividers placed within the feedback loop cause an opamp to have a multiplicative gain (division → multiplication). A circuit element exhibiting a logarithmic response, when placed within a negative feedback loop, will cause the amplifier to exhibit an exponential response (logarithm → exponent). Here, an element having a time-differentiating response, when placed inside the feedback loop, causes the amplifier to time-integrate (differentiation → integration). Since the opamp's output voltage must assume any value possible to maintain (nearly) zero differential voltage at the input terminals, placing a mathematical function in the feedback loop forces the output to assume the inverse of that function in order to "cancel out" its effects and achieve balance at the input terminals.

For example, if we were to input a constant DC voltage of $+1.7$ volts to this circuit with a resistor value of 81 kΩ and a capacitor value of 47 μF (i.e. $\tau_i = 3.807$ seconds), the output voltage would ramp at a constant rate of -0.447 volts per second[25]. If the output voltage were to begin at -3.0 volts and be allowed to ramp for exactly 12 seconds at this rate, it would reach a value of -8.359 volts at the conclusion of that time interval:

$$V_{out} = -\frac{1}{RC} \int_{t_0}^{t_f} V_{in}\, dt + V_0$$

$$V_{out} = -\left(\frac{1}{(81000\ \Omega)(47 \times 10^{-6}\ \text{F})}\right)\left(\int_0^{12} 1.7\ \text{V}\, dt\right) - 3\ \text{V}$$

$$V_{out} = -\left(\frac{1}{3.807\ \text{s}}\right)(20.4\ \text{V} \cdot \text{s}) - 3\ \text{V}$$

$$V_{out} = -5.359\ \text{V} - 3\ \text{V}$$

$$V_{out} = -8.359\ \text{V}$$

If, after ramping for some amount of time, the input voltage of this integrator circuit is brought to zero, the integrating action will cease. The circuit's output will simply hold at its last value until another non-zero input signal voltage appears.

Recall that the purpose of integral action in a PID controller is to eliminate offset between process variable and setpoint by calculating the error-time product (how far PV deviates from SP, and for how long). This circuit will fulfills that function if the input voltage is the error signal, and the output voltage contributes to the output signal of the controller.

---

[25]If this is not apparent, imagine a scenario where the $+1.7$ volt input existed for precisely one second's worth of time. However much the output voltage ramps in that amount of time must therefore be its rate of change in volts per second (assuming a linear ramp). Since we know the area accumulated under a constant value of 1.7 (high) over a time of 1 second (wide) must be 1.7 volt-seconds, and τ_i is equal to 3.807 seconds, the integrator circuit's output voltage must ramp 0.447 volts during that interval of time. If the input voltage is positive and we know this is an inverting opamp circuit, the direction of the output voltage's ramping must be negative, thus a ramping rate of -0.447 volts per second.

29.12.3 Full-PID circuit design

The following schematic diagram shows a full PID controller implemented using eight operational amplifiers, designed to input and output voltage signals representing PV, SP, and Output[26]:

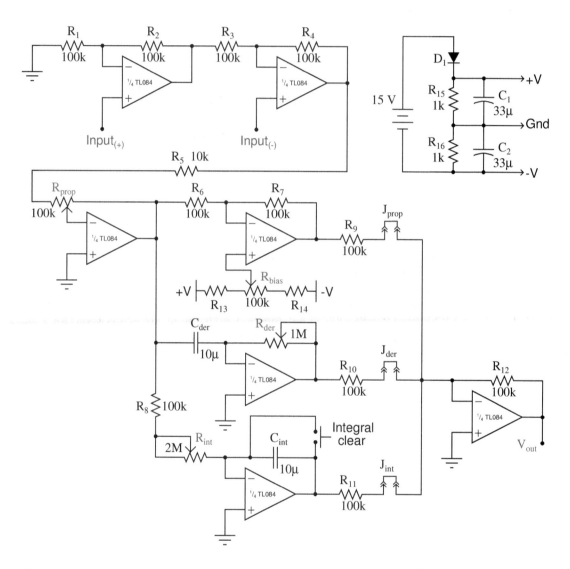

It is somewhat stunning to realize that such a controller, fully capable of controlling many industrial process types, may be constructed using only two integrated circuit "chips" (two "quad" operational amplifiers) and a handful of passive electronic components. The only significant engineering challenge in this simple circuit design is achieving slow enough time constants (in the

---

[26]The two input terminals shown, Input$_{(+)}$ and Input$_{(-)}$ are used as PV and SP signal inputs, the correlation of each depending on whether one desires direct or reverse controller action.

range of minutes rather than seconds) in the integrator and differentiator functions using non-polarized capacitors[27].

This controller implements the so-called *ideal* PID algorithm, with the proportional (gain) value distributing to the integral and derivative terms:

$$m = K_p \left(e + \frac{1}{\tau_i} \int e \, dt + \tau_d \frac{de}{dt} \right) \qquad \textbf{Ideal PID equation}$$

We may determine this from the schematic diagram by noting that the I and D functions each receive their input signals from the output of the proportional amplifier (the one with the R_{prop} potentiometer). Adjusting R_{prop} affects not only the controller's proportional gain, but also the sensitivity of τ_i and τ_d.

An actual implementation of this PID controller in printed circuit board form appears here:

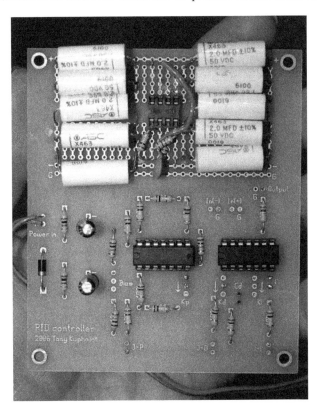

---

[27]This particular design has integral and derivative time value limits of 10 seconds, maximum. These relatively "quick" tuning values are the result of having to use non-polarized capacitors in the integrator and differentiator stages. The practical limits of cost and size restrict the maximum value of on-board capacitance to around 10 μF each.

It is possible to construct an analog PID controller with fewer components. An example is shown here:

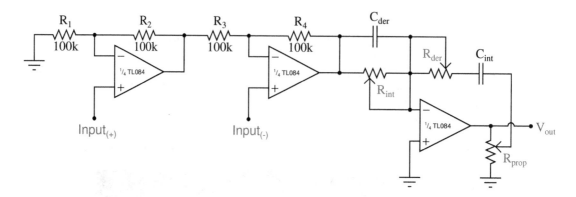

As you can see, a *single* operational amplifier does all the work of calculating proportional, integral, *and* derivative responses. The first two amplifiers do nothing but buffer the input signals and calculate error (PV − SP, or SP − PV, depending on the direction of action).

One of the consequences of consolidating all three control terms in a single amplifier is that those control terms interact with each other. The mathematical expression of this control action is shown here, called the *series* or *interacting* PID equation:

$$m = K_p \left[\left(\frac{\tau_d}{\tau_i} + 1 \right) e + \frac{1}{\tau_i} \int e \, dt + \tau_d \frac{de}{dt} \right] \qquad \textbf{Series or Interacting PID equation}$$

Not only does a change in gain (K_p) alter the relative responses of integral and derivative in the series equation (as it also does in the ideal equation), but changes in either integral or derivative time constants also have an effect on proportional response! This is especially noticeable when the integral time constant is set to some very small value, which is typically the case on fast-responding, self-regulating processes such as liquid flow or liquid pressure control.

It should be apparent that an analog controller implementing the series equation is simpler in construction than one implementing either the parallel or ideal PID equation. This also happens to be true for pneumatic PID controller mechanisms: the simplest analog controller designs all implement the series PID equation[28].

---

[28]An interesting example of engineering tradition is found in electronic PID controller designs. While it is not too terribly difficult to build an analog electronic controller implementing either the parallel or ideal PID equation (just a few more parts are needed), it is quite challenging to do the same in a pneumatic mechanism. When analog electronic controllers were first introduced to industry, they were often destined to replace old pneumatic controllers. In order to ease the transition from pneumatic to electronic control, manufacturers built their new electronic controllers to behave exactly the same as the old pneumatic controllers they would be replacing. The same legacy followed the advent of digital electronic controllers: many digital controllers were programmed to behave in the same manner as the old pneumatic controllers, for the sake of operational familiarity, not because it was easier to design a digital controller that way.

29.12.4 Single-loop analog controllers

One popular analog electronic controller was the Foxboro model 62H, shown in the following photographs. Like the model 130 pneumatic controller, this electronic controller was designed to fit into a rack next to several other controllers. Tuning parameters were adjustable by moving potentiometer knobs under a side-panel accessible by partially removing the controller from its rack:

The Fisher corporation manufactured a series of analog electronic controllers called the AC^2, which were similar in construction to the Foxboro model 62H, but very narrow in width so that many could be fit into a compact panel space. Here we see a pair of Fisher AC^2 controllers mounted side-by-side in the same rack, used to control liquid level in a pulping process:

Like the pneumatic panel-mounted controllers preceding, and digital panel-mount controllers to follow, the tuning parameters for a panel-mounted analog electronic controller were typically accessed on the controller's side. The controller could be slid partially out of the panel to reveal the P, I, and D adjustment knobs (as well as direct/reverse action switches and other configuration controls).

Indicators on the front of an analog electronic controller served to display the process variable (PV), setpoint (SP), and manipulated variable (MV, or output) for operator information. Many analog electronic controllers did not have separate meter indications for PV and SP, but rather used a single meter movement to display the *error signal*, or difference between PV and SP. On the Foxboro model 62H, a hand-adjustable knob provided both indication and control over SP, while a small edge-reading meter movement displayed the error. A negative meter indication showed that the PV was below setpoint, and a positive meter indication showed that the PV was above setpoint.

The Fisher AC^2 analog electronic controller used the same basic technique, cleverly applied in such a way that the PV was displayed in real engineering units. The setpoint adjustment was a large wheel, mounted so the edge faced the operator. Along the circumference of this wheel was a scale showing the process variable range, from the LRV at one extreme of the wheel's travel to the URV at the other extreme of the wheel's travel. The actual setpoint value was the middle of the wheel from the operator's view of the wheel edge. A single meter movement needle traced an arc along the circumference of the wheel along this same viewable range. If the error was zero (PV = SP), the needle would be positioned in the middle of this viewing range, pointed at the same value along the scale as the setpoint. If the error was positive, the needle would rise up to point to a larger (higher) value on the scale, and if the error was negative the needle would point to a smaller (lower) value on the scale. For any fixed value of PV, this error needle would therefore move in exact step with the wheel as it was rotated by the operator's hand. Thus, a single adjustment and a single meter movement displayed both SP and PV in very clear and unambiguous form.

Taylor manufactured a line of analog panel-mounted controllers that worked much the same way, with the SP adjustment being a graduated tape reeled to and fro by the SP adjustment knob. The middle of the viewable section of tape (as seen through a plastic window) was the setpoint value, and a single meter movement needle pointed to the PV value as a function of error. If the error happened to be zero (PV = SP), the needle would point to the middle of this viewable section of tape, which was the SP value.

Another popular panel-mounted analog electronic controller was the Moore Syncro, which featured plug-in modules for implementing different control algorithms (different PID equations, nonlinear signal conditioning, etc.). These plug-in function modules were a hardware precursor to the software "function blocks" appearing in later generations of digital controllers: a simple way of organizing controller functionality so that technicians unfamiliar with computer programming could easily configure a controller to do different types of control functions. Later models of the Syncro featured fluorescent bargraph displays of PV and SP for easy viewing in low-light conditions.

Analog single-loop controllers are largely a thing of the past, with the exception of some low-cost or specialty applications. An example of the former is shown here, a simple analog temperature controller small enough to fit in the palm of my hand:

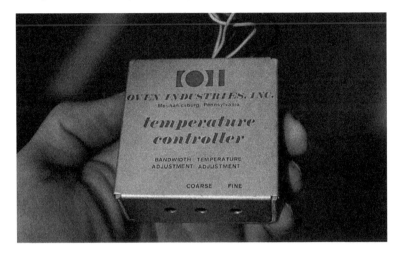

This particular controller happened to be part of a sulfur dioxide analyzer system, controlling the internal temperature of a gas regulator panel to prevent vapors in the sample stream from condensing in low spots of the tubing and regulator system. The accuracy of such a temperature control application was not critical – if temperature was regulated to ± 5 degrees Fahrenheit it would be more than adequate. This is an application where an analog controller makes perfect sense: it is very compact, simple, extremely reliable, and inexpensive. None of the features associated with digital PID controllers (programmability, networking, precision) would have any merit in this application.

29.12.5   Multi-loop analog control systems

In contrast to single-loop analog controllers, *multi-loop* systems control dozens or even hundreds of process loops at a time. Prior to the advent of reliable digital technology, the only electronic process control systems capable of handling the numerous loops within large industrial installations such as power generating plants, oil refineries, and chemical processing facilities were analog systems, and several manufacturers produced multi-loop analog systems just for these large-scale control applications.

One of the most technologically advanced analog electronic products manufactured for industrial control applications was the Foxboro SPEC 200 system[29]. Although the SPEC 200 system used panel-mounted indicators, recorders, and other interface components resembling panel-mounted control systems, the actual control functions were implemented in a separate equipment rack which Foxboro called a *nest*[30]. Printed circuit boards plugged into each "nest" provided all the control functions (PID controllers, alarm units, integrators, signal selectors, etc.) necessary, with analog signal wires connecting the various functions together with panel-mounted displays and with field instruments to form a working system.

Analog field instrument signals (4-20 mA, or in some cases 10-50 mA) were all converted to a 0-10 VDC range for signal processing within the SPEC 200 nest. Operational amplifiers (mostly the model LM301) formed the "building blocks" of the control functions, with a $\pm$ 15 VDC power supply providing DC power for everything to operate.

---

[29]Although the SPEC 200 system – like most analog electronic control systems – is considered "mature" (Foxboro officially declared the SPEC 200 and SPEC 200 Micro systems as such in March 2007), working installations may still be found at the time of this writing (2010). A report published by the Electric Power Research Institute (see References at the end of this chapter) in 2001 documents a SPEC 200 analog control system installed in a nuclear power plant in the United States as recently as 1992, and another as recently as 2001 in a Korean nuclear power plant.

[30]Foxboro provided the option of a self-contained, panel-mounted SPEC 200 controller unit with all electronics contained in a single module, but the split architecture of the display/nest areas was preferred for large installations where many dozens of loops (especially cascade, feedforward, ratio, and other multi-component control strategies) would be serviced by the same system.

The following photographs show a model 2AX+A4 proportional-integral (P+I) controller card for a SPEC 200 system inserted into a metal frame (called a "module" by Foxboro). This module was designed to fit into a slot in a SPEC 200 "nest" where it would reside alongside many other similar cards, each card performing its own control function:

 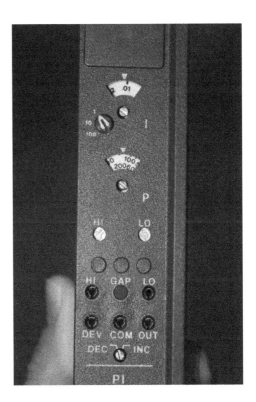

Tuning and alarm adjustments may be seen in the right-hand photograph. This particular controller is set to a proportional band value of approximately 170%, and an integral time constant of just over 0.01 minutes per repeat. A two-position rotary switch near the bottom of the card selected either reverse ("Dec") or direct ("Inc") control action.

The array of copper pins at the top of the module form the male half of a cable connector, providing connection between the control card and the front-panel instrument accessible to operations personnel. Since the tuning controls appear on the face of this controller card (making it a "card tuned" controller), they were not accessible to operators but rather only to the technical personnel with access to the nest area. Other versions of controller cards ("control station tuned") had blank places where the P and I potentiometer adjustments appear on this model, with tuning adjustments provided on the panel-mounted instrument displays for easier access to operators.

The set of ten screw terminals at the bottom of the module provided connection points for the input and output voltage signals. The following list gives the general descriptions of each terminal pair, with the descriptions for this particular P + I controller written in *italic* type:

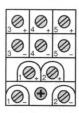

- Terminals (1+) and (1−): Input signal #1 (*Process variable input*)

- Terminals (2+) and (2−): Output signal #1 (*Manipulated variable output*)

- Terminals (3+) and (3−): Input #2, Output #4, or Option #1 (*Remote setpoint*)

- Terminals (4+) and (4−): Input #3, Output #3, or Option #2 (*Optional alarm*)

- Terminals (5+) and (5−): Input #4, Output #2, or Option #3 (*Optional 24 VAC*)

A photograph of the printed circuit board (card) removed from the metal module clearly shows the analog electronic components:

The following photograph shows a set of functioning Foxboro SPEC 200 controller modules residing in a "nest," used to control a flow loop and a pressure loop in a wood pulping process. Both active controller modules are P + I units. An unused PID module resides just to the left of the flow controller module:

Foxboro engineers went to great lengths in their design process to maximize reliability of the SPEC 200 system, already an inherently reliable technology by virtue of its simple, analog nature. As a result, the reliability of SPEC 200 control systems is the stuff of legend[31].

29.13 Digital PID controllers

The vast majority of PID controllers in service today are digital in nature. Microprocessors executing PID algorithms provide many advantages over any form of analog PID control (pneumatic or electronic), not the least of which being the ability to network with personal computer workstations and other controllers over wired or wireless (radio) networks.

---

[31]I once encountered an engineer who joked that the number "200" in "SPEC 200" represented the number of years the system was designed to continuously operate. At another facility, I encountered instrument technicians who were a bit afraid of a SPEC 200 system running a section of their plant: the system had *never suffered a failure of any kind* since it was installed decades ago, and as a result no one in the shop had any experience troubleshooting it. As it turns out, the entire facility was eventually shut down and sold, with the SPEC 200 nest running faithfully until the day its power was turned off! The functioning SPEC 200 controllers shown in the photograph were in continuous use at British Columbia Institute of Technology at the time of the photograph, taken in December of 2014.

29.13.1 Stand-alone digital controllers

If the internal components of a panel-mounted pneumatic or analog electronic controller (such as the Foxboro models 130 or 62, respectively) were completely removed and replaced by all-digital electronic componentry, the result would be a *stand-alone digital PID controller*. From the outside, such a digital controller looks very similar its technological ancestors, but its capabilities are far greater.

An example of a popular panel-mounted digital controller is the Siemens model 353 (formerly the Moore Products model 353):

This particular controller, like many high-end digital controllers and larger digital control systems, is programmed in a function block language. Each function block in the controller is a software subroutine performing a specific function on input signals, generating at least one output signal. Each function block has a set of configuration parameters telling it how to behave. For example, the PID function block in a digital controller would have parameters specifying direct or reverse action, gain (K_p), integral time constant (τ_i), derivative time constant (τ_d), output limits, etc.

Even the "stock" configuration for simple, single-loop PID control is a collection of function blocks linked together:

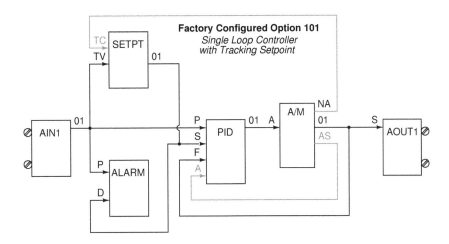

The beauty of function block programming is that the same blocks may be easily re-linked to implement custom control strategies. Take for instance the following function block program written for a Siemens model 353 controller to provide a pulse-width-modulation (PWM, or time-proportioned) output signal instead of the customary 4-20 mA DC analog output signal. The application is for an electric oven temperature control system, where the oven's heating element could only be turned on and off fully rather than continuously varied:

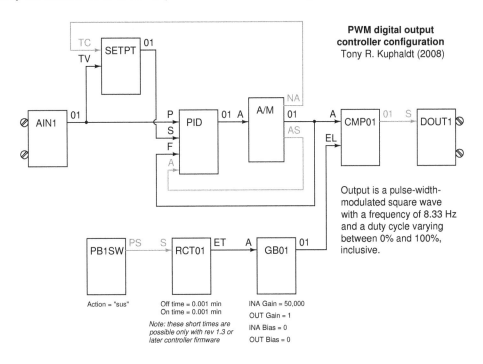

In order to specify links between function blocks, each of the used lettered block inputs is mapped to the output channel of another block. In the case of the time-proportioned function block program, for example, the "P" (process variable) input of the PID function block is set to get its signal from the "01" output channel of the AIN1 (analog input 1) function block. The "TV" (tracking value) input of the SETPT (setpoint) function block is also set to the "01" output channel of the AIN1 function block, so that the setpoint value generator has access to the process variable value in order to implement setpoint tracking. Any function block output may drive an unlimited number of function block inputs (fan-out), but each function block input may receive a signal from *only one* function block output. This is a rule followed within all function block languages to prevent multiple block output signals from conflicting (attempting to insert different signal values into the same input).

In the Siemens controllers, function block programming may be done by entering configuration data using the front-panel keypad, or by using graphical software running on a personal computer networked with the controller.

For applications not requiring so much capability, and/or requiring a smaller form factor, other panel-mounted digital controllers exist. The Honeywell model UDC3000 is a popular example of a 1/4 DIN (96 mm × 96 mm) size digital controller:

Even smaller panel-mounted controllers are produced by a wide array of manufacturers for applications requiring minimum functionality: 1/8 DIN (96 mm × 48 mm), 1/16 DIN (48 mm × 48 mm), and even 1/32 DIN (48 mm × 24 mm) sizes are available.

One of the advantageous capabilities of modern stand-alone controllers is the ability to exchange data over digital networks. This provides operations and maintenance personnel alike the ability to remotely monitor and even control (adjust setpoints, switch modes, change tuning parameters, etc.) the process controller from a computer workstation. The Siemens model 353 controller (with appropriate options) has the ability to digitally network over Ethernet, a very common and robust digital network standard. The following photographs show three such controllers connected to a network through a common 4-port Ethernet "hub" device:

Special software (in this case, Siemens *Procidia*) running on a computer workstation connected to the same Ethernet network acquires data from and sends data to the networked controllers. Screenshots of this software show typical displays allowing complete control over the function of the process controllers:

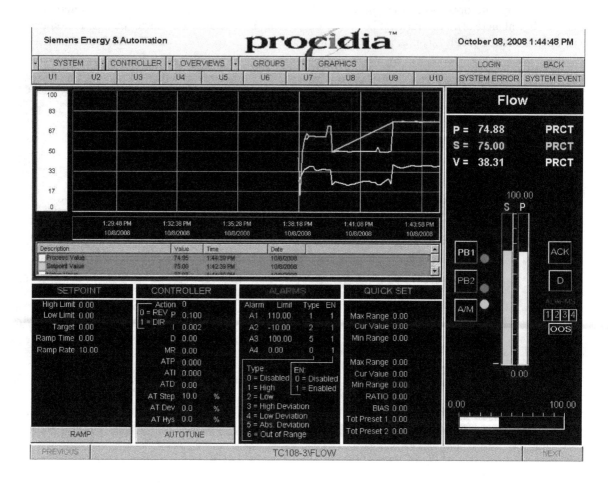

29.13.2 Direct digital control (DDC)

A microprocessor operating at sufficient clock speed is able to execute more than one PID control algorithm for a process loop, by "time-sharing" its calculating power: devoting slices of time to the evaluation of each PID equation in rapid succession. This not only makes multiple-loop digital control possible for a single microprocessor, but also makes it very attractive given the microprocessor's natural ability to manage data archival, transfer, and networking. A single computer is able to execute PID control for multiple loops, and also make that loop control data accessible between loops (for purposes of cascade, ratio, feedforward, and other control strategies) and accessible on networks for human operators and technicians to easily access.

Such *direct digital* control (DDC) has been applied with great success to the problem of building automation, where temperature and humidity controls for large structures benefit from large-scale data integration. The following photograph shows a Siemens APOGEE building automation system with multiple I/O (input/output) cards providing interface between analog instrument signals and the microprocessor's digital functions:

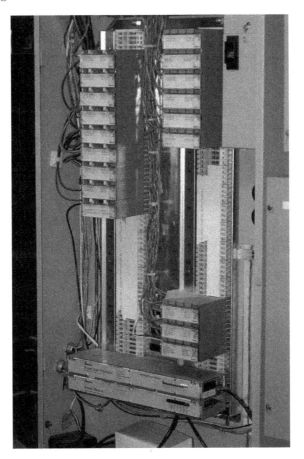

A close-up view of an APOGEE processor shows the device handling all mathematical calculations for the PID control:

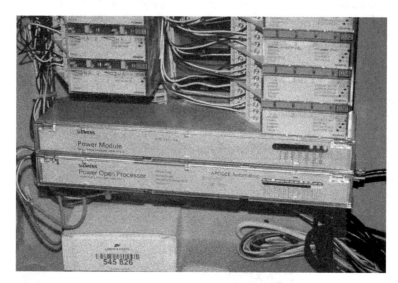

Other than a few LEDs, there is no visual indication in this panel of what the system is doing at any particular time. Operators, engineers, and technicians alike must use software running on a networked personal computer to access data in this control system.

An example of the HMI (Human-Machine Interface) software one might see used in conjunction with a DDC controller is shown here, also from a Siemens APOGEE building control system:

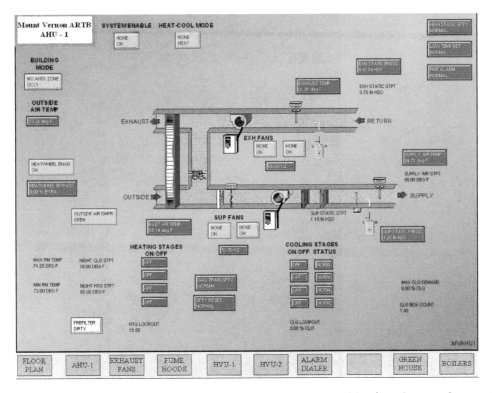

This particular screenshot shows monitored and controlled variables for a heat exchanger ("heat wheel") used to exchange heat between outgoing and incoming air for the building.

A smaller-scale example of a DDC system is the Delta model DSC-1280 controller, an example shown in the following photograph:

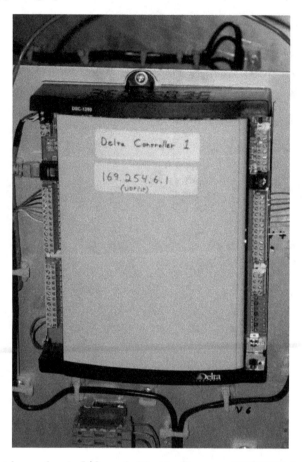

This system does not have plug-in I/O cards like the Siemens APOGEE, but instead is monolithic in design, with all inputs and outputs part of one large "motherboard" PCB. The model DSC-1280 controller has 12 input channels and 8 output channels (hence the model number "1280"). An Ethernet cable (RJ-45 plug) is seen in the upper-left corner of this unit, through which a remotely-located personal computer communicates with the DDC using a high-level protocol called BACnet. In many ways, BACnet is similar to Modbus, residing at layer 7 of the OSI Reference Model (the so-called *Application Layer*), unconcerned with the details of data communication at the Physical or Data Link layers. This means, like Modbus, BACnet commands may be sent and received over a variety of lower-level network standards, with Ethernet being the preferred[32] option at the time of this writing.

---

[32]Thanks to the explosion of network growth accompanying personal computers in the workplace, Ethernet is ubiquitous. The relatively high speed and low cost of Ethernet communications equipment makes it an attractive network standard over which a great many high-level industrial protocols communicate.

Another example of a small-scale DDC is this Distech model ECP-410 unit:

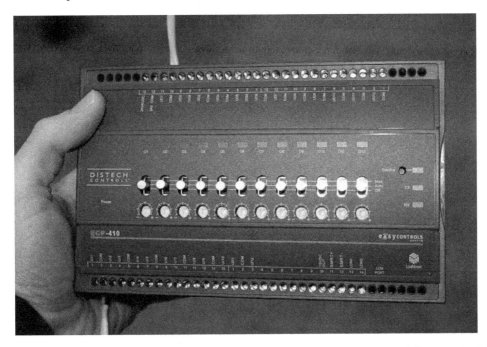

This controller uses *LonWorks* as its communication protocol rather than BACnet. Like BACnet, LonWorks is an upper-layer protocol and may be transported over Ethernet as well as simpler serial communication formats. The model ECP-410 DDC controller has a two-wire connection for its LonWorks network.

Programming of DDC controllers ranges from text-based languages (similar to BASIC) to function-block programming. The Delta DSC-1280 is an example of a controller programmed in text, while the Distech ECP-410 supports function blocks in addition to text-based programming.

The application-specific nature of DDC (environmental controls for the interior of large buildings and other facilities) lends itself to controller units pre-programmed to perform well-defined tasks rather than general-purpose controllers designed to be user-programmable for any task. An example of such a "fixed" program controller is the Distech model EC-RTU-L, designed to control a "rooftop unit" for air handling:

A more common application of industrial DDC is the use of programmable logic controllers (PLCs) to control multiple loops. PLCs were originally invented for on/off (discrete) process control functions, but have subsequently grown in speed and capability to execute analog PID control functions as well. This next photograph shows an Allen-Bradley (Rockwell) ControlLogix PLC used to control the operation of a gas turbine engine. The PLC may be seen in the upper-left corner of the enclosure, with the rest of the enclosure devoted to terminal blocks and accessory components:

A strong advantage of using PLCs for analog loop control is the ability to easily integrate discrete controls with the analog controls. It is quite easy, for example, to coordinate the sequential start-up and shut-down functions necessary for intermittent operation with the analog PID controls necessary for continuous operation, all within one programmable logic controller. It should be noted, however, that many early PLC implementations of PID algorithms were crude at best, lacking the finesse of stand-alone PID controllers. Even some modern PLC analog functions are mediocre[33] at the time of this writing (2008).

---

[33]An aspect common to many PLC implementations of PID control is the use of the "parallel" PID algorithm instead of the superior "ISA" or "non-interacting" algorithm. The choice of algorithm may have a profound effect on tuning, and on tuning procedures, especially when tuning parameters must be re-adjusted to accommodate changes in transmitter range.

29.13.3   SCADA and telemetry systems

A similar control system architecture to Direct Digital Control (DDC) – assigning a single microprocessor to the task of managing multiple control functions, with digital communication between the microprocessor units – is used for the management of systems which are by their very nature spread over wide geographical regions. Such systems are generally referred to as *SCADA*, which is an acronym standing for *Supervisory Control And Data Acquisition*.

The typical SCADA system consists of multiple *Remote Terminal Unit* (RTU) devices connected to process transmitters and final control elements, implementing basic control functions such as motor start/stop and PID loop control. These RTU devices communicate digitally to a *Master Terminal Unit* (MTU) device at a central location where human operators may monitor the process and issue commands.

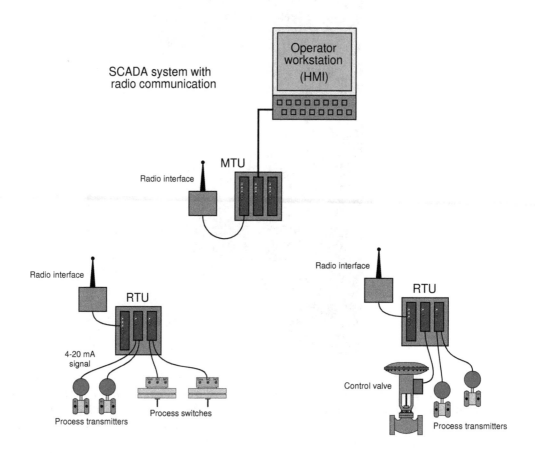

A photograph of an RTU "rack" operating at a large electric power substation is shown here:

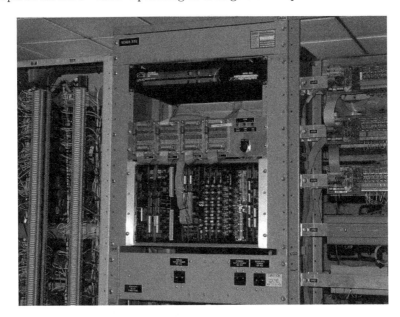

Some RTU hardware, such as the substation monitoring system shown above, is custom-manufactured for the application. Other RTU hardware is more general in purpose, intended for the monitoring and control of natural gas and oil production wells, but applicable to other applications as well.

The Fisher ROC 800 – shown in the photograph below – is an example of an RTU designed to operate with a minimum of electrical power, so that a single solar panel and battery will be sufficient for year-round operation in remote environments. The particular unit shown is installed in a natural gas metering station, where is monitors gas pressure, temperature, and flow rate, and also controls the injection of an "odorizing" compound into the gas to give it a bad smell:

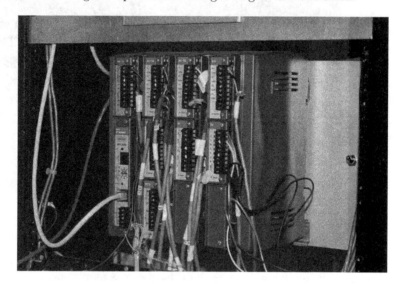

Standard programmable logic controllers (PLCs) are ideal candidates for use as RTU devices. Modern PLCs have all the I/O, networking, and control algorithm capability necessary to function as remote terminal units. Commercially available Human-Machine Interface (HMI) software allowing personal computers to display PLC variable values potentially turns every PC into a Master Terminal Unit (MTU) where operators can view process variables, change setpoints, and issue other commands for controlling the process.

A photograph of such HMI software used to monitor a SCADA system for a set of natural gas compressors is shown here:

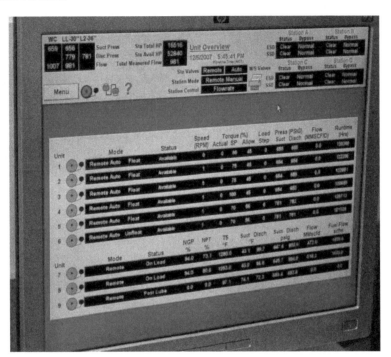

Another photograph of a similar system used to monitor and control drinking water reservoirs for a city is shown here:

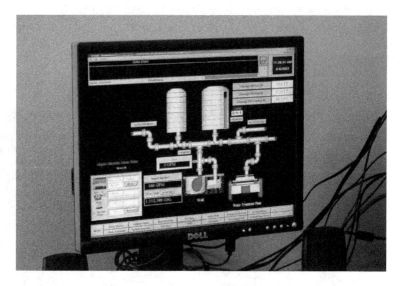

A concept closely related to SCADA is that of *telemetry*, the word literally meaning "distance measuring" (i.e. measuring something over a distance). The acronym SCADA, by containing the word "control," implies two-way communication (measurement and control) between the master location and the remote location. In applications where the flow of information is strictly one-way (simplex) from the remote location to the master location, "telemetry" is a more apt description.

Telemetry systems find wide application in scientific research. Seismographs, river and stream flowmeters, weather stations, and other remotely-located measurement instruments connected (usually by radio links) to some centralized data collection center are all examples of telemetry. Any industrial measurement (-only) application spanning a large distance could likewise be classified as a telemetry system, although you will sometimes find the term "SCADA" applied even when the communication is simplex in nature.

29.13.4 Distributed Control Systems (DCS)

A radically new concept appeared in the world of industrial control in the mid-1970's: the notion of *distributed* digital control. Direct digital control during that era[34] suffered a substantial problem: the potential for catastrophic failure if the single digital computer executing *multiple* PID control functions were to ever halt. Digital control brings many advantages, but it isn't worth the risk if the entire operation will shut down (or catastrophically fail!) following a hardware or software failure within that one computer.

Distributed control directly addressed this concern by having multiple control computers – each one responsible for only a handful of PID loops – distributed throughout the facility and networked together to share information with each other and with operator display consoles. With individual process control "nodes" scattered throughout the campus, each one dedicated to controlling just a few loops, there would be less concentration of liability as there would be with a single-computer DDC system. Such distribution of computing hardware also shortened the analog signal wiring, because now the hundreds or thousands of analog field instrument cables only had to reach as far as the distributed nodes, not all the way to a centralized control room. Only the networking cable had to reach that far, representing a drastic reduction in wiring needs. Furthermore, distributed control introduced the concept of *redundancy* to industrial control systems: where digital signal acquisition and processing hardware units were equipped with "spare" units designed to automatically take over all critical functions in the event of a primary failure.

---

[34]Modern DDC systems of the type used for building automation (heating, cooling, security, etc.) almost always consist of networked control nodes, each node tasked with monitoring and control of a limited area. The same may be said for modern PLC technology, which not only exhibits advanced networking capability (fieldbus I/O networks, Ethernet, Modbus, wireless communications), but is often also capable of redundancy in both processing and I/O. As technology becomes more sophisticated, the distinction between a DDC (or a networked PLC system) and a DCS becomes more ambiguous.

The following illustration shows a typical distributed control system (DCS) architecture:

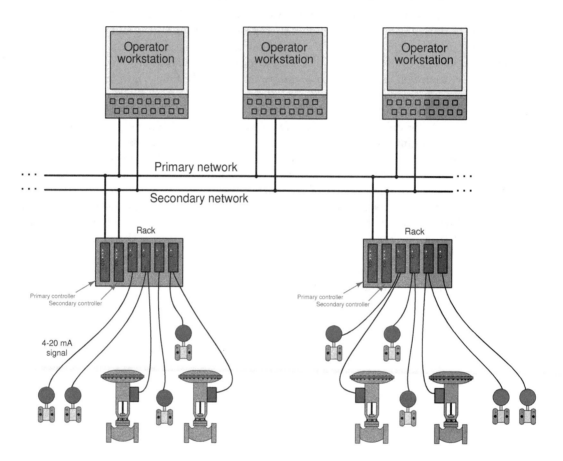

Each "rack" contains a microprocessor to implement all necessary control functions, with individual I/O (input/output) "cards" for converting analog field instrument signals into digital format, and vice-versa. Redundant processors, redundant network cables, and even redundant I/O cards address the possibility of component failure. DCS processors are usually programmed to perform routine self-checks[35] on redundant system components to ensure availability of the spare components in the event of a failure.

If there ever was a total failure in one of the "control racks" where the redundancy proved insufficient for the fault(s), the only PID loops faulted will be those resident in that rack, not any of the other loops throughout the system. Likewise, if ever the network cables become severed or otherwise faulted, only the information flow between those two points will suffer; the rest of the system will continue to communicate data normally. Thus, one of the "hallmark" features of a DCS is its tolerance to serious faults: even in the event of severe hardware or software faults, the impact to process control is minimized by design.

---

[35]An example of such a self-check is scheduled switching of the networks: if the system has been operating on network cable "A" for the past four hours, it might switch to cable "B" for the next four hours, then back again after another four hours to continually ensure both cables are functioning properly.

One of the very first distributed control systems in the world was the Honeywell TDC2000 system[36], introduced in 1975. By today's standards, the technology was crude[37], but the concept was revolutionary.

Each rack (called a "box" by Honeywell) consisted of an aluminum frame holding several large printed circuit boards with card-edge connectors. A "basic controller" box appears in the left hand photograph. The right-hand photograph shows the termination board where the field wiring (4-20 mA) connections were made. A thick cable connected each termination board to its respective controller box:

Controller redundancy in the TDC2000 DCS took the form of a "spare" controller box serving as a backup for up to eight other controller boxes. Thick cables routed all analog signals to this spare controller, so that it would have access to them in the event it needed to take over for a failed controller. The spare controller would become active on the event of *any* fault in any of the other controllers, including failures in the I/O cards. Thus, this redundancy system provided for processor failures as well as I/O failures. All TDC2000 controllers communicated digitally by means of a dual coaxial cable network known as the "Data Hiway." The dual cables provided redundancy in network communications.

---

[36]To be fair, the Yokogawa Electric Corporation of Japan introduced their CENTUM distributed control system the same year as Honeywell. Unfortunately, while I have personal experience maintaining and using the Honeywell TDC2000 system, I have zero personal experience with the Yokogawa CENTUM system, and neither have I been able to obtain technical documentation for the original incarnation of this DCS (Yokogawa's latest DCS offering goes by the same name). Consequently, I can do little in this chapter but mention its existence, despite the fact that it deserves just as much recognition as the Honeywell TDC2000 system.

[37]Just to give some perspective, the original TDC2000 system used whole-board processors rather than microprocessor chips, and magnetic core memory rather than static or dynamic RAM circuits! Communication between controller nodes and operator stations occurred over thick coaxial cables, implementing master/slave arbitration with a separate device (a "Hiway Traffic Director" or HTD) coordinating all communications between nodes. Like Bob Metcalfe's original version of Ethernet, these coaxial cables were terminated at their end-points by termination resistors, with coaxial "tee" connectors providing branch points for multiple nodes to connect along the network.

A typical TDC2000 operator workstation appears in the next photograph:

Over the years following its 1975 introduction, the Honeywell system grew in sophistication with faster networks (the "Local Control Network" or LCN), more capable controller racks (the "Process Manager" or PM series), and better operator workstations. Many of these improvements were incremental, consisting of add-on components that could work with existing TDC2000 components so that the entire system need not be replaced to accept the new upgrades.

Other control equipment manufacturers responded to the DCS revolution started by Honeywell and Yokogawa by offering their own distributed control systems. The Bailey Network 90 (Net90) DCS, Bailey Infi90 DCS, and the Fisher Provox systems are examples. Foxboro, already an established leader in the control system field with their SPEC 200 analog system, first augmented the SPEC 200 with digital capabilities (the VIDEOSPEC workstation consoles, FOX I/A computer, INTERSPEC and FOXNET data networks), then developed an entirely digital distributed control system, the SPECTRUM.

Some modern distributed control systems offered at the time of this writing (2008) include:

- ABB *800xA*

- Emerson *DeltaV* and *Ovation*

- Foxboro (Invensys) *I/A*

- Honeywell *Experion PKS*

- Yokogawa *CENTUM VP* and *CENTUM CS*

For a visual comparison with the Honeywell TDC2000 DCS, examine the following photograph of an Emerson DeltaV DCS rack, with processor and multiple I/O modules:

A photograph of an Emerson Ovation DCS rack shows a vertically-oriented backplane accepting multiple I/O modules:

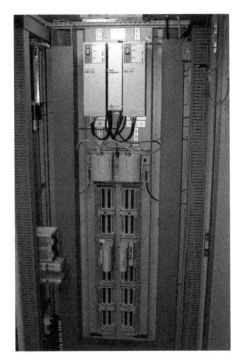

Many modern distributed control systems such as the Emerson DeltaV use regular personal computers rather than proprietary hardware as operator workstations. This cost-saving measure leverages existing computer and display technologies without sacrificing control-level reliability (since the control hardware and software is still industrial-grade):

As previously mentioned in the Direct Digital Control (DDC) subsection, programmable logic controllers (PLCs) are becoming more and more popular as PID control platforms due to their ever-expanding speed, functionality, and relatively low cost. It is now possible with modern PLC hardware and networking capabilities to build a truly distributed control system with individual PLCs as the processing nodes, and with redundancy built into each of those nodes so that any single failure does not interrupt critical control functions. Such a system may be purchased at a fraction of the up-front cost of a fully-fledged DCS.

However, what is currently lacking in the PLC world is the same level of hardware and software integration necessary to build a functional distributed control system that comes as ready-to-use as a system pre-built by a DCS manufacturer. In other words, if an enterprise chooses to build their own distributed control system using programmable logic controllers, they must be prepared to do a *lot* of programming work in order to emulate the same level of functionality and power as a pre-engineered DCS[38]. Any engineer or technician who has experienced the power of a modern DCS – with its self-diagnostic, "smart" instrument management, event auditing, advanced control strategy, pre-engineered redundancy, data collection and analysis, and alarm management capabilities – realizes these features are neither luxuries nor are they trivial to engineer. Woe to anyone who thinks these critical features may be created by incumbent staff at a lesser cost!

---

[38]I know of a major industrial manufacturing facility (which shall remain nameless) where a PLC vendor promised the same technical capability as a full DCS at approximately one-tenth the installed cost. Several years and several tens of thousands of man-hours later, the sad realization was this "bargain" did not live up to its promise, and the decision was made to remove the PLCs and go with a complete DCS from another manufacturer. *Caveat emptor!*

29.13.5 Fieldbus control

The DCS revolution started in the mid-1970's was fundamentally a moving of control system "intelligence" from a centralized location to distributed locations. Rather than have a single computer (or a panel full of single-loop controllers) located in a central control room implement PID control for a multitude of process loops, many (smaller) computers located closer to the process areas would execute the PID and other control functions, with network cables shuttling data between those distributed locations and the central control room.

Beginning in the late 1980's, the next logical step in this evolution of control architecture saw the relocation of control "intelligence" to the field instruments themselves. In other words, the new idea was to equip individual transmitters and control valve positioners with the necessary computational power to implement PID control all on their own, using digital networks to carry process data between the field instruments and any location desired. This is the fundamental concept of *fieldbus*.

"Fieldbus" as a technical term has multiple definitions. Many manufacturers use the word "fieldbus" to describe any digital network used to transport data to and from field instruments. In this subsection, I use the word "fieldbus" to describe a design philosophy where field instruments possess all the necessary "intelligence" to control the process, with no need for separate centralized (or even distributed) control hardware. *FOUNDATION Fieldbus* is the first standard to embody this fully-distributed control concept, the technical details of this open standard maintained and promoted by the *Fieldbus Foundation*. The aim of this Foundation is to establish an open, technical standard for *any* manufacturer to follow in the design of their fieldbus instruments. This means a FOUNDATION Fieldbus (FF) transmitter manufactured by Smar will work seamlessly with a FF control valve positioner manufactured by Fisher, communicating effortlessly with a FF-aware host system manufactured by ABB, and so on. This may be thought of in terms of being the digital equivalent of the 3-15 PSI pneumatic signal standard or the 4-20 mA analog electronic signal standard: so long as all instruments "talk" according to the same standard, brands and models may be freely interchanged to build any control system desired.

To illustrate the general fieldbus concept, consider this flow control system:

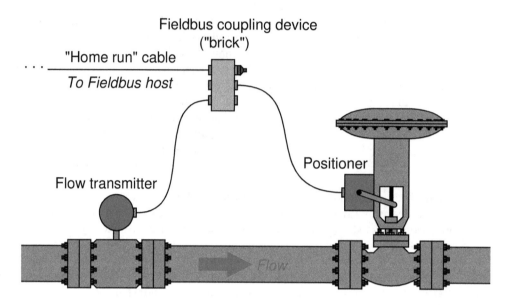

Here, a fieldbus *coupling device* provides a convenient junction point for cables coming from the transmitter, valve positioner, and host system. FOUNDATION Fieldbus devices both receive DC power and communicate digitally over the same twisted-pair cables. In this case, the host system provides DC power for the transmitter and positioner to function, while communication of process data occurs primarily between the transmitter and positioner (with little necessary involvement of the host system[39]).

As with distributed control systems, FOUNDATION Fieldbus instruments are programmed using a function block language. In this case, we must have an analog input (for the transmitter's measurement), a PID function block, and an analog output (for the valve positioner) to make a complete flow control system:

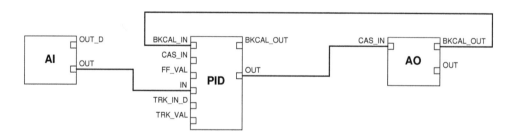

---

[39]Although it is customary for the host system to be configured as the *Link Active Scheduler* (LAS) device to schedule and coordinate all fieldbus device communications, this is not absolutely necessary. Any suitable field instrument may also serve as the LAS, which means a host system is not even necessary except to provide DC power to the instruments, and serve as a point of interface for human operators, engineers, and technicians.

The analog input (AI) block must reside in the transmitter, and the analog output (AO) block must reside in the valve positioner, since those blocks necessarily relate to the measured and controlled variables, respectively. However, the PID block may reside in *either* field device:

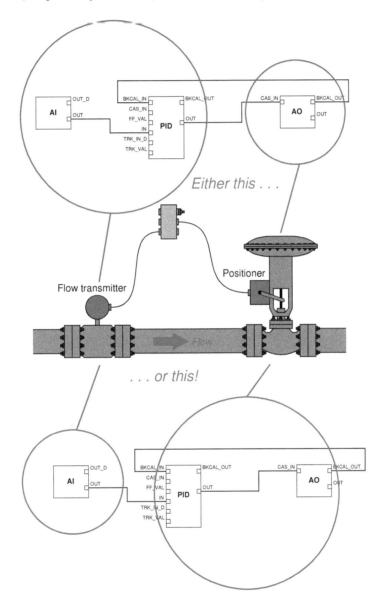

Practical reasons do exist for choosing one location of the PID function block over the other, most notably the difference in communication loading between the two options[40]. However, there

---

[40]With the PID function block programmed in the flow transmitter, there will be twice as many scheduled communication events per macrocycle than if the function block is programmed into the valve positioner. This is evident by the number of signal lines connecting circled block(s) to circled block(s) in the above illustration.

is no *conceptual* limitation to the location of the PID function block. In a fieldbus control system where the control "intelligence" is distributed all the way to the field instrument devices themselves, there are no limits to system flexibility.

29.14   Practical PID controller features

In order for any PID controller to be practical, it must be able to do more than just implement the PID equation. This section identifies and explains some of the basic features found on most (but not all!) modern PID controllers:

- Manual versus Automatic mode

- Output tracking

- Setpoint tracking

- Alarming

- PV characterization and damping

- Setpoint limits

- Output limits

- PID tuning security

29.14.1 Manual and automatic modes

When a controller continually calculates output values based on PV and SP values over time, it is said to be operating in *automatic* mode. This mode, of course, is what is necessary to regulate any process. There are times, however, when it is desirable to allow a human operator to manually "override" the automatic action of the PID controller. Applicable instances include process start-up and shut-down events, emergencies, and maintenance procedures. A controller that is being "overridden" by a human being is said to be in *manual mode*.

A very common application of manual mode is during maintenance of the sensing element or transmitter. If an instrument technician needs to disconnect a process transmitter for calibration or replacement, the controller receiving that transmitter's signal cannot be left in automatic mode. If it is, then the controller may[41] take sudden corrective action the moment the transmitter's signal goes dead. If the controller is first placed in manual mode before the technician disconnects the transmitter, however, the controller will ignore any changes in the PV signal, letting its output signal be adjusted at will by the human operator. If there is another indicator of the same process variable as the one formerly reported by the disconnected transmitter, the human operator may elect to read that other indicator and play the part of a PID controller, manually adjusting the final control element to maintain the alternate indicator at setpoint while the technician completes the transmitter's maintenance.

An extension of this "mode" concept applies to controllers configured to receive a setpoint from another device (called a *remote* or *cascaded* setpoint). In addition to an automatic and a manual mode selection, a third selection called *cascade* exists to switch the controller's setpoint from human operator control to remote (or "cascade") control.

---

[41]The only reason I say "may" instead of "will" is because some modern digital controllers are designed to automatically switch to manual-mode operation in the event of a sensor or transmitter signal loss. Any controller not "smart" enough to shed its operating mode to manual in the event of PV signal loss will react dramatically when that PV signal dies, and this is not a good thing for an operating loop!

29.14.2   Output and setpoint tracking

The provision of manual and automatic operating modes creates a set of potential problems for the PID controller. If, for example, a PID controller is switched from automatic to manual mode by a human operator, and then the output is manually adjusted to some new value, what will the output value do when the controller is switched *back* to automatic mode? In some crude PID controller designs, the result would be an immediate "jump" back to the output value calculated by the PID equation while the controller was in manual. In other words, some controllers never stop evaluating the PID equation – even while in manual mode – and will default to that automatically-calculated output value when the operating mode is switched from manual to automatic.

This can be very frustrating to the human operator, who may wish to use the controller's manual mode as a way to change the controller's bias value. Imagine, for example, that a PD controller (no integral action) is operating in automatic mode at some low output value, which happens to be too low to achieve the desired setpoint. The operator switches the controller to manual mode and then raises the output value, allowing the process variable to approach setpoint. When PV nearly equals SP, the operator switches the controller's mode back to automatic, expecting the PID equation to start working again from this new starting point. In a crude controller, however, the output would jump back to some lower value, right where the PD equation would have placed it for these PV and SP conditions.

A feature designed to overcome this problem – which is so convenient that I consider it an essential feature of any controller with a manual mode – is called *output tracking*. With output tracking, the bias value of the controller shifts every time the controller is placed into manual mode and the output value manually changed. Thus, when the controller is switched from manual mode to automatic mode, the output does *not* immediately jump to some previously-calculated value, but rather "picks up" from the last manually-set value and begins to control from that point as dictated by the PID equation. In other words, output tracking allows a human operator to arbitrarily offset the output of a PID controller by switching to manual mode, adjusting the output value, and then switching back to automatic mode. The output will continue its automatic action from this new starting point instead of the old starting point.

A very important application of output tracking is in the manual correction of integral wind-up (sometimes called *reset windup* or just *windup*). This is what happens to a controller with integral action if for some reason the process variable *cannot* achieve setpoint no matter how far the output signal value is driven by integral action. An example might be on a temperature controller where the source of heat for the process is a steam system. If the steam system shuts down, the temperature controller *cannot* warm the process up to the temperature setpoint value no matter how far open the steam valve is driven by integral action. If the steam system is shut down for too long, the result will be a controller output saturated at maximum value in a futile attempt to warm the process. If and when the steam system starts back up, the controller's saturated output will now send *too much heating steam* to the process, causing the process temperature to overshoot setpoint until integral action drives the output signal back down to some reasonable level. This situation may be averted, however, if the operator switches the temperature controller to manual mode as soon as the steam system shuts down. Even if this preventive step is not taken, the problem of overshoot may be averted upon steam system start-up if the operator uses output tracking by quickly switching the controller into manual mode, adjusting the output down to a reasonable level, and then switching back into automatic mode so that the controller's output value is no longer "wound up" at a high level[42].

A similar feature to output tracking – also designed for the convenience of a human operator switching a PID controller between automatic and manual modes – is called *setpoint tracking*. The purpose of setpoint tracking is to equalize SP and PV while the controller is in manual mode, so that when the controller gets switched back into automatic mode, it will begin its automatic operation with no error (PV = SP).

This feature is most useful during system start-ups, where the controller may have difficulty controlling the process in automatic mode under unusual conditions. Operators often prefer to run certain control loops in manual mode from the time of initial start-up until such time that the process is near normal operating conditions. At that point, when the operator is content with the stability of the process, the controller is assigned the responsibility of maintaining the process at setpoint. With setpoint tracking present in the controller, the controller's SP value will be held equal to the PV value (whatever that value happens to be) for the entire time the controller is in manual mode. Once the operator decides it is proper to switch the controller into automatic mode, the SP value freezes at that last manual-mode PV value, and the controller will continue to control the PV at that SP value. Of course, the operator is free to adjust the SP value to any new value while the controller is in automatic mode, but this is at the operator's discretion.

Without setpoint tracking, the operator would *have to* make a setpoint adjustment either before or after switching the controller from manual mode to automatic mode, in order to ensure the controller was properly set up to maintain the process variable at the desired value. With setpoint tracking, the setpoint value will default to the process variable value when the controller was last in manual mode, which (it is assumed) will be close enough to the desired value to suffice for continued operation.

Unlike output tracking, for which there is virtually no reason not to have the feature present in a PID controller, there may very well be applications where we do not wish to have setpoint tracking. For some processes[43], the setpoint value *should* remain fixed at all times, and as such it would be undesirable to have the setpoint value drift around with the process variable value every time the controller was placed into manual mode.

---

[42]I once had the misfortune of working on an analog PID controller for a chlorine-based wastewater disinfection system that lacked output tracking. The chlorine sensor on this system would occasionally fail due to sample system plugging by algae in the wastewater. When this happened, the PV signal would fail low (indicating abnormally low levels of chlorine gas dissolved in the wastewater) even though the actual dissolved chlorine gas concentration was adequate. The controller, thinking the PV was well below SP, would ramp the chlorine gas control valve further and further open over time, as integral action attempted to reduce the error between PV and SP. The error never went away, of course, because the chlorine sensor was plugged with algae and simply could not detect the actual chlorine gas concentration in the wastewater. By the time I arrived to address the "low chlorine" alarm, the controller output was already wound up to 100%. After cleaning the sensor, and seeing the PV value jump up to some outrageously high level, the controller would take a long time to "wind down" its output because its integral action was very slow. I could not use manual mode to "unwind" the output signal, because this controller lacked the feature of output tracking. My "work-around" solution to this problem was to re-tune the integral term of the controller to some really fast time constant, watch the output "wind down" in fast-motion until it reached a reasonable value, then adjust the integral time constant back to its previous value for continued automatic operation.

[43]Boiler steam drum water level control, for example, is a process where the setpoint really should be left at a 50% value at all times, even if there maybe legitimate reasons for occasionally switching the controller into manual mode.

29.14.3   Alarm capabilities

A common feature on many instrument systems is the ability to alert personnel to the onset of abnormal process conditions. The general term for this function is *alarm*. Process alarms may be triggered by process switches directly sensing abnormal conditions (e.g. high-temperature switches, low-level alarms, low-flow alarms, etc.), in which case they are called *hard alarms*. A *soft alarm*, by contrast, is an alarm triggered by some continuous measurement (i.e. a signal from a process transmitter rather than a process switch) exceeding a pre-programmed alarm limit value.

Since PID controllers are designed to input continuous process measurements, it makes sense that a controller could be equipped with programmable alarm limit values as well, to provide "soft" alarm capability without adding additional instruments to the loop[44]. Not only is PV alarming easy to implement in most PID controllers, but *deviation* alarming is easy to implement as well. A "deviation alarm" is a soft alarm triggered by excessive deviation (error) between PV and SP. Such an event indicates control problems, since a properly-operating feedback loop should be able to maintain reasonable agreement between PV and SP at all times.

Alarm capabilities find their highest level of refinement in modern distributed control systems (DCS), where the networked digital controllers of a DCS provide convenient access and advanced management of hard and soft alarms alike. Not only can alarms be accessed from virtually any location in a facility in a DCS, but they are usually time-stamped and archived for later analysis, which is an *extremely* important feature for the analysis of emergency events, and the continual improvement of process safety.

29.14.4   Output and setpoint limiting

In some process applications, it may not be desirable to allow the controller to automatically manipulate the final control element (control valve, variable-speed motor, heater) over its full 0% - 100% range. In such applications, a useful controller feature is an *output limit*. For example, a PID flow controller may be configured to have a minimum output limit of 5%, so that it is not able to close the control valve any further than the 5% open position in order to maintain "minimum flow" through a pump. The valve may still be fully closed (0% stem position) in manual mode, but just not in automatic mode[45].

Similarly, setpoint values may be internally limited in some PID controllers, such that an operator cannot adjust the setpoint above some limiting value or below some other limiting value. In the event that the process variable *must* be driven outside these limits, the controller may be placed in manual mode and the process "manually" guided to the desired state by an operator.

---

[44]It is very important to note that soft alarms are not a replacement for hard alarms. There is much wisdom in maintaining both hard and soft alarms for a process, so there will be redundant, non-interactive levels of alarming. Hard and soft alarms should complement each other in any critical process.

[45]Some PID controllers limit manual-mode output values as well, so be sure to check the manufacturer's documentation for output limiting on your particular PID controller!

29.14.5 Security

There is justifiable reason to prevent certain personnel from having access to certain parameters and configurations on PID controllers. Certainly, operations personnel need access to setpoint adjustments and automatic/manual mode controls, but it may be unwise to grant those same operators unlimited access to PID tuning constants and output limits. Similarly, instrument technicians may require access to a PID controller's tuning parameters, but perhaps should be restricted from editing configuration programs maintained by the engineering staff.

Most digital PID controllers have some form of security access control, allowing for different levels of permission in altering PID controller parameters and configurations. Security may be crude (a hidden switch located on a printed circuit board, which only the maintenance personnel should know about), sophisticated (login names and passwords, like a multi-user computer system), or anything in between, depending on the level of development invested in the feature by the controller's manufacturer.

An interesting solution to the problem of security in the days of analog control systems was the architecture of Foxboro's SPEC 200 analog electronic control system. The controller displays, setpoint adjustments, and auto/manual mode controls were located on the control room panel where anyone could access them. All other adjustments (PID settings, alarm settings, limit settings) could be located in the *nest* area where all the analog circuit control cards resided. Since the "nest" racks could be physically located in a room separate from the control room, personnel access to the nest room served as access security to these system parameters.

At first, the concept of controller parameter security may seen distrustful and perhaps even insulting to those denied access, especially when the denied persons possess the necessary knowledge to understand the functions and consequences of those parameters. It is not uncommon for soft alarm values to be "locked out" from operator access despite the fact that operators understand very well the purpose and functions of these alarms. At some facilities, PID tuning is the exclusive domain of process engineers, with instrument technicians and operators alike barred from altering PID tuning constants even though some operators and many technicians may well understand PID controller tuning.

When considering security access, there is more to regard than just knowledge or ability. At a fundamental level, security is a task of limiting access commensurate with *responsibility*. In other words, security restrictions exist to exclude those not charged with particular responsibilities. Knowledge and ability are necessary conditions of responsibility (i.e. one cannot reasonably be held responsible for something beyond their knowledge or control), but they are not *sufficient* conditions of responsibility (i.e. knowing how to, and being able to perform a task does not confer responsibility for that task getting completed). An operator may very well understand how and why a soft alarm on a controller works, but the responsibility for altering the alarm value may reside with someone else whose job description it is to ensure the alarm values correspond to plant-wide policies.

29.15 Digital PID algorithms

Instrument technicians should not have to concern themselves over the programming details internal to digital PID controllers. Ideally, a digital PID controller should simply perform the task of executing PID control with all the necessary features (setpoint tracking, output limiting, etc.) without the end-user having to know anything about those details. However, in my years of experience I have seen enough examples of poor PID implementation to warrant an explanatory section in this book, both so instrumentation professionals may recognize poor PID implementation when they see it, and also so those with the responsibility of designing PID algorithms may avoid some common mistakes.

29.15.1 Introduction to pseudocode

In order to show digital algorithms, I will use a form of notation called *pseudocode*: a text-based language instructing a digital computing device to implement step-by-step procedures. "Pseudocode" is written to be easily read and understood by human beings, yet similar enough in syntax and structure to real computer programming languages for a human programmer to be able to easily translate to a high-level programming language such as BASIC, C++, or Java. Since pseudocode is not a formal computer language, we may use it to very efficiently describe certain algorithms (procedures) without having to abide by strict "grammatical" rules as we would if writing in a formal language such as BASIC, C++, or Java.

Program loops

Each line of text in the following listing represents a command for the digital computer to follow, one by one, in order from top to bottom. The LOOP and ENDLOOP markers represent the boundaries of a program *loop*, where the same set of encapsulated commands are executed over and over again in cyclic fashion:

Pseudocode listing [46]

```
LOOP
    PRINT "Hello World!"    // This line prints text to the screen
    OUTPUT audible beep on the speaker  // This line beeps the speaker
ENDLOOP
```

In this particular case, the result of this program's execution is a continuous printing of the words "Hello World!" to the computer's display with a single "beep" tone accompanying each printed line. The words following a double-slash (//) are called *comments*, and exist only to provide explanatory text for the human reader, not the computer. Admittedly, this example program would be both impractical and annoying to actually run in a computer, but it does serve to illustrate the basic concept of a program "loop" shown in pseudocode.

---

[46] I have used a typesetting convention to help make my pseudocode easier for human beings to read: all formal commands appear in bold-faced blue type, while all comments appear in italicized red type. All other text appears as normal-faced black type. One should remember that the computer running any program cares not for how the text is typeset: all it cares is that the commands are properly used (i.e. no "grammatical" or "syntactical" errors).

Assigning values

For another example of pseudocode, consider the following program. This code causes a variable (x) in the computer's memory to alternate between two values of 0 and 2 indefinitely:

Pseudocode listing

```
DECLARE x to be an integer variable
SET x = 2   // Initializing the value of x

LOOP
    // This SET command alternates the value of x with each pass
    SET x = 2 - x
ENDLOOP
```

The first instruction in this listing declares the type of variable x will be. In this case, x will be an *integer* variable, which means it may only represent whole-number quantities and their negative counterparts – no other values (e.g. fractions, decimals) are possible. If we wished to limit the scope of x even further to represent just 0 or 1 (i.e. a single bit), we would have to declare it as a *Boolean* variable. If we required x to be able to represent fractional values as well, we would have to declare it as a *floating-point* variable. Variable declarations are important in computer programming because it instructs the computer how much space in its random-access memory to allocate to each variable, which necessarily limits the range of numbers each variable may represent.

The next instruction initializes x to a value of two. Like the declaration, this instruction need only happen once at the beginning of the program's execution, and never again so long as the program continues to run. The single SET statement located between the LOOP and ENDLOOP markers, however, repeatedly executes as fast as the computer's processor allows, causing x to rapidly alternate between the values of two and zero.

It should be noted that the "equals" sign ($=$) in computer programming often has a different meaning from that commonly implied in ordinary mathematics. When used in conjunction with the SET command, an "equals" sign *assigns* the value of the right-hand quantity to the left-hand variable. For example, the command SET x = 2 $-$ x tells the computer to first calculate the quantity $2 - x$ and then set the variable x to this new value. It definitely does *not* mean to imply x is actually equal in value to $2 - x$, which would be a mathematical contradiction. Thus, you should interpret the SET command to mean "set equal to . . ."

Testing values (conditional statements)

If we mean to simply test for an equality between two quantities, we may use the same symbol (=) in the context of a different command, such as "IF":

Pseudocode listing

```
DECLARE x to be an integer variable

LOOP

  // (other code manipulating the value of x goes here)

  IF x = 5 THEN
    PRINT "The value of the number is 5"
    OUTPUT audible beep on the speaker
  ENDIF
ENDLOOP
```

This program repeatedly tests whether or not the variable x is equal to 5, printing a line of text and producing a "beep" on the computer's speaker if that test evaluates as true. Here, the context of the IF command tells us the equals sign is a test for equality rather than a command to assign a new value to x. If the condition is met ($x = 5$) then all commands contained within the IF/ENDIF set are executed.

Some programming languages draw a more explicit distinction between the operations of equality test versus assignment by using different symbol combinations. In C and C++, for example, a single equals sign (=) represents assignment while a double set of equals signs (==) represents a test for equality. In Structured Text (ST) PLC programming, a single equals sign (=) represents a test for equality, while a colon plus equals sign (:=) represents assignment. The combination of an exclamation point and an equals sign (!=) represents "not equal to," used as a test condition to check for *inequality* between two quantities.

Branching and functions

A very important feature of any programming language is the ability for the path of execution to change (i.e. the program "flow" to *branch* in another direction) rather than take the exact same path every time. We saw shades of this with the IF statement in our previous example program: the computer would print some text and output a beep sound if the variable x happened to be equal to 5, but would completely skip the PRINT and OUTPUT commands if x happened to be any other value.

An elegant way to modularize a program into separate pieces involves writing portions of the program as separate *functions* which may be "called" as needed by the main program. Let us examine how to apply this concept to the following conditional program:

Pseudocode listing

```
DECLARE x to be an integer variable

LOOP

   // (other code manipulating the value of x goes here)

   IF x = 5 THEN
     PRINT "The value of the number is 5"
     OUTPUT audible beep on the speaker
   ELSEIF x = 7 THEN
     PRINT "The value of the number is 7"
     OUTPUT audible beep on the speaker
   ELSEIF x = 11 THEN
     PRINT "The value of the number is 11"
     OUTPUT audible beep on the speaker
   ENDIF
ENDLOOP
```

This program takes action (printing and outputting beeps) if ever the variable x equals either 5, 7, or 11, but not for any other values of x. The actions taken with each condition are quite similar: print the numerical value of x and output a single beep. In fact, one might argue this code is ugly because we have to keep repeating one of the commands verbatim: the OUTPUT command for each condition where we wish to computer to output a beep sound.

We may streamline this program by placing the PRINT and OUTPUT commands into their own separate "function" written outside the main loop, and then *call* that function whenever we need it. The boundaries of this function's code are marked by the BEGIN and END labels shown near the bottom of the listing:

Pseudocode listing

```
DECLARE n to be an integer variable
DECLARE x to be an integer variable
DECLARE PrintAndBeep to be a function

LOOP

  // (other code manipulating the value of x goes here)

  IF x = 5 OR x = 7 OR x = 11 THEN
    CALL PrintAndBeep(x)
  ENDIF
ENDLOOP


BEGIN PrintAndBeep (n)
  PRINT "The value of the number is" (n) "!"
  OUTPUT audible beep on the speaker
  RETURN
END PrintAndBeep
```

The main program loop is much shorter than before because the repetitive tasks of printing the value of x and outputting beep sounds has been moved to a separate function. In older computer languages, this was known as a *subroutine*, the concept being that flow through the main program (the "routine") would branch to a separate sub-program (a "subroutine") to do some specialized task and then return back to the main program when the sub-program was done with its task.

Note that the program execution flow never reaches the PrintAndBeep function unless x happens to equal 5, 7, or 11. If the value of x never matches any of those specific conditions, the program simply keeps looping between the LOOP and ENDLOOP markers.

Note also how the value of x gets *passed* on to the PrintAndBeep function, then read inside that function under another variable name, n. This was not strictly necessary for the purpose of printing the value of x, since x is the only variable in the main program. However, the use of a separate ("local") variable within the PrintAndBeep function enables us at some later date to use that function to act on other variables within the main program while avoiding conflict. Take this program for example:

Pseudocode listing

```
DECLARE n to be an integer variable
DECLARE x to be an integer variable
DECLARE y to be an integer variable
DECLARE PrintAndBeep to be a function

LOOP

   // (other code manipulating the value of x and y goes here)

  IF x = 5 OR x = 7 OR x = 11 THEN
    CALL PrintAndBeep(x)
  ENDIF
  IF y = 0 OR y = 2 THEN
    CALL PrintAndBeep(y)
  ENDIF
ENDLOOP



BEGIN PrintAndBeep (n)
  PRINT "The value of the number is" (n) "!"
  OUTPUT audible beep on the speaker
  RETURN
END PrintAndBeep
```

Here, the PrintAndBeep function gets used to print certain values of x, then re-used to print certain values of y. If we had used x within the PrintAndBeep function instead of its own variable (n), the function would only be useful for printing the value of x. Being able to pass values to functions makes those functions more useful.

A final note on branching and functions: most computer languages allow a function to call itself if necessary. This concept is known as *recursion* in computer science.

29.15.2   Position versus velocity algorithms

The canonical "ideal" or "ISA" variety of PID equation takes the following form:

$$m = K_p \left(e + \frac{1}{\tau_i} \int e \, dt + \tau_d \frac{de}{dt} \right)$$

Where,

  m = Controller output
  e = Error (SP − PV or PV − SP, depending on controller action being direct or reverse)
  K_p = Controller gain
  τ_i = Integral (reset) time constant
  τ_d = Derivative (rate) time constant

The same equation may be written in terms of "gains" rather than "time constants" for the integral and derivative terms. This re-writing exhibits the advantage of consistency from the perspective of PID tuning, where each tuning constant has the same (increasing) effect as its numerical value grows larger:

$$m = K_p \left(e + K_i \int e \, dt + K_d \frac{de}{dt} \right)$$

Where,

  m = Controller output
  e = Error
  K_p = Controller gain
  K_i = Integral (reset) gain (repeats per unit time)
  K_d = Derivative (rate) gain

However the equation is written, there are two major ways in which it is commonly implemented in a digital computer. One way is the *position* algorithm, where the result of each pass through the program "loop" calculates the actual output value. If the final control element for the loop is a control valve, this value will be the position of that valve's stem, hence the name *position algorithm*. The other way is the so-called *velocity* algorithm, where the result of each pass through the program "loop" calculates the amount the output value will *change*. Assuming a control valve for the final control element once again, the value calculated by this algorithm is the distance the valve stem will travel *per scan of the program*. In other words, the magnitude of this value describes how *fast* the valve stem will travel, hence the name *velocity algorithm*.

Mathematically, the distinction between the position and velocity algorithms is a matter of differentials: the position equation solves for the output value (m) directly while the velocity equation solves for small increments (differentials) of m, or dm.

A comparison of the position and velocity equations shows both the similarities and the differences:

$$m = K_p \left(e + K_i \int e\, dt + K_d \frac{de}{dt} \right) \qquad \text{Position equation}$$

$$dm = K_p \left(de + K_i e\, dt + K_d \frac{d^2 e}{dt} \right) \qquad \text{Velocity equation}$$

Of the two approaches to implementing PID control, the position algorithm makes the most intuitive sense and is the easiest to understand.

We will begin our exploration of both algorithms by examining their application to proportional-only control. This will be a simpler and "gentler" introduction than showing how to implement full PID control. The two respective proportional-only control equations we will consider are shown here:

$$m = K_p e + \text{Bias} \qquad \text{Position equation for P-only control}$$

$$dm = K_p de \qquad \text{Velocity equation for P-only control}$$

You will notice how a "bias" term is required in the position equation to keep track of the output's "starting point" each time a new output value is calculated. No such term is required in the velocity equation, because the computer merely calculates *how far the output moves from its last value* rather than the output's value from some absolute reference.

First, we will examine a simple pseudocode program for implementing the proportional-only equation in its "position" form:

Pseudocode listing for a "position algorithm" proportional-only controller

```
DECLARE PV, SP, and Out to be floating-point variables
DECLARE K_p, Error, and Bias to be floating-point variables
DECLARE Action, and Mode to be boolean variables

LOOP
  SET PV = analog_input_channel_N    // Update PV
  SET K_p = operator_input_channel_Gain    // From operator interface

  IF Action = 1 THEN
    SET Error = SP - PV        // Calculate error assuming reverse action
  ELSE THEN
    SET Error = PV - SP        // Calculate error assuming direct action
  ENDIF

  IF Mode = 1 THEN          // Automatic mode (if Mode = 1)
    SET Out = K_p * Error + Bias
    SET SP = operator_input_channel_SP    // From operator interface
  ELSE THEN                 // Manual mode (if Mode = 0)
    SET Out = operator_input_channel_Out  // From operator interface
    SET SP = PV      // Setpoint tracking
    SET Bias = Out   // Output tracking
  ENDIF
ENDLOOP
```

The first SET instructions within the loop update the PV to whatever value is being measured by the computer's analog input channel (channel N in this case), and the K_p variable to whatever value is entered by the human operator through the use of a keypad, touch-screen interface, or networked computer. Next, a set of IF/THEN conditionals determines which way the error should be calculated: Error = SP − PV if the control action is "reverse" (Action = 1) and Error = PV − SP if the control action is "direct" (Action = 0).

The next set of conditional instructions determines what to do in automatic versus manual modes. In automatic mode (Mode = 1), the output value is calculated according to the position equation and the setpoint comes from a human operator's input. In manual mode (Mode = 0), the output value is no longer calculated by an equation but rather is obtained from the human operator's input, the setpoint is forced equal to the process variable, and the Bias value is continually made equal to the value of the output. Setting SP = PV provides the convenient feature of *setpoint tracking*, ensuring an initial error value of zero when the controller is switched back to automatic mode. Setting the Bias equal to the output provides the essential feature of *output tracking*, where the controller begins automatic operation at an output value precisely equal to the last manual-mode output value.

Next, we will examine a simple pseudocode program for implementing the proportional-only equation in its "velocity" form:

Pseudocode listing for a "velocity algorithm" proportional-only controller

```
DECLARE PV, SP, and Out to be floating-point variables
DECLARE K_p, Error, and last_Error to be floating-point variables
DECLARE Action, and Mode to be boolean variables

LOOP
  SET PV = analog_input_channel_N    // Update PV
  SET K_p = operator_input_channel_Gain    // From operator interface
  SET last_Error = Error

  IF Action = 1 THEN
    SET Error = SP - PV        // Calculate error assuming reverse action
  ELSE THEN
    SET Error = PV - SP        // Calculate error assuming direct action
  ENDIF

  IF Mode = 1 THEN          // Automatic mode (if Mode = 1)
    SET Out = Out + (K_p * (Error - last_Error))
    SET SP = operator_input_channel_SP    // From operator interface
  ELSE THEN               // Manual mode (if Mode = 0)
    SET Out = operator_input_channel_Out  // From operator interface
    SET SP = PV   // Setpoint tracking
  ENDIF
ENDLOOP
```

The code for the velocity algorithm is mostly identical to the code for the position algorithm, with just a few minor changes. The first difference we encounter in reading the code from top to bottom is that we calculate a new variable called "last_Error" immediately prior to calculating a new value for Error. The reason for doing this is to provide a way to calculate the differential *change* in error (*de*) from scan to scan of the program. The variable "last_Error" remembers the value of Error during the previous scan of the program. Thus, the expression "Error − last_Error" is equal to the amount the error has changed from last scan to the present scan.

When the time comes to calculate the output value in automatic mode, we see the SET command calculating the change in output (K_p multiplied by the change in error), then adding this change in output to the existing output value to calculate a new output value. This is how the program translates calculated output increments into an actual output value to drive a final control element. The mathematical expression "K_p * (Error − last_Error)" defines the incremental change in output value, and this increment is then added to the current output value to generate a new output value.

From a human operator's point of view, the position algorithm and the velocity algorithm are identical with one exception: how each controller reacts to a sudden change in gain (K_p). To understand this difference, let us perform a "thought experiment" where we imagine a condition of constant error between PV and SP. Suppose the controller is operating in automatic mode, with a setpoint of 60% and a (steady) process variable value of 57%. We should not be surprised that a constant error might exist for a proportional-only controller, since we should be well aware of the phenomenon of *proportional-only offset*.

How will this controller react if the gain is suddenly increased in value while operating in automatic mode? If the controller executes the position algorithm, the result of a sudden gain change will be a sudden change in its output value, since output is a direct function of error and gain. However, if the controller executes the velocity algorithm, the result of a sudden gain change will be no change to the output at all, so long as the error *remains constant*. Only when the error begins to change will there be any noticeable difference in the controller's behavior compared to how it acted before the gain change. This is because the velocity algorithm is a function of gain and *change in error*, not error directly.

Comparing the two responses, the velocity algorithm's response to changes in gain is regarded as "better-mannered" than the position algorithm's response to changes in gain. When tuning a controller, we would rather not have the controller's output suddenly jump in response to simple gain changes[47], and so the velocity algorithm is generally preferred. If we allow the gain of the algorithm to be set by another process variable[48], the need for "stable" gain-change behavior becomes even more important.

---

[47]It should be noted that this is precisely what happens when you change the gain in a pneumatic or an analog electronic controller, since all analog PID controllers implement the "position" equation. Although the choice between "position" and "velocity" algorithms in a digital controller is arbitrary, it is *much* easier to build an analog mechanism or circuit implementing the position algorithm than it is to build an analog "velocity" controller.

[48]We call this an *adaptive gain* control system.

29.16   Note to students

PID control can be a frustrating subject for many students, even those with previous knowledge of calculus. At times it can seem like an impossibly abstract concept to master.

Thankfully, there is a relatively simple way to make PID control more "real," and that is hands-on experience with a real PID controller[49]. I advise you acquire an electronic single-loop PID controller[49] and set it up with an adjustable DC power supply and milliammeter as such:

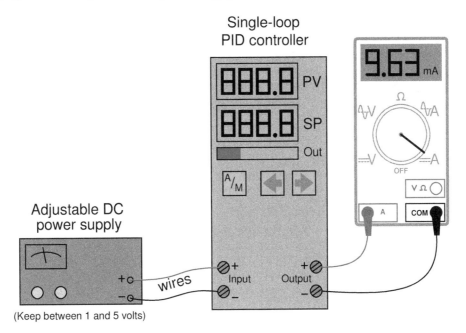

Most electronic controllers input a 1 to 5 VDC signal for the process variable (often with a 250 ohm resistor connected across the input terminals to generate a 1-5 VDC drop from a 4-20 mA current signal, which you will not need here). By adjusting the DC power supply between 1 and 5 volts DC, you will simulate a transmitter signal to the controller's input between 0% and 100%.

The milliammeter reads current output by the controller, 4 mA representing a 0% output signal and 20 mA representing a 100% output signal. With the power supply and milliammeter both connected to the appropriate terminals on the controller, you are all set to simulate input conditions and watch the controller's output response.

This arrangement does not simulate a process, and so there will be no feedback for you to observe. The purpose of this setup is to simply learn how the controller is supposed to respond to different PV and SP conditions, so that you may gain an intuitive "feel" for the PID algorithm to supplement your theoretical understanding of it. Experimentation with a real process (or even a simulated process) comes later (see section 30.6 beginning on page 2477).

---

[49]Many instrument manufacturers sell simple, single-loop controllers for reasonable prices, comparable to the price of a college textbook. You need to get one that accepts 1-5 VDC input signals and generates 4-20 mA output signals, and has a "manual" mode of operation in addition to automatic – these features are *very important!* Avoid controllers that can only accept thermocouple inputs, and/or only have time-proportioning (PWM) outputs.

Once you have all components connected, you should check to see that everything works:

- Set power supply to 1, 3, and then 5 volts DC. The controller's PV display should read 0%, 50%, and 100%, respectively. The PV display should follow closely to the power supply voltage signal over time. If the display seems to "lag" behind the power supply adjustment, then it means the controller has damping configured for the input signal. You should keep the damping set to the minimum possible value, so the controller is as responsive as it can be.

- Put the controller in manual mode and set the output to 0%, 50%, and then 100%. The milliammeter should register 4 mA, 12 mA, and 20 mA, respectively.

After checking these basic functions, you may proceed to do the following experiments. For each experiment, I recommend setting the PV input signal to 3 volts DC (50%), and manually setting the output to 50% (12 mA on the milliammeter). When you are ready to test the P,I,D responses of the controller, place the controller into automatic mode and then observe the results.

29.16.1   Proportional-only control action

1. Set the controller PV input to 50% (3 volts) and the output value to 50% in manual mode.

2. Configure the controller for reverse action (this is typically the default setting).

3. Configure the PID settings for proportional action only. This may be done by setting the gain equal to 1 (P.B. = 100%), the integral setting to zero repeats per minute (maximum minutes per repeat), and the derivative setting to zero minutes. Some controllers have the ability to switch to a "proportional-only" algorithm – if your controller has that ability, this is the best way to get set up for this exercise.

4. Switch the controller mode to "automatic."

5. Adjust the PV signal to 75% (4 volts) and observe the output. How far does the output signal move from its starting value of 50%? How does the magnitude of this step relate to the magnitude of the PV step? Does the output signal drift or does it remain the same when you stop changing the PV signal?

6. Adjust the PV signal to 25% (2 volts) and observe the output. How far does the output signal move from its starting value of 50%? How does the magnitude of this step relate to the magnitude of the PV step? Does the output signal drift or does it remain the same when you stop changing the PV signal?

7. Change the controller's gain setting to some different value and repeat the previous two steps. How does the output step magnitude relate to the input step-changes in each case? Do you see the relationship between controller gain and how the output responds to changes in the input?

8. Smoothly vary the input signal back and forth between 0% and 100% (1 and 5 volts). How does the output respond when you do this? Try changing the gain setting again and re-checking.

9. Switch the controller's action from *reverse* to *direct*, then repeat the previous step. How does the output respond now?

29.16.2 Integral-only control action

1. Set the controller PV input to 50% (3 volts) and the output value to 50% in manual mode.

2. Configure the controller for reverse action (this is typically the default setting).

3. Configure the PID settings for integral action only. If the controller has an "I-only" mode, this is the best way to get set up for this exercise. If there is no way to completely turn off proportional action, then I recommend setting the gain value to the minimum non-zero value allowed, and setting the integral constant to an aggressive value (many repeats per minute, or fractions of a minute per repeat). If your controller does have an integral-only option, I recommend setting the integral time constant at 1 minute. Set derivative action at zero minutes.

4. Switch the controller mode to "automatic."

5. Adjust the PV signal to 75% (4 volts) and observe the output. Which way does the output signal move? Does the output signal drift or does it remain the same when you stop changing the PV signal? How does this action compare with the proportional-only test?

6. Adjust the PV signal to 25% (2 volts) and observe the output. Which way does the output signal move? Does the output signal drift or does it remain the same when you stop changing the PV signal? How does this action compare with the proportional-only test?

7. Change the controller's integral setting to some different value and repeat the previous two steps. How does the rate of output ramping relate to the input step-changes in each case? Do you see the relationship between the integral time constant and how the output responds to changes in the input?

8. Smoothly vary the input signal back and forth between 0% and 100% (1 and 5 volts). How does the output respond when you do this? Try changing the integral setting again and re-checking.

9. Where must you adjust the input signal to get the output to stop moving? When the output finally does settle, is its value consistent (i.e. does it always settle at the same value, or can it settle at different values)?

10. Switch the controller's action from *reverse* to *direct*, then repeat the previous two steps. How does the output respond now?

29.16.3   Proportional plus integral control action

1. Set the controller PV input to 50% (3 volts) and the output value to 50% in manual mode.

2. Configure the controller for reverse action (this is typically the default setting).

3. Configure the PID settings with a proportional (gain) value of 1 (P.B. = 100%) and an integral value of 1 repeat per minute (or 1 minute per repeat). Set derivative action at zero minutes.

4. Switch the controller mode to "automatic."

5. Adjust the PV signal to 75% (4 volts) and observe the output. Which way does the output signal move? Does the output signal drift or does it remain the same when you stop changing the PV signal? How does this action compare with the proportional-only test and with the integral-only test?

6. Adjust the PV signal to 25% (2 volts) and observe the output. Which way does the output signal move? Does the output signal drift or does it remain the same when you stop changing the PV signal? How does this action compare with the proportional-only test and with the integral-only test?

7. Change the controller's gain setting to some different value and repeat the previous two steps. Can you tell which aspect of the output signal's response is due to proportional action and which aspect is due to integral action?

8. Change the controller's integral setting to some different value and repeat those same two steps. Can you tell which aspect of the output signal's response is due to proportional action and which aspect is due to integral action?

9. Smoothly vary the input signal back and forth between 0% and 100% (1 and 5 volts). How does the output respond when you do this? Try changing the gain and/or integral settings again and re-checking.

10. Switch the controller's action from *reverse* to *direct*, then repeat the previous two steps. How does the output respond now?

29.16.4 Proportional plus derivative control action

1. Set the controller PV input to 50% (3 volts) and the output value to 50% in manual mode.

2. Configure the controller for reverse action (this is typically the default setting).

3. Configure the PID settings with a proportional (gain) value of 1 (P.B. = 100%) and a derivative value of 1 minute. Set integral action at zero repeats per minute (maximum number of minutes per repeat).

4. Switch the controller mode to "automatic."

5. Adjust the PV signal to 75% (4 volts) and observe the output. Which way does the output signal move? How does the output signal value compare while you are adjusting the input voltage versus after you reach 4 volts and take your hand off the adjustment knob? How does this action compare with the proportional-only test?

6. Adjust the PV signal to 25% (2 volts) and observe the output. Which way does the output signal move? How does the output signal value compare while you are adjusting the input voltage versus after you reach 4 volts and take your hand off the adjustment knob? How does this action compare with the proportional-only test?

7. Change the controller's gain setting to some different value and repeat the previous two steps. Can you tell which aspect of the output signal's response is due to proportional action and which aspect is due to derivative action?

8. Change the controller's derivative setting to some different value and repeat those same two steps. Can you tell which aspect of the output signal's response is due to proportional action and which aspect is due to derivative action?

9. Smoothly vary the input signal back and forth between 0% and 100% (1 and 5 volts). How does the output respond when you do this? Try changing the derivative setting again and re-checking.

10. Switch the controller's action from *reverse* to *direct*, then repeat the previous two steps. How does the output respond now?

29.16.5   Full PID control action

1. Set the controller PV input to 50% (3 volts) and the output value to 50% in manual mode.

2. Configure the controller for reverse action (this is typically the default setting).

3. Configure the PID settings with a proportional (gain) value of 1 (P.B. = 100%), an integral value of 1 repeat per minute (or 1 minute per repeat), and a derivative action of 1 minute.

4. Switch the controller mode to "automatic."

5. Adjust the PV signal to 75% (4 volts) and observe the output. Which way does the output signal move? Does the output signal drift or does it remain the same when you stop changing the PV signal? How does magnitude of the output signal compare while you are changing the input voltage, versus when the input signal is steady?

6. Adjust the PV signal to 25% (2 volts) and observe the output. Which way does the output signal move? Does the output signal drift or does it remain the same when you stop changing the PV signal? How does magnitude of the output signal compare while you are changing the input voltage, versus when the input signal is steady?

7. Change the controller's gain setting to some different value and repeat the previous two steps. Can you tell which aspect of the output signal's response is due to proportional action, which aspect is due to integral action, and which aspect is due to derivative action?

8. Change the controller's integral setting to some different value and repeat the same two steps. Can you tell which aspect of the output signal's response is due to proportional action, which aspect is due to integral action, and which aspect is due to derivative action?

9. Change the controller's derivative setting to some different value and repeat the same two steps. Can you tell which aspect of the output signal's response is due to proportional action, which aspect is due to integral action, and which aspect is due to derivative action?

10. Smoothly vary the input signal back and forth between 0% and 100% (1 and 5 volts). How does the output respond when you do this? Try changing the gain, integral, and/or derivative settings again and re-checking.

11. Switch the controller's action from *reverse* to *direct*, then repeat the previous two steps. How does the output respond now?

29.17 Review of fundamental principles

Shown here is a partial listing of principles applied in the subject matter of this chapter, given for the purpose of expanding the reader's view of this chapter's concepts and of their general inter-relationships with concepts elsewhere in the book. Your abilities as a problem-solver and as a life-long learner will be greatly enhanced by mastering the applications of these principles to a wide variety of topics, the more varied the better.

- **Linear equations**: any function represented by a straight line on a graph may be represented symbolically by the slope-intercept formula $y = mx + b$. Relevant to proportional control algorithms.

- **Zero shift**: any shift in the offset of an instrument is fundamentally additive, being represented by the "intercept" (b) variable of the slope-intercept linear formula $y = mx + b$. Relevant to controller tuning: adjusting the "bias" of a loop controller always adds to or subtracts from its output signal.

- **Span shift**: any shift in the gain of an instrument is fundamentally multiplicative, being represented by the "slope" (m) variable of the slope-intercept linear formula $y = mx + b$. Relevant to controller tuning: adjusting the "gain" of a loop controller always multiplies or divides the response of its output for a given input change.

- **Negative feedback**: when the output of a system is degeneratively fed back to the input of that same system, the result is decreased (overall) gain and greater stability. Relevant to loop controller action: in order for a control system to be stable, the feedback must be negative.

- **Self-balancing pneumatic mechanisms**: all self-balancing pneumatic instruments work on the principle of negative feedback maintaining a nearly constant baffle-nozzle gap. Force-balance mechanisms maintain this constant gap by balancing force against force with negligible motion, like a tug-of-war. Motion-balance mechanisms maintain this constant gap by balancing one motion with another motion, like two dancers moving in unison.

- **Self-balancing opamp circuits**: all self-balancing operational amplifier circuits work on the principle of negative feedback maintaining a nearly zero differential input voltage to the opamp. Making the "simplifying assumption" that the opamp's differential input voltage is exactly zero assists in circuit analysis, as does the assumption that the input terminals draw negligible current.

- **Amplification**: the control of a relatively large signal by a relatively small signal. Relevant to the role of loop controllers exerting influence over a process variable at the command of a measurement signal. In behaving as amplifiers, loop controllers may oscillate if certain criteria are met.

- **Barkhausen criterion**: is overall loop gain is unity (1) or greater, and phase shift is 360°, the loop will sustain oscillations. Relevant to control system stability, explaining why the loop will "cycle" (oscillate) if gain is too high.

- **Time constant**: (τ), defined as the amount of time it takes a system to change 63.2% of the way from where it began to where it will eventually stabilize. The system will be within 1%

of its final value after 5 time constants' worth of time has passed (5τ). Relevant to process control loops, where natural lags contribute to time constants, usually of multiple order.

References

"FOUNDATION Fieldbus", document L454 EN, Samson AG, Frankfurt, Germany, 2000.

"Identification and Description of Instrumentation, Control, Safety, and Information Systems and Components Implemented in Nuclear Power Plants", EPRI, Palo Alto, CA: 2001. 1001503.

Kernighan, Brian W. and Ritchie, Dennis M., *The C Programming Language*, Bell Telephone Laboratories, Incorporated, Murray Hill, NJ, 1978.

Lavigne, John R., *Instrumentation Applications for the Pulp and Paper Industry*, The Foxboro Company, Foxboro, MA, 1979.

Lipták, Béla G. et al., *Instrument Engineers' Handbook – Process Control Volume II*, Third Edition, CRC Press, Boca Raton, FL, 1999.

Mollenkamp, Robert A., *Introduction to Automatic Process Control*, Instrument Society of America, Research Triangle Park, NC, 1984.

"Moore 353 Process Automation Controller User's Manual", document UM353-1, Revision 11, Siemens Energy and Automation, 2003.

Shinskey, Francis G., *Energy Conservation through Control*, Academic Press, New York, NY, 1978.

Shinskey, Francis G., *Process-Control Systems – Application / Design / Adjustment*, Second Edition, McGraw-Hill Book Company, New York, NY, 1979.

"SPEC 200 Systems", technical information document TI 200-100, Foxboro, 1980.

"SPEC 200 System Configuration", technical information document TI 200-105, Foxboro, January 1975.

"SPEC 200 System Wiring", technical information document TI 200-260, Foxboro, 1972.

Ziegler, J. G., and Nichols, N. B., "Optimum Settings for Automatic Controllers", *Transactions of the American Society of Mechanical Engineers* (ASME), Volume 64, pages 759-768, Rochester, NY, November 1942.

Chapter 30

Process dynamics and PID controller tuning

To *tune* a feedback control system means to adjust parameters in the controller to achieve robust control over the process. "Robust" in this context is usually defined as stability of the process variable despite changes in load, fast response to changes in setpoint, minimal oscillation following either type of change, and minimal offset (error between setpoint and process variable) over time.

"Robust control" is far easier to define than it is to achieve. With PID (Proportional-Integral-Derivative) control being the most common feedback control algorithm used in industry, it is important for all instrumentation practitioners to understand how to tune these controllers effectively and with a minimum investment of time.

Different types of processes, having different dynamic (time-dependent) behaviors, require different levels of proportional, integral, and derivative control action to achieve stability and robust response. It is therefore imperative for anyone seeking to tune a PID controller to understand the dynamic nature of the process being controlled. For this reason, the chapter begins with an exploration of common process characteristics before introducing techniques useful in choosing practical P, I, and D tuning parameter values.

30.1   Process characteristics

Perhaps the most important rule of controller tuning is to *know the process before attempting to adjust the controller's tuning*. Unless you adequately understand the nature of the process you intend to control, you will have little hope in actually controlling it well. This section of the book is dedicated to an investigation of different process characteristics and how to identify each.

Quantitative PID tuning methods (see section 30.3 beginning on page 2445) attempt to map the characteristics of a process so good PID parameters may be chosen for the controller. The goal of this section is for you to understand various process types by observation and qualitative analysis so you may comprehend why different tuning parameters are necessary for each type, rather than mindlessly following a step-by-step PID tuning procedure.

The three major classifications of process response are *self-regulating*, *integrating*, and *runaway*. Each of these process types is defined by its response to a step-change in the manipulated variable (e.g. control valve position or state of some other final control element). A "self-regulating" process responds to a step-change in the final control element's status by settling to a new, stable value. An "integrating" process responds by ramping either up or down at a rate proportional to the magnitude of the final control element's step-change. Finally, a "runaway" process responds by ramping either up or down at a rate that increases over time, headed toward complete instability without some form of corrective action from the controller.

Self-regulating, integrating, and runaway processes have very different control needs. PID tuning parameters that may work well to control a self-regulating process, for example, will *not* work well to control an integrating or runaway process, no matter how similar any of the other characteristics of the processes may be[1]. By first identifying the characteristics of a process, we may draw some general conclusions about the P, I, and D setting values necessary to control it well.

Perhaps the best method for testing a process to determine its natural characteristics is to place the controller in *manual mode* and introduce a step-change to the controller output signal. It is critically important that the loop controller be in manual mode whenever process characteristics are being explored. If the controller is left in the automatic mode, the response seen from the process to a setpoint or load change will be partly due to the natural characteristics of the process itself *and* partly due to the corrective action of the controller. The controller's corrective action thus interferes with our goal of exploring process characteristics. By placing the controller in "manual" mode, we turn off its corrective action, effectively removing its influence by breaking the feedback loop between process and controller, controller and process. In manual mode, the response we see from the process to an output (manipulated variable) or load change is *purely* a function of the natural process dynamics, which is precisely what we wish to discern.

A test of process characteristics with the loop controller in manual mode is often referred to as an *open-loop* test, because the feedback loop has been "opened" and is no longer a complete loop. Open-loop tests are the fundamental diagnostic technique applied in the following subsections.

---

[1]To illustrate, self-regulating processes require significant integral action from a controller in order to avoid large offsets between PV and SP, with minimal proportional action and no derivative action. Integrating processes, in contrast, may be successfully controlled primarily on proportional action, with minimal integral action to eliminate offset. Runaway processes absolutely require derivative action for dynamic stability, but derivative action alone is not enough: some integral action will be necessary to eliminate offset. Even if knowledge of a process's dominant characteristic does not give enough information for us to quantify P, I, or D values, it will tell us which tuning constant will be most important for achieving stability.

30.1.1 Self-regulating processes

If a liquid flow-control valve is opened in a step-change fashion, flow through the pipe tends to self-stabilize at a new rate very quickly. The following illustration shows a typical liquid flow-control installation, with a process trend showing the flow response following a manual-mode (also known as "open-loop") step-change in valve position:

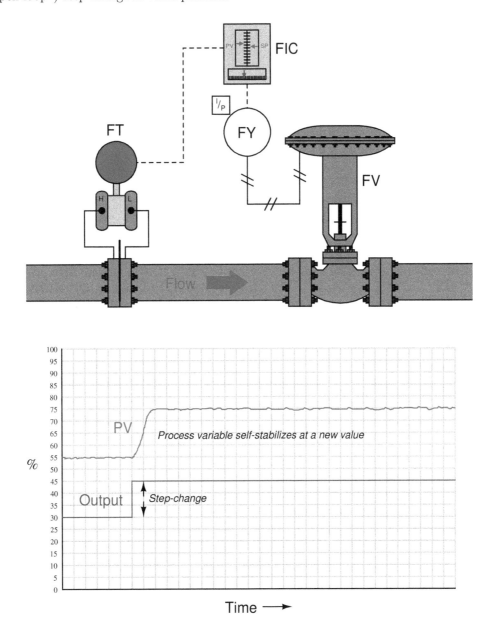

The defining characteristic of a self-regulating process is its inherent ability to settle at a new process variable value without any corrective action on the part of the controller. In other words, a self-regulating process will exhibit a unique process variable value for each possible output (valve) value. The inherently fast response of a liquid flow control process makes its self-regulating nature obvious: the self-stabilization of flow generally takes place within a matter of seconds following the valve's motion. Many other processes besides liquid flow are self-regulating as well, but their slow response times require patience on the part of the observer to tell that the process will indeed self-stabilize following a step-change in valve position.

A corollary to the principle of self-regulation is that a unique output value *will be required* to achieve a new process variable value. For example, to achieve a greater flow rate, the control valve must be opened further and held at that further-open position for as long as the greater flow rate is desired. This presents a fundamental problem for a proportional-only controller. Recall the formula for a proportional-only controller, defining the output value (m) by the error (e) between process variable and setpoint multiplied by the gain (K_p) and added to the bias (b):

$$m = K_p e + b$$

Where,

    m = Controller output

    e = Error (difference between PV and SP)

    K_p = Proportional gain

    b = Bias

Suppose we find the controller in a situation where there is no error (PV = SP), and the flow rate is holding steady at some value. If we then increase the setpoint value (calling for a greater flow rate), the error will increase, driving the valve further open. As the control valve opens further, flow rate naturally increases to match. This increase in process variable drives the error back toward zero, which in turn causes the controller to decrease its output value back toward where it was before the setpoint change. However, the error can never go all the way back to zero because if it did, the valve would return to its former position, and that would cause the flow rate to self-regulate back to its original value before the setpoint change was made. What happens instead is that the control valve begins to close as flow rate increases, and eventually the process finds some equilibrium point where the flow rate is steady at some value less than the setpoint, creating just enough error to drive the valve open just enough to maintain that new flow rate. Unfortunately, due to the need for an error to exist, this new flow rate will fall shy of our setpoint. We call this error *proportional-only offset*, or *droop*, and it is an inevitable consequence of a proportional-only controller attempting to control a self-regulating process.

For any fixed bias value, there will be only one setpoint value that is perfectly achievable for a proportional-only controller in a self-regulating process. Any other setpoint value will result in some degree of offset in a self-regulating process. If dynamic stability is more important than absolute accuracy (zero offset) in a self-regulating process, a proportional-only controller may suffice. A great many self-regulating processes in industry have been and still are controlled by proportional-only controllers, despite some inevitable degree of offset between PV and SP.

The amount of offset experienced by a proportional-only controller in a self-regulating process may be minimized by increasing the controller's gain. If it were possible to increase the gain of a proportional-only controller to infinity, it would be able to achieve any setpoint desired with zero offset! However, there is a practical limit to the extent we may increase the gain value, and that limit is *oscillation*. If a controller is configured with too much gain, the process variable will begin to oscillate over time, never stabilizing at any value at all, which of course is highly undesirable for any automatic control system. Even if the gain is not great enough to cause sustained oscillations, excessive values of gain will still cause problems by causing the process variable to oscillate with decreasing amplitude for a period of time following a sudden change in either setpoint or load. Determining the optimum gain value for a proportional-only controller in a self-regulating process is, therefore, a matter of compromise between excessive offset and excessive oscillation.

Recall that the purpose of integral (or "reset") control action was the elimination of offset. Integral action works by ramping the output of the controller at a rate determined by the magnitude of the offset: the greater the difference between PV and SP for an integral controller, the faster that controller's output will ramp over time. In fact, the output will stabilize at some value *only* if the error is diminished to zero (PV = SP). In this way, integral action works tirelessly to eliminate offset.

It stands to reason then that a self-regulating process *absolutely requires* some amount of integral action in the controller in order to achieve zero offset for all possible setpoint values. The more aggressive (faster) a controller's integral action, the sooner offset will be eliminated. Just how much integral action a self-regulating process can tolerate depends on the magnitudes of any time lags in the system. The faster a process's natural response is to a manual step-change in controller output, the better it will respond to aggressive integral controller action once the controller is placed in automatic mode. Aggressive integral control action in a slow process, however, will result in oscillation due to integral wind-up[2].

It is not uncommon to find self-regulating processes being controlled by *integral-only* controllers. An "integral-only" process controller is an instrument lacking proportional or derivative control modes. Liquid flow control is a nearly ideal candidate process for integral-only control action, due to its self-regulating and fast-responding nature.

Summary:

- Self-regulating processes are characterized by their natural ability to stabilize at a new process variable value following changes in the control element value or load(s).

- Self-regulating processes *absolutely require* integral controller action to eliminate offset between process variable and setpoint, because only integral action is able to create a different controller output value once the error returns to zero.

- Faster integral controller action results in quicker elimination of offset.

- The amount of integral controller action tolerable in a self-regulating process depends on the degree of time lag in the system. Too much integral action will result in oscillation, just like too much proportional control action.

---

[2]Recall that wind-up is what happens when integral action "demands" more from a process than the process can deliver. If integral action is too aggressive for a process (i.e. fast integral controller action in a process with slow time lags), the output will ramp too quickly, causing the process variable to overshoot setpoint which then causes integral action to wind the other direction. As with proportional action, too much integral action will cause a self-regulating process to oscillate.

30.1.2 Integrating processes

A good example of an integrating process is liquid level control, where either the flow rate of liquid into or out of a vessel is constant and the other flow rate varies. If a control valve is opened in a step-change fashion, liquid level in the vessel ramps at a rate proportional to the difference in flow rates in and out of the vessel. The following illustration shows a typical liquid level-control installation, with a process trend showing the level response to a step-change in valve position (with the controller in manual mode, for an "open-loop" test):

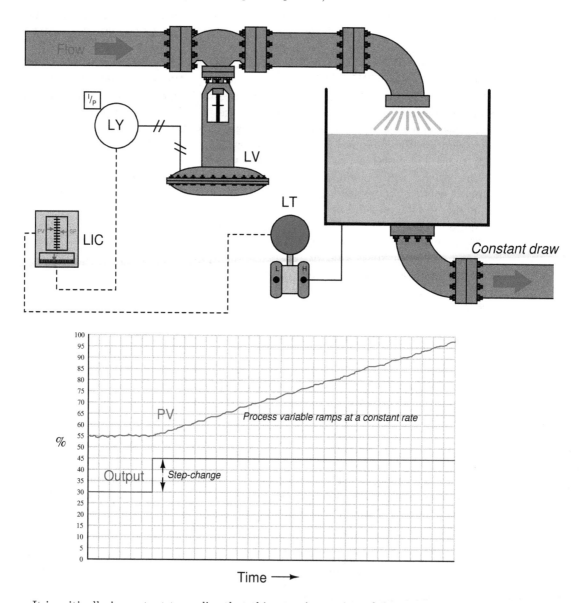

It is critically important to realize that this ramping action of the process variable over time is a characteristic of the process itself, not the controller. When liquid flow rates in and out of a vessel are mis-matched, the liquid level within that vessel will change at a rate proportional to the difference in flow rates. The trend shown here reveals a fundamental characteristic of the process, not the controller (this should be obvious once it is realized that the step-change in output is something that would only ever happen with the controller in *manual* mode).

Mathematically, we may express the integrating nature of this process using calculus notation. First, we may express the *rate of change* of volume in the tank over time ($\frac{dV}{dt}$) in terms of the flow rates in and out of the vessel:

$$\frac{dV}{dt} = Q_{in} - Q_{out}$$

For example, if the flow rate of liquid going into the vessel was 450 gallons per minute, and the constant flow rate drawn out of the vessel was 380 gallons per minute, the volume of liquid contained within the vessel would increase over time at a rate equal to 70 gallons per minute: the difference between the in-flow and the out-flow rates.

Another way to express this mathematical relationship between flow rates and liquid volume in the vessel is to use the calculus function of *integration*:

$$\Delta V = \int_0^T (Q_{in} - Q_{out}) \, dt$$

The amount of liquid volume accumulated in the vessel (ΔV) between time 0 and time T is equal to the sum ($\int$) of the products (multiplication) of difference in flow rates in and out of the vessel ($Q_{in} - Q_{out}$) during infinitesimal increments of time (dt).

In the given scenario of a liquid level control system where the out-going flow is held constant, this means the level will be stable only at one in-coming flow rate (where $Q_{in} = Q_{out}$). At any other controlled flow rate, the level will either be increasing over time or decreasing over time.

This process characteristic perfectly matches the characteristic of a proportional-only controller, where there is one unique output value when the error is zero (PV = SP). We may illustrate this by performing a "thought experiment" on the liquid level-control process shown earlier having a constant draw out the bottom of the vessel. Imagine this process controlled by a proportional-only controller in automatic mode, with the bias value (b) of the controller set to the exact value needed by the control valve to make in-coming flow exactly equal to the constant out-going flow (draw). This means that when the process variable is precisely equal to setpoint (PV = SP), the liquid level will hold constant. If now an operator were to increase the setpoint value (with the controller in automatic mode), it would cause the valve to open further, adding liquid at a faster rate to the vessel. The naturally integrating nature of the process will result in an increasing liquid level. As level increases, the amount of error in the controller decreases, causing the valve to approach its original (bias) position. When the level reaches the new setpoint, the controller output will have returned to its original (bias) value, which will make the control valve go to its original position and hold the level constant once again. Thus, a proportional-only controller will achieve the new setpoint with absolutely no offset ("droop").

The more aggressive the controller's proportional action, the sooner the integrating process will reach new setpoints. Just how much proportional action (gain) an integrating process can tolerate depends on the magnitudes of any time lags in the system as well as the magnitude of noise in the process variable signal. Any process system with time lags will oscillate if the controller has sufficient gain. Noise is a problem because proportional action directly reproduces process variable noise on the output signal: too much gain, and just a little bit of PV noise translates into a control valve whose stem position constantly jumps around.

Purely integrating processes do not require integral control action to eliminate offset as is the case with self-regulating processes, following a setpoint change. The natural integrating action of the process eliminates offset that would otherwise arise from setpoint changes. More than that, the presence of any integral action in the controller will actually force the process variable to overshoot setpoint following a setpoint change in a purely integrating process! Imagine a controller with integral action responding to a step-change in setpoint for the liquid level control process shown earlier. As soon as an error develops, the integral action will begin "winding up" the output value, forcing the valve to open more than proportional action alone would demand. By the time the liquid level reaches the new setpoint, the valve will have reached a position greater than where it originally was before the setpoint change[3], which means the liquid level will *not* stop rising when it reaches setpoint, but in fact will overshoot setpoint. Only after the liquid level has spent sufficient time above setpoint will the integral action of the controller "wind" back down to its previous level, allowing the liquid level to finally achieve the new setpoint.

This is not to say that integral control action is completely unnecessary in integrating processes – far from it. If the integrating process is subject to *load* changes, only integral action can return the PV back to the SP value (eliminate offset). Consider, in our level control example, if the out-going flow rate were to change. Now, a new valve position will be required to achieve stable (unchanging) level in the vessel. A proportional-only controller is able to generate a new valve position *only* if an error develops between PV and SP. Without at least some degree of integral action configured in the controller, that error will persist indefinitely. Or consider if the liquid supply pressure upstream of the control valve were to change, resulting in a different rate of incoming flow for the same valve stem position as before. Once again, the controller would have to generate a different output value to compensate for this process change and stabilize liquid level, and the only way a proportional-only controller could do that is to let the process variable drift a bit from setpoint (the definition of an error or offset).

The example of an integrating process used here is just one of many possible processes where we are dealing with either a *mass balance* or an *energy balance* problem. "Mass balance" is the accounting of all mass into and out of a process. Since the Law of Mass Conservation states the impossibility of mass creation or destruction, all mass into and out of a process must be accounted for. If the mass flow rate into a process does not equal the mass flow rate out of a process, the process must be either gaining or losing an internal store of mass. The same may be said for energy: all energy flowing into and out of a process must be accounted for, since the Law of Energy Conservation states the impossibility of energy creation or destruction. If the energy flow rate (input power) into a process does not equal the energy flow rate (output power) out of a process, the process must be either gaining or losing an internal store of energy.

---

[3]In a proportional-only controller, the output is a function of error (PV − SP) and bias. When PV = SP, bias alone determines the output value (valve position). However, in a controller with integral action, the zero-offset output value is determined by *how long* and *how far* the PV has previously strayed from SP. In other words, there is no fixed bias value anymore. Thus, the output of a controller with integral action will *not* return to its previous value once the new SP is reached. In a purely integrating process, this means the PV will *not* reach stability at the new setpoint, but will continue to rise until all the "winding up" of integral action is un-done.

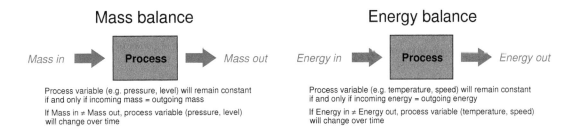

Common examples of integrating processes include the following:

- Liquid level control – *mass balance* – when the flow of liquid either into or out of a vessel is manipulated, and the other flows in or out of the vessel are constant

- Gas pressure control – *mass balance* – when the flow of gas either into or out of a vessel is manipulated, and the other flows in or out of the vessel are constant

- Storage bin level control – *mass balance* – when the conveyor feed rate into the bin is manipulated, and the draw from the bin is constant

- Temperature control – *energy balance* – when the flow of heat into or out of a process is manipulated, and all other heat flows are constant

- Speed control – *energy balance* – when the force (linear) or torque (angular) applied to a mass is manipulated, and all other loads are constant in force or torque

In a self-regulating process, the control element (valve) exerts control over *both* the in-flow and the out-flow of either mass or energy. In the previous subsection, where liquid flow control was the process example, the mass balance consisted of liquid flow into the valve and liquid flow out of the valve. Since the piping was essentially a "series" path for an incompressible fluid, where input flow must equal output flow at any given time, mass in and mass out were *guaranteed* to be in a state of balance, with one valve controlling both. This is why a change in valve position resulted in an almost immediate change and re-stabilization of flow rate: the valve exerts immediate control over both the incoming and the outgoing flow rates, with both in perfect balance. Therefore, nothing "integrates" over time in a liquid flow control process because there can never be an imbalance between in-flow and out-flow.

In an integrating process, the control element (valve) exerts control over *either* the in-flow *or* the out-flow of mass or energy, but never both. Thus, changing valve position in an integrating process causes an imbalance of mass flow and/or energy flow, resulting in the process variable ramping over time as either mass or energy accumulates in (or depletes from) the process.

CHAPTER 30. PROCESS DYNAMICS AND PID CONTROLLER TUNING

Our "simple" example of an integrating (level-control) process becomes a bit more complicated if the outgoing flow depends on level, as is the case with a gravity-drained vessel where the outgoing flow is a function of liquid level in the vessel rather than being fixed at a constant rate as it was in the previous example:

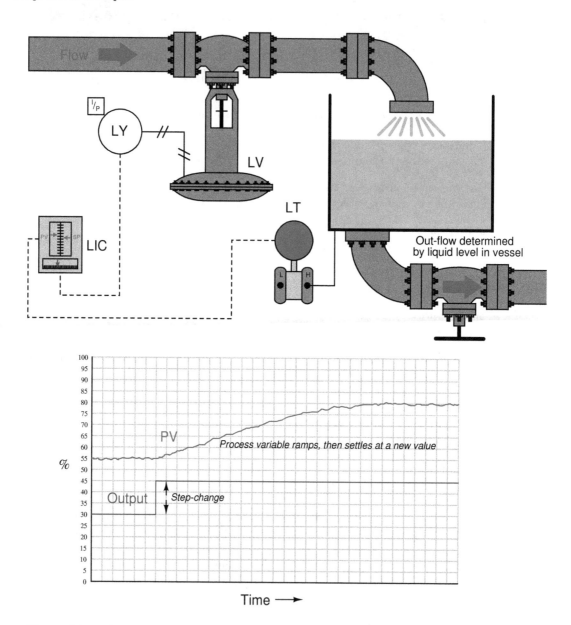

If we subject the control valve to a manual step-change increase, the flow rate of liquid into the vessel immediately increases. This causes an imbalance of incoming and outgoing flow, resulting in the liquid level rising over time. As level rises, however, increasing hydrostatic pressure across the manual valve at the vessel outlet causes the outgoing flow rate to increase. This causes the mass imbalance rate to be less than it was before, resulting in a decreased integration rate (rate of level rise). Thus, the liquid level still rises, but at a slower and slower rate as time goes on. Eventually, the liquid level will become high enough that the pressure across the manual valve forces a flow rate out of the vessel equal to the flow rate into the vessel. At this point, with matched flow rates, the liquid level stabilizes with no corrective action from the controller (remember, the step-change in output was made in manual mode!). Note the final result of letting the outgoing flow be a function of liquid level: *what used to be an integrating process has now become a self-regulating process,* albeit one with a substantial lag time.

Many processes ideally categorized as integrating actually behave in this manner. Although the manipulated variable may control the flow rate into or out of a process, the other flow rates often change with the process variable. Returning to our list of integrating process examples, we see how a PV-variable load in each case can make the process self-regulate:

- Liquid level control – *mass balance* – if the in-flow naturally decreases as liquid level rises and/or the out-flow naturally increases as liquid level rises, the vessel's liquid level will tend to self-regulate instead of integrate

- Gas pressure control – *mass balance* – if in-flow naturally decreases as pressure rises and/or the out-flow naturally increases as pressure rises, the vessel's pressure will tend to self-regulate instead of integrate

- Storage bin level control – *mass balance* – if the draw from the bin increases with bin level (greater weight pushing material out at a faster rate), the bin's level will tend to self-regulate instead of integrate

- Temperature control – *energy balance* – if the process naturally loses heat at a faster rate as temperature increases and/or the process naturally takes in less heat as temperature rises, the temperature will tend to self-regulate instead of integrate

- Speed control – *energy balance* – if drag forces on the object increase with speed (as they usually do for any fast-moving object), the speed will tend to self-regulate instead of integrate

We may generalize all these examples of integrating processes turned self-regulating by noting the one aspect common to all of them: some natural form of *negative feedback* exists internally to bring the system back into equilibrium. In the mass-balance examples, the physics of the process ensure a new balance point will eventually be reached because the in-flow(s) and/or out-flow(s) naturally change in ways that oppose any change in the process variable. In the energy-balance examples, the laws of physics again conspire to ensure a new equilibrium because the energy gains and/or losses naturally change in ways that oppose any change in the process variable. The presence of a control system is, of course, the ultimate example of negative feedback working to stabilize the process variable. However, the control system may not be the *only* form of negative feedback at work in a process. All self-regulating processes are that way because they intrinsically possess some degree of negative feedback acting as a sort of natural, proportional-only control system.

This one detail completely alters the fundamental characteristic of a process from integrating to self-regulating, and therefore changes the necessary controller parameters. Self-regulation guarantees at least some integral controller action is *necessary* to attain new setpoint values. A purely integrating process, by contrast, requires no integral controller action at all to achieve new setpoints, and in fact is *guaranteed* to suffer overshoot following setpoint changes if the controller is programmed with any integral action at all! Both types of processes, however, need some amount of integral action in the controller in order to recover from *load* changes.

Summary:

- Integrating processes are characterized by a ramping of the process variable in response to a step-change in the control element value or load(s).

- This integration occurs as a result of either *mass flow imbalance* or *energy flow imbalance* in and out of the process.

- Integrating processes are ideally controllable with proportional controller action alone.

- Integral controller action guarantees setpoint overshoot in a purely integrating process.

- Some integral controller action will be required in integrating processes to compensate for load changes.

- The amount of proportional controller action tolerable in an integrating process depends on the degree of time lag and process noise in the system. Too much proportional action will result in oscillation (time lags) and/or erratic control element motion (noise).

- An integrating process will become self-regulating if sufficient negative feedback is naturally introduced. This usually takes the form of loads varying with the process variable.

30.1.3 Runaway processes

A classic "textbook" example of a runaway process is an inverted pendulum: a vertical stick balanced on its end by moving the bottom side-to-side. Inverted pendula are typically constructed in a laboratory environment by fixing a stick to a cart by a pivot, then equipping the cart with wheels and a reversible motor to give it lateral control ability. A sensor (usually a potentiometer) detects the stick's angle from vertical, reporting that angle to the controller as the process variable. The cart's motor is the final control element:

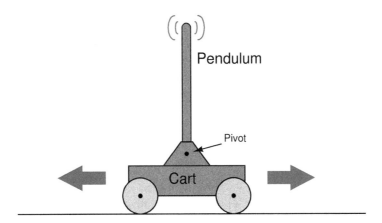

The defining characteristic of a runaway process is its tendency to accelerate away from a condition of stability with no corrective action applied. Viewed on a process trend, a runaway process tends to respond as follows to an open-loop step-change:

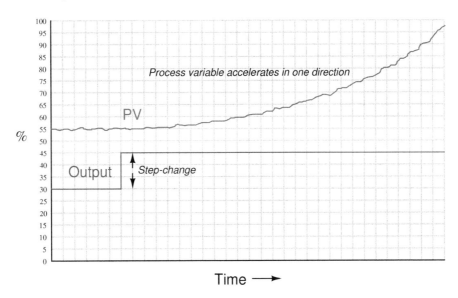

A synonym for "runaway" is *negative self-regulation* or *negative lag*, because the process variable curve over time for a runaway process resembles the mathematical inverse of a self-regulating curve with a lag time: it races away from the horizontal, while a self-regulating process variable draws closer and closer to the horizontal over time.

The "Segway^{TM}" personal transport device is a practical example of an inverted pendulum, with wheel motion controlled by a computer attempting to maintain the body of the vehicle in a vertical position. As the human rider leans forward, it causes the controller to spin the wheels with just the right amount of acceleration to maintain balance. There are many examples of runaway processes in motion-control applications, especially automated controls for vertical-flight vehicles such as helicopters and vectored-thrust aircraft such as the Harrier military fighter jet.

Some chemical reaction processes are runaway as well, especially *exothermic* (heat-releasing) reactions. Most chemical reactions increase in rate as temperature rises, and so exothermic reactions tend to accelerate with time (either becoming hotter or becoming colder) unless checked by some external influence. This poses a significant challenge to process control, as many exothermic reactions used to manufacture products must be temperature-controlled to ensure efficient production of the desired product. Off-temperature chemical reactions may not "favor" production of the desired products, producing unwanted byproducts and/or failing to efficiently consume the reactants. Furthermore, safety concerns usually surround exothermic chemical reactions, as no one wants their process to melt down or explode.

What makes a runaway process behave as it does is internal *positive feedback*. In the case of the inverted pendulum, gravity pulls works to pull an off-center pendulum even farther off center, accelerating it until it falls down completely. In the case of exothermic chemical reactions, the direct relationship between temperature and reaction rate forms a positive feedback loop: the hotter the reaction, the faster it proceeds, releasing even more heat, making it even hotter. It should be noted that *endothermic* chemical reactions (absorbing heat rather than releasing heat) tend to be self-regulating for the exact same reason exothermic reactions tend to be runaway: reaction rate usually has a positive correlation with reaction temperature.

It is easy to demonstrate for yourself how challenging a runaway process can be to control. Simply try to balance a long stick vertically in the palm of your hand. You will find that the only way to maintain stability is to react swiftly to any changes in the stick's angle – essentially applying a healthy dose of *derivative* control action to counteract any motion from vertical.

Fortunately, runaway processes are less common in the process industries. I say "fortunately" because these processes are notoriously difficult to control and usually pose more danger than inherently self-regulating processes. Many runaway processes are also nonlinear, making their behavior less intuitive to human operators.

Just as integrating processes may be forced to self-regulate by the addition of (natural) negative feedback, intrinsically runaway processes may also be forced to self-regulate given the presence of sufficient natural negative feedback. An interesting example of this is a pressurized water nuclear fission reactor.

Nuclear fission is a process by which the nuclei of specific types of atoms (most notably uranium 235 and plutonium-239) undergo spontaneous disintegration upon the absorption of an extra neutron, with the release of significant thermal energy and additional neutrons. A quantity of fissile material such as ^{235}U or ^{239}Pu is subjected to a source of neutron particle radiation, which initiates the fission process, releasing massive quantities of heat which may then be used to boil water into steam and drive steam turbine engines to generate electricity. The "chain reaction" of neutrons splitting fissile atoms, which then eject more neutrons to split more fissile atoms, is inherently exponential in nature. The more atoms split, the more neutrons are released, which then proceed to split even more atoms. The rate at which neutron activity within a fission reactor grows or decays over time is determined by the *multiplication factor*[4], and this factor is easily controlled by the insertion of neutron-absorbing *control rods* into the reactor core.

Thus, a fission chain-reaction naturally behaves as an inverted pendulum. If the multiplication factor is greater than 1, the reaction grows exponentially. If the multiplication factor is less than 1, the reaction dies exponentially. In the case of a nuclear weapon, the desired multiplication factor is as large as physically possible to ensure explosive reaction growth. In the case of an operating nuclear power plant, the desired multiplication factor is exactly 1 to ensure stable power generation.

---

[4]When a nucleus of uranium or plutonium undergoes fission ("splits"), it releases more neutrons capable of splitting additional uranium or plutonium nuclei. The ratio of new nuclei "split" versus old nuclei "split" is the multiplication factor. If this factor has a value of one (1), the chain reaction will sustain at a constant power level, with each new generation of atoms "split" equal to the number of atoms "split" in the previous generation. If this multiplication factor exceeds unity, the rate of fission will increase over time. If the factor is less than one, the rate of fission will decrease over time. Like an inverted pendulum, the chain reaction has a tendency to "fall" toward infinite activity or toward no activity, depending on the value of its multiplication factor.

A simplified diagram of a pressurized-water reactor (PWR) is shown here:

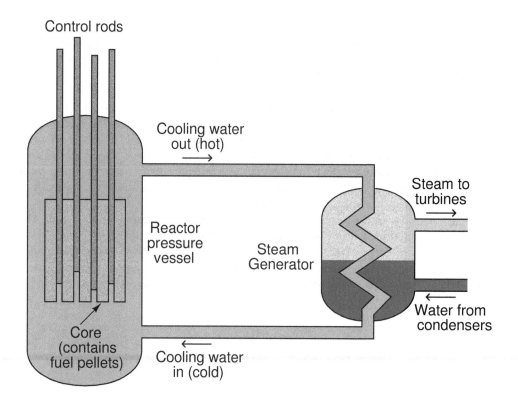

Water under high pressure (too high of pressure to boil) circulates through the reactor vessel, carrying heat away from the nuclear core, then transferring the heat energy to a heat exchanger ("steam generator") where a second water loop is allowed to boil into steam and drive turbine engines to spin electrical generators. Control rods inserted into the core by linear actuators adjust the multiplication factor of the reactor.

If the multiplication factor of a fission reactor were solely controlled by the positions of these control rods, it would be a classic "runaway" process, with the reactor's power level tending to increase toward infinity or decrease toward zero if the rods were at any position other than one yielding a multiplication factor of precisely unity (1). This would make nuclear reactors extremely difficult (if not impossible) to safely control. Fortunately, there are ways to engineer negative feedback directly into the design of the reactor core so that neutron activity *naturally* self-stabilizes without active control rod action. In water-cooled reactors, the water itself achieves this goal. Pressurized water plays a dual role in a fission reactor: it not only transfers heat out of the reactor core and into a boiler to produce steam, but it also offsets the multiplication factor inversely proportional to temperature. As the reactor core heats up, the water's density changes, affecting the probability[5] of neutrons being captured by fissile nuclei. This is called a *negative temperature*

---

[5]The mechanism by which this occurs varies with the reactor design, and is too detailed to warrant a full explanation here. In pressurized light-water reactors – the dominant design in the United States of America – this action occurs

coefficient for the reactor, and it forces the otherwise runaway process of nuclear fission to become self-regulating.

With this self-regulating characteristic in effect, control rod position essentially determines the reactor's steady-state temperature. The further the control rods are withdrawn from the core, the hotter the core will run. The cooling water's natural negative temperature coefficient prevents the fission reaction from "running away" either to destruction or to shutdown.

Some nuclear fission reactor designs are capable of "runaway" behavior, though. The ill-fated reactor at Chernobyl (Ukraine, Russia) was of a design where its power output could "run away" under certain operating conditions, and that is exactly what happened on April 26, 1986. The Chernobyl reactor used solid graphite blocks as the main neutron-moderating substance, and as such its cooling water did not provide enough natural negative feedback to overcome the intrinsically runaway characteristic of nuclear fission. This was especially true at low power levels where the reactor was being tested on the day of the accident. A combination of poor management decisions, unusual operating conditions, and unstable design characteristics led to the reactor's destruction with massive amounts of radiation released into the surrounding environment. It stands at the time of this writing as the world's worst nuclear accident[6].

Summary:

- Runaway processes are characterized by an exponential ramping of the process variable in response to a step-change in the control element value or load(s).

- This "runaway" occurs as a result of some form of *positive feedback* happening inside the process.

- Runaway processes cannot be controlled with proportional or integral controller action alone, and always requires derivative action for stability.

- Some integral controller action will be required in runaway processes to compensate for load changes.

- A runaway process will become self-regulating if sufficient negative feedback is naturally introduced, as is the case with water-moderated fission reactors.

---

due to the water's ability to *moderate* (slow down) the velocity of neutrons. Slow neutrons have a greater probability of being "captured" by fissile nuclei than fast neutrons, and so the water's moderating ability will have a direct effect on the reactor core's multiplication factor. As a light-water reactor core increases temperature, the water becomes less dense and therefore less effective at moderating (slowing down) fast neutrons emitted by "splitting" nuclei. These fast(er) neutrons then "miss" the nuclei of atoms they would have otherwise split, effectively reducing the reactor's multiplication factor without any need for regulatory control rod motion. The reactor's power level therefore self-stabilizes as it warms, rather than "running away" to dangerously high levels, and may thus be classified as a *self-regulating* process.

[6]Discounting, of course, the intentional discharge of nuclear weapons, whose sole design purpose is to self-destruct in a "runaway" chain reaction.

30.1.4 Steady-state process gain

When we speak of a controller's *gain*, we refer to the aggressiveness of its proportional control action: the ratio of output change to input change. However, we may go a step further and characterize each component within the feedback loop as having its own gain (a ratio of output change to input change):

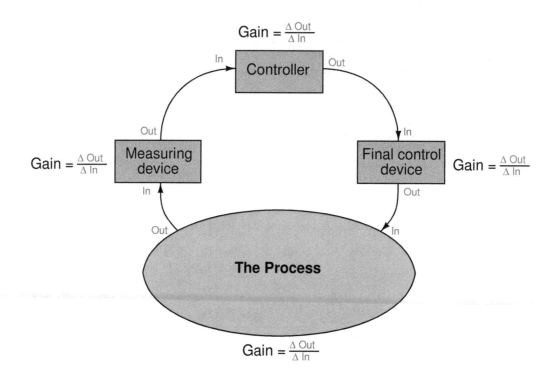

The gains intrinsic to the measuring device (transmitter), final control device (e.g. control valve), and the process itself are all important in helping to determine the necessary controller gain to achieve robust control. The greater the combined gain of transmitter, process, and valve, the less gain is needed from the controller. The less combined gain of transmitter, process, and valve, the more gain will be needed from the controller. This should make some intuitive sense: the more "responsive" a process appears to be, the less aggressive the controller needs to be in order to achieve stable control (and vice-versa).

These combined gains may be empirically determined by means of a simple test performed with the controller in manual mode, also known as an "open-loop" test. By placing the controller in manual mode (and thus disabling its automatic correction of process changes) and adjusting the output signal by some fixed amount, the resulting change in process variable may be measured and compared. If the process is self-regulating, a definite ratio of PV change to controller output change may be determined.

For instance, examine this process trend graph showing a manual "step-change" and process variable response:

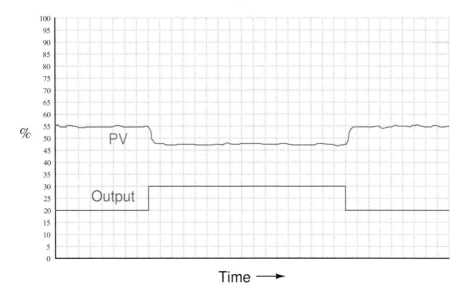

Here, the output step-change is 10% of scale, while the resulting process variable step-change is about 7.5%. Thus, the "gain" of the process[7] (together with transmitter and final control element) is approximately 0.75, or 75% (Gain = $\frac{7.5\%}{10\%}$). Incidentally, it is irrelevant that the PV steps *down* in response to the controller output stepping *up*. All this means is the process is reverse-responding, which necessitates *direct* action on the part of the controller in order to achieve negative feedback. When we calculate gains, we usually ignore directions (mathematical signs) and speak in terms of absolute values.

We commonly refer to this gain as the *steady-state gain* of the process, because the determination of gain is made after the PV settles to its self-regulating value.

Since from the controller's perspective the individual gains of transmitter, final control element, and physical process meld into one over-all gain value, the process may be made to appear more or less responsive (more or less steady-state gain) just by altering the gain of the transmitter and/or the gain of the final control element.

Consider, for example, if we were to reduce the span of the transmitter in this process. Suppose this was a flow control process, with the flow transmitter having a calibrated range of 0 to 200 liters per minute (LPM). If a technician were to re-range the transmitter to a new range of 0 to 150 LPM, what effect would this have on the apparent process gain?

---

[7]The general definition of gain is the ratio of output change over input change ($\frac{\Delta \text{Out}}{\Delta \text{In}}$). Here, you may have noticed we calculate process gain by dividing the process variable change (7.5%) by the controller output change (10%). If this seems "inverted" to you because we placed the *output* change value in the denominator of the fraction instead of the numerator, you need to keep in mind the perspective of our gain measurement. We are not calculating the gain of the controller, but rather the gain of the *process*. Since the output of the controller is the "input" to the process, it is entirely appropriate to refer to the 10% manual step-change as the change of *input* when calculating process gain.

To definitively answer this question, we must re-visit the process trend graph for the old calibrated range:

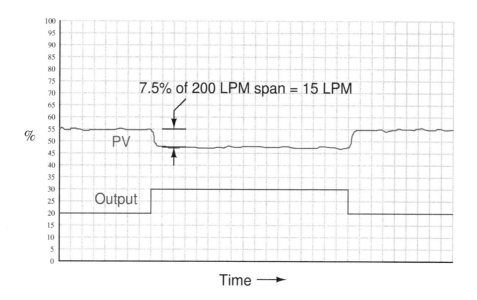

We see here that the 7.5% PV step-change equates to a change of 15 LPM given the flow transmitter's span of 200 LPM. However, if a technician re-ranges the flow transmitter to have just three-quarters that amount of span (150 LPM), the exact same amount of output step-change will *appear* to have a more dramatic effect on flow, even though the physical response of the process has the same as it was before:

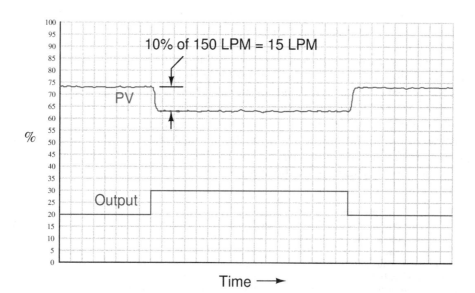

From the controller's perspective – which only "knows" percent[8] of signal range – the process gain appears to have increased from 0.75 to 1, with nothing more than a re-ranging of the transmitter. Since the process is now "more responsive" to controller output signals than it was before, there may be a tendency for the loop to oscillate in automatic mode even if it did not oscillate previously with the old transmitter range. A simple fix for this problem is to decrease the controller's gain by the same factor that the process gain increased: we need to make the controller's gain $\frac{3}{4}$ what it was before, since the process gain is now $\frac{4}{3}$ what it was before.

The exact same effect occurs if the final control element is re-sized or re-ranged. A control valve that is replaced with one having a different C_v value, or a variable-frequency motor drive that is given a different speed range for the same 4-20 mA control signal, are two examples of final control element changes which will result in different overall gains. In either case, a given change in controller output signal percentage results in a different amount of influence on the process thanks to the final control element being more or less influential than it was before. Re-tuning of the controller may be necessary in these cases to preserve robust control.

If and when re-tuning is needed to compensate for a change in loop instrumentation, all control modes should be proportionately adjusted. This is automatically done if the controller uses the *Ideal* or *ISA* PID equation, or if the controller uses the *Series* or *Interacting* PID equation[9]. All that needs to be done to an Ideal-equation controller in order to compensate for a change in process gain is to change that controller's proportional (P) constant setting. Since this constant directly affects all terms of the equation, the other control modes (I and D) will be adjusted along with the proportional term. If the controller happens to be executing the *Parallel* PID equation, you will have to manually alter all three constants (P, I, and D) in order to compensate for a change in process gain.

---

[8]While this is true of analog-signal transmitters, it is not necessarily true of digital-signal transmitters such as Fieldbus or wireless (digital radio). The reason for this distinction is that in a digital-signal transmitter, the reported process variable value is scaled in engineering units rather than percent. Applied to this case, if the flow transmitter gets re-ranged from 0-200 LPM to 0-150 LPM, the controller sees no change in process gain because a change of 10 LPM is still reported as a change in 10 LPM regardless of the transmitter's range.

[9]For more information on different PID equations, refer to Section 29.10 beginning on page 2302.

A very important aspect of process gain is how *consistent* it is over the entire measurement range. It is entirely possible (and in fact very likely) that a process may be more responsive (have higher gain) in some areas of control than in others. Take for instance this hypothetical trend showing process response to a series of manual-mode step-changes:

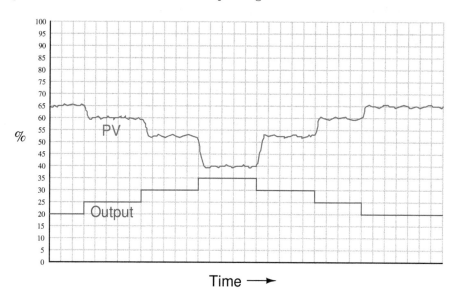

Note how the PV changes about 5% for the first 5% step-change in output, corresponding to a process gain of 1. Then, the PV changes about 7.5% for the next 5% output step-change, for a process gain of 1.5. The final increasing 5% step-change yields a PV change of about 12.5%, a process gain of 2.5. Clearly, the process being controlled here is not equally responsive throughout the measurement range. This is a concern to us in tuning the PID controller because any set of tuning constants that work well to control the process around a certain setpoint may not work as well if the setpoint is changed to a different value, simply because the process may be more or less responsive at that different process variable value.

Inconsistent process gain is a problem inherent to many different process types, which means it is something you will need to be aware of when investigating a process prior to tuning the controller. The best way to reveal inconsistent process gain is to perform a series of step-changes to the controller output while in manual mode, "exploring" the process response throughout the safe range of operation.

Compensating for inconsistent process gain is much more difficult than merely detecting its presence. If the gain of the process continuously grows from one end of the range to the other (e.g. low gain at low output values and high gain at high output values, or vice-versa), a control valve with a different characteristic may be applied to counter-act the process gain.

If the process gain follows some pattern more closely related to PV value rather than controller output value, the best solution is a type of controller known as an *adaptive gain controller*. In an adaptive gain controller, the proportional setting is made to vary in a particular way as the process changes, rather than be a fixed constant set by a human technician or engineer.

30.1.5 Lag time

If a square-wave signal is applied to an RC passive integrator circuit, the output signal will appear to have a "sawtooth" shape, the crisp rising and falling edges of the square wave replaced by damped curves:

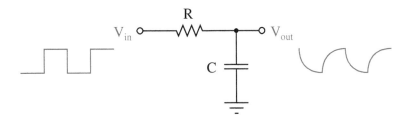

In a word, the output signal of this circuit *lags* behind the input signal, unable to keep pace with the steep rising and falling edges of the square wave.

Most mechanical and chemical processes exhibit a similar tendency: an "inertial" opposition to rapid changes. Even instruments themselves naturally[10] damp sudden stimuli. We could have just as easily subjected a pressure transmitter to a series of pressure pulses resembling square waves, and witnessed the output signal exhibit the same damped response:

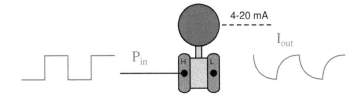

---

[10]It is also possible to *configure* many instruments to deliberately damp their response to input conditions. This is called *damping*, and it is covered in more detail in section 18.4 beginning on page 1261.

The gravity-drained level-control process highlighted in an earlier subsection exhibits a very similar response to a sudden change in control valve position:

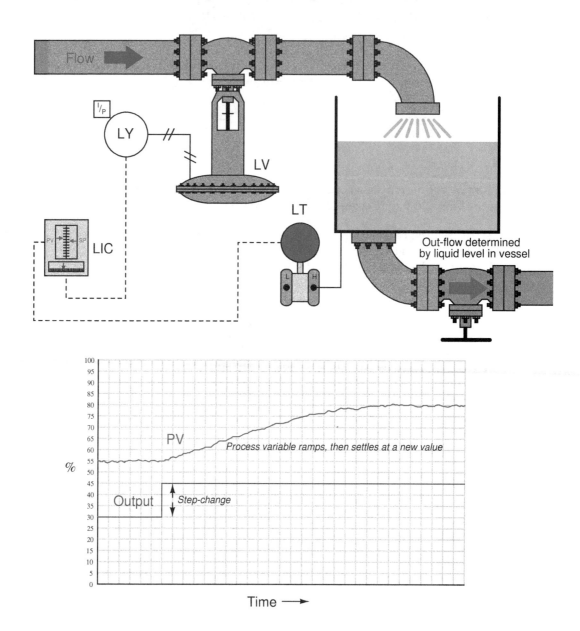

For any particular flow rate into the vessel, there will be a final (self-regulating) point where the liquid level "wants" to settle[11]. However, the liquid level does not *immediately* achieve that new

---

[11]Assuming a constant discharge valve position. If someone alters the hand valve's position, the relationship between incoming flow rate and final liquid level changes.

level if the control valve jumps to some new position, owing to the "capacity" of the vessel and the dynamics of gravity flow.

Any physical behavior exhibiting the same "settling" behavior over time may be said to illustrate a *first-order lag*. A classic "textbook" example of a first-order lag is the temperature of a cup of hot liquid, gradually equalizing with room temperature. The liquid's temperature drops rapidly at first, but then slows its approach to ambient temperature as time progresses. This natural tendency is described by *Newton's Law of Cooling*, mathematically represented in the form of a *differential equation* (an equation containing a variable along with one or more of its derivatives). In this case, the equation is a *first-order* differential equation, because it contains the variable for temperature (T) and the first derivative of temperature ($\frac{dT}{dt}$) with respect to time:

$$\frac{dT}{dt} = -k(T - T_{ambient})$$

Where,

T = Temperature of liquid in cup

$T_{ambient}$ = Temperature of the surrounding environment

k = Constant representing the thermal conductivity of the cup

t = Time

All this equation tells us is that the rate of cooling ($\frac{dT}{dt}$) is directly proportional ($-k$) to the difference in temperature between the liquid and the surrounding air ($T - T_{ambient}$). The hotter the temperature, the faster the object cools (the faster rate of temperature fall):

Newton's Cooling Law

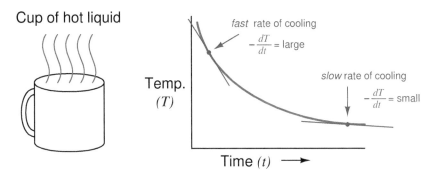

The proportionality constant in this equation (k) represents how readily thermal energy escapes the hot cup. A cup with more thermal insulation, for example, would exhibit a smaller k value (i.e. the rate of temperature loss $\frac{dT}{dt}$ will be less for any given temperature difference between the cup and ambient $T - T_{ambient}$).

A general solution to this equation is as follows:

$$T = (T_{initial} - T_{final}) \left(e^{-\frac{t}{\tau}}\right) + T_{final}$$

Where,

T = Temperature of liquid in cup at time t
$T_{initial}$ = Starting temperature of liquid ($t = 0$)
T_{final} = Ultimate temperature of liquid (ambient)
e = Euler's constant
τ = "Time constant" of the system

This mathematical analysis introduces a descriptive quantity of the system: something called a *time constant*. The "time constant" of a first-order system is the amount of time necessary for the system to come to within 36.8% (e^{-1}) of its final value (i.e. the time required for the system to go 63.2% of the way from the starting point to its ultimate settling point: $1 - e^{-1}$). After two time-constants' worth of time, the system will have come to within 13.5% (e^{-2}) of its final value (i.e. gone 86.5% of the way: $1 - e^{-2}$); after three time-constants' worth of time, to within 5% (e^{-3}) of the final value, (i.e. gone 95% of the way: $1 - e^{-3}$). After five time-constants' worth of time, the system will be within 1% (e^{-5}, rounded to the nearest whole percent) of its final value, which is often close enough to consider it "settled" for most practical purposes.

Time	Percent of final value	Percent change remaining
0	0.000%	100.000%
τ	63.212%	36.788%
2τ	86.466%	13.534%
3τ	95.021%	4.979%
4τ	98.168%	1.832%
5τ	99.326%	0.674%
6τ	99.752%	0.248%
7τ	99.909%	0.091%
8τ	99.966%	0.034%
9τ	99.988%	0.012%
10τ	99.995%	0.005%

The concept of a "time constant" may be shown in graphical form for both falling and rising variables:

Falling variable

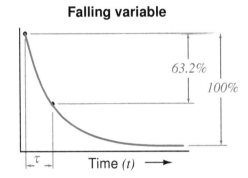

Rising variable

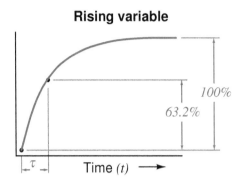

One "time constant" (τ) is the amount of time required
for the variable to change 63.2% of the way from its
starting point to its ultimate (terminal) value

Students of electronics will immediately recognize this concept, since it is widely used in the analysis and application of capacitive and inductive circuits. However, you should recognize the fact that the concept of a "time constant" for capacitive and inductive electrical circuits is only one case of a more general phenomenon. Literally *any* physical system described by the same first-order differential equation may be said to have a "time constant." Thus, it is perfectly valid for us to speak of a hot cup of coffee as having a time constant (τ), and to say that the coffee's temperature will be within 1% of room temperature after five of those time constants have elapsed.

In the world of process control, it is more customary to refer to this as a *lag time* than as a *time constant*, but these are really interchangeable terms. The term "lag time" makes sense if we consider a first-order system *driven* to achieve a constant rate of change. For instance, if we subjected our RC circuit to a ramping input voltage rather than a "stepped" input voltage – such that the output ramped as well instead of passively settling at some final value – we would find that the amount of time separating equal input and output voltage values was equal to this time constant (in an RC circuit, $\tau = RC$):

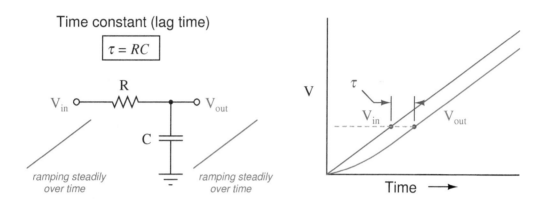

Lag time is thus defined as the difference in time between when the process variable ramps to a certain value and when it *would have* ramped to that same value were it not for the existence of first-order lag in the system. The system's output variable *lags behind* the ramping input variable by a fixed amount of time, regardless of the ramping rate. If the process in question is an RC circuit, the lag time will still be the product of ($\tau = RC$), just as the "time product" defined for a stepped input voltage. Thus, we see that "time constant" and "lag time" are really the exact same concept, merely manifesting in different forms as the result of two different input conditions (*stepped* versus *ramped*).

When an engineer or a technician describes a process being "fast" or "slow," they generally refer to the magnitude of this lag time. This makes lag time very important to our selection of PID controller tuning values. Integral and derivative control actions in particular are sensitive to the amount of lag time in a process, since both those actions are time-based. "Slow" processes (i.e. process types having large lag times) cannot tolerate aggressive integral action, where the controller "impatiently" winds the output up or down at a rate that is too rapid for the process to respond to. Derivative action, however, is generally useful on processes having large lag times.

30.1.6 Multiple lags (orders)

Simple, self-regulating processes tend to be first-order: that is, they have only one mechanism of lag. More complicated processes often consist of multiple sub-processes, each one with its own lag time. Take for example a convection oven, heating a potato. Being instrumentation specialists in addition to cooks, we decide to monitor both the oven temperature and the potato temperature using thermocouples and remote temperature indicators:

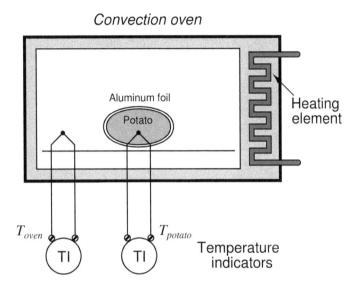

The oven itself is a first-order process. Given enough time and sufficiently thorough air circulation, the oven's air temperature will eventually self-stabilize at the heating element's temperature. If we graph its temperature over time as the heater power is suddenly stepped up to some fixed value[12], we will see a classic first-order response:

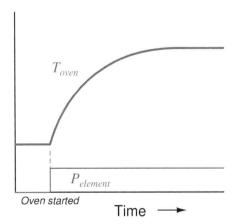

---

[12]We will assume here the heating element reaches its final temperature immediately upon the application of power, with no lag time of its own.

The potato forms another first-order process, absorbing heat from the air within the oven (heat transfer by convection), gradually warming up until its temperature (eventually) reaches that of the oven[13]. From the perspective of the heating element to the oven air temperature, we have a first-order process. From the perspective of the heating element to the potato, however, we have a *second*-order process.

Intuition might lead you to believe that a second-order process is just like a first-order process – except slower – but that intuition would be wrong. Cascading two first-order lags creates a fundamentally different time dynamic. In other words, two first-order lags do not simply result in a *longer* first-order lag, but rather a *second-order* lag with its own unique characteristics.

If we superimpose a graph of the potato temperature with a graph of the oven temperature (once again assuming constant power output from the heating element, with no thermostatic control), we will see that the *shape* of this second-order lag is different. The curve now has an "S" shape, rather than a consistent downward concavity:

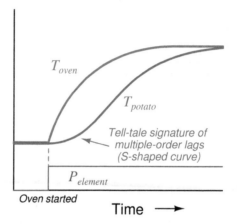

This, in fact, is one of the tell-tale signature of multiple lags in a process: an "S"-shaped curve rather than the characteristically abrupt initial rise of a first-order curve.

---

[13]Given the presence of water in the potato which turns to steam at 212 °F, things are just a bit more complicated than this, but let's ignore the effects of water in the potato for now!

Another tell-tale signature of multiple lags is that the lagging variable does not immediately reverse its direction of change following a reversal in the final control element signal. We can see this effect by cutting power to the heating element before either the oven air or potato temperatures have reached their final values:

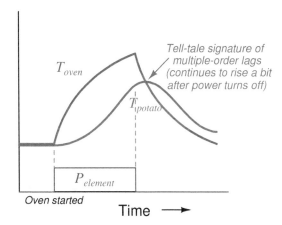

Note how the air temperature trend *immediately* reverses direction following the cessation of power to the heating element, but how the potato temperature trend continues to rise for a short amount of time[14] before reversing direction and cooling. Here, the contrast between first-order and second-order lag responses is rather dramatic – the second-order response is clearly not just a longer version of the first-order response, but rather something quite distinct unto itself.

This is why multiple-order lag processes have a greater tendency to *overshoot* their setpoints while under automatic control: the process variable exhibits a sort of "inertia" whereby it fails to switch directions simultaneously with the controller output.

---

[14]The amount of time the potato's temperature will continue to rise following the down-step in heating element power is equal to the time it takes for the oven's air temperature to equal the potato's temperature. The reason the potato's temperature keeps rising after the heating element turns off is because the air inside the oven is (for a short time) still hotter than the potato, and therefore the potato continues to absorb thermal energy from the air for a time following power-off.

If we were able to ramp the heater power at a constant rate and graph the heater element, air, and potato temperatures, we would clearly see the separate lag times of the oven and the potato as offsets in time at any given temperature:

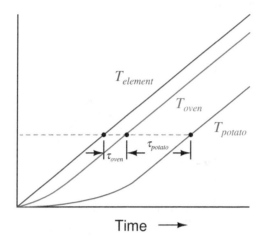

As another example, let us consider the control of level in three cascaded, gravity-drained vessels:

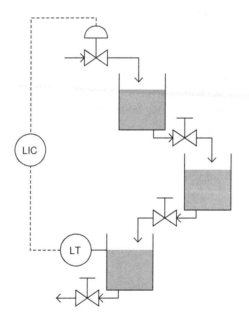

From the perspective of the level transmitter on the last vessel, the control valve is driving a *third-order* process, with three distinct lags cascaded in series. This would be a challenging process to control, and not just because of the possibility of the intermediate vessels overflowing (since their levels are not being measured)!

When we consider the dynamic response of a process, we are usually concerned primarily with the physical process itself. However, the instruments attached to that process also influence lag orders and lag times. As discussed in the previous subsection, almost every physical function exhibits some form of lag. Even the instruments we use to measure process variables have their own (usually very short) lag times. Control valves may have substantial lag times, measured in the tens of seconds for some large valves. Thus, a "slow" control valve exerting control over a first-order process effectively creates a second-order loop response. Thermowells used with temperature sensors such as thermocouples and RTDs can also introduce lag times into a loop (especially if the sensing element is not fully contacting the bottom of the well!).

This means it is nearly impossible to have a control loop with a purely first-order response. Many real loops come close to being first-order, but only because the lag time of the physical process swamps (dominates) the relatively tiny lag times of the instruments. For inherently fast processes such as liquid flow and liquid pressure control, however, the process response is so fast that even short time lags in valve positioners, transmitters, and other loop instruments significantly alter the loop's dynamic character.

Multiple-order lags are relevant to the issue of PID loop tuning because they encourage oscillation. The more lags there are in a system, the more delayed and "detached" the process variable becomes from the controller's output signal.

A system with lag time exhibits *phase shift* when driven by a sinusoidal stimulus: the outgoing waveform lags behind the input waveform by a certain number of degrees at one frequency. The exact amount of phase shift depends on frequency – the higher the frequency, the more phase shift (to a maximum of -90° for a first-order lag):

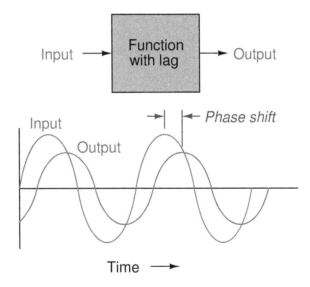

The phase shifts of multiple, cascaded lag functions (or processes, or physical effects) add up. This means each lag in a system contributes an additional negative *phase shift* to the loop. This can be detrimental to negative feedback, which by definition is a 180° phase shift. If sufficient lags exist in a system, the total loop phase shift may approach 360°, in which case the feedback becomes *positive* (regenerative): a necessary[15] condition for oscillation.

---

[15]The so-called *Barkhausen criterion* for oscillation in a feedback system is that the total loop gain is at least unity

It is worthy to note that multiple-order lags are constructively applied in electronics when the express goal is to create oscillations. If a series of RC "lag" networks are used to feed the output of an inverting amplifier circuit back to its input with sufficient signal strength intact[16], and those networks introduce another 180 degrees of phase shift, the total loop phase shift will be 360° (i.e. positive feedback) and the circuit will self-oscillate. This is called an *RC phase-shift oscillator* circuit:

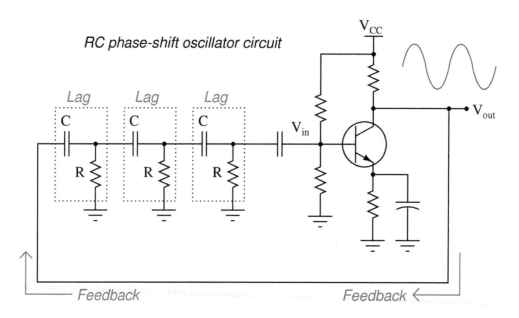

The amplifier works just like a proportional-only process controller, with action set for negative feedback. The resistor-capacitor networks act like the lags inherent to the process being controlled. Given enough controller (amplifier) gain, the cascaded lags in the process (RC networks) create the perfect conditions for self-oscillation. The amplifier creates the first 180° of phase shift (being inverting in nature), while the RC networks collectively create the other 180° of phase shift to give a total phase shift of 360° (positive, or *regenerative* feedback).

In theory, the most phase shift a single RC network can create is −90°, but even that is not practical[17]. This is why more than two RC phase-shifting networks are required for successful operation of an RC phase-shift oscillator circuit.

---

(1) and the total loop phase shift is 360°.

[16]The conditions necessary for self-sustaining oscillations to occur is a total phase shift of 360° *and* a total loop gain of 1. Merely having positive feedback *or* having a total gain of 1 or more will not guarantee self-sustaining oscillations; both conditions must simultaneously exist. As a measure of how close any feedback system is to this critical confluence of conditions, we may quantify a system's *phase margin* (how many degrees of phase shift the system is away from 360° while at a loop gain of 1) and/or a system's *gain margin* (how many decibels of gain the system is away from 0 dB while at a phase shift of 360°). The less phase or gain margin a feedback system has, the closer it is to a condition of instability.

[17]At maximum phase shift, the gain of any first-order RC network is zero. Both phase shift and attenuation in an RC lag network are frequency-dependent: as frequency increases, phase shift grows larger (from 0° to a maximum of −90°) and the output signal grows weaker. At its theoretical maximum phase shift of exactly −90°, the output signal would be reduced to nothing!

As an illustration of this point, the following circuit is incapable[18] of self-oscillation. Its lone RC phase-shifting network cannot create the -180° phase shift necessary for the overall loop to have positive feedback and oscillate:

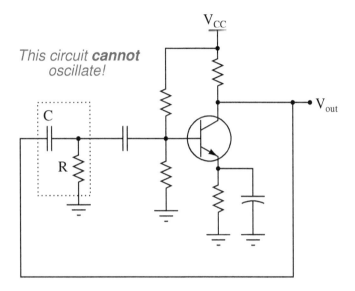

The RC phase-shift oscillator circuit design thus holds a very important lesson for us in PID loop tuning. It clearly illustrates how multiple orders of lag are a more significant obstacle to robust control than a single lag time of *any* magnitude. A purely first-order process will tolerate enormous amounts of controller gain without ever breaking into oscillations, because it lacks the phase shift necessary to self-oscillate. This means – barring any other condition limiting our use of high gain, such as process noise – we may use very aggressive proportional-only action (e.g. gain values of 20 or more) to achieve robust control on a first-order process[19]. Multiple-order processes are less forgiving of high controller gains, because they *are* capable of generating enough phase shift to self-oscillate.

---

[18]In its pure, theoretical form at least. In practice, even a single-lag circuit may oscillate given enough gain due to the unavoidable presence of parasitic capacitances and inductances in the wiring and components causing multiple orders of lag (and even some dead time). By the same token, even a "pure" first-order process will oscillate given enough controller gain due to unavoidable lags and dead times in the field instrumentation (especially the control valve). The point I am trying to make here is that there is more to the question of stability (or instability) than loop gain.

[19]Truth be told, the same principle holds for purely integrating processes as well. A purely integrating process *always* exhibits a phase shift of -90° at any frequency, because that is the nature of integration in calculus. A purely first-order lag process will exhibit a phase shift anywhere from 0° to -90° depending on frequency, but never more lagging than -90°, which is not enough to turn negative feedback into positive feedback. In either case, so long as we don't have process noise to deal with, we can increase the controller's gain all the way to *eleven*. If that last sentence (a joke) does not make sense to you, be sure to watch the 1984 movie *This is Spinal Tap* as soon as possible. Seriously, I have used controller gains as high as *50* on low-noise, first-order processes such as furnace temperature control. With such high gain in the controller, response to setpoint and load changes is quite swift, and integral action is almost unnecessary because the offset is naturally so small.

30.1.7   Dead time

Lag time refers to a damped response from a process, from a change in manipulated variable (e.g. control valve position) to a measured change in process variable: the initial effect of a change in controller output is immediately seen, but the final effect takes time to develop. *Dead time*, by contrast, refers to a period of time during which a change in manipulated variable produces *no effect whatsoever* in the process variable: the process appears "dead" for some amount of time before showing a response. The following graph contrasts first-order and multiple-order lag times against pure dead time, as revealed in response to a manual step-change in the controller's output (an "open-loop" test of the process characteristics):

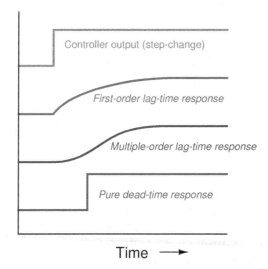

Although the first-order response does takes some time to settle at a stable value, there is no time delay between when the output steps up and the first-order response *begins* to rise. The same may be said for the multiple-order response, albeit with a slower rate of initial rise. The dead-time response, however, is actually delayed some time after the output makes its step-change. There is a period of time where the dead-time response does *absolutely nothing* following the output step-change.

Dead time is also referred to as *transport delay*, because the mechanism of dead time is often a time delay caused by the transportation of material at finite speed across some distance. The following cookie-baking process has dead time by virtue of the time delay inherent to the cookies' journey from the oven to the temperature sensor:

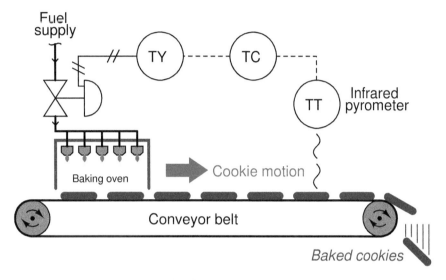

Dead time is a far worse problem for feedback control systems than lag time. The reason why is best understood from the perspective of phase shift: the delay (measured in degrees of angular displacement) between input and output for a system driven by a sinusoidal stimulus. Excessive phase shift in a feedback system makes possible self-sustaining oscillations, turning what is supposed to be negative feedback into positive feedback. Systems with lag produce phase shift that is frequency-dependent (the greater the frequency, the more the output "lags" behind the input), but this phase shift has a natural limit. For a first-order lag function, the phase shift has an absolute maximum value of -90°; second-order lag functions have a theoretical maximum phase shift of -180°; and so on. Dead time functions also produce phase shift that increases with frequency, but there is no ultimate limit to the amount of phase shift. This means a single dead-time element in a feedback control loop is capable of producing *any* amount of phase shift given the right frequency[20]. What is more, the gain of a dead time function usually does not diminish with frequency, unlike the gain of a lag function.

Recall that a feedback system will self-oscillate if two conditions are met: a total phase shift of 360° (or -360°: the same thing) and a total loop gain of at least one. Any feedback system meeting these criteria[21] will oscillate, be it an electronic amplifier circuit or a process control loop. In the interest of achieving robust process control, we need to prevent these conditions from ever occurring simultaneously.

---

[20]A sophisticated way of saying this is that a dead-time function has no *phase margin*, only *gain margin*. All that is needed in a feedback system with dead time is sufficient gain to make the system oscillate.

[21]Sometimes referred to as the *Barkhausen criterion*.

A visual comparison between the phase shifts and gains exhibited by lag versus dead time functions may be seen here, the respective functions modeled by the electrical entities of a simple RC network (lag time) and an LC "delay line" network (dead time):

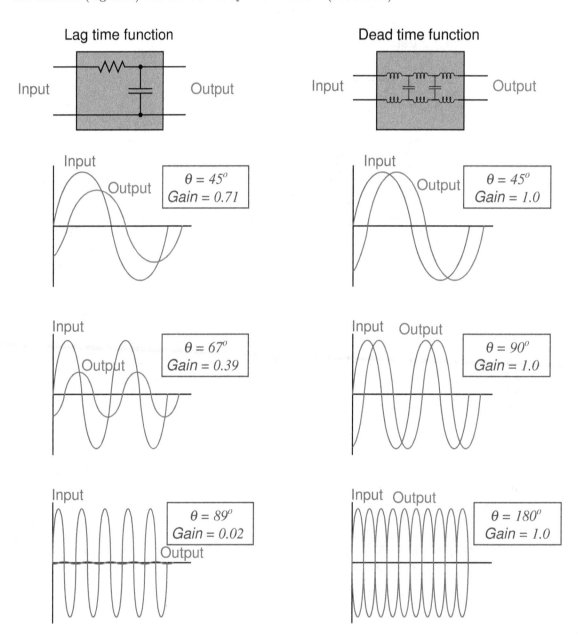

As frequency increases, the lag time function's phase shift asymptotically approaches -90° while its attenuation asymptotically approaches zero. Ultimately, when the phase shift reaches its maximum of -90°, the output signal amplitude is reduced to nothing. By contrast, the dead time function's phase shift grows linearly with frequency (to -180° and beyond!) while its attenuation remains unchanged. Clearly, dead time better fulfills the dual criteria of sufficient phase shift and sufficient loop gain needed for feedback oscillation than lag time, which is why dead time invites oscillation in a control loop more than lag time.

Pure dead-time processes are rare. Usually, an industrial process will exhibit at least some degree of lag time in addition to dead time. As strange as it may sound, this is a fortunate for the purpose of feedback control. The presence of lag(s) in a process guarantees a degradation of loop gain with frequency increase, which may help avoid oscillation. The greater the ratio between dead time and lag time in a loop, the more unstable it tends to be.

The appearance of dead time may be created in a process by the cascaded effect of multiple lags. As mentioned in an earlier subsection, multiple lags create a process response to step-changes that is "S"-shaped, responding gradually at first instead of immediately following the step-change. Given enough lags acting in series, the beginning of this "S" curve may be so flat that it appears "dead:"

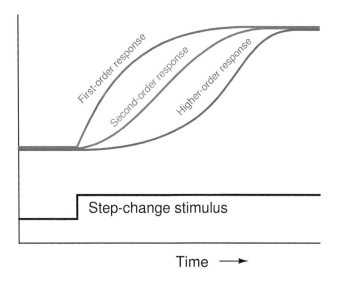

While dead time may be impossible to eliminate in some processes, it should be minimized wherever possible due to its detrimental impact on feedback control. Once an open-loop (manual-mode step-change) test on a process confirms the existence of dead time, the source of dead time should be identified and eliminated if at all possible.

One technique applied to the control of dead-time-dominant processes is a special variation of the PID algorithm called *sample-and-hold*. In this variation of PID, the controller effectively alternates between "automatic" and "manual" modes according to a pre-programmed cycle. For a short period of time, it switches to "automatic" mode in order to "sample" the error (PV − SP) and calculate a new output value, but then switches right back into "manual" mode ("hold") so as to give time for the effects of those corrections to propagate through the process dead time before taking another sample and calculating another output value. This sample-and-hold cycle of course slows the controller's response to changes such as setpoint adjustments and load variations, but it does allow for more aggressive PID tuning constants than would otherwise work in a continuously sampling controller because it effectively blinds[22] the controller from "seeing" the time delays inherent to the process.

All digital instruments exhibit dead time due to the nature of their operation: processing signals over discrete time periods. Usually, the amount of dead time seen in a modern digital instrument is too short to be of any consequence, but there are some special cases meriting attention. Perhaps the most serious case is the example of *wireless* transmitters, using radio waves to communicate process information back to a host system. In order to maximize battery life, a wireless transmitter must transmit its data sparingly. Update times (i.e. dead time) measured in minutes are not uncommon for battery-powered wireless process transmitters.

---

[22]An interesting analogy is that of a narcoleptic human operator manually controlling a process with a lot of dead time. If we imagine this person helplessly falling asleep at periodic intervals, then waking up to re-check the process variable and make another valve adjustment before falling asleep again, we see that the dead time of the process disappears from the perspective of the operator. The operator never realizes the process even has dead time, because they don't remain awake long enough to notice. So long as the poor operator's narcolepsy occurs at just the right intervals (i.e. not too short so as to notice dead time, and not too long so as to miss important changes in setpoint or load), good control of the process is possible.

30.1.8 Hysteresis

A detrimental effect to feedback control is a characteristic known as *hysteresis*: a lack of responsiveness to a change in direction. Although hysteresis typically resides with instruments rather than the physical process they connect to, it is most easily detected by a simple open-loop ("step-change") test with the controller in manual mode just like all the important process characteristics (self-regulating versus integrating, steady-state gain, lag time, dead time, etc.).

The most common source of hysteresis is found in pneumatically-actuated control valves possessing excess stem friction. The "jerky" motion of such a valve to smooth increases or decreases in signal is sometimes referred to as *stiction*. Similarly, a pneumatically-actuated control valve with excess friction will be unresponsive to small reversals in signal direction. To illustrate, this means the control valve's stem position will not be the same at a *rising* signal value of 50% (typically 12 mA, or 9 PSI) as it will be at a *falling* signal value of 50%.

Control valve stiction may be quite severe in valves with poor maintenance histories, and/or lacking positioners to correct for deviations between controller signal value and actual stem position. I have personally encountered control valves with hysteresis values in excess of 10%[23], and have heard of even more severe cases.

Detecting hysteresis in a control loop is as simple as performing "up-and-down" tests of the controller output signal in manual mode. The following trend shows how hysteresis might appear in a self-regulating process such as liquid flow control:

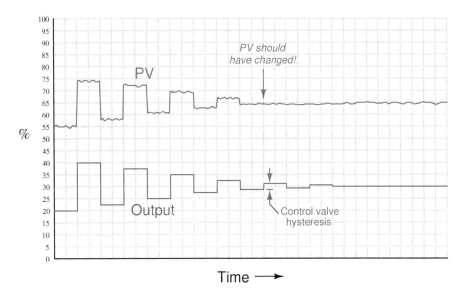

Note how the PV responds to large up-and-down output step-changes, but stops responding as soon as the magnitude of these open-loop step-changes falls below a threshold equal to the control valve's hysteresis.

---

[23]A 10% hysteresis value means that the signal must be changed by 10% following a reversal of direction before any movement is seen from the valve stem.

Applied to an integrating process such as liquid level control, the same type of test reveals the control valve's hysteresis by the largest step-change that does not alter the PV's slope:

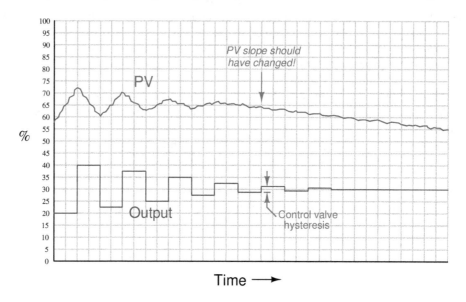

It is not as simple to perform this test on a process with slow lag or dead times, of course, or on a process possessing a "runaway" (rather than self-regulating or integrating) characteristic, in which case a better test for valve hysteresis would be to monitor valve stem position rather than the PV when executing the step-changes.

Hysteresis is a problem in feedback control because it essentially acts like a variable dead time. Recall that "dead time" was defined as a period of time during which a change in manipulated variable produces no effect in the process variable: the process appears "dead" for some amount of time before showing a response. If a change in controller output (manipulated variable) is insufficient to overcome the hysteresis inherent to a control valve or other component in a loop, the process variable will not respond to that output signal change at all. Only when the manipulated variable signal continues to change sufficiently to overcome hysteresis will there be a response from the process variable, and the time required for that to take place depends on how soon the controller's output happens to reach that critical value. If the controller's output moves quickly, the "dead time" caused by hysteresis will be short. If the controller's output changes slowly over time, the "dead time" caused by hysteresis will be longer.

Another problem caused by hysteresis in a feedback loop occurs in combination with integral action, whether it be programmed into the controller or is inherent to the process (i.e. an *integrating* process). It is highly unlikely that a "sticky" control valve will happen to "stick" at exactly the right stem position required to equalize PV and SP. Therefore, the probability at any time of an error developing between PV and SP, or of an offset developing between the valve position and the equilibrium position required by an integrating process, is very great. This leads to a condition of guaranteed instability. For a self-regulating process with integral action in the controller, the almost guaranteed existence of PV − SP error means the controller output will ceaselessly ramp up and down as the valve first slips and sticks to give a positive error, then slips and sticks to give a negative error. For an integrating process with proportional action in the controller, the process variable will ceaselessly ramp up and down as the valve first sticks too far open, then too far closed to equalize process in-flow and out-flow which is necessary to stabilize the process variable. In either case, this particular form of instability is called a *slip-stick cycle*.

The following process trend shows a slip-stick cycle in a self-regulating process, controlled by an integral-only controller:

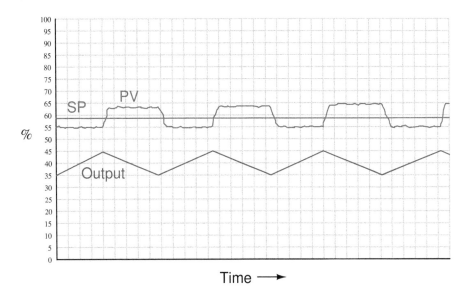

Note how the output ceaselessly ramps in a futile attempt to drive the process variable to setpoint. Once sufficient pressure change accumulates in the valve actuator to overcome static stem friction, the valve "slips to and sticks at" a new stem position where the PV is unequal to setpoint, and the controller's integral action begins to ramp the output in the other direction.

The next trend shows a slip-stick cycle in an integrating process, controlled by a proportional-only controller:

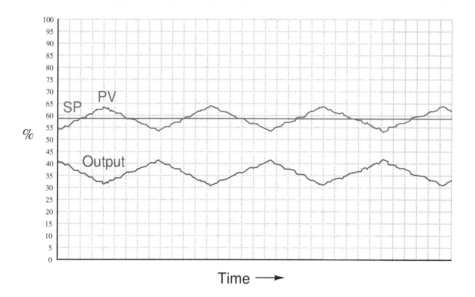

Time $\longrightarrow$

Note how the process variable's slope changes every time the valve "slips to and sticks at" a new stem position unequal to the balance point for this integrating process. The process's natural integrating action then causes the PV to ramp, causing the controller's proportional action to similarly ramp the output signal until the valve has enough accumulated force on its stem to jump to a new position.

It is very important to note that the problems created by a "sticky" control valve *cannot* be completely overcome by controller tuning[24]. For instance, if one were to de-tune the integral-only controller (i.e. longer time constant, or fewer repeats per minute) in the self-regulating process, it would *still* exhibit the same slip-stick behavior, only over a longer period (lower frequency) than before. If one were to de-tune the proportional-only controller (i.e. greater proportional band, or less gain) in the integrating process, the exact same thing would happen: a decrease in cycle frequency, but no elimination of the cycling. Furthermore, de-tuning the controller in either process would also result in less responsive (poorer) corrective action to setpoint and load changes. The only solution[25] to either one of these problems is to reduce or eliminate the friction inside the control valve.

---

[24]Some integral controllers are equipped with a useful feature called *integral deadband* or *reset deadband*. This is a special PID function inhibiting integration whenever the process variable comes close enough to setpoint, the "deadband" value specifying how close the PV must come to SP before integration stops. If this deadband value is set equal to or wider than the error caused by the valve's stiction, the controller will stop its integral-driven cycling. The trade-off, of course, is that the controller will no longer work to eliminate all error, but rather will be content with an error equal to or less than the specified deadband.

[25]An alternate solution is to install a positioner on the control valve, which acts as a secondary (cascaded) controller seeking to equalize stem position with the loop controller's output signal at all times. However, this just "shifts" the problem from one controller to another. I have seen examples of control valves with severe packing friction which will *self-oscillate* their own positioners (i.e. the positioner will "hunt" back and forth for the correct valve stem position given a constant signal from the loop controller)! If valve stem friction can be minimized, it should be minimized.

30.2 Before you tune . . .

Much has been written about the benefits of robust PID control. Increased productivity, decreased equipment strain, and increased process safety are some of the advantages touted of proper PID tuning. What is often overlooked, though, are the negative consequences of poor PID controller tuning. If robust PID control can increase productivity, then poor PID control can decrease productivity. If a well-tuned system helps equipment run longer and safer, then a poorly tuned system may increased failure frequency and safety incidents. The instrumentation professional should be mindful of this dichotomy when proceeding to tune a PID control system. One should never think there is "nothing to lose" by trying different PID settings. Tuning a PID controller is as serious a matter as reconfiguring any field instrument.

PID tuning parameters are easy to access, which makes them a tempting place to begin for technicians looking to improve the performance of a feedback loop. Another temptation driving technicians to focus on controller tuning as a first step is the prestige associated with being able to tame an unruly feedback loop with a few adjustments to the controller's PID tuning constants. For those who do not understand PID control (and this constitutes the vast majority of the human population, even in the industrial world), there is something "magic" about being able to achieve robust control behavior simply by making small adjustments to numbers in a computer (or to knobs in an analog controller). The reality, though, is that many poorly-behaving control systems are that way not due (at least purely) to a deficit of proper PID tuning values, but rather to problems external to the controller which no amount of "tuning" will solve. Adjusting PID tuning constants *as a first step* is almost always a bad idea.

This section aims to describe and explain some of the recommended considerations prior to making adjustments to the tuning of a loop controller. These considerations include:

- Identifying operational needs (i.e. "How do the operators want the system to respond?")

- Identifying process and system hazards before manipulating the loop

- Identifying whether it is a tuning problem, a field instrument problem, and/or a design problem

30.2.1   Identifying operational needs

As defined elsewhere in this book, "robust" control is a stability of the process variable despite changes in load, fast response to changes in setpoint, minimal oscillation following either type of change, and minimal offset (error between setpoint and process variable) over time. However, these criteria are not equally valued in all processes, and neither are they equally attainable with simple PID control in all processes. It may be critical, for example, in a boiler water level control process to have fast response to changes in load, but minimal offset over time is not as important. It may be completely permissible to have a persistent 5% error between PV and SP in such a system, so long as the water level does not deviate much over 20% for any length of time due to load changes. In another process, such as liquid level control inside one stage ("effect") of a multi-stage ("multi-effect") evaporator system, a priority may be placed upon relatively steady flow control through the valve rather than steady level in the process. A level controller tuned for aggressive response to setpoint changes will cause large fluctuations in liquid flow rate to all successive stages ("effects") of the evaporator process in the event of a sudden load or setpoint change, which would be more detrimental to product quality than some deviation from setpoint in that one effect.

Thus, we must first determine what the operational needs of a control system are before we aim to adjust the performance of that control system. The operations personnel (operators, unit managers, process engineers) are your best resources here. Ultimately, they are your "internal customers." Your task is to give the customers the system performance they need to do their jobs best.

Keep in mind the following process control objectives, knowing that it will likely be impossible to achieve *all* of them with any particular PID tuning. Try to rank the relative importance of these objectives, then concentrate on achieving those most important, at the expense of those least important:

- Minimum change in PV (dynamic stability) with changes in load

- Fast response to setpoint changes (minimum dynamic error)

- Minimum overshoot/undershoot/oscillation following sudden load or setpoint changes

- Minimum error (PV − SP) over time

- Minimum valve velocity (i.e. minimal effect to upstream or downstream processes)

The control actions best suited for rapid response to load and/or setpoint changes are proportional and derivative. Integral action takes effect only *after* error has had time to develop, and as such cannot act as immediately as either proportional or derivative.

If the priority is to minimize overshoot, undershoot, and/or oscillations, the controller's response will likely need to be more sluggish than is typical. New setpoint values will take longer to achieve, and load changes will not be responded to with quite the same vigor. Derivative action may be helpful in some applications to "tame" the oscillatory tendencies of proportional and integral actions.

Minimum error over time can really only be addressed by integral action. No other controller action pays specific "attention" to the magnitude and duration of error. This is not to say that the process will work well on integral-only control, but rather that integral action will be absolutely necessary (i.e. a P-only or PD controller will not suffice).

Minimum valve velocity is a priority in processes where the manipulated variable has an effect on some *other* process in the system. For example, liquid level control in a multi-stage (multi-"effect") evaporator system where the discharge flow from one evaporator becomes the incoming flow for another evaporator, is a system where sharp changes in the manipulated variable of one control loop can upset downstream processes:

A multi-effect evaporator system

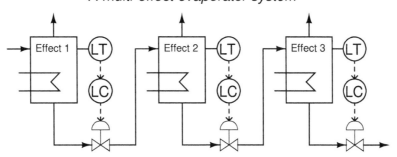

In other words, an aggressively-tuned level controller on an upstream evaporator (e.g. Effect 1) may achieve its goal of holding liquid level very steady in that evaporator by varying its out-going flow, but it will do so at the expense of causing level variations in all downstream evaporators. Cases such as this call for controller tuning (at least in the upstream effects) responding slowly to errors. Proportional action will very likely be limited to low gain values (high proportional band values), and derivative action (if any is used at all) should be set to respond only to the process variable, not to error (PV − SP). This leaves the main work of stabilizing the loop to integral action, even though we know that integral action tends to overshoot following setpoint changes in an integrating process such as liquid level control. Understand that tuning a PID loop with the goal of minimizing valve motion *will* result in longer deviations from setpoint than if the controller were tuned to respond faster to process or setpoint changes.

30.2.2   Identifying process and system hazards

When students practice PID control in an Instrumentation program, they usually do so using computer simulation software and/or "toy" processes constructed in a lab environment. A potential disadvantage to this learning environment is a failure to recognize real problems that may develop when tuning an actual production process. Rarely will you find a completely isolated feedback loop in industry: generally there are interactive effects between control loops in a process, which means one cannot proceed to tune a loop with impunity.

A very important question to ask the operations personnel before tuning a loop is, "How far and how fast am I allowed to let the process variable increase and decrease?" Processes and process equipment may become dangerously unstable, for example, if certain temperatures become to high (or too low, as is the case in process liquids that solidify when cold). It is not uncommon for certain control loops in a process to be equipped with alarms, either hard or soft, that automatically *shut down* equipment if exceeded. Clearly, these "shutdown" limits must be avoided during the tuning of the process loop.

One should also examine the control strategy before proceeding to tune. Is this a cascaded loop? If so, the slave controller needs to be tuned before the master. Does this loop incorporate feedforward action to act on load changes? If so, the effectiveness of that feedforward loop (gain, dynamic compensation) should be checked and adjusted before the feedback loop is tuned. Are there limits in this loop? Is this a selector or override control strategy? If so, you need to be able to clearly tell which loop components are selected, and which signals are being limited, at any given time.

Another consideration is whether or not the process is in a "normal" condition before you attempt to improve its performance. Ask the operations personnel if this is a typical day, or if there is some abnormal condition in effect (equipment shutdown, re-routing of flows, significantly different production rates, etc.) that might skew the response of the process loop to be tuned. Once again we see a need for input from the operations personnel, because they know the day-to-day behavior of the system better than anyone else.

30.2.3 Identifying the problem(s)

One of the questions I advise instrument technicians to ask of operators when diagnosing any process problem is simply, "How long has this problem existed?" The age of a problem can be a very important indicator of possible causes. If you were told that a problem suddenly developed after the last night shift, you would be inclined to suspect an equipment failure, or something else that could happen *suddenly* (e.g. a hand valve someone opened or shut when they shouldn't have). Alternatively, if you were told a problem has been in existence since the day the process was constructed, you would be more inclined to suspect an issue with system design or improper installation. This same diagnostic technique – obtaining a "history" of the "patient" – applies to loop tuning as well. A control loop that suddenly stopped working as it should might be suffering from an instrument failure (or an unauthorized change of controller parameters), whereas a chronically misbehaving loop would more likely be suffering from poor design, bad instrument installation, or a controller that was never tuned properly.

In either case, poor control is just as likely to be caused by field instrument problems as it is by incorrect PID tuning parameters. No PID settings can fully compensate for faulty field instrumentation, but it is possible for some instrument problems to be "masked" by controller tuning. Your first step in actually manipulating the control loop should be a check of instrument health. Thankfully, this is relatively easy to do by performing a series of "step-change" tests with the controller in manual mode. By placing the controller in manual and making small changes in output signal (remember to check with operations to see how far you are allowed to move the output, and how far you can let the PV drift!), you can determine much about the process and the loop instrumentation, including:

- Whether the PV signal is "noisy" (first turn off all damping in the controller and transmitter)

- How much "stiction" is in the control valve

- Whether the process is integrating, runaway, or self-regulating

- Process gain (and whether this gain is stable or if it changes as PV changes)

- Process lag time and lag degree (first-order versus multiple-order)

- Process dead time

Such an open-loop test might reveal potential problems without pinpointing the exact nature or location of those problems. For example, a large lag time may be intrinsic to the process, or it may be the result of a poorly-installed sensor (e.g. a thermocouple not pushed to touch the bottom of its thermowell) or even a control valve in need of a volume booster or positioner. Dead time measured in an open-loop test may also be intrinsic to the process (transport delay), intrinsic to the sensor (e.g. a gas chromatograph where each analysis cycle takes several *minutes* of time), or it could be the result of stiction in the valve. The only way to definitively identify the problem is to test the instruments themselves, ideally in the field location.

An indispensable tool for identifying loop problems is a *trend recorder*, showing all the relevant variables in a control loop graphed over time. In order to obtain the best "view" of the process, you need to make sure the graphing trend display has sufficient resolution and responsiveness. If the trend fails to show fine details such as noise in the process, it is possible that the graphing device will be insufficient for your needs.

If this is the case, you may still perform response tests of the loop, but you will have to use some other instrument(s) to graph the controller and process actions. A modern tool useful for this purpose is a portable computer with a data acquisition device connected, giving the computer the ability to read instrument signal voltages. Many data graphing programs exist for taking acquired data and plotting it over the time domain. Data acquisition modules with sample rates in the thousands of samples per second are available for very modest prices.

30.2.4   Final precautions

Be prepared to document your work! This means capturing and recording "screen shot" images of process trend graphs, both for the initial open-loop tests and the closed-loop PID trials. It also means documenting the original PID settings of the controller, and all PID setting values attempted during the tuning process (linked to their respective trend graphs, so it will be easy to tell which sets of PID constants produced which process responses). If there are any instrument configuration settings (e.g. damping time values in process transmitters) changed during the tuning exercise, both the original values and all your changes need to be documented as well.

As a final word, I would like to cast a critical vote against auto-tuning controllers. With all due respect for the engineers who work hard to make controllers "intelligent" enough to adjust their own PID settings, there is no controller in the world able to account for all the factors described in this "Before you tune . . ." section. Feel free to use the automatic tuning feature of a controller, but only *after* you have ensured all instrument and process problems are corrected, and *after* you have confirmed the tuning goal of the controller matches the behavioral goal of the control loop as defined by the operators (e.g. fast response versus minimum overshoot, etc.). Some people in the automation business are over-confident with regard to the capabilities of auto-tuning controllers. We would all do well to recognize this feature as a *tool*, and just like any other tool it is only as useful as the person handling it is knowledgeable regarding how and why it works. Wielding any tool in ignorance is a recipe for disaster.

30.3 Quantitative PID tuning procedures

A *quantitative* PID tuning procedure is a step-by-step approach leading directly to a set of numerical values to be used in a PID controller. These procedures may be split into two categories: *open loop* and *closed loop*. An "open loop" tuning procedure is implemented with the controller in manual mode: introducing a step-change to the controller output and then mathematically analyzing the results of the process variable response to calculate appropriate PID settings for the controller to use when placed into automatic mode. A "closed loop" tuning procedure is implemented with the controller in automatic mode: adjusting tuning parameters to achieve an easily-defined result, then using those PID parameter values and information from a graph of the process variable over time to calculate new PID parameters.

Quantitative PID tuning got its start with a paper published in the November 1942 *Transactions of the American Society of Mechanical Engineers* written by two engineers named Ziegler and Nichols. "Optimum Settings For Automatic Controllers" is a seminal paper, and deserves to be read by every serious student of process control theory. That Ziegler's and Nichols' recommendations for PID controller settings may still be found in modern references more than 60 years after publication is a testament to its impact in the field of industrial control. Although dated in its terminology and references to pneumatic controller technology (some controllers mentioned as not even having adjustable proportional response, and others as having only discrete degrees of reset adjustment rather than continuously variable settings!), the PID algorithm described by its authors and the effects of P, I, and D adjustments on process control behavior are as valid today as they were then.

This section is devoted to a discussion of quantitative PID tuning procedures in general, and the "Ziegler-Nichols" methods in specific. It is the opinion of this author that the Ziegler-Nichols tuning methods are useful primarily as historical references, and indeed suffer from serious practical impediments. The most serious reservation I have with the Ziegler-Nichols methods (and in fact any algorithmic procedure for PID tuning) is that these methods tend to absolve the practitioner of responsibility for understanding the process they intend to tune. Any time you provide people with step-by-step instructions to perform complex tasks, there will be a great many readers of those instructions tempted to mindlessly follow the instructions, even to their doom. PID tuning is one of these "complex tasks," and there is significant likelihood for a person to do more harm than good if all they do is implement a step-by-step approach rather than understand what they are doing, why they are doing it, and what it means if the results do not meet with satisfaction. Please bear this in mind as you study any PID tuning procedure, Ziegler-Nichols or otherwise.

30.3.1   Ziegler-Nichols closed-loop ("Ultimate Gain")

Closed-loop refers to the operation of a control system with the controlling device in "automatic" mode, where the flow of the information from sensing element to transmitter to controller to control element to process and back to sensor represents a continuous ("closed") feedback loop. If the total amount of signal amplification provided by the instruments is too much, the feedback loop will self-oscillate at the system's natural (resonant) frequency. While oscillation is almost always considered undesirable in a control system, it may be used as an exploratory test of process dynamics if the controller acts purely on proportional action (no integral or derivative action): providing data useful for calculating effective PID controller settings. Thus, a "closed-loop" PID tuning procedure entails disabling any integral or derivative actions in the controller, then raising the gain value of the controller just far enough that self-sustaining oscillations ensue. The minimum amount of controller gain necessary to sustain sinusoidal oscillations is called the *ultimate sensitivity* (S_u) or *ultimate gain* (K_u) of the process, while the time (period) between successive oscillation peaks is called the *ultimate period* (P_u) of the process. We may then use the measured values of K_u and P_u to calculate reasonable controller tuning parameter values (K_p, τ_i, and/or τ_d).

When performing such a test on a process loop, it is important to ensure the oscillation peaks do not reach the limits of the instrumentation, either measurement or final control element. In other words, in order for the oscillation to accurately reveal the process characteristics of ultimate sensitivity and ultimate period, the oscillations must be naturally limited and not artificially limited by either the transmitter or the control valve saturating. Oscillations characterized by either the transmitter or the final control element reaching their range limits should be avoided in order to obtain the best closed-loop oscillatory test results. An illustration is shown here as a model of what to avoid:

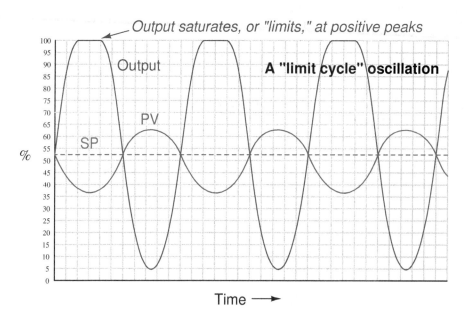

Here the controller gain is set too high, the result being saturation at the positive peaks of the output waveform. The controller gain should be decreased until symmetrical, sinusoidal waves result.

If the controller in question is proportional-only (i.e. capable of providing no integral or derivative control actions), Ziegler and Nichols' recommendation is to set the controller gain[26] to one-half the value of the ultimate sensitivity determined in the closed-loop test, which I will call ultimate *gain* (K_u) from now on:

$$K_p = 0.5K_u$$

Where,

K_p = Controller gain value that you should enter into the controller for good performance

K_u = "Ultimate" gain determined by increasing controller gain until self-sustaining oscillations are achieved

Generally, a controller gain of one-half the experimentally determined "ultimate" gain results in reasonably quick response to setpoint and process load changes. Oscillations of the process variable following such setpoint and load changes typically damp with each successive wave peak being approximately one-quarter the amplitude of the one preceding. This is known as *quarter-wave damping*. While certainly not ideal, it is a compromise between fast response and stability.

The following process trend shows what "quarter-wave damping" looks like with the controller in automatic mode, with the process variable (PV) exhibiting decaying oscillations following a step-change in setpoint (SP):

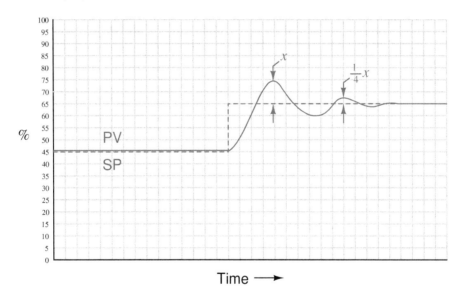

Ziegler and Nichols were careful to qualify quarter-wave damping as less than optimal for some applications. In their own words (page 761):

> "The statement that a sensitivity setting of one half the ultimate with attendant 25 per cent amplitude ratio gives optimum control must be modified in some cases. For example, the actual level maintained by a liquid-level controller might not be nearly as important as the effect of sudden valve movements on further portions of the process. In this case the sensitivity should be lowered to reduce the amplitude ratio even though the offset is increased by so doing. On the other hand, a pressure-control application giving oscillations with very short period could be set to give an 80 or 90 per cent amplitude ratio. Due to the short period, a disturbance would die out in reasonable time, even though there were quite a few oscillations. The offset would be reduced somewhat though it should be kept in mind that it can never be reduced to less than one half of the amount given at our previously defined optimum sensitivity of one half the ultimate."

Some would argue (myself included) that quarter-wave damping exercises the control valve needlessly, causing undue stem packing wear and consuming large quantities of compressed air over time. Given the fact that all modern process controllers have integral (reset) capability, unlike the simple pneumatic controllers of Ziegler and Nichols' day, there is really no need to tolerate prolonged offset (failure of the process variable to exactly equalize with setpoint over time) as a necessary cost of avoiding valve oscillation.

If the controller in question has integral (reset) action in addition to proportional, Ziegler and Nichols' recommendation is to set the controller gain to slightly less than one-half the value of the ultimate sensitivity, and to set the integral time constant[27] to a value slightly less than the ultimate period:

$$K_p = 0.45 K_u$$

$$\tau_i = \frac{P_u}{1.2}$$

Where,

K_p = Controller gain value that you should enter into the controller for good performance

K_u = "Ultimate" gain determined by increasing controller gain until self-sustaining oscillations are achieved

τ_i = Controller integral setting that you should enter into the controller for good performance (minutes per repeat)

P_u = "Ultimate" period of self-sustaining oscillations determined when the controller gain was set to K_u (minutes)

---

[27]Either minutes per repeat or seconds per repeat. If the controller's integral rate is expressed in units of repeats per minute (or second), the formula would be $K_i = \frac{1.2}{P_u}$.

If the controller in question has all three control actions present (full PID), Ziegler and Nichols' recommendation is to set the controller tuning constants as follows:

$$K_p = 0.6K_u$$

$$\tau_i = \frac{P_u}{2}$$

$$\tau_d = \frac{P_u}{8}$$

Where,

K_p = Controller gain value that you should enter into the controller for good performance

K_u = "Ultimate" gain determined by increasing controller gain until self-sustaining oscillations are achieved

τ_i = Controller integral setting that you should enter into the controller for good performance (minutes per repeat)

τ_d = Controller derivative setting that you should enter into the controller for good performance (minutes)

P_u = "Ultimate" period of self-sustaining oscillations determined when the controller gain was set to K_u (minutes)

An important caveat with any tuning procedure based on ultimate gain is the potential to cause trouble in a process while experimentally determining the ultimate gain. Recall that "ultimate" gain is the amount of controller gain (proportional action) resulting in self-sustaining oscillations of constant amplitude. In order to precisely determine this gain setting, one must spend some time provoking the process with sudden setpoint changes (to induce oscillation) and experimenting with greater and greater gain settings until constant oscillation amplitude is achieved. Any more gain than the "ultimate" value, of course, leads to ever-*growing* oscillations which may be brought under control only by decreasing controller gain or switching to manual mode (thereby stopping all feedback in the system). The problem with this is, one never knows for certain when ultimate gain is achieved until this critical value has been exceeded, as evidenced by ever-growing oscillations. In other words, *the system must be brought to the brink of total instability in order to determine its ultimate gain value*. Not only is this time-consuming to achieve – especially in systems where the natural period of oscillation is long, as is the case with many temperature and composition control applications – but potentially hazardous to equipment and certainly detrimental to process quality[28]. In fact, one might argue that any process tolerant of such abuse probably doesn't need to be well-tuned at all!

Despite its practical limitations, the rules given by Ziegler and Nichols do shed light on the relationship between realistic P, I, and D tuning parameters and the operational characteristics of the process. Controller gain should be some fraction of the gain necessary for the process to self-oscillate. Integral time constant should be proportional to the process time constant; i.e. the

---

[28]Imagine informing the lead operations manager or a unit supervisor in a chemical processing facility you wish to over-tune the temperature controller in the main reaction furnace or the pressure controller in one of the larger distillation columns until it nearly oscillates out of control, and that doing so may necessitate hours of unstable operation before you find the perfect gain setting. Consider yourself fortunate if your declaration of intent does not result in security personnel escorting you out of the control room.

"slower" the process is to respond, the "slower" (less aggressive) the controller's integral response should be. Derivative time constant should likewise be proportional to the process time constant, although this has the opposite meaning from the perspective of aggressiveness: a "slow" process deserves a long derivative time constant; i.e. *more aggressive* derivative action.

30.3.2 Ziegler-Nichols open-loop

In contrast to the first tuning technique presented by Ziegler and Nichols in their landmark 1942 paper where the process was made to oscillate using proportional-only automatic control and the parameters of that oscillation served to define PID tuning parameters, their second tuning technique did not even rely on the presence of a controller. Instead, this second technique consisted of making a manual "step-change" of the control element (valve) and analyzing the resulting effect on the process variable, much the same way as described in the Process Characterization section of this chapter (section 30.1 beginning on page 2396).

After making the step-change in output signal with the controller in manual mode, the process variable trend is closely analyzed for two salient features: the *dead time* and the *reaction rate*. Dead time (L)[29] is the amount of time delay between the output step-change and the first indication of process variable change. Reaction rate is the maximum rate at which the process variable changes following the output step-change (the maximum time-derivative of the process variable):

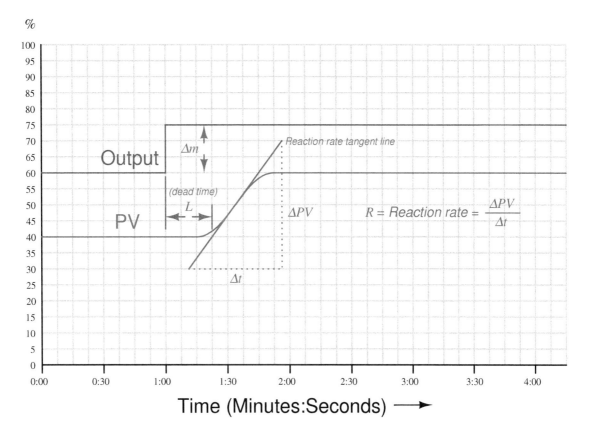

Dead time and reaction rate are responses common to self-regulating and integrating processes alike. Whether or not the process variable ends up stabilizing at some new value, its rate of rise will

---

[29]Unfortunately, Ziegler and Nichols chose to refer to dead time by the word *lag* in their paper. In modern technical parlance, "lag" refers to a first-order inverse-exponential function, which is fundamentally different from dead time.

reach some maximum value following the output step-change, and this will be the reaction rate of the process[30]. The unit of measurement for reaction rate is *percent per minute*:

$$R = \frac{\Delta \text{PV}}{\Delta t} = \frac{[\text{Percent rise}]}{[\text{Minutes run}]}$$

While dead time in a process tends to be constant regardless of the output step-change magnitude, reaction rate tends to vary directly with the magnitude of the output step-change. For example, an output step-change of 10% will generally cause the PV to rise at a rate twice as steep compared to the effects of a 5% output step-change. In order to ensure our predictive calculations capture only what is inherent to the process and not our own arbitrary open-loop tuning actions, we must include the output step-change magnitude (Δm) in those calculations as well[31].

If the controller in question is proportional-only (i.e. capable of providing no integral or derivative control actions), Ziegler and Nichols' recommendation is to set the controller gain as follows:

$$K_p = \frac{\Delta m}{RL}$$

Where,
  K_p = Controller gain value that you should enter into the controller for good performance
  Δm = Output step-change magnitude made while testing in open-loop (manual) mode (percent)
  R = Process reaction rate = $\frac{\Delta \text{PV}}{\Delta t}$ (percent per minute)
  L = Process dead time (minutes)

If the controller in question has integral (reset) action in addition to proportional, Ziegler and Nichols' recommendation is to set the controller gain to 90% of the proportional-only value, and to set the integral time constant to a value just over three times the measured dead time value:

$$K_p = 0.9 \frac{\Delta m}{RL}$$

$$\tau_i = 3.33L$$

Where,
  K_p = Controller gain value that you should enter into the controller for good performance
  Δm = Output step-change magnitude made while testing in open-loop (manual) mode (percent)
  R = Process reaction rate = $\frac{\Delta \text{PV}}{\Delta t}$ (percent per minute)
  L = Process dead time (minutes)
  τ_i = Controller integral setting that you should enter into the controller for good performance (minutes per repeat)

---

[30]Right away, we see a weakness in the Ziegler-Nichols open-loop method: it makes absolutely no distinction between self-regulating and integrating process types. We know this is problematic from the analysis of each process type in sections 30.1.1 and 30.1.2.

[31]Ziegler and Nichols' approach was to define a normalized reaction rate called the *unit reaction rate*, equal in value to $\frac{R}{\Delta m}$. I opt to explicitly include Δm in all the tuning parameter equations in order to avoid the possibility of confusing reaction rate with unit reaction rate.

If the controller has full PID capability, Ziegler and Nichols' recommendation is to set the controller gain to 120% of the proportional-only value, to set the integral time constant to twice the measured dead time value, and to set the derivative time constant to one-half the measured dead time value:

$$K_p = 1.2 \frac{\Delta m}{RL}$$

$$\tau_i = 2L$$

$$\tau_d = 0.5L$$

Where,

K_p = Controller gain value that you should enter into the controller for good performance

Δm = Output step-change magnitude made while testing in open-loop (manual) mode (percent)

R = Process reaction rate = $\frac{\Delta \text{PV}}{\Delta t}$ (percent per minute)

L = Process dead time (minutes)

τ_i = Controller integral setting that you should enter into the controller for good performance (minutes per repeat)

τ_d = Controller derivative setting that you should enter into the controller for good performance (minutes)

As you can see, the Ziegler-Nichols open-loop tuning method relies heavily on dead time (L) as a descriptive parameter for the process. This may be problematic in processes having insubstantial dead time, as the small L values obtained during the open-loop test will predict large controller gain (K_p) and aggressive integral (τ_i) time constant values, often too large to be practical. The open-loop method, however, is less disruptive to an operating process than the closed-loop method (which necessitated over-tuning the controller to the brink of total instability).

Another limitation, common to both the closed-loop and open-loop tuning methods, is that other factors in the process such as noise and hysteresis are completely overlooked. Noise is troublesome for large controller gain values (because the controller's proportional action reproduces that noise on the output) and is especially troublesome for derivative action which amplifies any noise it sees. Hysteresis causes integral action to continually "hunt" up and down, leading to cycling of the process variable. The lesson here is that no algorithmic PID tuning method can replace informed judgment on the part of the person tuning the loop. The methods proposed by Ziegler and Nichols (and others!) are merely starting points, and should never be taken as a definitive answer for controller tuning.

30.4 Heuristic PID tuning procedures

In contrast to quantitative tuning procedures where definite numerical values for P, I, and D controller settings are obtained through data collection and analysis, a *heuristic* tuning procedure is one where general rules are followed to obtain approximate or qualitative results. The majority of PID loops in the world have been tuned with such methods, for better or for worse. My goal in this section is to optimize the effectiveness of such tuning methods.

When I was first educated on the subject of PID tuning, I learned this rather questionable tuning procedure:

1. Configure the controller for proportional action only (integral and derivative control actions set to minimum effect), setting the gain near or at 1.

2. Increase controller gain until self-sustaining oscillations are achieved, "bumping" the setpoint value up or down as necessary to provoke oscillations.

3. When the ultimate gain is determined, reduce the aggressiveness of proportional action by a factor of two.

4. Repeat steps 2 and 3, this time adjusting integral action instead of proportional.

5. Repeat steps 2 and 3, this time adjusting derivative action instead of proportional.

The first three steps of this procedure are identical to the steps recommended by Ziegler and Nichols for closed-loop tuning. The last two steps are someone else's contribution. The results of this method are generally poor, *and I strongly recommend against using it!*

While this particular procedure is crude and ineffective, it does illustrate a useful principle in trial-and-error PID tuning: we can tune a PID-controlled process by incrementally adjusting the aggressiveness of a controller's P, I, and/or D actions until we see oscillations (suggesting the action has become too aggressive), then reducing the aggressiveness of the action until stable control is achieved. The "trick" to doing this effectively and efficiently is knowing which action(s) to focus on, which action(s) to avoid, and how to tell which of the actions is too aggressive when things do begin to oscillate. The following portions of this subsection describe the utility of each control action, the limitations of each, and how to recognize an overly-aggressive condition.

Much improvement may be made to any "trial-and-terror" PID tuning procedure if one is aware of the process characteristics and recognizes the applicability of P, I, and D actions to those process characteristics. Random experimentation with P, I, and D parameter values is tedious at best and dangerous at worst! As always, the key is to *understand* the role of each action, their applicability to different process types, and their limitations. The competent loop-tuner should be able to visually analyze trends of PV, SP, and Output, and be able to discern the degrees of P, I, and D action in effect at any given time in that trend.

Here is an improved PID tuning technique employing heuristics (general rules) regarding P, I, and D actions. It is assumed that you have taken all necessary safety precautions (e.g. you know the hazards of the process and the limits you are allowed to change it) and other steps recommended in the "Before You Tune . . ." section (30.2) beginning on page 2439:

1. Perform open-loop (manual-mode) tests of the process to determine its natural characteristics (e.g. self-regulating versus integrating versus runaway, steady-state gain, noisy versus calm, dead time, time constant, lag order) and to ensure no field instrument or process problems exist (e.g. control valve with excessive friction, inconsistent process gain, large dead time). *Correct all problems before proceeding*[32].

2. Identify any controller actions that may be problematic (e.g. derivative action on a noisy process), noting to use them sparingly or not at all.

3. Identify whether the process will depend mostly on proportional action or integral action for stability. This will be the controller's "dominant" action when tuned. You may find the chart on page 2457 useful for this.

4. Start with all terms of the controller set for minimal response (minimal P, minimal I, no D).

5. Set the dominant action to some safe value[33] (e.g. gain less than 1, integral time much longer than time constant of process) and check the loop's response to setpoint and/or load changes in automatic mode.

6. Increase aggressiveness of this action until a point is reached where any more causes excessive overshoot or oscillation.

7. Increase aggressiveness of the other action(s) as needed to achieve the best compromise between stability and quick response.

8. If the loop ever shows signs of being too aggressive (e.g. oscillations), use the technique of phase-shift comparison between PV and Output trends (see page 2458) to identify which controller action to attenuate.

9. Repeat the last three steps as often as needed.

Note: the loop should *never* "porpoise" (see page 2462) when responding to a setpoint or load change! Some oscillation around setpoint may be tolerable (especially when optimizing the tuning for fast response to setpoint and load changes), but "porpoising" is always something to avoid.

---

[32]This is very important: no degree of controller "tuning" will fix a poor control valve, noisy transmitter, or ill-designed process. If your open-loop tests reveal significant process problems, you must remedy them before attempting to tune the controller.

[33]It is important to know which PID equation your controller implements in order to adjust just one action (P, I, or D) of the controller without affecting the others. Most PID controllers, for example, implement either the "Ideal" or "Series" equations, where the gain value (K_p) multiplies every action in the controller including integral and derivative. If you happen to be tuning such a controller for integral-dominant control, you cannot set the gain to zero (in order to minimize proportional action) because this will nullify integral action too! Instead, you must set K_p to some value small enough that the proportional action is minimal while allowing integral action to function.

30.4.1   Features of P, I, and D actions

Purpose of each action

- **Proportional action** is the "universal" control action, capable of providing at least marginal control quality for any process.

- **Integral action** is useful for eliminating offset caused by load variations and process self-regulation.

- **Derivative action** is useful for canceling lags, but useless by itself.

Limitations of each action

- **Proportional action** will cause oscillations if sufficiently aggressive, in the presence of lags and/or dead time. The more lags (higher-order), the worse the problem. It also directly reproduces process noise onto the output signal.

- **Integral action** will cause oscillation if sufficiently aggressive, in the presence of lags and/or dead time. Any amount of integral action will guarantee overshoot following setpoint changes in purely integrating processes.

- **Derivative action** dramatically amplifies process noise, and will cause oscillations in fast-acting processes.

Special applicability of each action

- **Proportional action** works exceptionally well when aggressively applied to processes lacking the phase shift necessary to oscillate: self-regulating processes dominated by first-order lag, and purely integrating processes.

- **Integral action** works exceptionally well when aggressively applied to fast-acting, self-regulating processes. Has the unique ability to ignore process noise.

- **Derivative action** works exceptionally well to speed up the response of processes dominated by large lag times, and to help stabilize runaway processes. Small amounts of derivative action will sometimes allow more aggressive P and/or I actions to be used than otherwise would be possible without unacceptable overshoot.

Gain and phase shift of each action

- **Proportional action** acts on the *present*, adding no phase shift to a sinusoidal signal. Its gain is constant for any signal frequency.

- **Integral action** acts on the *past*, adding a -90° phase shift to a sinusoidal signal. Its gain decreases with increasing frequency.

- **Derivative action** acts on the *future*, adding a $+90^o$ phase shift to a sinusoidal signal. Its gain increases with increasing frequency.

30.4.2 Tuning recommendations based on process dynamics

Knowing which control actions to focus on first is a matter of characterizing the process (identifying whether it is self-regulating, integrating, runaway, noisy, has lag or dead time, or any combination of these traits based on an open-loop response test[34]) and then selecting the best actions to fit those characteristics. The following table shows some general recommendations for fitting PID tuning to different process characteristics

	Pure self-regulating	*May be controlled with aggressive integral action, and perhaps with a bit of proportional action. Use absolutely no derivative action!*
	Self-reg w/ pure 1st order lag	*Responds well to aggressive proportional action, with integral action needed only for recovery from load changes.*
	Self-reg w/ multiple lags	*Proportional action needed for quick response to setpoint changes, integral action needed for recovery from load changes, and derivative needed to prevent overshoot. Proportional and integral actions are limited by tendency to oscillate.*
	Integrating w/ lag(s)	*Proportional action should be aggressive as possible without generating oscillations. Integral action needed only for recovery from load changes.*
	Pure integrating	*Responds well to aggressive proportional action, with integral action needed only for recovery from load changes.*

General rules:

- Use no derivative action if the process signal is "noisy"

- Use proportional action sparingly if the process signal is "noisy"

- The slower the time lag(s), the less integral action to use (a good approximation is to set the integration time τ_i equal to the measured lag time of the process)

- The higher-order the time lag(s), the less proportional action (gain) to use

- Self-regulating processes *need* integral action

- Integrating processes *need* proportional action

---

[34]Recall that an open-loop response test consists of placing the loop controller in manual mode, introducing a step-change to the controller output (manipulated variable), and analyzing the time-domain response of the process variable as it reacts to that perturbation.

• Dead time requires a reduction of all PID constants below what would normally work

Once you have determined the basic character of the process, and understand from that characterization what the needs of the process will be regarding P, I, and/or D control actions, you may "experiment" with different tuning values of P, I, and D until you find a combination yielding robust control.

30.4.3 Recognizing an over-tuned controller by phase shift

When performing heuristic tuning of a PID controller, it is important to be able to identify a condition where one or more of the "actions" (P, I, or D) is configured too aggressively for the process. The characteristic indication of over-tuning is the presence of sinusoidal oscillations. At best, this means damped oscillations following a sudden setpoint or load change. At worst, this means oscillations that *never* decay:

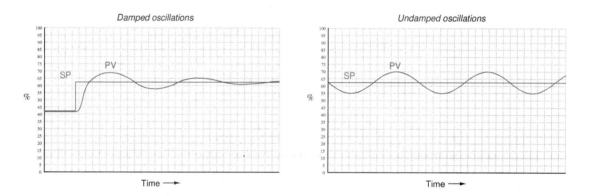

At this point, the question is: *which action of the controller is causing this oscillation?* We know all three actions (P, I, and D) are fully capable of causing process oscillation if set too aggressive, so in the lack of clarifying information it could be any of them (or some combination!).

One clue is the trend of the controller's *output* compared with the trend of the process variable (PV). Modern digital control systems all have the ability to trend manipulated variables as well as process variables, allowing personnel to monitor exactly how a loop controller is managing a process. If we compare the two trends, we may examine the *phase shift* between PV and output of a loop controller to discern what it's dominant action is.

For example, the following trend graphs show the PV and output signals for a loop controller with proportional-dominant response. Both direct- and reverse-acting versions are shown:

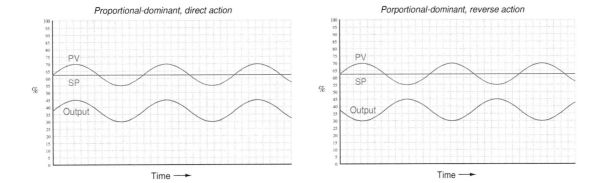

Since we know proportional action is *immediate*, there should be no phase shift[35] between the PV and output waveforms. This makes sense from a mathematical perspective: if we substitute a sine function for the error variable in a proportional-only controller equation, we see that different gain values (K_p) simply result in an output signal of the same phase, just amplified or attenuated:

$$m = K_p e + b$$

$$m = K_p(\sin t) + b$$

If ever you see a process oscillating like this (sinusoidal waveforms, with the PV and output signals in-phase), you know that the controller's response to the process is dominated by proportional action, and that the gain needs to be reduced (i.e. increase proportional band) to achieve stability.

---

[35]For reverse-acting controllers, I am ignoring the obvious 180° phase shift necessary for negative feedback control when I say "no phase shift" between PV and output waveforms. I am also ignoring dead time resulting from the scanning of the PID algorithm in the digital controller. For some controllers, this scan time may be significant enough to see on a trend!

Integral and derivative actions, however, introduce phase shift between the PV waveform and the output waveform. The direction of phase shift will reveal which time-based action (either I or D) dominates the controller's response and is therefore most likely the cause of oscillation.

For example, the following trend graph shows the PV and output signals for a loop controller with integral-dominant response. Both direct- and reverse-acting versions are shown:

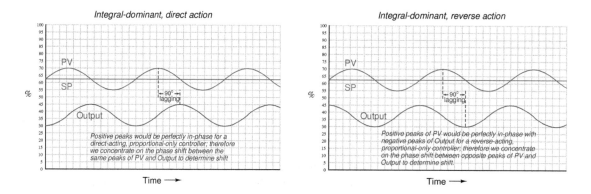

Here, the output waveform is phase-shifted 90° behind (lagging) the PV waveform compared to what it would be by proportional action alone. This -90° phase shift is most difficult to see for reverse-acting controllers, since the natural 180° phase shift caused by reverse action makes the additional -90° shift look like a $+90^o$ shift (i.e. a -270° shift). One method I use for discerning the direction of phase shift for reverse acting controllers is to imagine what the output waveform would look like for a proportional-only controller, and compare to that. Another method is to perform a "thought experiment" looking at the PV waveform, noting the velocity of the controller's output at points of maximum error (when integral action would be expected to move the output signal fastest). In the previous example of an integral-dominant reverse-acting controller, we see that the output signal is indeed descending at its most rapid pace when the PV is highest above setpoint (greatest positive error), which is exactly what we would expect reverse-acting integration to do.

Mathematically, this makes sense as well. Integration of a sine function results in a negative cosine waveform:

$$-\cos t = \int \sin t \, dt$$

Another way of stating this is to say integration always adds a -90° phase shift:

$$\sin(t - 90^o) = \int \sin t \, dt$$

If derivative action is set too aggressive for the needs of the process, the resulting output waveform will be phase-shifted $+90^o$ from that of a purely proportional response:

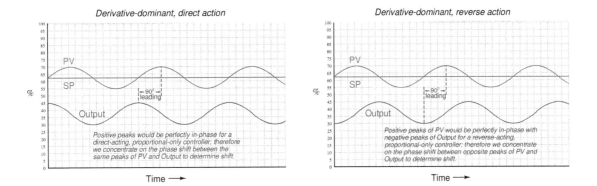

Once again, the direction of phase shift is easiest to discern in the direct-acting case, since we expect the PV and output sine waves to be perfectly in-phase for proportional-dominant response. A $+90^o$ phase shift is very clear to see in the direct-acting example because the peaks of the output waveform clearly precede the corresponding peaks of the PV waveform. The reverse-acting control example is more difficult due to the added 180° of phase shift intrinsic to reverse action.

The derivative of a sine function is always a cosine function (or alternatively stated, a sine function with a $+90^o$ phase shift):

$$\cos t = \frac{d}{dt} \sin t$$

$$\sin(t + 90^o) = \frac{d}{dt} \sin t$$

You may encounter cases where *multiple* PID terms are set too aggressively for the needs of the process, in which case the phase shift will be split somewhere between 0° and $\pm\ 90^o$ owing to the combined actions. In such cases one must "round up" or "round down" the phase shift to the nearest value of -90^0, 0°, or $+90^o$ in order to determine which of the three actions (I, P, or D, respectively) is most responsible for the oscillations.

30.4.4   Recognizing a "porpoising" controller

An interesting case of over-tuning is when the process variable "porpoises[36]" on its way to setpoint following a step-change in setpoint. The following trend shows such a response:

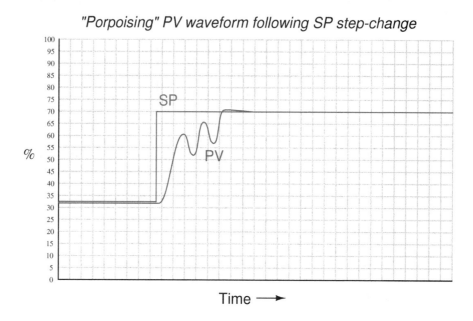

"Porpoising" is universally poor behavior for a loop, because it combines the negative consequences of over-tuning (instability and excessive valve travel) with the negative consequence of under-tuning (delay achieving setpoint). There is no practical purpose served by a loop "porpoising," and so this behavior should be avoided if at all possible.

Thankfully, identifying the cause of "porpoising" is rather easy to do. Only two control actions are capable of causing this response: proportional and derivative. Integral action simply *cannot* cause porpoising. In order for the process variable to "porpoise," the controller's output signal must reverse direction before the process variable ever reaches setpoint. Integral action, however, will always drive the output in a consistent direction when the process variable is on one side of setpoint. Only proportional and derivative actions are capable of producing a directional change in the output signal prior to reaching setpoint.

Solely examining the process variable waveform will not reveal whether it is proportional action, derivative action, or both responsible for the "porpoising" behavior. A trial reduction in the derivative[37] tuning parameter is one way to identify the culprit, as is phase-shift analysis between the PV and output waveforms during the "porpoising" period.

---

[36]The term "porpoise" comes from the movements of a porpoise swimming rapidly toward the water's surface as it chases along the bow of a moving ship. In order to generate speed, the animal undulates its body up and down to powerfully drive forward with its horizontal tail, tracing a sinusoidal path on its way up to breaching the surface of the water.

[37]You could try reducing the controller's gain as a first step, but if the controller implements the Ideal or Series algorithm, reduction in gain will *also* reduce derivative action, which may mask an over-tuned derivative problem.

30.5 Tuning techniques compared

In this section I will show screenshots from a process loop simulation program illustrating the effectiveness of Ziegler-Nichols open-loop ("Reaction Rate") and closed-loop ("Ultimate") PID tuning methods, and then contrast them against the results of my own heuristic tuning. As you will see in some of these cases, the results obtained by either Ziegler-Nichols method tends toward instability (excessive oscillation of the process variable following a setpoint change). This is not necessarily an indictment of Ziegler's and Nichols' recommendations as much as it is a demonstration of the power of understanding. Ziegler and Nichols presented a simple step-by-step procedure for obtaining *approximate* PID tuning constant values based on closed-loop and open-loop process responses, which could be applied by anyone regardless of their level of understanding PID control theory. If I were tasked with drafting a procedure to instruct anyone to quantitatively determine PID constant values without an understanding of process dynamics or process control theory, I doubt my effort would be an improvement. Ultimately, robust PID control is attainable only at the hands of someone who understands how PID works, what each mode does (and why), and is able to distinguish between intrinsic process characteristics and instrument limitations. The purpose of this section is to clearly demonstrate the limitations of ignorantly-followed procedures, and contrast this "mindless" approach against the results of simple experimentation directed by qualitative understanding.

Each of the examples illustrated in this section were simulations run on a computer program called *PC-ControLab* developed by Wade Associates, Inc. Although these are simulated processes, in general I have found similar results using both Ziegler-Nichols and heuristic tuning methods on real processes. The control criteria I used for heuristic tuning were fast response to setpoint changes, with minimal overshoot or oscillation.

30.5.1   Tuning a "generic" process

Ziegler-Nichols open-loop tuning procedure

The first process tuned in simulation was a "generic" process, unspecific in its nature or application. Performing an open-loop test (two 10% output step-changes made in manual mode, both increasing) on this process resulted in the following behavior:

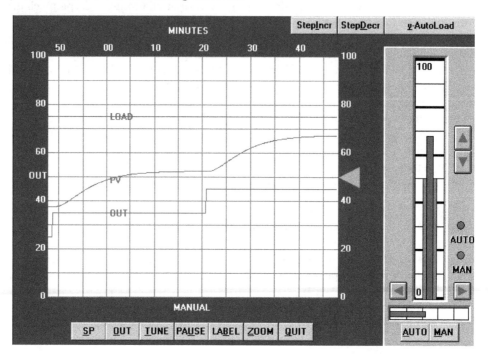

From the trend, we can see that this process is self-regulating, with multiple lags and some dead time. The reaction rate (R) is 20% over 15 minutes, or 1.333 percent per minute. Dead time (L) appears to be approximately 2 minutes. Following the Ziegler-Nichols recommendations for PID tuning based on these process characteristics (also including the 10% step-change magnitude Δm):

$$K_p = 1.2 \frac{\Delta m}{RL} = 1.2 \frac{10\%}{\frac{20\%}{15 \text{ min}} 2 \text{ min}} = 4.5$$

$$\tau_i = 2L = (2)(2 \text{ min}) = 4 \text{ min}$$

$$\tau_d = 0.5L = (0.5)(2 \text{ min}) = 1 \text{ min}$$

Applying the PID values of 4.5 (gain), 4 minutes per repeat (integral), and 1 minute (derivative) gave the following result in automatic mode (with a 10% setpoint change):

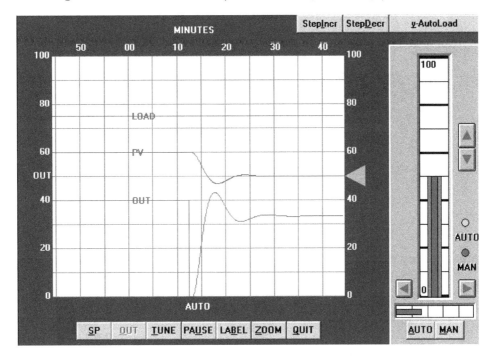

The result is reasonably good behavior with the PID values predicted by the Ziegler-Nichols open-loop equations, and would be acceptable for applications where some setpoint overshoot were tolerable.

We may tell from analyzing the phase shift between the PV and OUT waveforms that the dominant control action here is proportional: each negative peak of the PV lines up fairly close with each positive peak of the OUT, for this reverse-acting controller. If we were interested in minimizing overshoot and oscillation, the logical choice would be to reduce the gain value somewhat.

Ziegler-Nichols closed-loop tuning procedure

Next, the closed-loop, or "Ultimate" tuning method of Ziegler and Nichols was applied to this process. Eliminating both integral and derivative control actions from the controller, and experimenting with different gain (proportional) values until self-sustaining oscillations of consistent amplitude[38] were obtained, gave a gain value of 11:

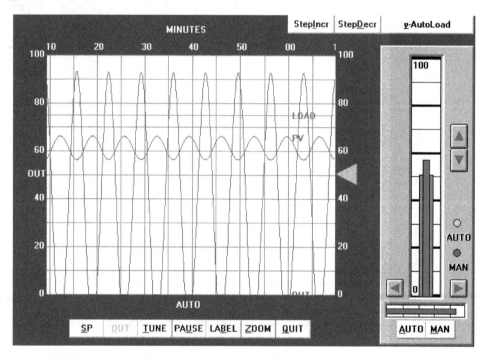

From the trend, we can see that the ultimate period (P_u) is approximately 7 minutes in length. Following the Ziegler-Nichols recommendations for PID tuning based on these process characteristics:

$$K_p = 0.6K_u = (0.6)(11) = 6.6$$

$$\tau_i = \frac{P_u}{2} = \frac{7 \text{ min}}{2} = 3.5 \text{ min}$$

$$\tau_d = \frac{P_u}{8} = \frac{7 \text{ min}}{8} = 0.875 \text{ min}$$

It should be immediately apparent that these tuning parameters will yield poor control. While the integral and derivative values are close to those predicted by the open-loop (Reaction Rate) method, the gain value calculated here is even larger than what was calculated before. Since we

---

[38]The astute observer will note the presence of some limiting (saturation) in the output waveform, as it attempts to go below zero percent. Normally, this is unacceptable while determining the ultimate gain of a process, but here it was impossible to make the process oscillate at consistent amplitude without saturating on the output signal. The gain of this process falls off quite a bit at the ultimate frequency, such that a high controller gain is necessary to sustain oscillations, causing the output waveform to have a large amplitude.

know proportional action was excessive in the last tuning attempt, and this one recommends an even higher gain value, we can expect our next trial to oscillate even worse.

Applying the PID values of 6.6 (gain), 3.5 minutes per repeat (integral), and 0.875 minute (derivative) gave the following result in automatic mode:

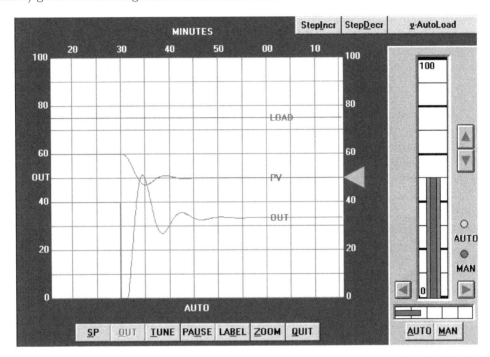

This time the loop stability is a bit worse than with the PID values given by the Ziegler-Nichols open-loop tuning equations, owing mostly to the increased controller gain value of 6.6 (versus 4.5). Proportional action is still the dominant mode of control here, as revealed by the minimal phase shift between PV and OUT waveforms (ignoring the 180 degrees of shift inherent to the controller's reverse action).

In all fairness to the Ziegler-Nichols technique, the excessive controller gain value probably resulted more from the saturated output waveform than anything else. This led to more controller gain being necessary to sustain oscillations, leading to an inflated K_p value.

Heuristic tuning procedure

From the initial open-loop (manual output step-change) test, we could see this process contains multiple lags in addition to about 2 minutes of dead time. Both of these factors tend to limit the amount of gain we can use in the controller before the process oscillates. Both Ziegler-Nichols tuning attempts confirmed this fact, which led me to try much lower gain values in my initial heuristic tests. Given the self-regulating nature of the process, I knew the controller needed integral action, but once again the aggressiveness of this action would be necessarily limited by the lag and dead times. Derivative action, however, would prove to be useful in its ability to help "cancel" lags, so I suspected my tuning would consist of relatively tame proportional and integral values, with a relatively aggressive derivative value.

After some experimenting, the values I arrived at were 1.5 (gain), 10 minutes (integral), and 5 minutes (derivative). These tuning values represent a proportional action only one-third as aggressive as the least-aggressive Ziegler-Nichols recommendation, an integral action less than half as aggressive as the Ziegler-Nichols recommendations, and a derivative action *five times* more aggressive than the most aggressive Ziegler-Nichols recommendation. The results of these tuning values in automatic mode are shown here:

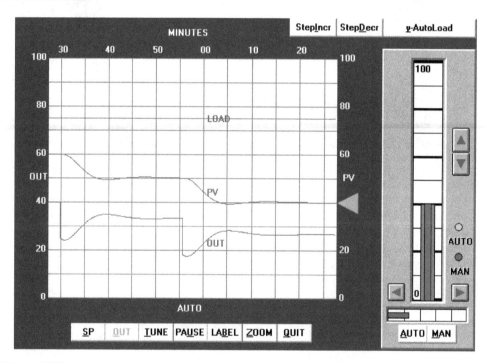

With this PID tuning, the process responded with much less overshoot of setpoint than with the results of either Ziegler-Nichols technique.

30.5.2 Tuning a liquid level process

Ziegler-Nichols open-loop tuning procedure

The next simulated process I attempted to tune was a liquid level-control process. Performing an open-loop test (one 10% increasing output step-change, followed by a 10% decreasing output step-change, both made in manual mode) on this process resulted in the following behavior:

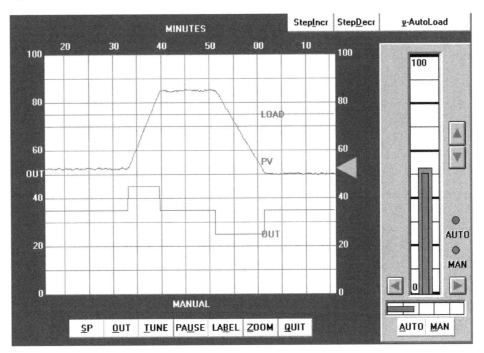

From the trend, the process appears to be purely integrating, as though the control valve were throttling the flow of liquid into a vessel with a constant out-flow. The reaction rate (R) on the first step-change is 50% over 10 minutes, or 5 percent per minute. Dead time (L) appears virtually nonexistent, estimated to be 0.1 minutes simply for the sake of having a dead-time value to use in the Ziegler-Nichols equations. Following the Ziegler-Nichols recommendations for PID tuning based on these process characteristics (also including the 10% step-change magnitude Δm):

$$K_p = 1.2\frac{\Delta m}{RL} = 1.2\frac{10\%}{\frac{50\%}{10\ \text{min}}0.1\ \text{min}} = 24$$

$$\tau_i = 2L = (2)(0.1\ \text{min}) = 0.2\ \text{min}$$

$$\tau_d = 0.5L = (0.5)(0.1\ \text{min}) = 0.05\ \text{min}$$

Applying the PID values of 24 (gain), 0.2 minutes per repeat (integral), and 0.05 minutes (derivative) gave the following result in automatic mode:

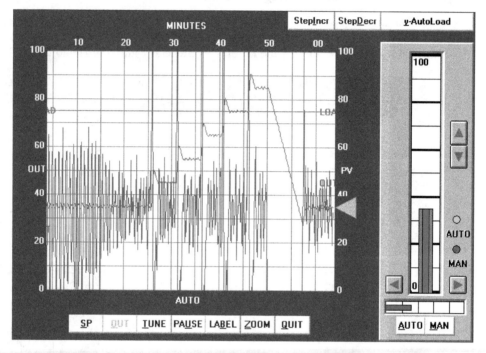

The process variable certainly responds rapidly to the five increasing setpoint changes and also to the one large decreasing setpoint change, but the valve action is hopelessly chaotic. Not only would this "jittery" valve motion prematurely wear out the stem packing, but it would also result in vast over-consumption of compressed air to continually stroke the valve from one extreme to the other. Furthermore, we see evidence of "overshoot" at every setpoint change, most likely from excessive integral action.

We can see from the valve's wild behavior even during periods when the process variable is holding at setpoint that the problem is not a loop oscillation, but rather the effects of process noise on the controller. The extremely high gain value of 24 is amplifying PV noise by that factor, and reproducing it on the output signal.

Ziegler-Nichols closed-loop tuning procedure

Next, I attempted to perform a closed-loop "Ultimate" gain test on this process, but I was not successful. Even the controller's maximum possible gain value would not generate oscillations, due to the extremely crisp response of the process (minimal lag and dead times) and its integrating nature (constant phase shift of -90°).

Heuristic tuning procedure

From the initial open-loop (manual output step-change) test, we could see this process was purely integrating. This told me it could be controlled primarily by proportional action, with very little integral action required, and no derivative action whatsoever. The presence of some process noise is the only factor limiting the aggressiveness of proportional action. With this in mind, I experimented with increasingly aggressive gain values until I reached a point where I felt the output signal noise was at a maximum acceptable limit for the control valve. Then, I experimented with integral action to ensure reasonable elimination of offset.

After some experimenting, the values I arrived at were 3 (gain), 10 minutes (integral), and 0 minutes (derivative). These tuning values represent a proportional action only one-eighth as aggressive as the Ziegler-Nichols recommendation, and an integral action *fifty times* less aggressive than the Ziegler-Nichols recommendation. The results of these tuning values in automatic mode are shown here:

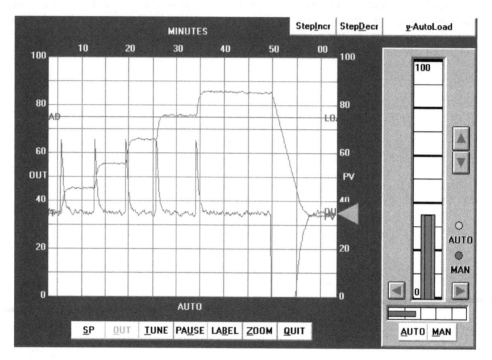

You can see on this trend five 10% increasing setpoint value changes, with crisp response every time, followed by a single 50% decreasing setpoint step-change. In all cases, the process response clearly meets the criteria of rapid attainment of new setpoint values and no overshoot or oscillation.

If it was decided that the noise in the output signal was too detrimental for the valve, we would have the option of further reducing the gain value and (possibly) compensating for slow offset recovery with more aggressive integral action. We could also attempt the insertion of a damping constant into either the level transmitter or the controller itself, so long as this added lag did not cause oscillation problems in the loop[39]. The best solution would be to find a way to isolate the level transmitter from noise, so that the process variable signal was much "quieter." Whether or not this is possible depends on the process and on the particular transmitter used.

---

[39]We would have to be *very* careful with the addition of damping, since the oscillations could create may not appear on the trend. Remember that the insertion of damping (low-pass filtering) in the PV signal is essentially an act of "lying" to the controller: telling the controller something that differs from the real, measured signal. If our PV trend shows us this damped signal and not the "raw" signal from the transmitter, it is possible for the process to oscillate and the PV trend to be deceptively stable!

30.5.3   Tuning a temperature process

Ziegler-Nichols open-loop tuning procedure

This next simulated process is a temperature control process. Performing an open-loop test (two 10% increasing output step-changes, both made in manual mode) on this process resulted in the following behavior:

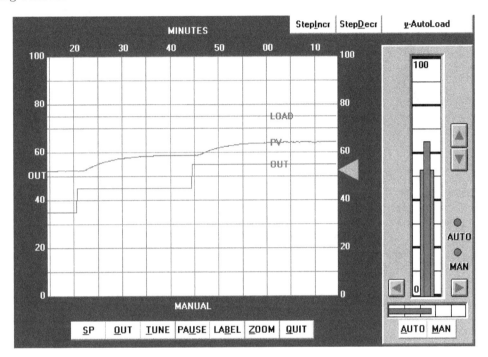

From the trend, the process appears to be self-regulating with a slow time constant (lag) and a substantial dead time. The reaction rate (R) on the first step-change is 30% over 30 minutes, or 1 percent per minute. Dead time (L) looks to be approximately 1.25 minutes. Following the Ziegler-Nichols recommendations for PID tuning based on these process characteristics (also including the 10% step-change magnitude Δm):

$$K_p = 1.2 \frac{\Delta m}{RL} = 1.2 \frac{10\%}{\frac{30\%}{30 \text{ min}} 1.25 \text{ min}} = 9.6$$

$$\tau_i = 2L = (2)(1.25 \text{ min}) = 2.5 \text{ min}$$

$$\tau_d = 0.5L = (0.5)(1.25 \text{ min}) = 0.625 \text{ min}$$

Applying the PID values of 9.6 (gain), 2.5 minutes per repeat (integral), and 0.625 minutes (derivative) gave the following result in automatic mode:

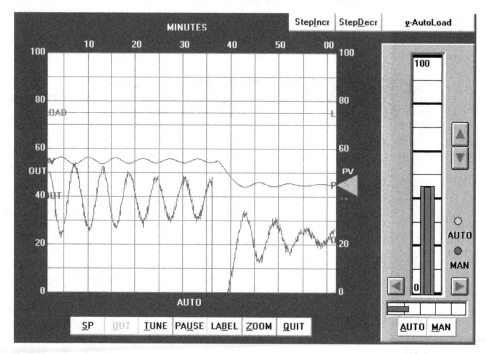

As you can see, the results are quite poor. The PV is still oscillating with a peak-to-peak amplitude of almost 20% from the last process upset at the time of the 10% downward SP change. Additionally, the output trend is rather noisy, indicating excessive amplification of process noise by the controller.

Ziegler-Nichols closed-loop tuning procedure

Next, the closed-loop, or "Ultimate" tuning method of Ziegler and Nichols was applied to this process. Eliminating both integral and derivative control actions from the controller, and experimenting with different gain (proportional) values until self-sustaining oscillations of consistent amplitude were obtained, gave a gain value of 15:

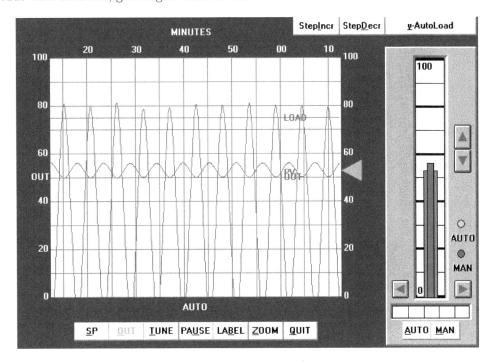

From the trend, we can see that the ultimate period (P_u) is approximately 5.2 minutes in length. Following the Ziegler-Nichols recommendations for PID tuning based on these process characteristics:

$$K_p = 0.6K_u = (0.6)(15) = 9$$

$$\tau_i = \frac{P_u}{2} = \frac{5.2 \text{ min}}{2} = 2.6 \text{ min}$$

$$\tau_d = \frac{P_u}{8} = \frac{5.2 \text{ min}}{8} = 0.65 \text{ min}$$

These PID tuning values are quite similar to those predicted by the open loop ("Reaction Rate") method, and so we would expect to see very similar results:

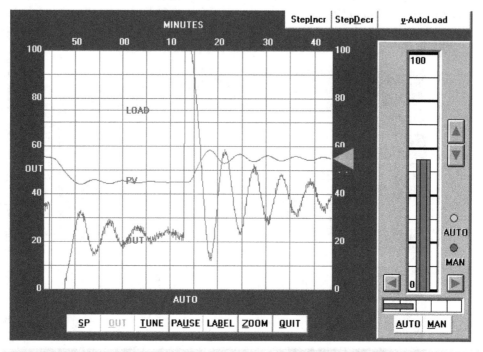

As expected, we still see excessive oscillation following a 10% setpoint change, as well as excessive "noise" in the output trend.

Heuristic tuning procedure

From the initial open-loop (manual output step-change) test, we could see this process was self-regulating with a slow lag and substantial dead time. The self-regulating nature of the process demands at least some integral control action to eliminate offset, but too much will cause oscillation given the long lag and dead times. The existence of over 1 minute of process dead time also prohibits the use of aggressive proportional action. Derivative action, which is generally useful in overcoming lag times, will cause problems here by amplifying process noise. In summary, then, we would expect to use mild proportional, integral, *and* derivative tuning values in order to achieve good control with this process. Anything too aggressive will cause problems for this process.

After some experimenting, the values I arrived at were 3 (gain), 5 minutes (integral), and 0.5 minutes (derivative). These tuning values represent a proportional action only one-third as aggressive as the Ziegler-Nichols recommendation, and an integral action about half as aggressive as the Ziegler-Nichols recommendation. The results of these tuning values in automatic mode are shown here:

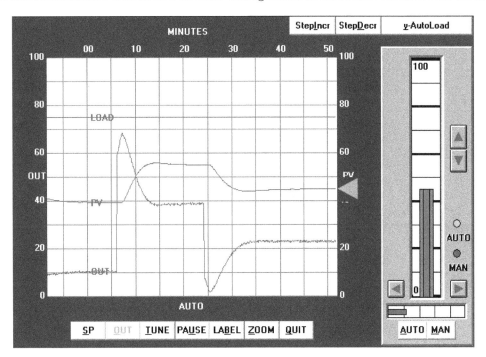

As you can see, the system's response has almost no overshoot (with either a 10% setpoint change or a 15% setpoint change) and very little "noise" on the output trend. Response to setpoint changes is relatively crisp considering the naturally slow nature of the process: each new setpoint is achieved within about 7.5 minutes of the step-change.

30.6 Note to students

Learning how to tune PID controllers is a skill born of much practice. Regardless of how thoroughly you may study the subject of PID control on paper, you really do not understand it until you have spent a fair amount of time actually tuning real controllers.

In order to gain this experience, though, you need access to working processes and the freedom to disturb those processes over and over again. If your school's lab has several "toy" processes built to facilitate this type of learning experience, that is great. However, your learning will grow even more if you have a way to practice PID tuning at your own convenience.

30.6.1   Electrically simulating a process

Thankfully, there is a relatively simple way to build your own "process" for PID tuning practice. First, you need to obtain an electronic single-loop PID controller[40] and connect it to a resistor-capacitor network such as this:

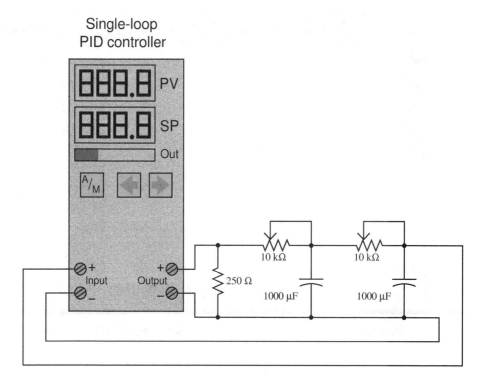

The 250 Ω resistor converts the controller's 4-20 mA signal into a 1-5 VDC signal, which then drives the passive integrator (lag) RC networks. The two stages of RC "lag" simulate a self-regulating process with a second-order lag and a steady-state gain of 1. The potentiometers establish the lag times for each stage, providing a convenient way to alter the process characteristics for more tuning practice. Feel free to extend the circuit with additional RC lag networks for even more delay (and an even harder-to-tune process!).

Since this simulated "process" is direct-acting (i.e. increasing manipulated variable signal results in an increasing process variable signal), the controller must be configured for *reverse* action (i.e. increasing process variable signal results in a decreasing manipulated variable signal) in order to achieve negative feedback. You are welcome to configure the controller for direct action just to see what the effects will be, but I assure you control will be impossible: the PV will saturate beyond 100% or below 0% no matter how the PID values are set.

---

[40]Many instrument manufacturers sell simple, single-loop controllers for reasonable prices, comparable to the price of a college textbook. You need to get one that accepts 1-5 VDC input signals and generates 4-20 mA output signals, and has a "manual" mode of operation in addition to automatic – these features are *very important!* Avoid controllers that can only accept thermocouple inputs, and/or only have time-proportioning (PWM) outputs. Additionally, I strongly recommend you take the time to experimentally learn the actions of proportional, integral, and derivative as outlined in section 29.16 beginning on page 2387 before you embark on any PID tuning exercises.

30.6.2 Building a "Desktop Process" unit

A more sophisticated approach to gaining hands-on experience tuning PID controllers is to actually build a working "process" that the controller can regulate. A relatively simple way to do this for students is to build what I like to call *Desktop Processes*, where a loop controller is used to control the speed of a motor/generator set made from small DC "hobby" electric motors. An illustration of a "Desktop Process" is shown here:

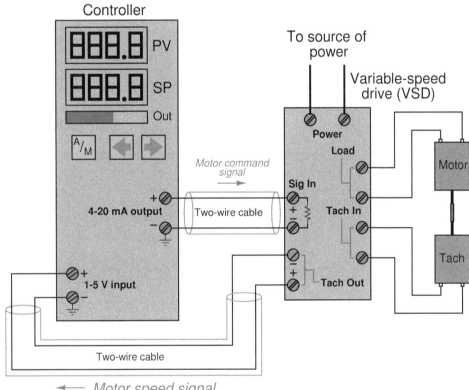

You must build your own variable-speed drive (VSD) circuit to convert the controller's 4-20 mA output signal into a DC voltage powerful enough to drive the motor. This same circuit should also contain components for "scaling" and filtering the tachogenerator's DC voltage signal so it may be read by the controller's input. Fortunately, the following circuit is a proven and simple design for doing just that:

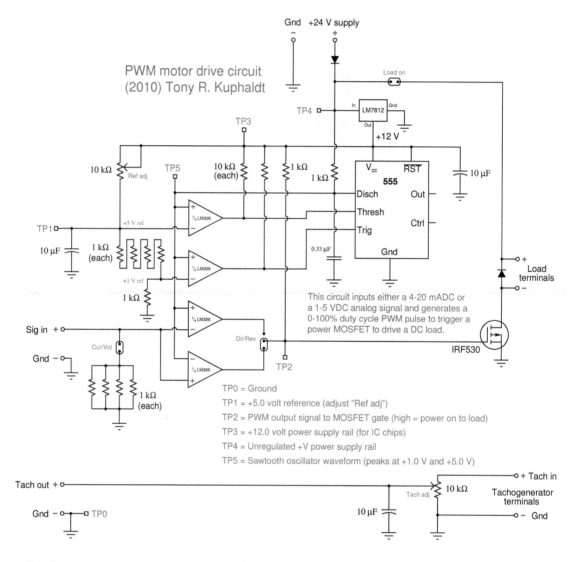

Diodes included in this design protect against reverse-polarity power supply connections and inductive "kickback" resulting from de-energizing inductive loads.

Photographs showing a complete "Desktop Process" unit in operation, including close-ups of the motor/generator set and the variable speed drive circuit board, appear here:

As you can see from the photograph, the motor and generator are held in a short length of split

PVC pipe. This is a simple way to clamp and align both machines so their shafts turn on the same centerline. The coupling between the two shafts is nothing more than a piece of rubber tube (or wire insulation, or heat-shrink tubing, or even electrical tape!).

An optional accessory to add to a Desktop Process is a data acquisition unit capable of measuring the DC voltage motor speed and controller output signals, plotting them on a computer display for further analysis. This becomes very useful when fine-tuning PID response, allowing students to visually recognize oscillation, overshoot, windup, and other phenomena of closed-loop control.

The controller model shown in these photographs happens to be a Siemens 353, but any loop controller capable of receiving a 1-5 volt DC input signal and generating 4-20 mA DC output signal will work just fine. In fact, I've connected this very same VSD and motor/generator set to different controllers[41] to compare operation.

Interesting experiments to perform with a Desktop Process – other than PID tuning practice – include the following:

- Introducing process loads by touching the spinning motor shaft (slowing it down using your finger) and compare the responses between the controller's "manual" and "automatic" modes. This proves to be a very effective way for students to comprehend the difference between these two modes of operation. I have yet to encounter a student who does not immediately grasp the concept after doing this experiment for themselves, feeling the motor's shaft speed respond to their finger load in both modes, also watching the controller's output response.

- Switching the controller mode from reverse action to direct action to see how a process "runs away" when the loop feedback is positive rather than negative.

- Switching the VSD action from direct to reverse, then reconfiguring the controller's action to complement it, maintaining negative feedback in the system.

- Try switching between auto and manual modes in the controller, comparing the response with and without the feature of *setpoint tracking*. Again, this is a concept many students struggle to grasp in theory, but immediately comprehend when they see it in action.

---

[41]Among these different controllers were a Distech ESP-410 building (HVAC) controller and a small PLC programmed with a custom PID control algorithm. In fact, a Desktop Process is ideal for courses where students create their own control algorithms in PLC or data acquisition hardware. The significance of controller scan rate becomes very easy to comprehend when controlling a process like this with such a short time constant. The contrast between a DDC controller with a 500 millisecond scan rate and a PLC with a 50 millisecond scan rate, for example, is marked.

30.6.3 Simulating a process by computer

A fascinating solution for realistic PID tuning in the classroom was offered to me by Blair MacNeil of Cape Breton University (located in the town of Sydney, on the island of Nova Scotia, Canada) in 2010 by way of email correspondence. Professor MacNeil uses Moore 353 loop controllers connected to analog computer I/O ("data acquisition") modules, with a personal computer running VisSim Realtime software to simulate the dynamics of a real process. With the power of a personal computer simulating the process, virtually any process dynamic (as well as any instrument fault) may be generated for the benefit of the loop controller to control:

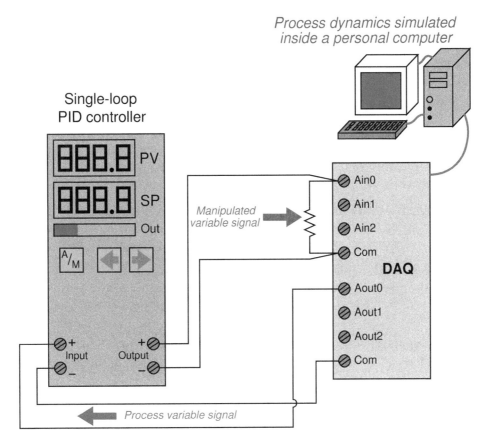

This approach provides realistic process dynamics for the loop controller to manage, yet requires little in the way of capital expense or physical space to implement. Different process models, instrument faults, and control strategies may be easily implemented in the personal computer's software, making it far more flexible as a teaching tool than any analog electronic simulation network or real process connected to the controller. It is also completely safe to operate, with absolutely no danger of harming anything in the event of a process "excursion" or other upset.

30.7   Review of fundamental principles

Shown here is a partial listing of principles applied in the subject matter of this chapter, given for the purpose of expanding the reader's view of this chapter's concepts and of their general inter-relationships with concepts elsewhere in the book. Your abilities as a problem-solver and as a life-long learner will be greatly enhanced by mastering the applications of these principles to a wide variety of topics, the more varied the better.

- **Conservation of mass**: mass is an intrinsic property of matter, and as such cannot be created or destroyed. Relevant to the *mass balance* of a process, meaning that all mass flowing into a process must equal all mass flowing out of a process (unless mass is being accumulated or released from the process). This is relevant to the determination of a process' characteristic as either self-regulating or integrating: whether the mass balance of the process naturally equalizes or not as the process variable changes value.

- **Conservation of energy**: energy cannot be created or destroyed, only converted between different forms. Relevant to the *energy balance* of a process, meaning that all energy flowing into a process must equal all energy flowing out of a process (unless energy is being accumulated or released from the process). This is relevant to the determination of a process' characteristic as either self-regulating or integrating: whether the energy balance of the process naturally equalizes or not as the process variable changes value.

- **Negative feedback**: when the output of a system is degeneratively fed back to the input of that same system, the result is decreased (overall) gain and greater stability. Relevant to loop controller action: in order for a control system to be stable, the feedback must be negative.

- **Amplification**: the control of a relatively large signal by a relatively small signal. Relevant to the role of loop controllers exerting influence over a process variable at the command of a measurement signal. In behaving as amplifiers, loop controllers may oscillate if certain criteria are met.

- **Barkhausen criterion**: is overall loop gain is unity (1) or greater, and phase shift is 360°, the loop will sustain oscillations. Relevant to control system stability, explaining why the loop will "cycle" (oscillate) if gain is too high.

- **Time constant**: (τ), defined as the amount of time it takes a system to change 63.2% of the way from where it began to where it will eventually stabilize. Also known as the *lag time* of a system. The system will be within 1% of its final value after 5 time constants' worth of time has passed (5τ). Relevant to process control loops, where natural lags contribute to time constants, usually of multiple order.

- **Phase shift**: the angular difference between two sinusoidal waves of the same frequency. Relevant to the analysis of controller trend graphs: zero phase shift between PV and Output is the hallmark of proportional action; lagging phase shift is the hallmark of integral action; leading phase shift is the hallmark of derivative action.

- **Differentiation (calculus)**: where one variable is proportional to the rate-of-change of two others. Differentiation always results in a positive phase shift if the input is a wave. Relevant to the output of a controller, for determining by (leading) phase shift whether derivative action is dominant.

- **Integration (calculus)**: where one variable is proportional to the accumulation of the product of two others. Integration always results in a negative phase shift if the input is a wave. Relevant to the output of a controller, for determining by (lagging) phase shift whether integral action is dominant.

References

Lipták, Béla G. et al., *Instrument Engineers' Handbook – Process Control Volume II*, Third Edition, CRC Press, Boca Raton, FL, 1999.

McMillan, Greg, "Is Wireless Process Control Ready for Prime Time?", *Control*, May 2009, pp. 54-57.

Mollenkamp, Robert A., *Introduction to Automatic Process Control*, Instrument Society of America, Research Triangle Park, NC, 1984.

Palm, William J., *Control Systems Engineering*, John Wiley & Sons, Inc., New York, NY, 1986.

Shinskey, Francis G., *Energy Conservation through Control*, Academic Press, New York, NY, 1978.

Shinskey, Francis G., *Process-Control Systems – Application / Design / Adjustment*, Second Edition, McGraw-Hill Book Company, New York, NY, 1979.

St. Clair, David W., *Controller Tuning and Control Loop Performance, a primer*, Straight-Line Control Company, Newark, DE, 1989.

Ziegler, J. G., and Nichols, N. B., "Optimum Settings for Automatic Controllers", *Transactions of the American Society of Mechanical Engineers* (ASME), Volume 64, pages 759-768, Rochester, NY, November 1942.

Chapter 31

Basic process control strategies

In a simple control system, a process variable (PV) is measured and compared with a setpoint value (SP). A manipulated variable (MV, or output) signal is generated by the controller and sent to a final control element, which then influences the process variable to achieve stable control. The algorithm by which the controller develops its output signal is typically PID (Proportional-Integral-Derivative), but other algorithms may be used as well:

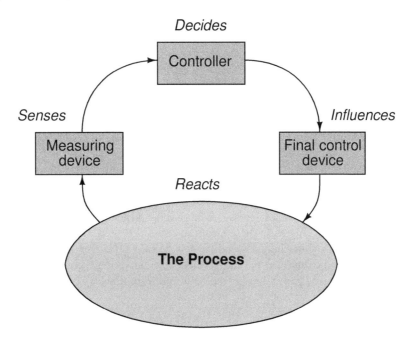

This form of simple control may be improved upon and expanded for a greater range of process applications by interconnecting multiple controllers and/or redirecting measurement and control signals in more complex arrangements. An exploration of some of the more common control system configurations is the subject of this chapter.

31.1   Supervisory control

In a manually-controlled process, a human operator directly actuates some form of final control element (usually a valve) to influence a process variable. Simple automatic ("regulatory") control relieves human operators of the need to continually adjust final control elements by hand, replacing this task with the occasional adjustment of setpoint values. The controller then manipulates the final control element to hold the process variable at the setpoint value determined by the operator.

The next step in complexity after simple automatic control is to automate the adjustment of the setpoint for a process controller. A common implementation of this concept is the automatic cycling of setpoint values according to a timed schedule. An example of this is a temperature controller for a heat-treatment furnace used to temper metal samples:

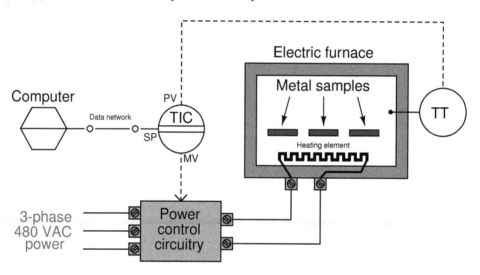

Here, a computer "supervises" the furnace's temperature by communicating setpoint values to the temperature indicating controller (TIC) over a digital network interface such as Ethernet. From the temperature controller's perspective, this is a *remote* setpoint signal, as opposed to a *local* setpoint value which would be set by a human operator at the controller faceplate. Since the heat-treatment of metals requires particular temperature ranges and rates of change over time, this control system relieves the human operator of having to manually adjust setpoint values again and again during heat-treatment cycles. Instead, the computer schedules different setpoint values at different times (even setpoint values that change steadily at a certain rate over a period of time) according to the needs of the particular metal type and treatment type. Such a control scheme is quite common for heat-treating processes, and it is referred to as *ramp and soak*[1].

---

[1]In honor of the system's ability to slowly "ramp" temperature up or down at a specified rate, then "soak" the metal at a constant temperature for set periods of time. Many single-loop process controllers have the ability to perform ramp-and-soak setpoint scheduling without the need of an external "supervisory" computer.

Process controllers configured for supervisory setpoint control typically have three operating modes:

- **Manual mode:** Controller takes no automatic action. Output value set by human operator.

- **Automatic mode with local SP:** Controller automatically adjusts its output to try to keep PV = SP. Setpoint value set "locally" by human operator.

- **Automatic mode with remote SP:** Controller automatically adjusts its output to try to keep PV = SP. Setpoint value set "remotely" by supervising computer.

Supervisory setpoint control is also used in the chemical processing industries to optimize production efficiencies by having a powerful computer provide setpoint adjustments to regulatory controls based on mathematical models of the process and optimization constraints. In simple terms, this means having a computer make setpoint adjustments to the normal PID loop controllers instead (or in addition to) human operators making setpoint changes. This forms a two-layer process control system: the "base" or "regulatory" layer of control (PID loop controllers) and the "high" or "supervisory" level of control (the powerful computer with the mathematical process models).

Such "optimizing" control systems are usually built over a digital network for reasons of convenience. A single network cable not only is able to communicate the frequent setpoint changes from the supervisory computer to the multitude of process loop controllers, but it may also carry process variable information from those controllers back to the supervisory computer so it has data for its optimization algorithms to operate on:

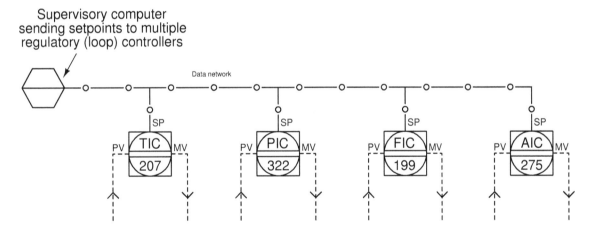

The complexity of these optimization algorithms is limited only by the computational power of the supervisory computer and the creativity of the programmers and engineers who implement it. A modern trend in process optimization for industries able to produce varying proportions of different products from the same raw material feed is to have computer algorithms select and optimize production not only for maximum cost efficiency, but also for maximum market sales and minimum storage of volatile product[2].

---

[2]I once attended a meeting of industry representatives where one person talked at length about a highly automated

31.2 Cascade control

A simple control system drawn in block diagram form looks like this:

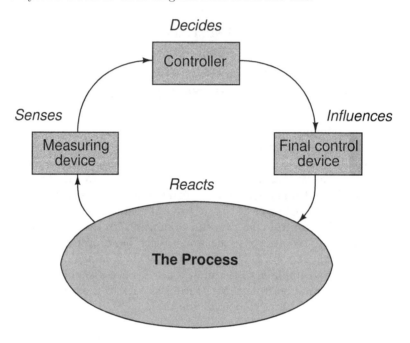

Information from the measuring device (e.g. transmitter) goes to the controller, then to the final control device (e.g. control valve), influencing the process which is sensed again by the measuring device. The controller's task is to inject the proper amount of negative feedback such that the process variable stabilizes over time. This flow of information is collectively referred to as a feedback control *loop*.

To *cascade* controllers means to connect the output signal of one controller to the setpoint of another controller, with each controller sensing a different aspect of the same process. The first controller (called the *primary*, or *master*) essentially "gives orders" to the second controller (called the *secondary* or *slave*) via a *remote setpoint* signal.

---

lumber mill where logs were cut into lumber not only according to minimum waste, but also according to the real-time market value of different board types and stored inventory. The joke was, if the market value of wooden toothpicks suddenly spiked up, the control system would shred every log into toothpicks in an attempt to maximize profit!

Thus, a cascade control system consists of two feedback control loops, one nested inside the other:

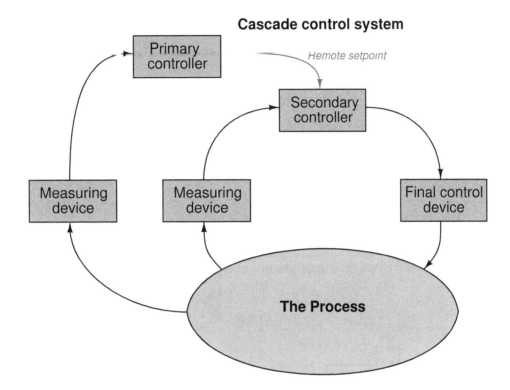

Cascade control system

A very common example of cascade control is a *valve positioner*, which receives a command signal from a regular process controller, and in turn works to ensure the valve stem position precisely matches that command signal. The control valve's stem position is the process variable (PV) for the positioner, just as the command signal is the positioner's setpoint (SP). Valve positioners therefore act as "slave" controllers to "master" process controllers controlling pressure, temperature, flow, or some other process variable.

The purpose of cascade control is to achieve greater stability of the primary process variable by regulating a secondary process variable in accordance with the needs of the first. An essential requirement of cascaded control is that the secondary process variable be faster-responding (i.e. shorter lag and dead times) than the primary process variable.

An analogy for understanding cascade control is that of *delegation* in a work environment. If a supervisor delegates some task to a subordinate, and that subordinate performs the task without further need of guidance or assistance from the supervisor, the supervisor's job is made easier. The subordinate takes care of all the little details that would otherwise burden the supervisor if the supervisor had no one to delegate to. This analogy also makes it clear why the secondary process variable must be faster-responding than the primary process variable: the supervisor-subordinate management structure fails to work if the supervisor does not maintain focus on long-term goals (i.e. longer-term than the completion time of the tasks given to subordinates). If a supervisor focuses on achieving goals that are shorter-term than the time required for subordinates to complete their assignments, the supervisor will inevitably call for "course changes" that are too quick for the subordinates to execute. This will lead to the subordinates "lagging" behind the supervisor's orders, to the detriment of everyone's satisfaction.

An example of cascade control applied to a real industrial process is shown here, for a *dryer* system where heated air is used to evaporate water from a granular solid. The primary process variable is the outlet air exiting the dryer, which should be maintained at a high enough temperature to ensure water will not remain in the upper layers of the solid material. This outlet temperature is fairly slow to react, as the solid material mass creates a large lag time:

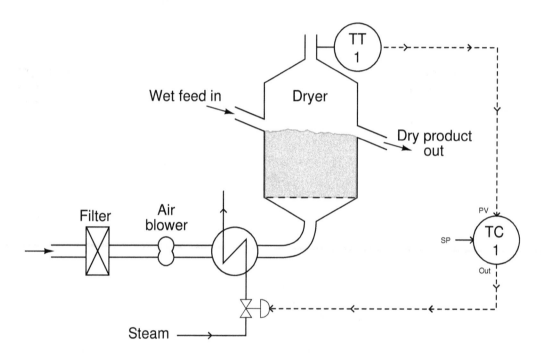

There are several parameters influencing the temperature of the outlet air other than the moisture content of the drying material. These include air flow, ambient air temperature, and variations in steam temperature. Each one of these variables is a *load* on the process variable we are trying to control (outlet air temperature). If any of these parameters were to suddenly change, the effect would be slow to register at the outlet temperature even though there would be immediate impact at the bottom of the dryer where the heated air enters. Correspondingly, the control system would be slow to correct for any of these changing loads.

We may better compensate for these loads by installing a second temperature transmitter at the inlet duct of the dryer, with its own controller to adjust steam flow at the command of the primary controller:

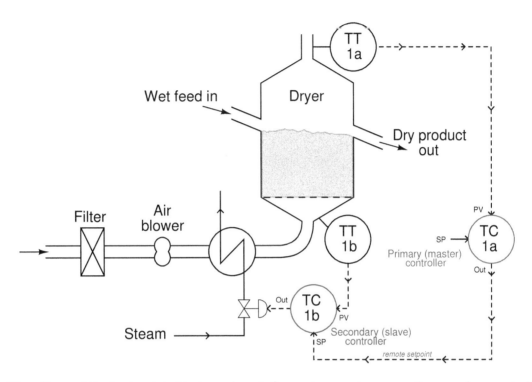

Now, if any of the loads related to incoming air flow or temperature vary, the secondary controller (TC-1b) will *immediately* sense the change in dryer inlet temperature and compensate by adjusting steam flow through the heat exchanger. Thus, the "slave" control loop (1b) helps stabilize the "master" control loop (1a) by reacting to load changes long before any effect might manifest at the dryer outlet.

A helpful way to think of this is to consider the slave controller as *shielding* the master controller from the loads previously mentioned (incoming air flow, ambient temperature, and steam temperature). Of course, these variables still act as loads to the slave controller, as it must continuously adjust the steam valve to compensate for changes in air flow, ambient air temperature, and steam temperature. However, so long as the slave controller does a good job of stabilizing the air temperature entering the dryer, the master controller will never "see" the effects of those load changes. Responsibility for incoming air temperature has been delegated to the slave controller, and as a result the master controller is conveniently isolated from the loads impacting that loop.

To re-emphasize an important point, one of the non-negotiable requirements for cascade control is that the secondary (slave) loop must be *faster-responding* than the primary (master) loop. Cascade control cannot function if this speed relationship is reversed. Temperature controller TC-1b is able to be a slave to controller TC-1a because the natural response time of the temperature at the dryer's bottom is much shorter than at the dryer's top with respect to any changes in steam valve position.

A common implementation of cascade control is where a flow controller receives a setpoint from some other process controller (pressure, temperature, level, analytical, etc.), fluid flow being one of the fastest-responding process types in existence. A feedwater control system for a steam boiler – shown here in pneumatic form – is a good example:

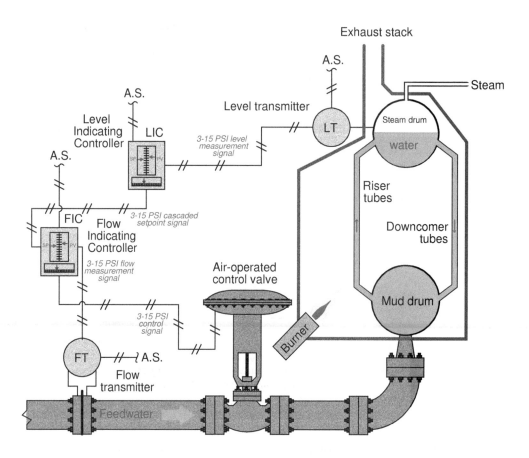

The "secondary" or "slave" flow controller works to maintain feedwater flow to the boiler at whatever flow rate is desired by the level controller. If feedwater pressure happens to increase or decrease, any resulting changes in flow will be quickly countered by the flow controller without the level controller having to react to a consequent upset in steam drum water level. Thus, cascade control works to guard against steam drum level instability resulting from changes in the feedwater flow caused by factors outside the boiler. As stated previously, the slave (flow) controller effectively *shields* the master (level) controller from loads in the feedwater supply system, so that master controller doesn't have to deal with those loads.

This level/flow cascade control system also embodies the principle of the secondary (slave) loop being faster-responding than the primary (master) loop. Water flow is an inherently fast process, the flow rate responding immediately to changes in valve position. By contrast, water level is a much slower-responding type of process. If you perform a "thought experiment" where the feedwater valve is suddenly opened fully, it is easy to see that the feedwater flow rate will immediately reach its full

(100%) value while the steam drum's water level will merely begin to rise, taking time to reach its full (100%) value.

It is worth noting that the inclusion of a flow control "slave" loop to this boiler water level control system also helps to overcome a potential problem of the control valve: nonlinear behavior. In the control valves chapter, we explore the phenomenon of *installed valve characteristics* (Section 27.13.1 beginning on page 2178), specifically noting how changes in pressure drop across a control valve influences its throttling behavior. The result of these pressure changes is a non-linearization of valve response, such that the valve tends to be more responsive near its closed position and less responsive near its open position. One of the benefits of cascaded flow control is that this problem becomes confined to the secondary (flow control) loop, and is effectively removed from the primary control loop. To phrase it simply, distorted valve response becomes "the flow controller's problem" rather than something the level controller must manage. The result is a level control system with more predictable response.

A classic example of cascade control strategy is found in *motion control* applications, where an electric motor is used as the final control element to precisely position a piece of machinery. In this capacity, the motor is usually called a *servo*. Robotic systems make extensive use of servo motors and cascaded control loops to modulate power to those motors. The following illustration shows a triple-cascade control system[3] for a motor-actuated elevator, precisely controlling the position of the elevator through cascaded velocity and motor current control:

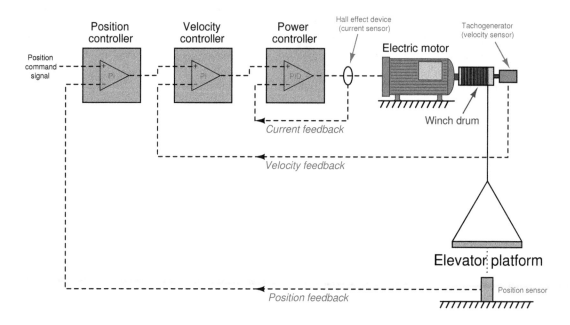

Hypothetically, the position of the elevator could be controlled with a single PID controller sensing platform position and directly sending power to the motor. However, much more precise control of the platform is achievable by sensing position, velocity, and motor current, and controlling each one of those variables with its own loop. In motion control systems, each successive variable is the time-derivative of its precursor. Velocity, for instance, is the time-derivative of position ($v = \frac{dx}{dt}$). Motor current, which is usually proportional to motor torque, which in turn is proportional to the angular acceleration (α) of the winch and consequently the linear acceleration of the platform (a), is the time-derivative of velocity ($a = \frac{dv}{dt}$). If it were not for cascading, a single PID controller would have to control position by manipulating the *acceleration* of the platform (i.e. motor current). This would make the process characteristic "runaway" in nature, as any fixed amount of current will cause the platform to accelerate[4].

---

[3]Interestingly, servo motor control is one application where *analog* loop controllers have historically been favored over digital loop controllers, simply for their superior speed. An opamp-based P, PI, or PID controller is lightning-fast because it has no need to digitize any analog process variables (analog-to-digital conversion) nor does it require time for a clock to sequence step-by-step through a written program as a microprocessor does. Servomechanism processes are inherently fast-responding, and so the controller(s) used to control servos must be faster yet.

[4]At one specific current level, the motor will develop just enough torque to hold the platform's weight, at which point the acceleration will be zero. Any amount of current above this value will cause an upward acceleration, while any amount of current below this value will cause a downward acceleration.

Here with servomechanisms we see how cascading not only has the effect of "shielding" certain load variables from the master controller's view, but it also simplifies the dynamic characteristics of the process from that same point of view. Instead of the position controller having to regulate an inherently "runaway" process, it now sees the process as having an "integrating" characteristic, since any constant output signal from the position controller results in the platform holding to a constant velocity (i.e. platform position will change at a constant rate over time, rather than at an accelerating rate).

A necessary step in implementing cascade control is to ensure the secondary ("slave") controller is well-tuned *before* any attempt is made to tune the primary ("master") controller. Just a moment's thought is all that is needed to understand why this precedence in tuning must be: it is a simple matter of dependence. The slave controller does not depend on good tuning in the master controller in order to control the slave loop. If the master controller were placed in manual (effectively turning off its automatic response), the slave controller would simply control to a constant setpoint. However, the master controller most definitely depends on the slave controller being well-tuned in order to fulfill the master's "expectations." If the slave controller were placed in manual mode, the master controller would not be able to exert any control over its process variable whatsoever. Clearly then, the slave controller's response is essential to the master controller being able to control its process variable, therefore the slave controller should be tuned first when initially commissioning or optimizing a cascade control system.

Just like supervisory control systems where a process controller receives a "remote" setpoint signal from some other system, the secondary ("slave") controller in a cascade system typically has three different operating modes:

- **Manual mode:** Controller takes no automatic action. Output value set by human operator.

- **Automatic mode:** Controller automatically adjusts its output to try to keep PV = SP. Setpoint value set "locally" by human operator.

- **Cascade mode:** Controller automatically adjusts its output to try to keep PV = SP. Setpoint value set "remotely" by primary (master) controller.

This means it is possible to defeat a cascade control system by placing the secondary controller in the wrong mode (automatic) just as it is possible to defeat any control system by placing the controller in manual mode. If a controller is "slaved" to another controller, it must be left in *cascade* mode in order for the control strategy to function as designed.

31.3   Ratio control

Most people reading this book have likely had the experience of adjusting water temperature using two hand valves as they took a shower: one valve controlling the flow of hot water and the other valve controlling the flow of cold water. In order to adjust water temperature, the *proportion* of one valve opening to the other must be changed. Increasing or decreasing total water flow rate without upsetting the outlet temperature is a matter of adjusting both valves in the same direction, maintaining that same proportion of hot to cold water flow.

Although you may not have given it much thought while taking your shower, you were engaged in a control strategy known as *ratio control*, where the ratio of one flow rate to another is controlled for some desired outcome. Many industrial processes also require the precise mixing of two or more ingredients to produce a desired product. Not only do these ingredients need to be mixed in proper proportion, but it is usually desirable to have precise control over the total flow rate as well.

A simple example of ratio control is in the production of paint, where a base liquid must be mixed with one or more pigments to achieve a desired consistency and color. A manually controlled paint mixing process, similar to the hot and cold water valve "process" in some home showers, is shown here. Two flowmeters, a ratio calculating relay, and a display provide the human operator with a live measurement of pigment-to-base ratio:

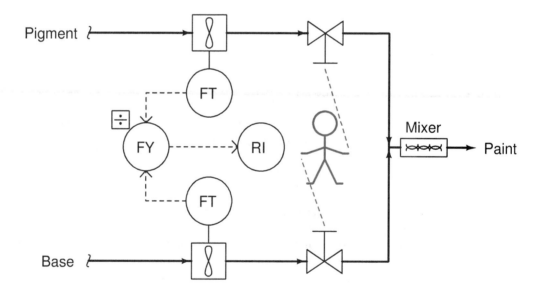

One alteration we could make to this mixing system is to link the two manual control valve handles together in such a way that the ratio of base to pigment was *mechanically* established. All the human operator needs to do now is move the one link to increase or decrease mixed paint production:

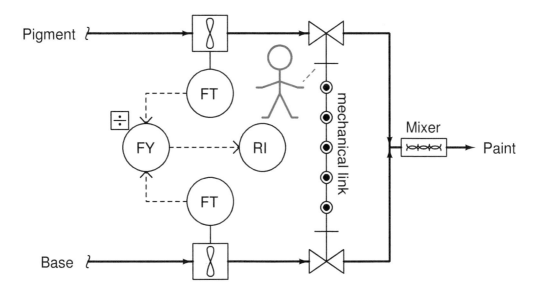

Adjusting the pigment-to-base ratio is now a matter of adjusting the linkage ratio, a task most likely performed by a mechanic or someone else skilled in the alignment of mechanical linkages. The convenience of total flow adjustment gained by the link comes at the price of inconvenient ratio adjustment.

Mechanical link ratio-control systems are commonly used to manage simple burners, proportioning the flow rates of fuel and air for clean, efficient combustion. A photograph of such a system appears here, showing how the fuel gas valve and air damper motions are coordinated by a single rotary actuator:

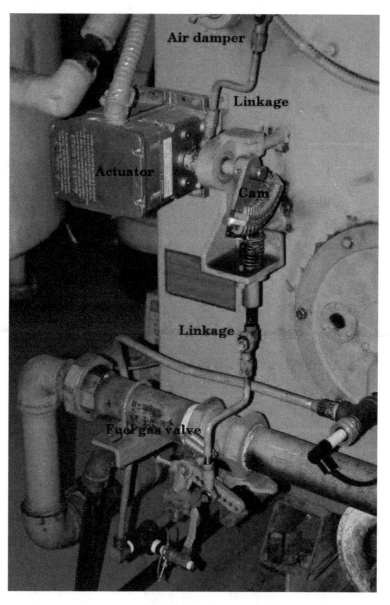

As you can see in this photo, the fuel gas valve is actuated by means of a cam, allowing precise "tuning" of the valve characteristics for consistent fuel/air ratio across a wide range of firing rates. Making ratio adjustments in such a linkage system is obviously a task for a skilled mechanic or technician.

A more automated approach to the general problem of ratio control involves the installation of a flow control loop on one of the lines and a flow-sensing transmitter on the other line. The signal coming from the uncontrolled flow transmitter becomes the setpoint for the flow control loop:

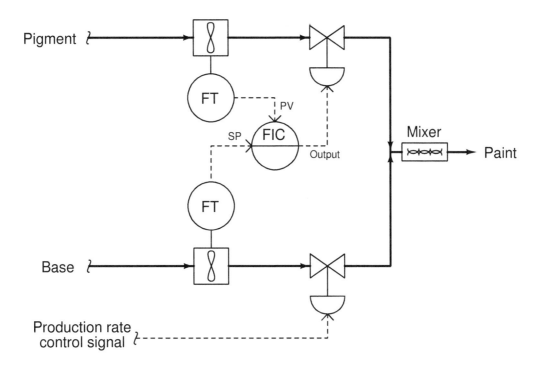

Here, the flow transmitter on the uncontrolled line measures the flow rate of base, sending a flow rate signal to the pigment flow controller which acts to match flow rates. If the calibrations of each flow transmitter are precisely equal to one another, the ratio of pigment to base will be 1:1 (equal). The flow of base liquid into the mixing system is called a *wild flow* or *wild variable*, since this flow rate is not controlled by the ratio control system. The only purpose served by the ratio control system is to match the pigment flow rate to the wild (base) flow rate, so the same ratio of pigment to base will always be maintained regardless of total flow rate. Thus, the flow rate of pigment will be held *captive* to match the "wild" base flow rate, which is why the controlled variable in a ratio system is sometimes called the *captive variable* (in this case, a captive *flow*).

As with the mechanically-linked manual ratio mixing system, this ratio control system provides convenient total flow control at the expense of convenient ratio adjustment. In order to alter the ratio of pigment to base, someone must re-range one or more flow transmitters. To achieve a 2:1 ratio of base to pigment, for example, the base flow transmitter's range would have to be double that of the pigment flow transmitter. This way, an equal percentage of flow registered by both flow transmitters (as the ratio controller strives to maintain equal percentage values of flow between pigment and base) would actually result in twice the amount of base flow than pigment flow.

We may incorporate convenient ratio adjustment into this system by adding another component (or function block) to the control scheme: a device called a *signal multiplying relay* (or alternatively, a *ratio station*). This device (or computer function) takes the flow signal from the base (wild) flow transmitter and multiplies it by some constant value (k) before sending the signal to the pigment (captive) flow controller as a setpoint:

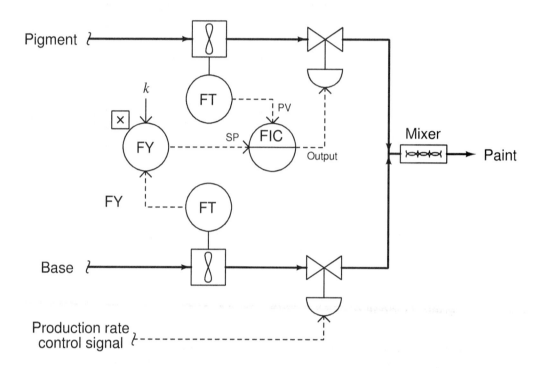

With identical flow range calibrations in both flow transmitters, this multiplying constant k directly determines the pigment-to-base ratio (i.e. the ratio will be 1:1 when $k = 1$; the ratio will be 2:1 when $k = 2$, etc.). If the k value is easily adjusted by a human operator, mixing ratio becomes a very simple parameter to change at will, just as the total production rate is easy to adjust by moving the base flow control valve.

Another example of ratio control at work is in a process whereby hydrocarbon gases (usually methane) are converted into hydrogen gas and carbon dioxide gas. This is known as the *steam-hydrocarbon reforming process*, and it is one of the more popular ways of generating hydrogen gas for industrial use. The overall reaction for this process with methane gas (CH_4) and steam (H_2O) as the reactants is as follows[5]:

$$CH_4 + 2H_2O \rightarrow 4H_2 + CO_2$$

---

[5]The conversion from hydrocarbon and steam to hydrogen and carbon dioxide is typically a two-stage process: the first (*reforming*) stage produces hydrogen gas and carbon monoxide, while a second (*water-gas-shift*) stage adds more steam to convert the carbon monoxide into carbon dioxide with more hydrogen liberated. Both reactions are endothermic, with the reforming reaction being more endothermic than the water-gas-shift reaction.

This is an *endothermic chemical reaction*, which means a net input of energy is required to make it happen. Typically, the hydrocarbon gas and steam are mixed together in a heated environment in the presence of a catalyst (to reduce the activation energy requirements of the reaction). This usually takes the form of catalyst-packed metal tubes inside a gas-fired furnace. It is important to control the proportion of gas to steam flow into this process. Too much hydrocarbon gas, and the result will be "coking" (solid hydrocarbon deposits) inside the heated tubes and on the surface of the catalyst beads, decreasing the efficiency of the process over time. Too much steam and the result is wasted energy as unreacted steam simply passes through the heater tubes, absorbing heat and carrying it away from the catalyst where it would otherwise do useful work.

One way to achieve the proper ratio of hydrocarbon gas to steam flow is to install a normal flow control loop on one of these two reactant feed lines, then use that process variable (flow) signal as a setpoint to a flow controller installed on the other reactant feed line. This way, the second controller will maintain a proper balance of flow to proportionately match the flow rate of the other reactant. An example P&ID is shown here, where the methane gas flow rate establishes the setpoint for steam flow control:

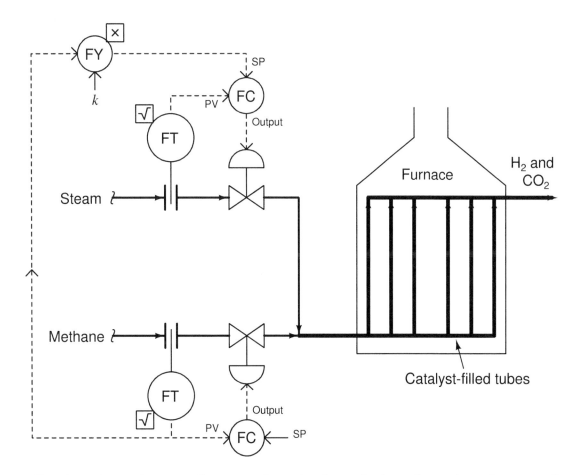

Note how the methane gas flow transmitter signal goes both to the methane flow controller and to a *multiplying relay* that multiplies this signal by a constant value (k) before passing it on to the steam flow controller as a setpoint. This k value sets the *ratio* of steam flow to methane flow. Although this might appear to be a cascade control system at first glance, it is actually quite different. In a cascade system, the *output* of one controller becomes the setpoint for another. Here in a ratio control system, the *process variable* of one controller becomes the setpoint for another, such that two process variables remain in constant proportion (ratio) to one another.

If the two flow transmitters are compensated to measure mass flow, the ideal value of k should be set such that two molecules of steam vapor (H_2O) enter the reforming furnace for every one molecule of methane (CH_4). With a 2-to-1 molecular ratio of steam to methane (2 moles of steam per one mole of methane), this equates to a 9-to-4 mass flow ratio once the formula weights of steam and methane are calculated[6]. Thus, if the methane and gas flowmeters are calibrated for equal mass flow ranges, the ideal value for k should be $\frac{9}{4}$, or 2.25. Alternatively, the flow transmitter calibrations could be set in such a way that the ideal ratio is intrinsic to those transmitters' ranges (i.e. the methane flow transmitter has 2.25 times the mass flow range of the steam flow transmitter), with k set to an ideal value of 1. This way a 9:4 ratio of methane mass flow to steam mass flow will result in equal percentage output values from both flow transmitters. In practice, the value for k is set a bit higher than ideal, in order to ensure just a little excess steam to guard against coking inside the reaction heater tubes[7].

---

[6]Steam has a formula weight of 18 amu per molecule, with two hydrogen atoms (1 amu each) and one oxygen atom (16 amu). Methane has a formula weight of 16 amu per molecule, with one carbon atom (12 amu) and four hydrogen atoms (1 amu each). If we wish to have a molecular ratio of 2:1, steam-to-methane, this makes a formula weight ratio of 36:16, or 9:4.

[7]It is quite common for industrial control systems to operate at ratios a little bit "skewed" from what is stoichiometrically ideal due to imperfect reaction efficiencies. Given the fact that no chemical reaction ever goes to 100% completion, a decision must be made as to which form of incompleteness is worse. In a steam-hydrocarbon reforming system, we must ask ourselves which is worse: excess (unreacted) steam at the outlet, or excess (unreacted) hydrocarbon at the outlet. Excess hydrocarbon content will "coke" the catalyst and heater tubes, which is very bad for the process over time. Excess steam merely results in a bit more operating energy loss, with no degradation to equipment life. The choice, then, is clear: it is better to operate this process "hydrocarbon-lean" (more steam than ideal) than "hydrocarbon-rich" (less steam than ideal).

We could add another layer of sophistication to this ratio control system by installing a gas analyzer at the outlet of the reaction furnace designed to measure the composition of the product stream. This analyzer's signal could be used to adjust the value of k so the ratio of steam to methane would automatically vary to ensure optimum production quality even if the feedstock composition (i.e. percentage concentration of methane in the hydrocarbon gas input) changes:

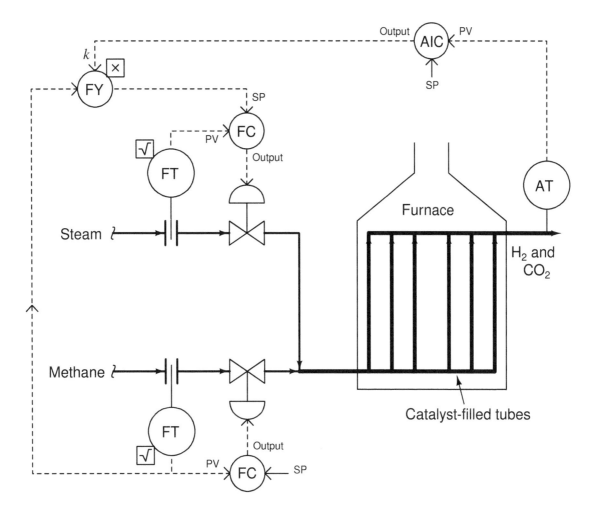

As we saw before, pure methane feed requires a 9-to-4 steam-to-methane mass flow ratio for the desired reaction to be stoichiometrically balanced. This mass ratio, however, is not balanced for hydrocarbons other than methane. Ethane (C_2H_6) processed in the same way requires a 12-to-5 steam-to-ethane mass flow ratio. Propane (C_3H_8) requires a 26-to-11 steam-to-propane mass flow ratio. If the hydrocarbon feed to the reforming furnace varies in composition, the steam flow ratio (k) must change accordingly for efficient reaction.

31.4 Relation control

A control strategy similar to ratio control is *relation* control. This is similar to ratio control in that a "wild" variable determines the setpoint for a captive variable, but with relation control the mathematical relationship between the wild and captive variables is one of addition (or subtraction) rather than multiplication (or division). In other words, a relation control system works to maintain a specific *difference* between wild and captive flow values, whereas a ratio control system works to maintain a specific *ratio* between wild and captive flow values.

An example of relation control appears here, where a temperature controller for a steam superheater on a boiler receives its setpoint from the biased output of a temperature transmitter sensing the temperature of saturated steam (that is, steam exactly at the boiling point of water) in the steam drum:

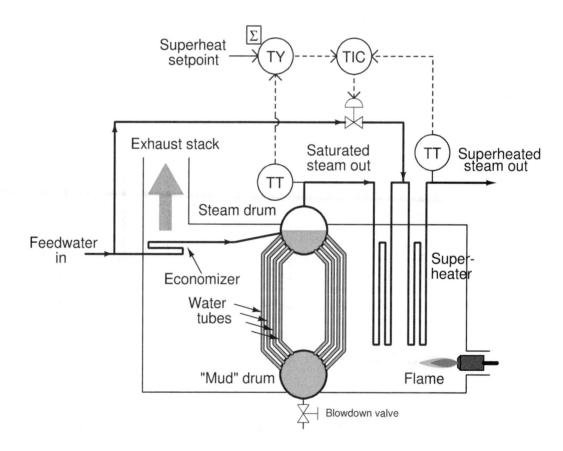

It is a basic principle of thermodynamics that the vapor emitted at the surface of a boiling liquid will be at the same temperature as that liquid. Furthermore, any heat lost from that vapor will cause at least some of that vapor to condense back into liquid. In order to ensure the vapor is "dry" (i.e. it may lose substantial heat energy without condensing), the vapor must be heated beyond the liquid's boiling point at some later stage in the process.

Steam within the steam drum of a boiler is *saturated* steam: at the same temperature as the boiling water. Any heat lost from saturated steam causes at least some of it to immediately condense back into water. In order to ensure "dry" steam output from the boiler, the saturated steam taken from the steam drum must be further heated through a set of tubes called a *superheater*. The resulting "dry" steam is said to be *superheated*, and the difference between its temperature and the temperature of the boiling water (saturated steam) is called *superheat*.

This control system maintains a set amount of superheat by measuring the saturated steam's temperature (within the steam drum), adding a "superheat setpoint" bias value to that signal, then passing the biased signal to the temperature indicating controller (TIC) where the superheated steam temperature is regulated by adding water[8] to the superheated steam. With this system in place, the boiler operator may freely define how much superheat is desired, and the controller attempts to maintain the superheated steam at that much higher temperature than the saturated steam in the drum, over a wide range of saturated steam temperatures.

A ratio control system would not be appropriate here, since what we desire in this process is a controlled *offset* (rather than a controlled *ratio*) between two steam temperatures. The control strategy looks very much like a ratio control, except for the substitution of a summing function instead of a multiplying function.

---

[8]This mixing of superheated steam and cold water happens in a specially-designed device called a *desuperheater*. The basic concept is that the water will absorb heat from the superheated steam, turning that injected water completely into steam and also reducing the temperature of the superheated steam. The result is a greater volume of steam than before, at a reduced temperature. So long as some amount of superheat remains, the de-superheated steam will still be "dry" (above its condensing temperature). The desuperheater control merely adds the appropriate amount of water until it achieves the desired superheat value.

31.5   Feedforward control

"Feedforward" is a rather under-used control strategy capable of managing a great many types of process problems. It is based on the principle of *preemptive load counter-action:* that if all significant loads on a process variable are monitored, and their effects on that process variable are well-understood, a control system programmed to take appropriate action based on load changes will shield the process variable from any ill effect. That is to say, the feedforward control system uses data from load sensors to predict when an upset is about to occur, then *feeds that information forward to the final control element* to counteract the load change before it has an opportunity to affect the process variable. Feedback control systems are *reactive*, taking action after to changes in the process variable occur. Feedforward control systems are *proactive*, taking action before changes to the process variable can occur.

This photograph shows a kind of feedforward strategy employed by human operators running a *retort*: a steam-powered machine used to pressure-treat wooden beams at a milled lumber operation. The sign taped to this control panel reminds the operator to warn the maintenance department of an impending steam usage:

The story behind this sign is that a sudden demand in retort steam causes the entire facility's steam supply pressure to sag if it happens at a time when the boiler is idling. Since the boiler's pressure control system can only react to deviations in steam pressure from setpoint, the boiler pressure controller will not take any action to compensate for sudden demand until *after* it sees the steam pressure fall, at which point it may be too late to fully recover. If operators give the maintenance personnel advance notice of the steam demand, though, the boiler may be fired up for extra steam capacity and thus will be prepared for the extra demand when it comes. The upset avoided here is abnormally low steam header pressure, with the predictive load being the retort operator's planned usage of steam. Crude as this solution might be, it illustrates the fundamental concept of feedforward control: information about a load change is "fed forward" to the final control element to preemptively stabilize the process variable.

As the following section explains, perfect feedforward control action is nearly impossible to achieve. However, even imperfect feedforward action is often far better than none at all, and so this control strategy is quite valuable in process control applications challenged by frequent and/or large variations in load.

31.5.1 Load Compensation

Feedback control works on the principle of information from the outlet of a process being "fed back" to the input of that process for corrective action. A block diagram of feedback control looks like a loop:

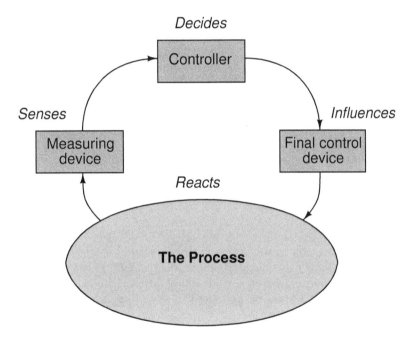

The reason any control system is necessary at all[9] to maintain a process variable at some stable value is the existence of something called a *load*. A "load" is a variable influencing a process that is not itself under direct control, and may be represented in the block diagram as an arrow entering the process, but not within the control loop:

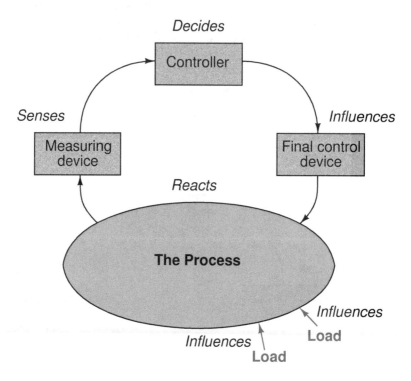

For example, consider the problem of controlling the speed of an automobile. In this scenario, vehicle speed is the process variable being measured and controlled, while the final control device is the accelerator pedal controlling engine power output. If it were not for the existence of hills and valleys, head-winds and tail-winds, air temperature changes, road surface variations, and a host of other "load" variables affecting car speed, maintaining a constant speed would be as simple as holding the accelerator pedal at a constant position.

However, the presence of these "load" variables makes necessitates a human driver (or a *cruise control* system) continually adjusting engine power to maintain constant speed. Using the car's measured speed as feedback, the driver (or cruise control) adjusts the accelerator pedal position as necessary based on whether or not the car's speed matches the desired "setpoint" value.

An inherent weakness of any feedback control system is that it can never be *proactive*. The best any feedback control system can ever do is *react* to detected disturbances in the process variable. This makes deviations from setpoint inevitable, even if only for short periods of time. In the context of our automobile cruise control system, this means the car can never maintain a *perfectly* constant

---

[9]This statement is true only for self-regulating processes. Integrating and "runaway" processes require control systems to achieve stability even in the complete absence of any loads. However, since self-regulation typifies the vast majority of industrial processes, we may conclude that the fundamental purpose of most control systems is to counteract the effects of loads.

speed in the face of loads because the control system does not have the ability to anticipate loads (e.g. hills, wind gusts, changes in air temperature, changes in road surface, etc.). At best, all the feedback cruise control system can do is react to changes in speed it senses *after* some load has disturbed it.

Feedforward control addresses this weakness by taking a fundamentally different approach, basing final control decisions on the states of load variables rather than the process variable. In other words, a feedforward control system monitors the factor(s) influencing a process and decides how to compensate *ahead of time* before the process variable deviates from setpoint. If all loads are accurately measured, and the control algorithm realistic enough to predict process response for these known load values, the process variable (ideally) need not be measured at all:

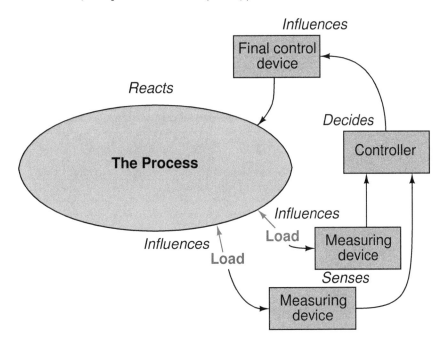

As was the case with cascade control, feedforward control also has an analogue in workplace management. If you consider a supervisor to be the "controller" of a work group (issuing orders to his or her subordinates to accomplish important tasks), a feedforward system would be when someone informs the supervisor of an important change that will soon impact the work group. By having this information "fed forward" to the supervisor, the supervisor may then take *preemptive* measures to better manage this change before its effects are fully felt. If this predictive information is accurate, and the supervisor's response appropriate, any negative impacts of the change will be minimized to the point where no reactive steps will be needed. Stated differently, good feedforward control action translates what would otherwise be a crisis into an insignificant event.

Returning to the cruise control application, a purely feedforward automobile cruise control system would be interfaced with topographical maps, real-time weather monitors, and road surface sensors to decide how much engine power was necessary at any given time to attain the desired speed[10].

---

[10]The load variables I keep mentioning that influence a car's speed constitute an incomplete list at best. Many

Assuming all relevant load variables are accounted for, the cruise control would be able to maintain constant speed regardless of conditions, and without the need to even monitor the car's speed.

This is the promise of feedforward control: a method of controlling a process variable so perfect in its predictive power that it eliminates the need to even measure that process variable. If you are skeptical of this feedforward principle and its ability to control a process variable without even measuring it, this is a good thing – you are thinking critically! In practice, it is nearly impossible to accurately account for *all* loads influencing a process and to both anticipate and counter-act their combined effects, and so *pure* feedforward control systems are rare[11]. Instead, the feedforward principle finds use as a supplement to normal feedback control. To understand feedforward control better, however, we will consider its pure application before exploring how it may be combined with feedback control.

---

other variables come into play, such as fuel quality, engine tuning, and tire pressure, just to name a few. In order for a purely feedforward (i.e. no feedback monitoring of the process variable) control system to work, *every single load variable* must be accurately monitored and factored into the system's output signal. This is impractical or impossible for a great many applications, which is why we usually find feedforward control used in conjunction with feedback control, rather than feedforward control used alone.

[11]In fact, the only pure feedforward control strategies I have ever seen have been in cases where the process variable was nearly impossible to measure and could only be inferred from other variables.

First, let us consider a liquid level control system on an open tank, where three different fluid ingredients (shown in the following P&ID simply as A, B, and C) are mixed to produce a final product. A level transmitter (LT) measures liquid level, while a level controller (LC) compares this level to a setpoint value, and outputs a signal calling for a certain amount of discharge flow. A cascaded (slave) flow controller (FC) senses outgoing flow via a flow transmitter (FT) and works to maintain whatever rate of flow is "asked" for by the level controller:

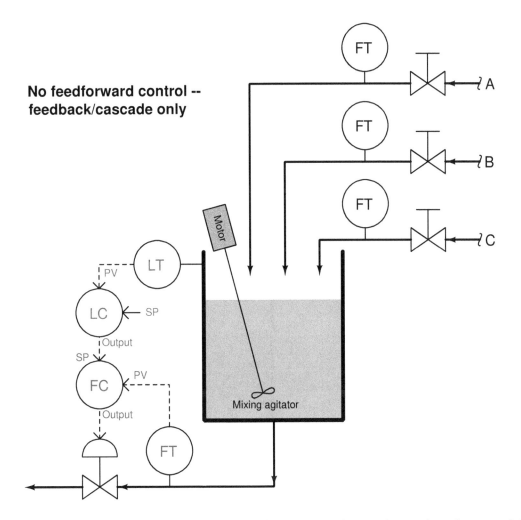

The level control system acts to keep liquid level constant in the vessel, ensuring adequate mixing of the three ingredients[12]. Being a feedback level control system, it adjusts the discharge flow rate in response to measured changes in liquid level. Like all feedback control systems, this one is *reactive* in nature: it can only take corrective action *after* a deviation between process variable (level) and setpoint is detected. As a result, temporary deviations from setpoint are guaranteed to occur with

---

[12]If the liquid level drops too low, there will be insufficient *retention time* in the vessel for the fluids to mix before they exit the product line at the bottom.

this control system every time the combined flow rate of the three ingredients increases or decreases.

Let us now change the control system strategy from feedback to feedforward. It is clear what the loads are in this process: the three ingredient flows entering the vessel. If we measure and sum these three flow rates[13], then use the total incoming flow signal as a setpoint for the discharge flow controller, the outlet flow should (ideally) match the inlet flow, resulting in a constant liquid level. Being a purely feedforward control system, there is no level transmitter (LT) any more, just flow transmitters measuring the three loads:

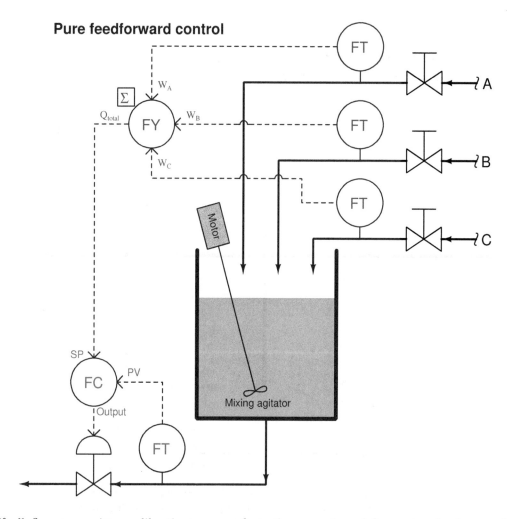

If all flow transmitter calibrations are perfect, the summing of flow rates flawless, and the flow controller's tuning robust, this level control system should control liquid level in the vessel by proactive effort ("thinking ahead") rather than reactive effort ("after the fact"). Any change

---

[13]The device or computer function performing the summation is shown in the P&ID as a bubble with "FY" as the label. The letter "F" denotes *Flow*, while the letter "Y" denotes a signal relay or transducer.

in the flow rate of ingredients A, B, and/or C is quickly matched by an equal adjustment to the discharge flow rate. So long as total volumetric flow out of the vessel is held equal to total volumetric flow into the vessel, the liquid level inside the vessel *cannot* change[14].

If this feedforward strategy reminds you of ratio control, you are thinking correctly: the ingredient flow sum signal is the *wild variable*, and the discharge flow signal is the *captive variable*. The flow controller simply maintains the discharge flow rate at a 1:1 ratio with the (total) ingredient flow rate. In fact, pure feedforward control is a variation of 1:1 ratio control, except that the real process variable (tank level) is neither the wild (total incoming flow) nor the captive variable (discharge flow) in the process.

An interesting property of feedforward and ratio control systems alike is that they cannot generate oscillations as is the case with an over-tuned (excessive gain) feedback system. Since a feedforward system does not monitor the effects of its actions, it cannot react to something it did to the process, which is the root cause of feedback oscillation. While it is entirely possible for a feedforward control system to be configured with too much gain, the effect of this will be *overcompensation* for a load change rather than oscillation. In the case of the mixing tank feedforward level control process, improper instrument scaling and/or offsets will merely cause the discharge and inlet flows to mismatch, resulting in a liquid level that either continues to increase or decrease over time ("integrate"). However, no amount of mis-adjustment can cause this feedforward system to produce *oscillations* in the liquid level.

In reality, this pure feedforward control system is impractical even if all instrument calibrations and control calculations are perfect. There are still loads unaccounted for: evaporation of liquid from the vessel, for example, or the occasional pipe fitting leak. Furthermore, since the control system has no "knowledge" of the actual liquid level, it cannot make adjustments to that level. If an operator, for instance, desired to decrease the liquid level in order to reduce the residence time (also known as "retention time")[15], he or she would have to manually drain liquid out of the vessel, or temporarily place the discharge flow controller in "manual" mode and increase the flow there (then place back into "cascade" mode where it follows the remote setpoint signal again). The advantage of proactive control and minimum deviation from setpoint over time comes at a fairly high price of impracticality and inconvenience.

---

[14]Incidentally, this is a good example of an *integrating* mass-balance process, where the rate of process variable change over time is proportional to the imbalance of flow rates in and out of the process. Stated another way, total accumulated (or lost) mass in a mass-balance system such as this is the time-integral of the difference between incoming and outgoing mass flow rates: $\Delta m = \int_0^T (W_{in} - W_{out})\, dt$.

[15]*Residence time* or *Retention time* is the average amount of time each liquid molecule spends inside the vessel. It is an important variable in chemical reaction processes, where adequate time must be given to the reactant molecules in order to ensure a complete reaction. It is also important for non-reactive mixing processes such as paint and food manufacturing, to ensure the ingredients are thoroughly mixed together and not stratified. For any given flow rate through a vessel, the residence time is directly proportional to the volume of liquid contained in that vessel: double the captive volume, and you double the residence time. For any given captive volume, the residence time is inversely proportional to the flow rate through the vessel: double the flow rate through the vessel, and you halve the residence time. In some mixing systems where residence time is critical to the thorough mixing of liquids, vessel level control may be coupled to measured flow rate, such that an increase in flow rate results in an increased level setpoint, thus maintaining a constant residence time despite changes in production rate.

For these reasons, feedforward control is most often found in conjunction with feedback control. To show how this would work in the liquid level control system, we will incorporate a level transmitter and level controller back into the system, the output of that level controller being summed with the feedforward flow signal (by the LY summing relay) before going to the cascaded setpoint input of the discharge flow controller:

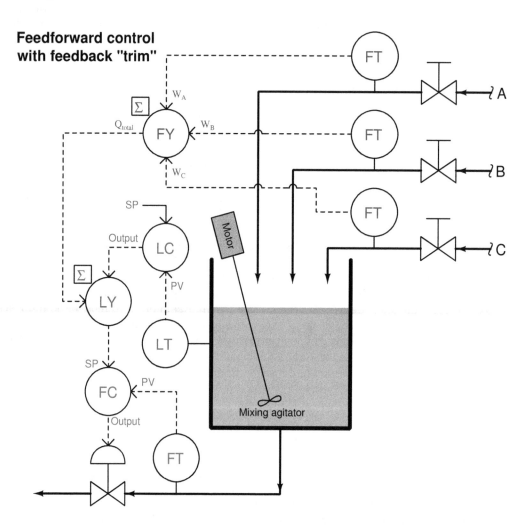

This hybrid control strategy is sometimes called *feedforward with trim*. In this context, "trim" refers to the level controller's (LC) output signal contributing to the discharge flow setpoint, helping to compensate for any unaccounted loads (evaporation, leaks) and provide for level setpoint changes. This "trim" signal should do very little of the control work in this system, the bulk of the liquid level stability coming from the feedforward signals provided by the incoming flow transmitters.

A very similar control strategy commonly used on large steam boilers for the precise control of steam drum water level goes by the name of *three-element feedwater control*. The following illustration shows an example of this control strategy implemented with pneumatic (3-15 PSI signal) instruments:

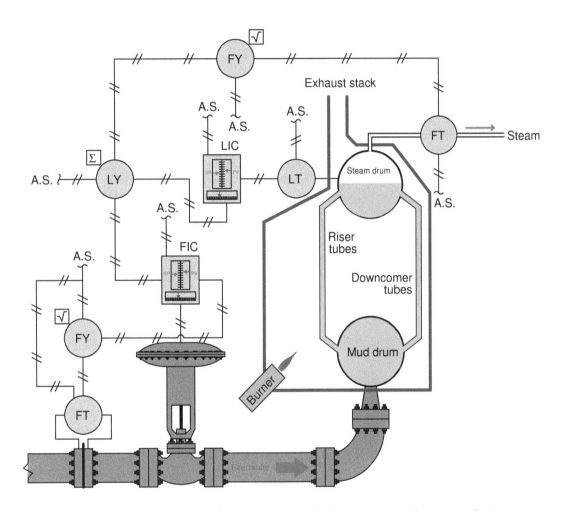

Such a control system is called "three-element" because it makes use of three process measurements:

- Feedwater flow rate

- Steam drum water level

- Steam flow rate

Feedwater flow is controlled by a dedicated flow controller (FIC), receiving a remote setpoint signal from a summing relay (LY). The summer receives two inputs: a steam flow signal and the output signal (trim) from the level controller (LIC). The feedforward portion of this system (steam flow feeding forward to water flow) is intended to match the mass flow rates of water into the boiler with steam flow out of the boiler. If steam demand suddenly increases, this feedforward portion of the system immediately calls for a matching increase in water flow into the boiler, since every molecule of steam exiting the boiler must come from one molecule of water entering the boiler. The level controller and transmitter act as a feedback control loop, supplementing the feedforward signal to the cascaded water flow controller to make up for ("trim") any shortcomings of the feedforward loop.

A three-element boiler feedwater control system is a good example of a feedforward strategy designed to ensure *mass balance*, defined as a state of equality between all incoming mass flow rates and all outgoing mass flow rates. The steam flow transmitter measures outgoing mass flow, its signal being used to adjust incoming water mass rate. Since mass cannot be created or destroyed (the Law of Mass Conservation), every unit of steam mass leaving the boiler must be accounted for as an equivalent unit of water mass entering the boiler. If the control system perfectly balances these mass flow rates, water level inside the boiler *cannot* change.

In processes where the process variable is affected by energy flow rates rather than mass, the balance maintained by a feedforward control system will be *energy balance* rather than mass balance. Like mass, energy cannot be created or destroyed (the Law of Energy Conservation), but must be accounted for. A feedforward control system monitoring all incoming energy flows into a process and adjusting the outgoing energy flow rate (or vice-versa) will ensure no energy is depleted from or accumulated within the process, thus ensuring the stability of the processes' internal energy state.

An example of energy-balance feedforward control appears in this heat exchanger temperature control system:

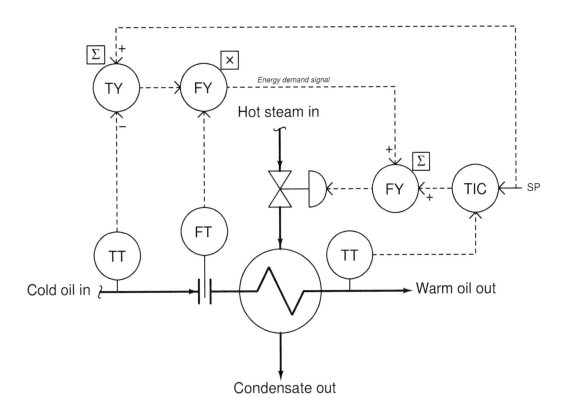

The two transmitters on the incoming (cold oil) line measure oil temperature and oil flow rate, respectively. The first "summing" function subtracts the incoming oil temperature from the setpoint (desired) temperature, and then the difference of these two temperatures is then multiplied by the flow rate signal to produce a signal representing the *energy demand*[16] of the incoming oil (i.e. how much energy will be required to elevate the oil flow's temperature to setpoint). The "energy demand" signal is summed with the temperature controller's output signal to set the steam valve position (adding energy to the process).

There do exist other loads in this process, such as ambient air temperature and chemical composition of the oil, but these variables are generally less influential on discharge temperature than feed temperature and flow rate. This illustrates a practical facet of feedforward control: although there may be a great many loads affecting our process variable, we must generally limit our application of feedforward to only the most dominant loads in order to limit control system cost. Simply put, we usually cannot justify the expense and complexity of a feedforward control system compensating for *every single load* in a system.

---

[16]Energy demand is an example of what is called an *inferred variable*: a physical quantity that we cannot measure directly but instead calculate from measurements made of other variables.

31.5.2 Proportioning feedforward action

Feedforward control works by directly modulating the manipulated variable in a control system according to changes sensed in the load(s). In order for feedforward to function optimally, it must adjust the manipulated variable in a manner that is proportionate to the need: no more, and no less. At this juncture it is appropriate to ask the question, "how do we know the amount of feedforward action that will be adequate for a process, and how do we adjust it if it is too much or too little?"

In processes where the feedforward control strategy attempts to achieve direct mass- or energy-balance, the question of adequate feedforward action is answered in the mathematics of measuring the incoming and outgoing flows. Consider the following mass-balance level control system where the combined sum of three inlet flows is routed to the setpoint of the exit flow control loop. In this diagram, the portions of the control strategy implemented as function blocks (algorithms in software) appear inside a yellow-colored bounded area, while all real physical instruments appear outside the yellow area:

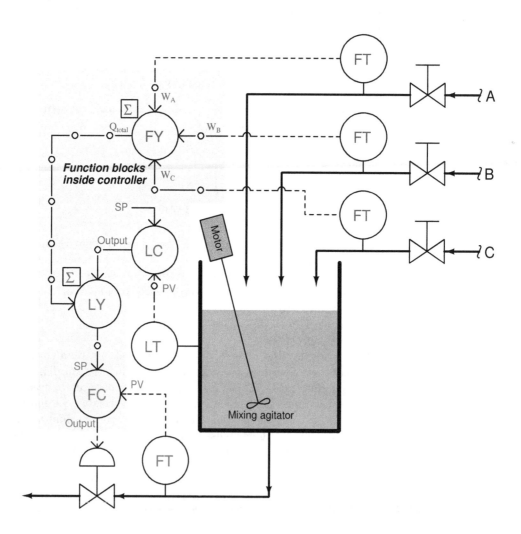

If all flowmeters are calibrated in pounds per minute, then the feedforward signal will likewise be scaled in pounds per minute, and so will the setpoint be for flow control loop. In a digital control system, it is quite customary to scale each and every analog input signal with some real "engineering unit" of measurement, so that the signal will be treated as a physical quantity throughout as opposed to being treated as some anonymous percentage value. Not only is this consistent scaling a standard feature in digital control systems, but it also helps the implementation of this feedforward control strategy, because we desire the out-going mass flow rate to precisely match the (total) in-coming flow rate. So long as all flowmeters and their associated scaling factors are accurate, the feedforward control's action *must* be exactly right: an increase of +5 pounds per minute in incoming flow rate will prompt an immediate increase of +5 pounds per minute in outgoing flow rate, simply by virtue of all these measured flows having been scaled in the same unit of measurement.

The situation is not as simple in systems where the feedforward control is not precisely balancing mass-flow or energy rates. By contrast, let us examine the following pH neutralization system equipped with feedforward control action. Here, the incoming liquid is alkaline (pH greater than 7), and the control system's job is to mix just enough acid reagent to "neutralize" the solution (bring the pH value down to 7):

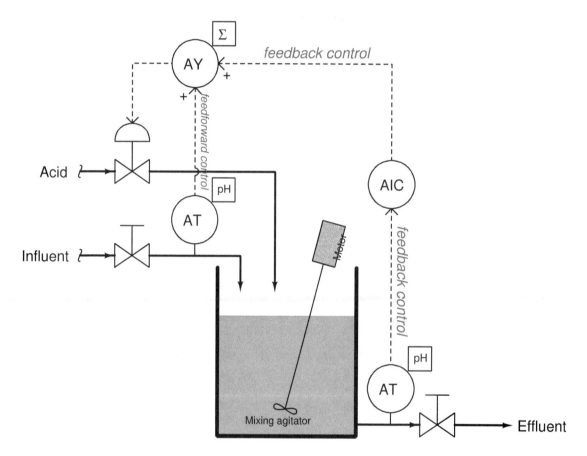

Controlling the pH (acidity/alkalinity) of a liquid solution is challenging for many reasons, not the least of which being the need to have adequate mixing time for the reagent to react with the influent. This mixing time translates to *dead time* in the feedback control system. If the influent's pH suddenly changes for any reason, the feedback control system will be slow to alter the reagent flow rate due to this dead time, causing long-lasting deviations from setpoint. The goal of the feedforward signal (from the influent pH transmitter to the summer) is to preemptively adjust reagent flow rate according to how alkaline the incoming flow is, countering any sudden changes in influent pH so the feedback control system doesn't have to take (delayed) action.

Once again, it is appropriate to ask the question, "how do we know the amount of feedforward action that will be adequate for this process, and how do we adjust it if it is too much or too little?" It would be blind luck if the system happened to work perfectly as shown, with the influent pH transmitter's signal going straight to the summing function to be added to the pH controller's output signal. Certainly, an increase in influent pH would cause more acid to be added to the mix thanks to feedforward action, but it would likely add either too much or too little acid than it should. The scale of the influent pH transmitter does not match the scale of the signal sent to the control valve, and so we do not have a neat "pound-for-pound" balance of mass flow as we did in the case of the level control system.

A neat solution to this problem is to add another function block to the feedforward portion of the control system. This block takes the influent pH transmitter signal and skews it using multiplication and addition, using the familiar linear equation $y = mx + b$ (where y is the output signal of the function and x is the input signal; m and b being constants). This function block is typically called a *gain and bias* block:

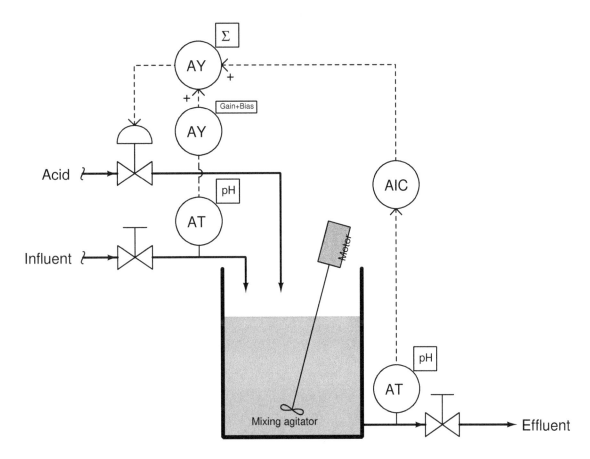

The gain adjustment (m) in this function block serves to amplify or attenuate the feedforward signal's magnitude, while the bias adjustment (b) offsets it.

Determining practical values for these "feedforward tuning constants" is relatively easy. First, place the feedback controller into manual mode[17] with the output value fixed at or near 50%. Next, introduce load changes to the process while watching the process variable's value after sufficient time has elapsed to see the effects of those load changes. Increase or decrease the gain value until step-changes in load cease to yield significant changes in the process variable. The following trends show what too much and too little feedforward gain would look like in this pH control system:

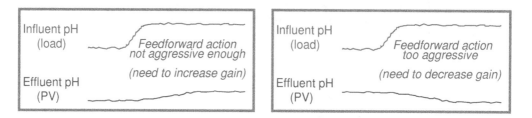

If the gain is properly set in the gain/bias function block, these load changes should have minimal effect on the process variable. Step-changes in influent pH should have little effect on the process variable after sufficient time has passed for the load change to have fully propagated through the process.

Once a good gain value has been found, change the bias value until the process variable approaches the normal setpoint value[18].

---

[17]Most control systems' feedforward function blocks are designed in such a way that both the feedback and the feedforward signal paths are disabled when the controller is placed into manual mode, in order to give the human operator 100% control over the final element (valve) in that mode. For the purpose of "tuning" the feedforward gain/bias function block, one must disable the feedback control *only* so that only pure feedforward response is seen. If simply switching the feedback controller to manual mode is not an option (which it usually is not), one may achieve the equivalent result by setting the gain value of the feedback controller to zero.

[18]This is why it was recommended to leave the feedback controller's output at or near 50%. The goal is to have the feedforward action adjusted such that the feedback controller's output is "neutral," and has room to swing either direction if needed to provide necessary trim to the process.

31.6 Feedforward with dynamic compensation

As we have seen, feedforward control is a way to improve the stability of a feedback control system in the face of changing loads. Rather than rely on feedback to make corrective changes to a process only *after* some load change has driven the process variable away from setpoint, feedforward systems monitor the relevant load(s) and use that information to preemptively make stabilizing changes to the final control element such that the process variable will not be affected. In this way, the feedback loop's role is to merely "trim" the process for factors lying outside the realm of the feedforward system.

At least, this is how feedforward control is *supposed* to work. One way feedforward controls commonly fail to live up to their promise is if the effects of load changes and of manipulated variable changes possess different time lags in their respective effects on the process variable. This is a problem in feedforward control systems because it means the corrective action called for in response to a change in load will not affect the process variable at the same time, or in the same way over time, as the load will. In order to correct this problem, we must intelligently insert time lags (or advancing time-based functions called *leads*) into the control system to equalize the time lags of load and feedforward correction. This is called *dynamic compensation.*

The following subsections will explore illustrative examples to make both the problem and the solution(s) clear.

A common area of confusion among students first approaching this topic is deciding where to place the dynamic compensation function in a feedforward control system. The answer to this question is surprisingly simple, although it may seem elusive at first glance. The key is found in the following principle: *the only time-dynamic we have the ability to alter with our control system is the dynamic of the final control element.* We cannot alter the time-dynamic of the load's effect on the process variable, as that is strictly a function of process physics. Therefore, when we test a process employing feedforward control with an eye toward incorporating dynamic compensation, we must measure the time lag of the load's effect on PV and also the time lag of our final control element's effect on PV, then compare those two time lags. If the final control element's time lag is shorter (quicker) than the load's, then we must add a delay or lag to the feedforward signal so that the final control element's preemptive action does not occur too soon. If the final control element's time lag is longer (slower) than the load's, then we must either find a way to alter the process itself to decrease the load's time lag, or add a "lead" function to the feedforward signal in order to advance the final control element's response and thereby ensure the preemptive action does not occur too late. *Remember, all we can do with dynamic compensation is alter how the final control element responds (i.e. how slowly or quickly the preemptive action of feedforward occurs). The load's effect on the process variable is fixed by the physics of the process and therefore lies beyond our direct control.*

31.6.1   Dead time compensation

Examine the following control system P&ID showing the addition of *flocculant* (a chemical compound used in water treatment to help suspended solids clump together for easier removal by filtering and/or gravity clarification) and *lime* for pH balance. Flocculant is necessary to expedite the removal of impurities from the water, but some flocculation compounds have the unfortunate effect of decreasing the pH value of the water (turning it more acidic). If the water's pH value is too low, the flocculant ironically loses its ability to function. Thus, lime (an alkaline substance – high pH value) must be added to the water to counter-act the flocculant's effect on pH to ensure efficient flocculation. Both substances are powders in this water pre-treatment system, metered by variable-speed screw conveyors and carried to the mixing tank by belt-style conveyors:

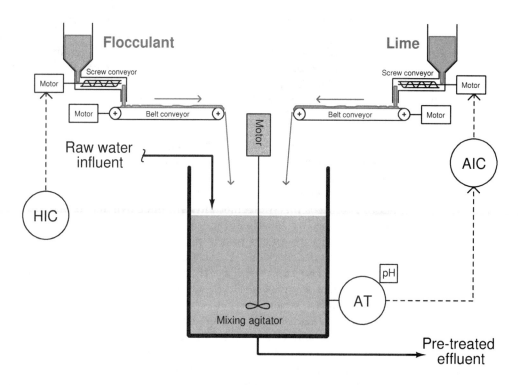

The control system shown in this P&ID consists of a pH analyzer (AT) transmitting a signal to a pH indicating controller (AIC), adjusting the speed of the lime screw conveyor. The flocculant screw conveyor speed is manually set by a *hand indicating controller* (HIC) – sometimes known as a *manual loading station* – adjusted when necessary by experienced water treatment operators who periodically monitor the effectiveness of flocculation in the system.

This simple feedback control system will work fine in steady-state conditions, but if the operator suddenly changes flocculant flow rate into the mixing vessel, there will be a temporary deviation of pH from setpoint before the pH controller is able to find the correct lime flow rate into the vessel to compensate for the change in flocculant flow. In other words, flocculant feed rate into the mixing tank is a *load* for which the pH control loop must compensate.

Dynamic response could be greatly improved with the addition of feedforward control to this system:

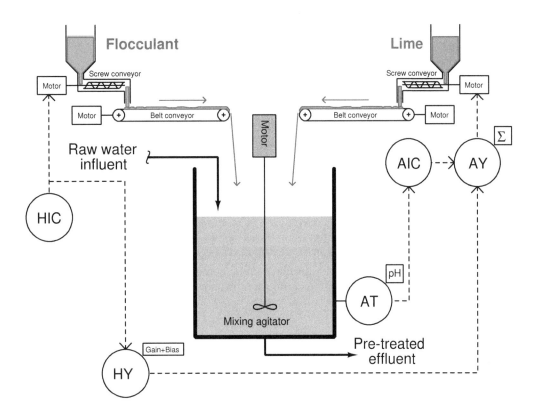

Here, the hand controller's signal gets added to the pH controller's output signal to directly influence lime feed rate in addition to acting as a control signal to the flocculant screw conveyor motor drive. If an operator changes the flocculant feed rate, the lime feed rate will immediately adjust to compensate, *before* any change in pH value takes place in the water. Ideally, the pH controller need only make minor "trim" adjustments to lime feed rate, while the feedforward signal does most of the work in maintaining a steady pH value. The proper proportioning and offset between flocculant and lime feed rates is established in the gain/bias function, which is "tuned"[19] to ensure the feedforward signal does not over- or under-react, calling for too much (or too little) lime to compensate.

Even if all components in the feedforward system have been calibrated and configured properly, however, a potential problem still lurks in this system which can cause the pH value to temporarily deviate from setpoint following flocculant feed rate changes. This problem is the *transport delay* – otherwise known as *dead time* – inherent to the two belt conveyors transporting both flocculant and lime powder from their respective screw conveyors to the mixing vessel. If the rotational speed of a

---

[19] Tuning this gain/bias block is done with the pH controller in manual mode with its output at 50%. The gain value is adjusted such that step-changes in flocculant feed rate have little long-term effect on pH. The bias value is adjusted until the pH approaches setpoint (even with the pH controller in manual mode).

screw conveyor changes, the flow rate of powder exiting that screw conveyor will immediately and proportionately change. However, the belt conveyor imposes a time delay before the new powder feed rate enters the mixing vessel. In other words, the water in the vessel will not "see" the effects of a change in flocculant or lime feed rate until after the *belt conveyor's* time delay has elapsed. This is not a problem if the dead times of both belt conveyors are exactly equal, since this means any compensatory change in lime feed rate initiated by the feedforward system will reach the water at exactly the same time the new flocculant rate reaches the water. So long as flocculant and lime feed rates are precisely balanced with one another at the point in time they reach the mixing vessel, pH should remain stable. But what if their arrival times are not coordinated – what will happen to pH then?

Let us engage in a "thought experiment" to explore the consequences of the flocculant conveyor belt moving much slower than, and/or being much longer than, the lime conveyor belt. Suppose the flocculant belt imposed a dead time of 60 seconds on flocculant powder making it to the vessel, while the lime belt only delayed lime powder transit by 5 seconds from screw conveyor to mixing tank. This would mean changes in flocculant flow (set by the hand controller) would compensate with changes in lime flow *55 seconds too soon*. Now imagine the human operator making a sudden increase to the flocculant powder feed rate. The lime feed rate would immediately increase thanks to the efforts of the feedforward system. However, since the increased flow rate of lime powder will reach the mixing vessel 55 seconds before the increased flow rate of flocculant powder, the effect will be a temporary increase in pH value beginning about 5 seconds after the operator's change, and then a settling of pH value back to setpoint[20], as shown in this timing diagram:

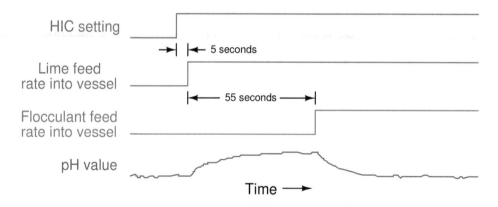

The obvious solution to this problem is to mechanically alter the belt conveyor systems for equal transport times of flocculant and lime powders. If this is impractical, we may achieve a similar result by incorporating another signal relay (or digital function block) inserting dead time into the feedforward control system. In other words, we can modify the control system in such a way to emulate what would be impractical to modify in the process itself.

---

[20]This "thought experiment" assumes no compensating action on the part of the feedback pH controller for the sake of simplicity. However, even if we include the pH controller's efforts, the problem does not go away. As pH rises due to the premature addition of extra lime, the controller will try to reduce the lime feed rate. This will initially reduce the degree to which pH deviates from setpoint, but then the reverse problem will occur when the increased flocculant enters the vessel 55 seconds later. Now, the pH will drop below setpoint, and the feedback controller will have to ramp up lime addition (to the amount it was before the additional lime reached the vessel) to achieve setpoint.

This new function will add a dead time of 55 seconds to the feedforward signal before it enters the summer, thus delaying the lime feed rate's response to feedforward effect by just the right amount of time such that any lime feed rate changes called for by feedforward action will arrive at the vessel *simultaneously* with the changed flocculant feed rate:

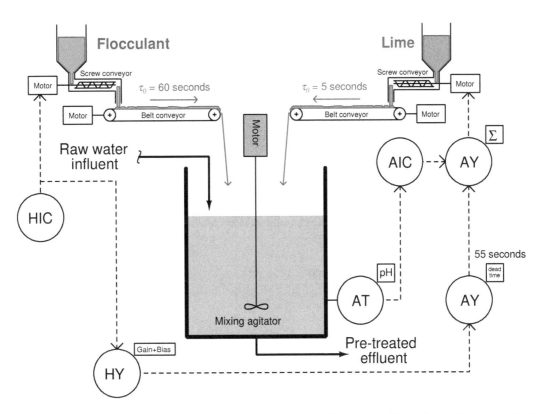

Adding time-based functions to a control system in order to equalize inherently unequal time delays in the physical process is called *dynamic compensation*. It is important to note that dynamic compensation cannot make physically unequal time lags equal – all it does is modify the feedforward system so that signal's effect arrives at the right time to compensate the load. In simple terms, we can only use dynamic compensation to modify the FCE's behavior, not the process behavior. Here, there is absolutely nothing the feedforward system can do to speed up the slower flocculant belt, so instead we chose to slow down the feedforward manipulation of lime flow to make it match the flocculant flow.

Note how the feedback pH controller's loop was purposely spared the effects of the added dead time function, by placing the function outside of that controller's feedback loop. This is important, as dead time in any form is the bane of feedback control. The more dead time within a feedback loop, the easier that loop will tend to oscillate. By strategically placing the dead time function before the summing relay rather than after (between the summer and the lime screw conveyor motor drive), the feedback control system still achieves minimum response time while only the feedforward signal gets delayed.

Let us now consider the same flocculant and lime powder control system, this time with transport delays reversed between the two belt conveyors. If the flocculant conveyor belt is now the fast one (5 seconds dead time) and the lime belt slow (60 seconds), the effects of flocculant feed rate changes will be reversed. An increase in flocculant powder feed rate to the vessel will result in a drop in pH beginning 5 seconds after the HIC setting change, followed by a rise in pH value after the additional lime feed rate finally reaches the vessel:

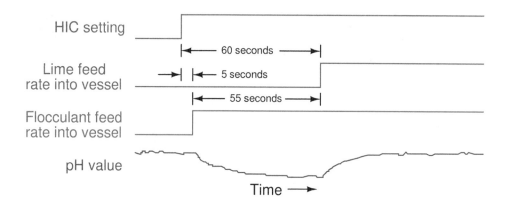

What is happening here is that the feedforward signal's effect is coming too late. In a perfect world our lime flow rate into the mixing vessel should change at the same time that the flocculant flow rate changes. Instead, our lime flow rate's necessary increase happens 55 seconds too late to prevent the pH from deviating.

It would be possible to compensate for the difference in conveyor belt transport times using a special relay in the same location of the feedforward signal path as before, if only there was such a thing as a relay that could *predict the future exactly 55 seconds in advance!*[21]. Since no such device exists (or ever will exist), we must apply dynamic compensation elsewhere in the feedforward control system.

If a time delay is the only type of compensation function at our disposal, then the only thing we can delay in this system to make the two dead times equal is the flocculation feed rate. Thus, we should place a 55-second dead time relay in the signal path between the hand indicating controller (HIC) and the flocculant screw conveyor motor drive.

---

[21]Let me know if you are ever able to invent such a thing. I'll even pay your transportation costs to Stockholm, Sweden so you can collect your Nobel prize. Of course, I will demand to see the prize before buying tickets for your travel, but with your time-travel device that should not be a problem for you.

This diagram shows the proper placement of the dead time function:

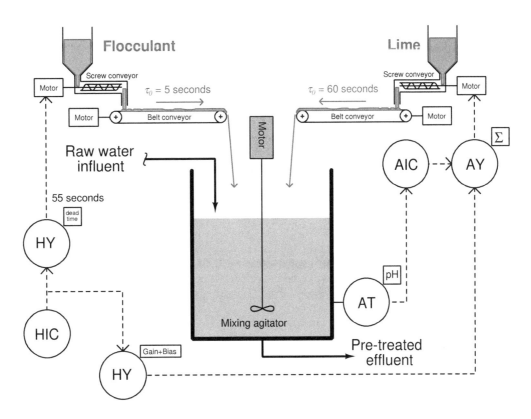

With this dead time relay in place, any change in flocculation feed rate initiated by a human operator will immediately adjust the feed rate of lime powder, but delay an adjustment to flocculant powder feed rate by 55 seconds, so the two powders' feed rate changes arrive at the mixing vessel simultaneously.

Bear in mind that this solution will really only work in a system like this where the major load happens to be controlled by the human operator. In most processes the load variable is not under anyone's control, making it difficult if not impossible to purposely delay its action. If the load has a shorter dead time than the compensation, there is usually little we can do within the control strategy to equalize those effects for better dynamic stability. Typically, the best solution is to *physically alter the process* (e.g. slow down the flocculant belt conveyor's speed) so that the load and compensation dead times are closer to being equal.

31.6.2   Lag time compensation

Process time delays characterized by pure transport delay (dead time) are less common in industry than other forms of time delays, most notably *lag times*[22]. A simple "lag" time is the characteristic exhibited by a low-pass RC filter circuit, where a step-change in input voltage results in an output voltage asymptotically rising to the new voltage value over time:

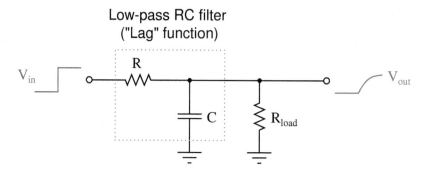

The *time constant* (τ) of such a system – be it an RC circuit or some other physical process – is the time required for the output to move 63.2% of the way to its final value ($1 - e^{-1}$). For an RC circuit such as the one shown, $\tau = RC$ (assuming $R_{load} >> R$ so the load resistance will have negligible effect on timing).

Lag times differ fundamentally from dead times. With a dead time, the effect is simply time-delayed by a finite amount from the cause, like an echo. With a lag time, the effect begins at the exact same time as the cause, but does not follow the same rapid change over time as the cause. Like dead times in a feedforward system, it is quite possible (and in fact usually the case) for loads and final control variables to have differing lag times regarding their respective effects on the process variable. This presents another form of the same problem we saw in the two-conveyor water pretreatment system, where an attempt at feedforward control was not completely successful because the corrective feedforward action did not occur with the same amount of time delay as the load.

---

[22]For a more detailed discussion of lag times and their meaning, see section 30.1.5 beginning on page 2417.

To illustrate, we will analyze a heat exchanger used to pre-heat fuel oil before being sent to a combustion furnace. Hot steam is the heating fluid used to pre-heat the oil in the heat exchanger. As steam gives up its thermal energy to the oil through the walls of the heat exchanger tubes, it undergoes a phase change to liquid form (water), where it exits the shell of the exchanger as "condensate" ready to be re-boiled back into steam.

A simple feedback control system regulates steam flow to the heat exchanger, maintaining the discharge temperature of the oil at a constant setpoint value:

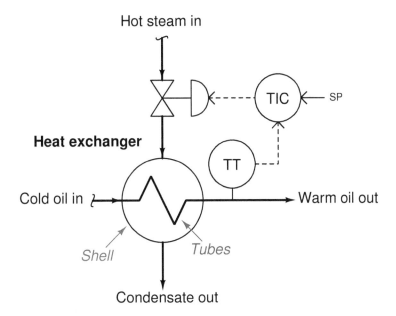

Once again, it should come as no surprise to us that the outlet temperature will suffer temporary deviations from setpoint if load conditions happen to change. The feedback control system may be able to *eventually* bring the exiting oil's temperature back to setpoint, but it cannot begin corrective action until *after* a load has driven the oil temperature off setpoint. What we need for improved control is *feedforward* action in addition to feedback action. This way, the control system can take corrective action in response to load changes *before* the process variable gets affected.

Suppose we know that the dominant load in this system is oil flow rate[23], caused by changes in demand at the combustion furnace where this oil is being used as fuel. Adapting this control system to include feedforward is as simple as installing an oil flow transmitter, a gain/bias function, and a summing function block:

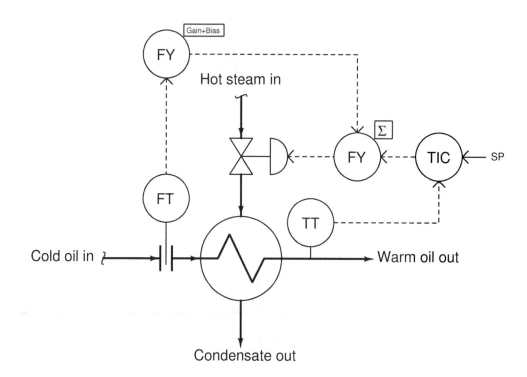

With feedforward control action in place, the steam flow rate will immediately change with oil flow rate, preemptively compensating for the increased or decreased heat demand of the oil. In other words, the feedforward system attempts to maintain *energy balance* in the process, with the goal of stabilizing the outlet temperature:

There is a problem of time delay in this system, however: a change in oil flow rate has a *faster* effect on outlet temperature than a proportional change in steam flow rate. This is due to the relative masses impacting the temperature of each fluid. The oil's temperature is primarily coupled to the temperature of the tubes, whereas the steam's temperature is coupled to both the tubes and the shell of the heat exchanger. So, the steam has a greater mass to heat than the oil has to cool, giving the steam a larger thermal time constant than the oil.

For the sake of illustration, we will assume transport delays are short enough to ignore[24], so we

---

[23]Knowing this allows us to avoid measuring the incoming cold oil temperature and just measure incoming cold oil flow rate as the feedforward variable. If the incoming oil's temperature were known to vary substantially over time, we would be forced to measure it as well as flow rate, combining the two variables together to calculate the *energy demand* and use this inferred variable as the feedforward variable.

[24]Transport delay (dead time) in heat exchanger systems can be a thorny problem to overcome, as they they tend to change with flow rate! For reasons of simplicity in our illustration, we will treat this process as if it only possessed lag times, not dead times.

are only dealing with different *lag* times between the oil flow's effect on temperature and the steam flow's effect on temperature.

This is what would happen to the heated oil temperature if steam flow were held constant and oil flow were suddenly increased:

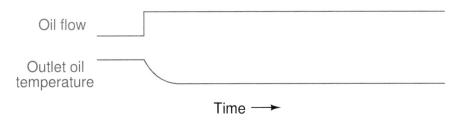

Increased oil flow convects heat away from the steam at a faster rate than before, resulting in decreased oil temperature. This drop in temperature is fairly quick, and is self-regulating.

By contrast, this is what would happen to the heated oil temperature if oil flow were held constant and steam flow were suddenly increased:

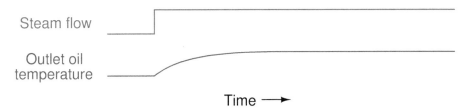

Increased steam flow convects heat into the oil at a faster rate than before, resulting in increased oil temperature. This rise in temperature is also self-regulating, but much slower than the temperature change resulting from a proportional adjustment in oil flow. In other words, the time constant (τ) of the process with regard to steam flow changes is greater than the time constant of the process with regard to oil flow changes ($\tau_{steam} > \tau_{oil}$).

If we superimpose these two effects, as will be the case when the feedforward system is working (without the benefit of feedback "trim" control), what we will see when oil flow suddenly increases is a "fight" between the cooling effect of the increased oil flow and the heating effect of the increased steam flow. However, it will not be a fair fight: the oil flow's effect will temporarily win over the steam's effect because of the oil's faster time constant. Another way of stating this is to say the feedforward action *temporarily under-compensates* for the change in load. The result will be a momentary dip in outlet temperature before the system achieves equilibrium again:

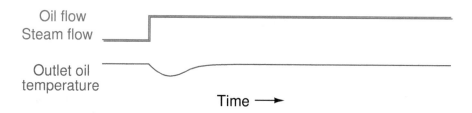

The solution to this problem is not unlike the solution we applied to the water treatment system: we must somehow equalize these two lag times so their superimposed effects will directly cancel, resulting in an undisturbed process variable. An approximate solution for equalizing two different lag times is to cascade two lags together in order to emulate one larger lag time[25]. This may be done by inserting a lag time relay or function block in the feedforward system.

When we look at our P&ID, though, a problem is immediately evident. The lag time we need to slow down is the lag time of the oil flow's effect on temperature. In this system, oil flow is a wild variable, not something we have the ability to control (or delay at will). Our feedforward control system can only manipulate the steam valve position in response to oil flow, not influence oil flow in order to give the steam time to "catch up."

If we cannot slow down the time constant inherent to the wild variable (oil flow), then the best we can do is speed up the time constant of the variable we do have influence over (steam flow). The solution is to insert something called a *lead function* into the feedforward signal driving the steam valve. A "lead" is the mathematical inverse of a lag. If a lag is modeled by an RC low-pass filter circuit, then a "lead" is modeled by an RC high-pass filter circuit:

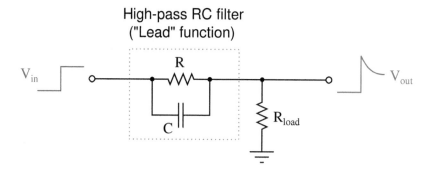

High-pass RC filter
("Lead" function)

---

[25]Technically, two cascaded lag times is not the same as one large lag time, no matter the time constant values. Two first-order lags in series with one another create a *second-order lag*, which is a different effect. However imperfect as the added lag solution is, it is still better than nothing at all!

Being mathematical inverses of each other, a lead function should perfectly cancel a lag function when the output of one is fed to the input of the other, and when the time constants of each are equal. If the time constants of lead and lag are not equal, their cascaded effect will be a partial cancellation. In our heat exchanger control application, this is what we need to do: partially cancel the steam valve's slow time constant so it will be more equal with the oil flow's time constant. Therefore, we need to insert a lead function into the feedforward signal path.

A lead function will take the form of either a physical signal relay or (more likely with modern technology) a function block executed inside a digital control system. The proper place for the lead function is between the oil flow transmitter and the summation function:

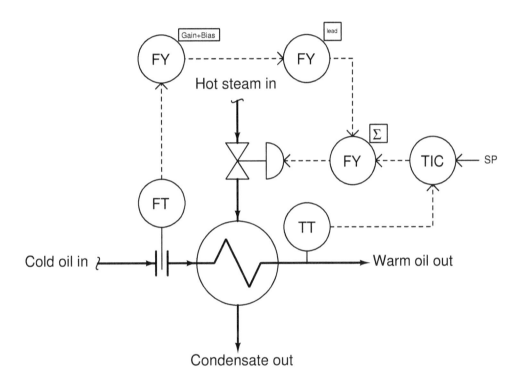

Now, when the oil flow rate to this heat exchanger suddenly increases, the lead function will add a "surge" to the feedforward signal before it goes to the summing function, quickly opening the steam valve further than usual and sending a surge of steam to the exchanger to help overcome the naturally sluggish response of the oil temperature to changes in steam flow. The feedforward action won't be perfect with this lead function added, but it will be substantially better than if there was no dynamic compensation added to the feedforward signal.

31.6.3   Lead/Lag and dead time function blocks

The addition of dynamic compensation in a feedforward control system may require a lag function, a lead function, and/or a dead time function, depending on the nature of the time delay differences between the relevant process load and the system's corrective action. Modern control systems provide all these functions as digital *function blocks*. In the past, these functions could only be implemented in the form of individual instruments with these time characteristics, called *relays*. As we have already seen, lead and lag functions may be rather easily implemented as simple RC filter circuits. Pneumatic equivalents also exist, which were the only practical solution in the days of pneumatic transmitters and controllers. Dead time is notoriously difficult to emulate using analog components of any kind, and so it was common to use lag-time elements (sometimes more than one connected in series) to provide an approximation of dead time.

With digital computer technology, all these dynamic compensation functions are easy to implement and readily available in a control system. Some single-loop controllers even have these capabilities programmed within, ready to use when needed.

A dead time function block is most easily implemented using the concept of a *first-in, first-out shift register*, sometimes called a *FIFO*. With this concept, successive values of the input variable are stored in a series of registers (memory), their progression to the output delayed by a certain amount of time:

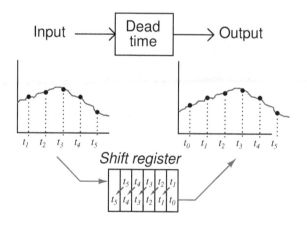

Lead and lag functions are also implemented digitally in modern controllers and control systems, but they are actually easier to comprehend in their analog (RC circuit) forms. The most common way lead and lag functions are found in modern control systems is in combination as the so-called *lead/lag function*, merging both lead and lag characteristics in a single function block (or relay):

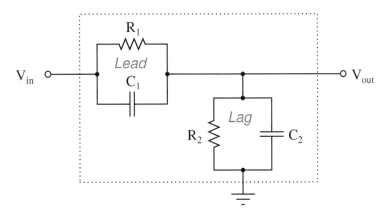

Each parallel RC subcircuit represents a time constant (τ), one for lead and one for lag. The overall behavior of the network is determined by the relative magnitudes of these two time constants. Which ever time constant is larger, determines the overall characteristic of the network.

If the two time constant values are equal to each other ($\tau_{lead} = \tau_{lag}$), then the circuit performs no dynamic compensation at all, simply passing the input signal to the output with no change except for some attenuation:

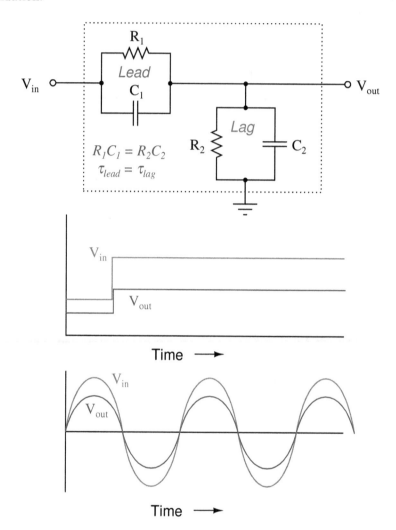

A square wave signal entering this network will exit the network as a square wave. If the input signal is sinusoidal, the output will also be sinusoidal and in-phase with the input.

If the lag time constant exceeds the lead time constant ($\tau_{lag} > \tau_{lead}$), then the overall behavior of the circuit will be to introduce a first-order lag to the voltage signal:

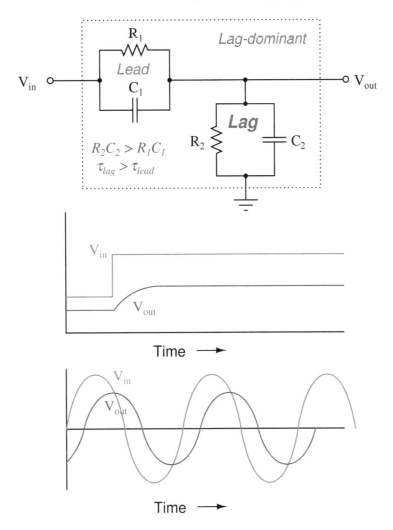

A square wave signal entering the network will exit the network as a sawtooth-shaped wave. A sinusoidal input will emerge sinusoidal, but with a lagging phase shift. This, in fact, is where the *lag* function gets its name: from the negative ("lagging") phase shift it imparts to a sinusoidal input.

Conversely, if the lead time constant exceeds the lag time constant ($\tau_{lead} > \tau_{lag}$), then the overall behavior of the circuit will be to introduce a first-order lead to the voltage signal (a step-change voltage input will cause the output to "spike" and then settle to a steady-state value):

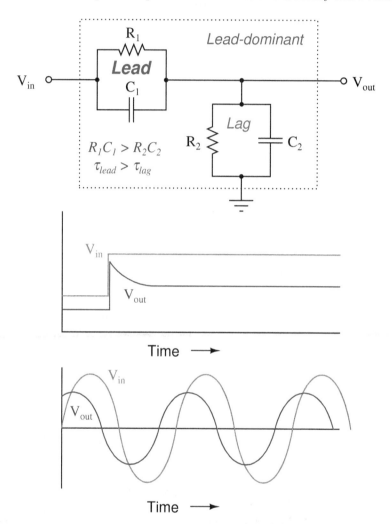

A square wave signal entering the network will exit the network with sharp transients on each leading edge. A sinusoidal input will emerge sinusoidal, but with a leading phase shift. Not surprisingly, this is where the *lead* function gets its name: from the positive ("leading") phase shift it imparts to a sinusoidal input.

   This exact form of lead/lag circuit finds application in a context far removed from process control: compensation for coaxial cable capacitance in a ×10 oscilloscope probe. Such probes are used to extend the voltage measurement range of standard oscilloscopes, and/or to increase the impedance of the instrument for minimal loading effect on sensitive electronic circuits. Using a ×10 probe, an oscilloscope will display a waveform that is $\frac{1}{10}$ the amplitude of the actual signal, and present ten times the normal impedance to the circuit under test.

   If a 9 MΩ resistor is connected in series with a standard oscilloscope input (having an input impedance of 1 MΩ) to create a 10:1 voltage division ratio, problems will result from the cable capacitance connecting the probe to the oscilloscope input. What should display as a square-wave input instead looks "rounded" by the effect of capacitance in the coaxial cable and at the oscilloscope input:

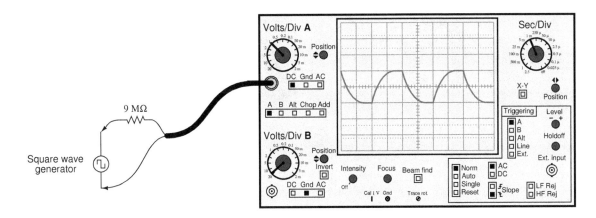

   The reason for this is signal distortion the combined effect of the 9 MΩ resistor and the cable's natural capacitance forming an RC network:

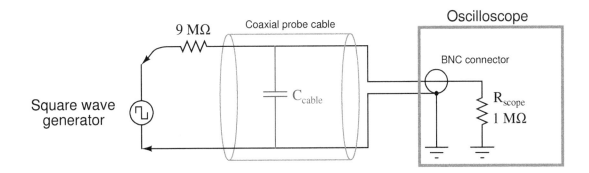

A simple solution to this problem is to build the 10:1 probe with a variable capacitor connected in parallel across the 9 MΩ resistor. The combination of the 9 MΩ resistor and this capacitor creates a lead network to cancel out the effects of the lag caused by the cable capacitance and 1 MΩ oscilloscope impedance in parallel. When the capacitor is properly adjusted, the oscilloscope will accurately show the shape of any waveform at the probe tip, including square waves:

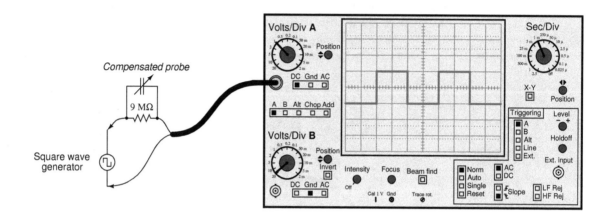

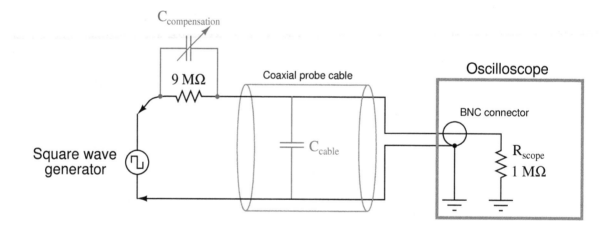

If we re-draw this compensated probe circuit to show the resistor pair and the capacitor pair both working as 10:1 voltage dividers, it becomes clearer to see how the two divider circuits work in parallel with each other to provide the same 10:1 division ratio *only* if the component ratios are properly proportioned:

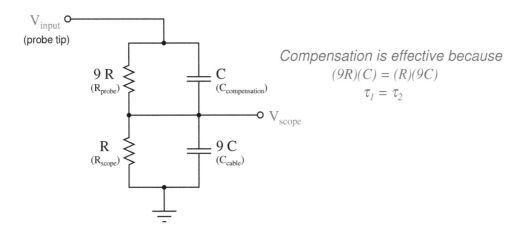

The series resistor pair forms an obvious 10:1 division ratio, with the smaller of the two resistors being in parallel with the oscilloscope input. The upper resistor, being 9 times larger in resistance value, drops 90% of the voltage applied to the probe tip. The series capacitor pair forms a less obvious 10:1 division ratio, with the cable capacitance being the larger of the two. Recall that the *reactance* of a capacitor to an AC voltage signal is inversely related to capacitance: a larger capacitor will have fewer ohms of reactance for any given frequency according to the formula $X_C = \frac{1}{2\pi f C}$. Thus, the capacitive voltage divider network has the same 10:1 division ratio as the resistive voltage divider network, even though the capacitance ratios may look "backward" at first glance.

If the compensation capacitor is adjusted to an excessive value, the probe will *overcompensate* for lag (too much lead), resulting in a "spiked" waveform on the oscilloscope display with a perfect square-wave input:

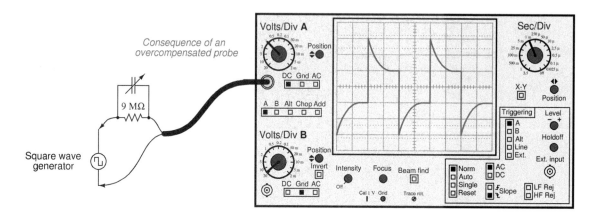

With the probe's compensating capacitor exhibiting an excessive amount of capacitance, the capacitive voltage divider network has a voltage division ratio *less than* 10:1. This is why the waveform "spikes" on the leading edges: the capacitive divider dominates the network's response in the short term, producing a voltage pulse at the oscilloscope input greater than it should be (divided by some ratio less than 10). Soon after the leading edge of the square wave passes, the capacitors' effects will wane, leaving the resistors to establish the voltage division ratio on their own. Since the two resistors have the proper 10:1 ratio, this causes the oscilloscope's signal to "settle" to its proper value over time. Thus, the waveform "spikes" too far at each leading edge and then decays to its proper amplitude over time.

While undesirable in the context of oscilloscope probes, this is precisely the effect we desire in a process control *lead* function. The purpose of a lead/lag function is to provide a signal gain that begins at some initial value, then "settles" at another value over time. This way, sudden changes in the feedforward signal will either be amplified or attenuated for a short duration to compensate for lags in other parts of the control system, while the steady-state gain of the feedforward loop remains at some other value necessary for long-term stability. For a lag function, the initial gain is less than the final gain; for a lead function, the initial gain exceeds the final gain. If the lead and lag time constants are set equal to each other, the initial and final gains will likewise be equal, with the function exhibiting a constant gain at all times.

Although lead-lag functions for process control systems may be constructed from analog electronic components, modern systems implement the function arithmetically using digital microprocessors. A typical time-domain equation describing a digital lead/lag function block's output response (y) to an input step-change from zero (0) to magnitude x over time (t) is as follows:

$$y = x \left(1 + \frac{\tau_{lead} - \tau_{lag}}{\tau_{lag} \ e^{\frac{t}{\tau_{lag}}}} \right)$$

As you can see, if the two time constants are set equal to each other ($\tau_{lead} = \tau_{lag}$), the second term inside the parentheses will have a value of zero at all times, reducing the equation to $y = x$. If the lead time constant exceeds the lag time constant ($\tau_{lead} > \tau_{lag}$), then the fraction will begin with a positive value and decay to zero over time, giving us the "spike" response we expect from a lead function. Conversely, if the lag time constant exceeds the lead ($\tau_{lag} > \tau_{lead}$), the fraction will begin with a negative value at time = 0 (the beginning of the step-change) and decay to zero over time, giving us the "sawtooth" response we expect from a lag function.

It should also be evident from an examination of this equation that the "decay" time of the lead/lag function is set by the lag time constant (τ_{lag}). Even if we just need the function to produce a "lead" response, we must still properly set τ_{lag} in order for the lead response to decay at the correct rate for our control system. The intensity of the lead function (i.e. how far it "spikes" when presented with a step-change in input signal) varies with the ratio $\frac{\tau_{lead}}{\tau_{lag}}$, but the duration of the "settling" following that step-change is entirely set by τ_{lag}.

To summarize the behavior of a lead/lag function:

- If $\tau_{lead} = \tau_{lag}$, the lead/lag function will simply pass the input signal through to the output (no lead or lag action at all)

- If $\tau_{lead} = 0$, the lead/lag function will provide a pure lag response with a final gain of unity and a time constant of τ_{lag}

- If $\tau_{lead} = 2(\tau_{lag})$, the lead/lag function will provide a lead response with an initial gain of 2, a final gain of unity, and a time constant of τ_{lag}

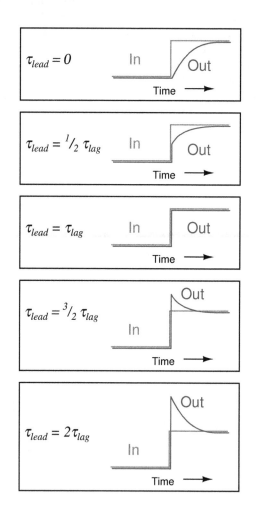

31.7 Limit, Selector, and Override controls

Another category of control strategies involves the use of signal relays or function blocks with the ability to switch between different signal values, or re-direct signals to new pathways. Such functions are useful when we need a control system to choose between multiple signals of differing value in order to make the best control decisions.

The "building blocks" of such control strategies are special relays (or function blocks in a digital control system) shown here:

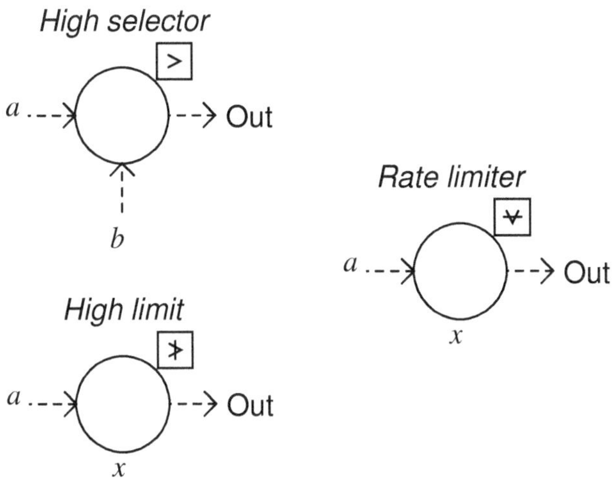

High-select functions output whichever input signal has the *greatest* value. *Low-select* functions do just the opposite: output whichever input signal has the *least* value. "Greater-than" and "Less than" symbols mark these two selector functions, respectively, and each type may be equipped to receive more than two input signals.

Sometimes you will see these relays represented in P&IDs simply by an inequality sign in the middle of the large bubble, rather than off to the side in a square. You should bear in mind that the location of the input lines has no relationship at all to the direction of the inequality symbol – e.g., it is not as though a high-select relay looks for the input on the left side to be greater than the input on the right. Note the examples shown below, complete with sample signal values:

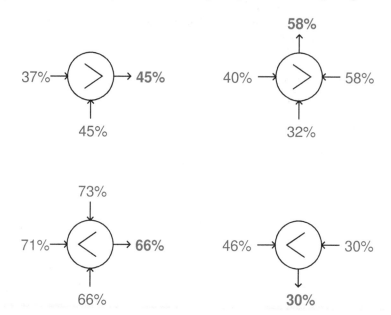

High-limit and *low-limit* functions are similar to high- and low-select functions, but they only receive one input each, and the limit value is a parameter programmed into the function rather than received from another source. The purpose of these functions is to place a set limit on how high or how low a signal value is allowed to go before being passed on to another portion of the control system. If the signal value lies within the limit imposed by the function, the input signal value is simply passed on to the output with no modification.

Like the select functions, limit functions may appear in diagrams with nothing more than the limit symbol inside the bubble, rather than being drawn in a box off to the side:

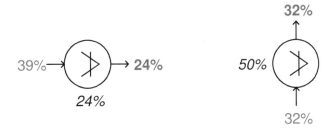

Rate limit functions place a maximum rate-of-change limit on the input signal, such that the output signal will follow the input signal precisely until and unless the input signal's rate-of-change over time $\left(\frac{dx}{dt}\right)$ exceeds the pre-configured limit value. In that case, the relay still produces a ramping output value, but the rate of that ramp remains fixed at the limit $\frac{dx}{dt}$ value no matter how fast the input keeps changing. After the output value "catches up" with the input value, the function once again will output a value precisely matching the input unless the input begins to rise or fall at too fast a rate again.

31.7.1 Limit controls

A common application for select and limit functions is in *cascade* control strategies, where the output of one controller becomes the setpoint for another. It is entirely possible for the primary (master) controller to call for a setpoint that is unreasonable or unsafe for the secondary (slave) to attain. If this possibility exists, it is wise to place a limit function between the two controllers to limit the cascaded setpoint signal.

In the following example, a cascade control system regulates the temperature of molten metal in a furnace, the output of the master (metal temperature) controller becoming the setpoint of the slave (air temperature) controller. A high limit function limits the maximum value this cascaded setpoint can attain, thereby protecting the refractory brick of the furnace from being exposed to excessive air temperatures:

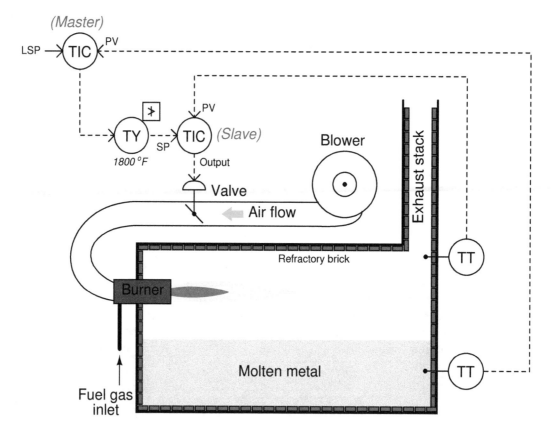

It should be noted that although the different functions are drawn as separate bubbles in the P&ID, it is possible for multiple functions to exist within one physical control device. In this example, it is possible to find a controller able to perform the functions of both PID control blocks (master and slave) and the high limit function as well. It is also possible to use a distributed technology such as FOUNDATION Fieldbus to place all control functions inside field instruments, so only three field instruments exist in the loop: the air temperature transmitter, the metal temperature transmitter, and the control valve (with a Fieldbus positioner).

This same control strategy could have been implemented using a low select function block rather than a high limit:

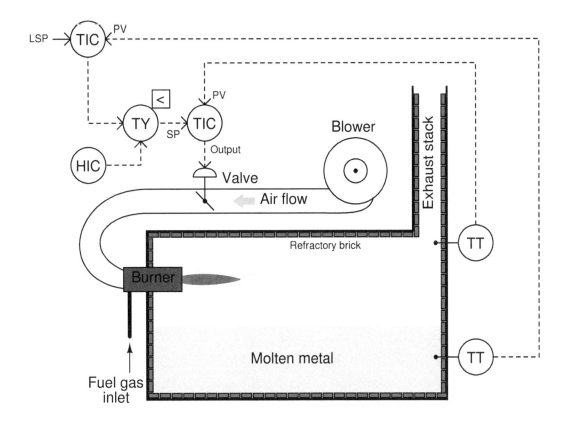

Here, the low-select function selects whichever signal value is lesser: the setpoint value sent by the master temperature controller, or the maximum air temperature limit value sent by the hand indicating controller (HIC – sometimes referred to as a *manual loading station*).

An advantage of this latter approach over the former might be ease of limit value changes. With a pre-configured limit value residing in a high-limit function, it might be that only qualified maintenance people have access to changing that value. If the decision of the operations department is to have the air temperature limit value easily adjusted by anyone, the latter control strategy's use of a manual loading station would be better suited[26].

Another detail to note in this system is the possibility of *integral windup* in the master controller in the event that the high setpoint limit takes effect. Once the high-limit (or low-select) function secures the slave controller's remote setpoint at a fixed value, the master controller's output is no

---

[26]I generally suggest keeping such limit values inaccessible to low-level operations personnel. This is especially true in cases such as this where the presence of a high temperature setpoint limit is intended for the longevity of the equipment. There is a strong tendency in manufacturing environments to "push the limits" of production beyond values considered safe or expedient by the engineers who designed the equipment. Limits are there for a reason, and should not be altered except by people with full understanding of and full responsibility over the consequences!

longer controlling anything: it has become decoupled from the process. If, when in this state of affairs, the metal temperature is still below setpoint, the master controller's integral action will "wind up" the output value over time with absolutely no effect, since the slave controller is no longer following its output signal. If and when the metal temperature reaches setpoint, the master controller's output will likely be saturated at 100% due to the time it spent winding up. This will cause the metal temperature to overshoot setpoint, as a positive error will be required for the master controller's integral action to wind back down from saturation.

A relatively easy solution to this problem is to configure the master controller to stop integral action when the high limit relay engages. This is easiest to do if the master PID and high limit functions both reside in the same physical controller. Many digital limit function blocks generate a bit representing the state of that block (whether it is passing the input signal to the output or limiting the signal at the pre-configured value), and some PID function blocks have a boolean input used to disable integral action. If this is the case with the function blocks comprising the high-limit control strategy, it may be implemented like this:

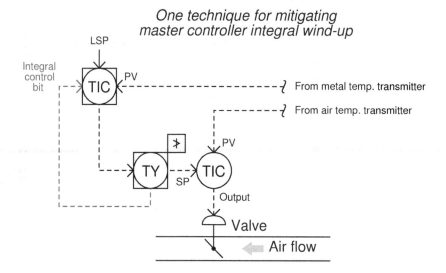

One technique for mitigating master controller integral wind-up

Another method used to prevent integral windup is to make use of the *feedback* input available on some PID function blocks. This is an input used to calculate the integral term of the PID equation. In the days of pneumatic PID controllers, this option used to be called *external reset*. Normally connected to the output of the PID block, if connected to the output of the high-limit function it will let the controller know whether or not any attempt to wind up the output is having an effect. If the output has been de-selected by the high-limit block, integral windup will cease:

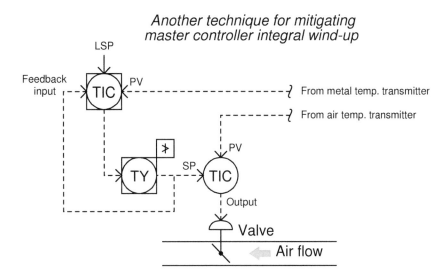

Limit control strategies implemented in FOUNDATION Fieldbus instruments use the same principle, except that the concept of a "feedback" signal sending information backwards up the function block chain is an aggressively-applied design philosophy throughout the FOUNDATION Fieldbus standard. Nearly every function block in the Fieldbus suite provides a "back calculation" output, and nearly every function block accepts a "back calculation" input from a downstream block. The "Control Selector" (CS) function block specified in the FOUNDATION Fieldbus standard provides the limiting function we need between the master and slave controllers. The BKCAL_OUT signal of this selector block connects to the master controller's BKCAL_IN input, making the master controller aware of its selection status. If ever the Control Selector function block de-selects the master controller's output, the controller will immediately know to halt integral action:

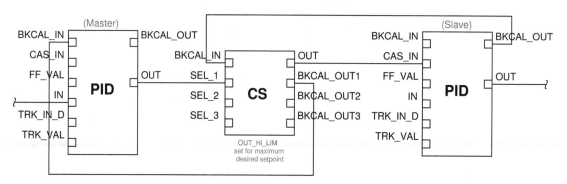

This same "back calculation" philosophy – whereby the PID algorithm is aware of how another function is limiting or over-riding its output – is also found in some programmable logic controller (PLC) programming conventions. The Allen-Bradley Logix5000 series of PLCs, for example, provides a *tieback* variable to force the PID function's output to track the overriding function. When the "tieback" variable is properly used, it allows the PID function to bumplessly transition from the "in-control" state to the "overridden" state.

31.7.2 Selector controls

In the broadest sense, a "selector" control strategy is where one signal is selected from multiple signals in a system to perform a measurement control function. In the context of this book and this chapter, I will use the term "selector" to reference automatic selection among multiple *measurement* or *setpoint* signals. Selection between multiple *controller output* signals will be explored in the next subsection, under the term "override" control.

Perhaps one of the simplest examples of a selector control strategy is where we must select a process variable signal from multiple transmitters. For example, consider this chemical reactor, where the control system must throttle the flow of coolant to keep the *hottest* measured temperature at setpoint, since the reaction happens to be exothermic (heat-releasing)[27]:

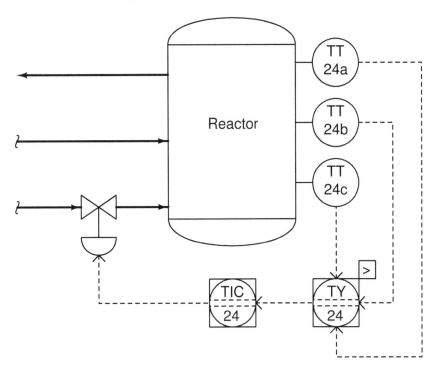

The high-select relay (TY-24) sends only the highest temperature signal from the three transmitters to the controller. The other two temperature transmitter signals are simply ignored.

Another use of selector relays (or function blocks) is for the determination of a *median* process measurement. This sort of strategy is often used on triple-redundant measurement systems, where three transmitters are installed to measure the exact same process variable, providing a valid measurement even in the event of transmitter failure.

---

[27]Only the coolant flow control instruments and piping are shown in this diagram, for simplicity. In a real P&ID, there would be many more pipes, valves, and other apparatus shown surrounding this process vessel.

The median select function may be implemented one of two ways using high- and low-select function blocks:

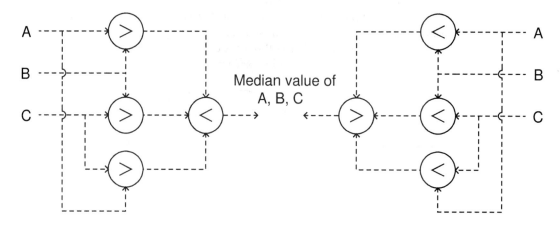

Two ways of obtaining a median signal value from three redundant inputs

The left-hand selector strategy selects the highest value from each pair of signals (A and B, B and C, A and C), then selects the lowest value of those three primary selections. The right-hand strategy is exactly opposite – first selecting the lowest value from each input pair, then selecting the highest of those values – but it still accomplishes the same function. Either strategy outputs the *middle* value of the three input signals[28].

Although either of these methods of obtaining a median measurement requires four signal selector functions, it is quite common to find function blocks available in control systems ready to perform the median select function all in a single block. The median-select function is so common to redundant sensor control systems that many control system manufacturers provide it as a standard function unto itself:

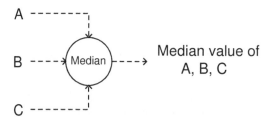

---

[28]In order to understand how this works, I advise you try a "thought experiment" for each function block network whereby you arbitrarily assign three different numerical values for A, B, and C, then see for yourself which of those three values becomes the output value.

This is certainly true in the FOUNDATION Fieldbus standard, where two standardized function blocks are capable of this function, the CS (Control Selector) and the ISEL (Input Selector) blocks:

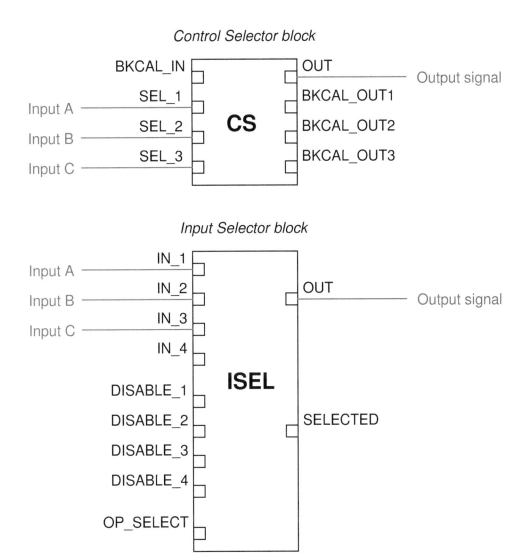

Of these two Fieldbus function blocks, the latter (ISEL) is expressly designed for selecting transmitter signals, whereas the former (CS) is best suited for selecting controller outputs with its "back calculation" facilities designed to modify the response of all de-selected controllers. Using the terminology of this book section, the ISEL function block is best suited for *selector* strategies, while the CS function block is ideal for *limit* and *override* strategies (discussed in the next section).

If receiving three "good" inputs, the ISEL function block will output the middle (median) value of the three. If one of the inputs carries a "bad" status[29], the ISEL block outputs the averaged value of the remaining two (good) inputs. Note how this function block also possesses individual "disable" inputs, giving external boolean (on/off) signals the ability to disable any one of the transmitter inputs to this block. Thus, the ISEL function block may be configured to de-select a particular transmitter input based on some programmed condition other than internal diagnostics.

If receiving four "good" inputs, the ISEL function block normally outputs the average value of the two middle (median) signal values. If one of the four inputs becomes "bad" is disabled, the block behaves as a normal three-input median select.

A general design principle for redundant transmitters is that you *never* install exactly two transmitters to measure the same process variable. Instead, you should install three (minimum). The problem with having two transmitters is a lack of information for "voting" if the two transmitters happen to disagree. In a three-transmitter system, the function blocks may select the median signal value, or average the "best 2 out of 3." If there are just two transmitters installed, and they do not substantially agree with one another, it is anyone's guess which one should be trusted[30].

---

[29]In FOUNDATION Fieldbus, each and every signal path not only carries the signal value, but also a "status" flag declaring it to be "Good," "Bad," or "Uncertain." This status value gets propagated down the entire chain of connected function blocks, to alert dependent blocks of a possible signal integrity problem if one were to occur.

[30]This principle holds true even for systems with no function blocks "voting" between the redundant transmitters. Perhaps the installation consists of two transmitters with remote indications for a human operator to view. If the two displays substantially disagree, which one should the operator trust? A set of *three* indicators would be much better, providing the operator with enough information to make an intelligent decision on which display(s) to trust.

A classic example of selectors in industrial control systems is that of a *cross-limited ratio control* strategy for air/fuel mixture applications. Before we explore the use of selector functions in such a strategy, we will begin by analyzing a simplified version of that strategy, where we control air and fuel flows to a precise ratio with no selector action at all:

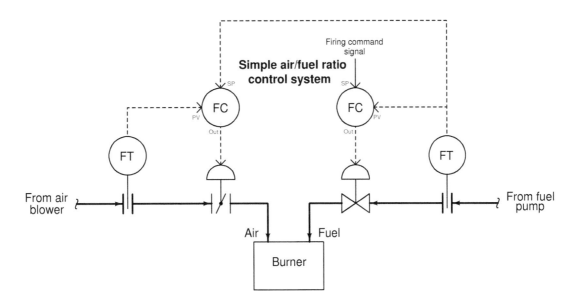

Here, the fuel flow controller receives its setpoint directly from the firing command signal, which may originate from a human operator's manual control or from the output of a temperature controller regulating temperature of the combustion-heated process. The air flow controller receives its setpoint from the fuel flow transmitter, with the calibrations of the air and fuel flow transmitters being appropriate to establish the proper air:fuel flow ratio when the transmitters register equally. From the perspective of the air flow controller, fuel flow is the *wild* flow while air flow is the *captive* flow.

There is a problem with this control system that may not be evident upon first inspection: the air:fuel ratio will tend to vary as the firing command signal increases or decreases in value over time. This is true even if the controllers are well-tuned and the air:fuel ratio remains well-controlled under steady-state conditions. The reason for this is linked to the roles of "wild" and "captive" flows, fuel and air flow respectively. Since the air flow controller receives its setpoint from the fuel flow transmitter, changes in air flow will always *lag* behind changes in fuel flow.

This sort of problem can be difficult to understand because it involves changes in multiple variables over time. A useful problem-solving technique to apply here is a "thought experiment," coupled with a time-based graph to display the results. Our thought experiment consists of imagining what would happen if the firing command signal were to *suddenly* jump in value, then sketching the results on a graph.

If the firing command signal suddenly increases, the fuel flow controller responds by opening up the fuel valve, which after a slight delay results in increased fuel flow to the burner. This increased fuel flow signal gets sent to the setpoint input of the air flow controller, which in turn opens up the air valve to increase air flow proportionately. If the firing command signal suddenly decreased, the same changes in flow would occur in reverse direction but in the same chronological sequence, since the fuel flow change still "leads" the subsequent air flow change:

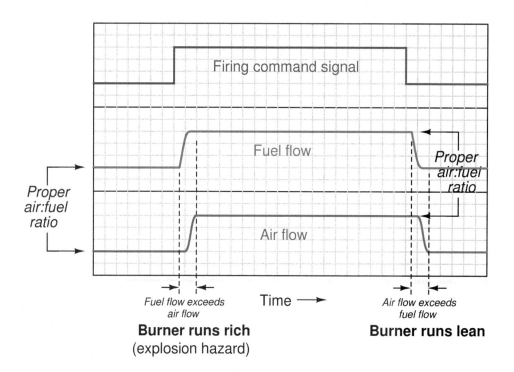

Inevitable delays in controller response, valve response, and flow transmitter response conspire to upset the air:fuel ratio during the times immediately following a step-change in firing command signal. When the firing command steps up, the fuel flow increases before the air flow, resulting in a short time when the burner runs "rich" (too much fuel, not enough air). When the firing command steps down, the fuel flow decreases before the air flow, resulting in a short time when the burner runs "lean" (too much air, not enough fuel). The scenario problem is dangerous because it may result in an explosion if an accumulation of unburnt fuel collects in a pocket of the combustion chamber and then ignites. The second scenario is generally not a problem unless the flame burns *so* lean that it risks blowing out.

The solution to this vexing problem is to re-configure the control scheme so that the air flow controller "takes the lead" whenever the firing command signal rises, and that the fuel flow controller "takes the lead" whenever the firing command signal falls. This way, any upsets in air:fuel ratio resulting from changes in firing command will always err on the side of a lean burn rather than a rich burn.

We may implement precisely this strategy by adding some signal selector functions to our ratio control system. The ratio is now *cross-limited*, where both measured flow rates serve as limiting variables to each other to ensure the air:fuel ratio can never be too rich:

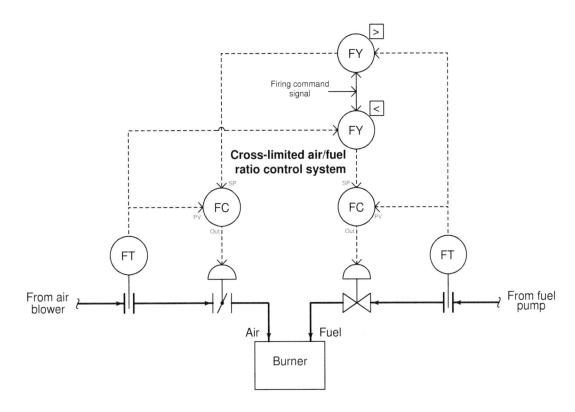

Now, the air flow controller receives its setpoint either directly from the firing command signal or from the fuel flow transmitter, whichever signal is *greater* in value. The fuel flow controller receives its setpoint either directly from the firing command signal or from the air flow transmitter, whichever signal is *least* in value. Thus, the air flow controller "takes the lead" and makes fuel flow the "captive" variable when the firing command signal rises. Conversely, the fuel flow controller "takes the lead" and makes air flow the "captive" variable when the firing command signal falls. Instead of having the roles of "wild" and "captive" flows permanently assigned, these roles switch depending on which way the firing command signal changes.

Examining the response of this cross-limited system to sudden changes in firing command signal, we see how the air flow controller takes the lead whenever the firing rate signal increases, and how the fuel flow controller takes the lead whenever the firing rate signal decreases:

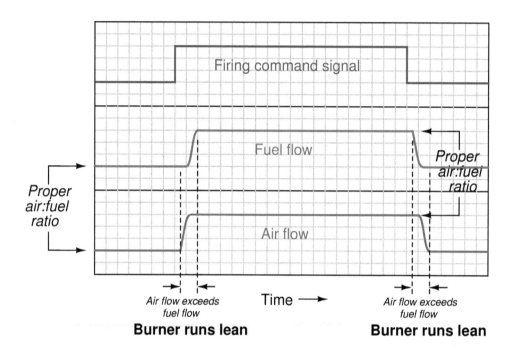

In both transient scenarios, the mixture runs lean (safe) rather than rich (dangerous). Of course, care must be taken to ensure the firing rate signal never steps up or down so quickly that the flame runs lean enough to blow out (i.e. the mixture becomes much too lean during a transient "step-change" of the firing rate signal). If this is a problem, we may fix it by installing *rate-limiting* functions in the firing command signal path, so that the firing command signal can never rise or fall too rapidly.

A realistic cross-limited ratio control system also incorporates a means to adjust the air:fuel ratio without having to re-range the air and/or fuel flow transmitters. Such ratio adjustment may be achieved by the insertion of a "multiplying" function between one of the selectors and a controller setpoint, plus a "dividing" function to return that scaled flow to a normalized value for cross-limiting.

The complete control strategy looks something like this, complete with cross-limiting of air and fuel flows, rate-limiting of the firing command signal, and adjustable air:fuel ratio:

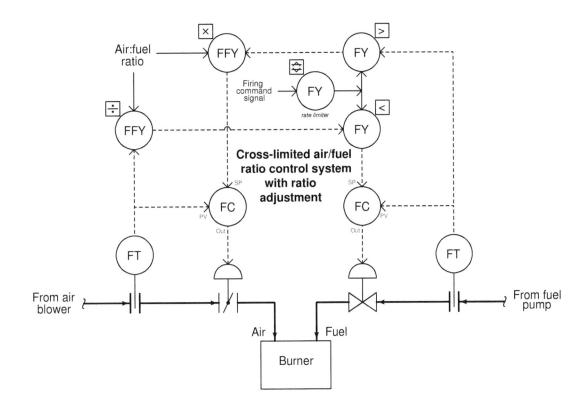

31.7.3   Override controls

An "override" control strategy involves a selection between two or more controller *output* signals, where only one controller at a time gets the opportunity to exert control over a process. All other "de-selected" controllers are thus *overridden* by the selected controller.

The general concept of override control is easily understood by appeal to a human example. Cargo truck drivers must monitor and make control decisions on a wide number of variables, including diesel engine operating parameters and road rules. A truck driver needs to keep a close watch on the exhaust gas temperature of the truck engine: a leading indicator of impending engine damage (if exhaust temperature exceeds a pre-determined limit established by the engine manufacturer). The same truck driver must also drive as fast as the law will allow on any given road in order to minimize shipping time and thereby maximize the amount of cargo transported over long periods of time. These two goals may become mutually exclusive when hauling heavy cargo loads up steep inclines, such as when ascending a mountain pass. The goal of avoiding engine damage necessarily overrides the goal of maintaining legal road speed in such conditions.

Imagine a diesel truck driver maintaining the legal speed limit on a highway, occasionally glancing at the EGT (Exhaust Gas Temperature) indicator in the instrument panel. Under normal operating conditions, the EGT should be well below the danger threshold for the engine. However, after pulling a full load up a mountain pass and noticing the EGT approach the high operating limit, the truck driver makes the decision to regulate the engine's power based on EGT rather than road speed. In other words, the legal speed limit is no longer the "setpoint" to control to, and EGT now is.

If we were to model the truck driver's decision-making processes in industrial instrumentation terms, it would look something like this:

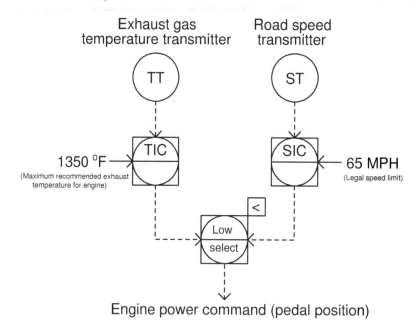

Which ever control decision calls for the least engine power output, "wins the vote" to control the engine's power.

As is the case with limit and selector control strategies, a "select" function is used to choose one signal from multiple signals. The difference here is that the signals being selected are both *controller outputs* rather than transmitter (measurement) or setpoint signals. Both controllers are still active, but only one at a time will have any actual control over the process.

This model maps well to the truck driver analogy. Despite having "overridden" the goal of maintaining legal road speed in favor of maintaining a safe engine exhaust temperature, the driver is still thinking about road speed. In fact, if the driver happens to be behind schedule, you can be absolutely sure the goal of maintaining the highway speed limit has not been forgotten! In fact, the driver may become impatient as the long incline wears on, eager to make up lost time as soon as the opportunity allows. This is a potential problem for all override control systems: making sure the de-selected controller does not "wind up" (with integral action still active) while it has no control over the process.

An municipal example of override control is seen in this water pumping system, where a water pump is driven by a variable-speed[31] electric motor to draw water from a well and provide constant water pressure to a customer:

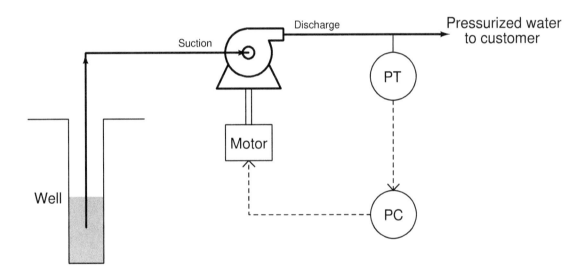

Incidentally, this is an excellent application for a variable-speed motor as the final control element rather than a control valve. Reducing pump speed in low-flow conditions will save a lot of energy over time compared to the energy that would be wasted by a constant-speed pump and control valve.

A potential problem with this system is the pump running "dry" if the water level in the well gets too low, as might happen during summer months when rainfall is low and customer demand is high. If the pump runs for too long with no water passing through it, the seals will become damaged. This will necessitate a complete shut-down and costly rebuild of the pump, right at the time customers need it the most.

---

[31]In most applications this takes the form of an AC induction motor receiving power from a *Variable Frequency Drive* or *VFD*. Since the rotational speed of an induction motor is a function of frequency, the VFD achieves motor speed control by electronically converting the fixed-frequency line power into variable-frequency power to drive the motor.

One solution to this problem would be to install a level switch in the well, sensing water level and shutting off the electric motor driving the pump if the water level ever gets too low:

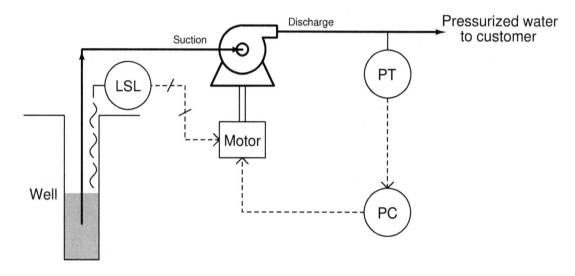

This may be considered a kind of "override" strategy, because the low-level switch over-rides the pressure controller's command for the pump to turn. It is also a crude solution to the problem, for while it protects the pump from damage, it does so at the cost of completely shutting off water to customers. One way to describe this control strategy would be to call it a *hard override* system, suggesting the uncompromising action it will take to protect the pump.

A better solution to the dilemma would be to have the pump merely slow down as the well water level approaches a low-level condition. This way at least the pump could be kept running (and some amount of pressure maintained), decreasing demand on the well while maintaining curtailed service to customers and still protecting the pump from dry-running. This would be termed a *soft override* system.

We may create just such a control strategy by replacing the well water level switch with a level *transmitter*, connecting the level transmitter to a level controller, and using a low-select relay or function block to select the lowest-valued output between the pressure and level controllers. The level controller's setpoint will be set at some low level above the acceptable limit for continuous pump operation:

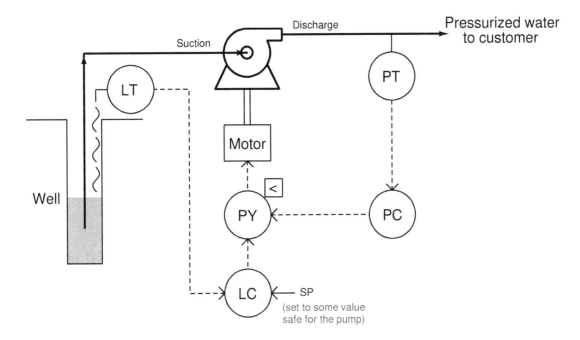

If ever the well's water level goes below this setpoint, the level controller will command the pump to slow down, even if the pressure controller is calling for a higher speed. The level controller will have *overridden* the pressure controller, prioritizing pump longevity over customer demand.

Bear in mind that the concept of a low-level switch completely shutting off the pump is not an entirely bad idea. In fact, it might be prudent to integrate such a "hard" shutdown control in the override control system, just in case something goes wrong with the level controller (e.g. an improperly adjusted setpoint or poor tuning) or the low-select function.

With two layers of safety control for the pump, this system provides both a "soft constraint" providing moderated action and a "hard constraint" providing aggressive action to protect the pump from dry operation:

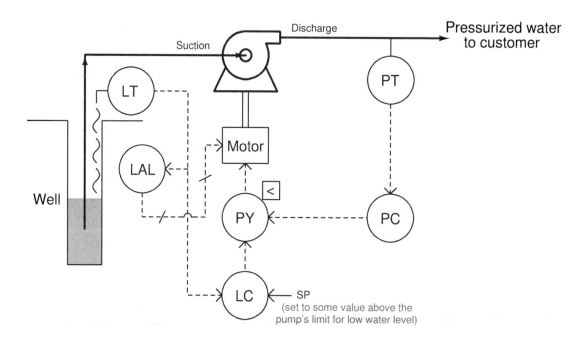

In order that these two levels of pump protection work in the proper order, the level controller's (LC) setpoint needs to be set to a higher value than the low level alarm's (LAL) trip point.

A very important consideration for any override control strategy is how to manage integral windup. Any time a controller with any integral (reset) action at all is de-selected by the selector function, the integral term of the controller will have the tendency to wind up (or wind down) over time. With the output of that controller de-coupled from the final control element, it can have no effect on the process variable. Thus, integral control action – the purpose of which being to constantly drive the output signal in the direction necessary to achieve equality between process variable and setpoint – will work in vain to eliminate an error it cannot influence. If and when control is handed back to that controller, the integral action will have to spend time "winding" the other way to un-do what it did while it was de-selected.

Thus, override controls demand some form of integral windup limits that engage when a controller is de-selected. Methods of accomplishing this function are discussed in an earlier section on limit controls (section 31.7.1 beginning on page 2554).

31.8 Techniques for analyzing control strategies

Control strategies such as cascade, ratio, feedforward, and those containing limit and selector functions can be quite daunting to analyze, especially for students new to the subject. As a teacher, I have seen first-hand where students tend to get confused on these topics, and have seen how certain problem-solving techniques work well to overcome these conceptual barriers. This section explores some of these techniques and the reasons why they work.

31.8.1 Explicitly denoting controller actions

The direction of action for a loop controller – either *direct* or *reverse* – at first seems like a very simple concept. It certainly is fundamental to the comprehension of any control strategy containing PID loop controllers, but this seemingly simple concept harbors an easy-to-overlook fact causing much confusion for students as they begin to analyze any control strategy where a loop controller receives a remote setpoint signal from some other device, most notably in cascade and ratio control strategies.

A *direct-acting* loop controller is defined as one where the output signal increases as the process variable signal increases. A *reverse-acting* controller is defined as one where the output signal decreases as the process variable signal increases. Both types of action are shown here:

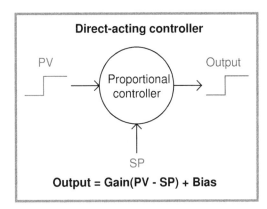
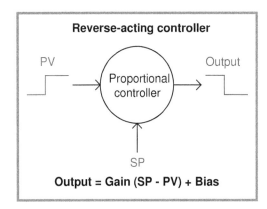

Let us apply this concept to a realistic application, in this case the control of temperature in a steam-heated chemical reactor vessel:

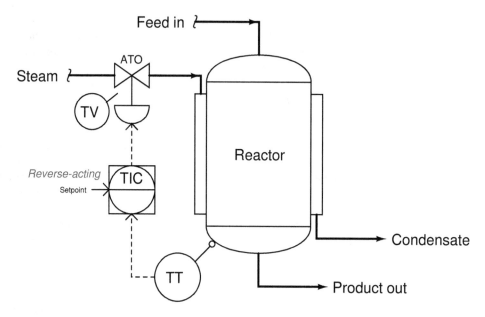

As the reactor vessel's temperature increases, we need the temperature controller (TIC) to reduce the amount of hot steam entering the jacket in order to stabilize that temperature. Since the steam control valve is air-to-open (ATO), this means we need the controller to output a *decreasing* signal as the process variable (temperature) signal increases. This, by definition, is a *reverse-acting* controller. This example also showcases the utility of the problem-solving technique known as a "thought experiment," whereby we imagine a certain condition changing (in this case, the reactor temperature increasing) and then we mentally model the desired response of the system (in this case, closing the steam valve) in order to determine the necessary controller action.

So far, this example poses no confusion. But suppose we were to perform another thought experiment, this time supposing the *setpoint* signal increases rather than the reactor temperature increases. How will the controller respond now?

Many students will conclude that the controller's output signal will once again decrease, because we have determined this controller's action to be *reverse*, and "reverse" implies the output will go the opposite direction as the input. However, this is not the case: the controller output will actually *increase* if its setpoint signal is increased. This, in fact, is precisely how any reverse-acting controller should respond to an increase in setpoint.

The reason for this is evident if we take a close look at the characteristic equation for a reverse-acting proportional controller. Note how the gain is multiplied by the difference between setpoint and process variable. Note how the process variable has a negative sign in front of it, while setpoint does not.

Reverse-acting controller

Output = Gain (SP - PV) + Bias

Direct effect on Output   *Reverse effect on Output*

An increase in process variable (PV) causes the quantity inside the parentheses to become more negative, or less positive, causing the output to decrease toward 0%. Conversely, an increase in setpoint (SP) causes the quantity inside the parentheses to become more positive, causing the output to increase toward 100%. This is precisely how any loop controller should respond: with the setpoint having the opposite effect of the process variable, because those two quantities are always being *subtracted* from one another in the proportional controller's equation.

Where students get confused is the single label of either "direct" or "reverse" describing a controller's action. We define a controller as being either "direct-acting" or "reverse-acting" based on how it responds to changes in process variable, but it is easy to overlook the fact that the controller's setpoint input must necessarily have the *opposite* effect. What we really need is a way to more clearly denote the respective actions of a controller's two inputs than a single word.

Thankfully, such a convention already exists in the field of electronics[32], where we must denote the "actions" of an operational amplifier's two inputs. In the case of an opamp, one input has a direct effect on the output (i.e. a change in signal at that input drives the output the same direction) while the other has a reverse effect on the output (i.e. a change in signal at that input drives the output in the opposite direction). Instead of calling these inputs "direct" and "reverse", however, they are conventionally denoted as *noninverting* and *inverting*, respectively. If we draw a proportional controller as though it were an opamp, we may clearly denote the actions of both inputs in this manner:

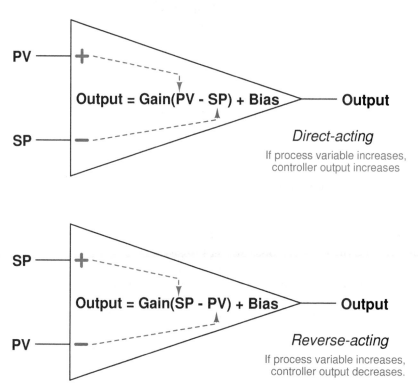

I strongly recommend students label the loop controllers in any complex control strategy in the same manner, with "+" and "−" labels next to the PV and SP inputs for each controller, in order to unambiguously represent the effects of each signal on a controller's output. This will be far more informative, and far less confusing, than merely labeling each controller with the word "direct" or "reverse".

---

[32]Some differential pressure transmitter manufacturers, such as Bailey, apply the same convention to denote the actions of a DP transmitter's two pressure ports: using a "+" label to represent direct action (i.e. increasing pressure at this port drives the output signal up) and a "−" symbol to represent reverse action (i.e. increasing pressure at this port drives the output signal down).

Let us return to our example of the steam-heated reactor to apply this technique, labeling the reverse-acting controller's process variable input with a "−" symbol and its setpoint input with a "+" symbol:

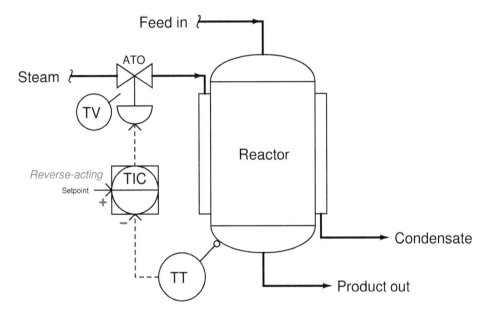

With these labels in place we can see clearly how an increase in temperature going into the "−" (inverting) input of the temperature controller will drive the valve signal down, counter-acting the change in temperature and thereby stabilizing it. Likewise, we can see clearly how an increase in setpoint going into the "+" (noninverting) input of the temperature controller will drive the valve signal up, sending more steam to the reactor to achieve a greater temperature.

While this technique of labeling the PV and SP inputs of a loop controller as though it were an operational amplifier is helpful in single-loop controller systems, it is incredibly valuable when analyzing more complex control strategies where the setpoint to a controller is a live signal rather than a static value set by a human operator. In fact, it is for this very reason that many students do not begin to have trouble with this concept until they begin to study cascade control, where one controller provides a live ("remote") setpoint value to another controller. Up until that point in their study, they never rarely had to consider the effects of a setpoint change on a control system because the setpoint value for a single-loop controller is usually static.

Let us modify our steam-heated reactor control system to include a cascade strategy, where the temperature controller drives a setpoint signal to a "slave" steam flow controller:

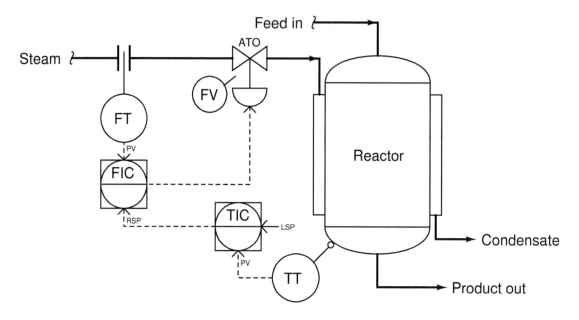

In order to determine the proper actions for each controller in this system, it is wise to begin with the slave controller (FIC), since the master controller (TIC) depends on the slave controller being properly configured in order to do its job properly. Just as we would first tune the slave controller in a cascade system prior to tuning the master controller, we should first determine the correct action for the slave controller prior to determining the correct action for the master controller.

Once again we may apply a "thought experiment" to this system in order to choose the appropriate slave controller action. If we imagine the steam flow rate suddenly increasing, we know we need the control valve to close off in order to counter-act this change. Since the valve is still air-to-open, this requires a decrease in the output signal from the FIC. Thus, the FIC must be reverse-acting. We shall denote this with a "$-$" label next to the process variable (PV) input, and a "$+$" label next to the remote setpoint (RSP) input:

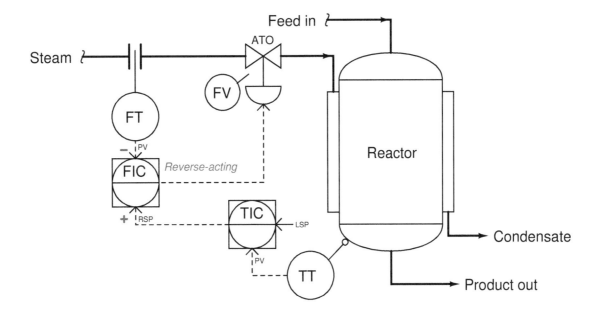

Now that we know the slave controller must be reverse-acting, we may choose the action of the master controller. Applying another "thought experiment" to this system, we may imagine the reactor temperature suddenly increasing. If this were to happen, we know we would need the control valve to close off in order to counter-act this change: sending less steam to a reactor that is getting too hot. Since the valve is air-to-open, this requires a decrease in the output signal from the FIC. Following the signal path backwards from the control valve to the FIC to the TIC, we can see that the TIC must output a decreasing signal to the FIC, calling for less steam flow. A decreasing output signal at the TIC enters the FIC's noninverting ("+") input, causing the FIC output signal to also decrease. Thus, we need the TIC to be reverse-acting as well. We shall denote this with a "−" label next to the process variable (PV) input, and a "+" label next to the local setpoint (LSP) input:

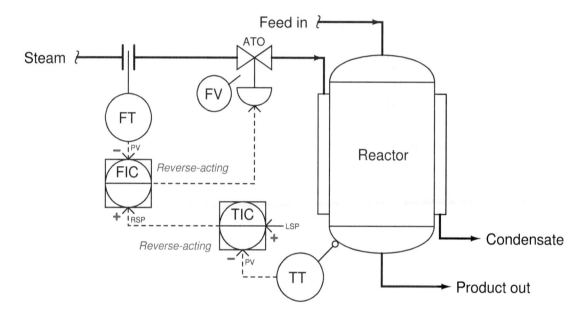

With these unambiguous labels in place at each controller's inputs, we are well-prepared to qualitatively analyze the response of this cascade control system to process upsets, to instrument failure scenarios, or to any other change. No longer will we be led astray by the singular label of "reverse-acting", but instead will properly recognize the different directions of action associated with each input signal to each controller.

31.8.2 Determining the design purpose of override controls

Override control strategies are a source of much confusion for students first learning the concept. Perhaps the most fundamental question students find difficult to answer when faced with an override strategy is how to determine the intended purpose for that strategy if no explanation is given.

Take for example this surge tank level/flow control system. While it may be obvious that the flow controller is there for the purpose of regulating flow out of the tank, it is not so clear what the two level controllers are doing, or what purposes are served by the two selector functions:

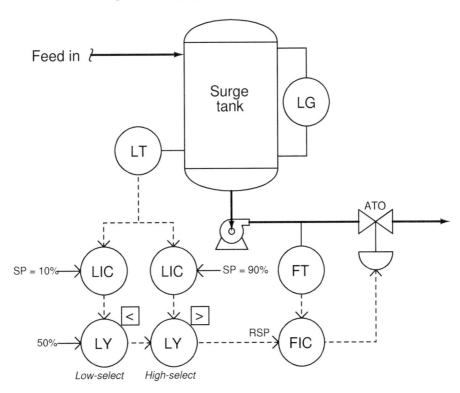

A good starting point in our analysis is to first determine the proper directions of action for each controller. This is wise because the selector functions perform their tasks based on the relative values of the controller output signals: controllers become selected or de-selected on the basis of their output signals being greater or less than some other signal. Therefore, before we may be able to determine the purpose of a selector function, we must know how the loop controller feeding that selector function is supposed to react to process conditions. Once we have determined each controller's proper action, we may then interpret each selector's function in light of what process conditions will cause a particular controller to become selected.

When choosing the proper action for each controller, we must consider each controller in this system – one at a time – as though it were the one being selected. In other words, we may give ourselves license to ignore the selector functions and just concentrate for the time being on how each controller needs to act in order to do its job when selected. Looking at the system from this perspective, we see that each level controller (when selected) acts as a master to the flow (slave) controller. Thus, what we have here is a cascade level/flow control system, with two master controllers selected on the basis of their output signals.

The flow controller (FIC) needs to be reverse-acting, because in order to counter-act an increase in flow rate it must close off the valve (i.e. decreasing output with increasing input = reverse action). Each level controller needs to be direct-acting, because in order to counter-act an increase in level it must call for more flow exiting the tank (i.e. increasing output with increasing input = direct action). Denoting these actions using "+" and "−" labels at each PV and SP input line:

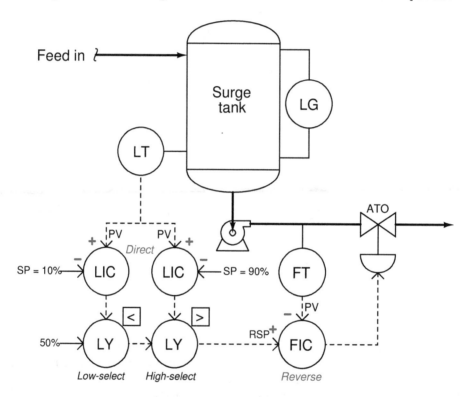

Only now are we prepared to analyze the purpose of each selector function. Let's begin with the low-select first. It selects the lowest of two values, either a fixed value of 50% or the output of the level controller with the 10% setpoint. Since we know this level controller is direct-acting, we may conclude that it will be selected if it sees a *low level* at its PV input. More specifically, it will be surely be selected if the measured tank level drops significantly below the setpoint value of 10%. Thus, we may conclude that the purpose of this level controller is to take over control if the tank level reaches or drops below the 10% mark.

Next, let us analyze the purpose of the other level controller (connected to the high-select function). Since the high-select function will select this level controller only if its output signal exceeds the signal passed on by the low-select function, and we know that this controller is direct-acting, we may conclude that it will be selected if it sees a *high level* at its PV input. More specifically, it will surely be selected if the measured tank level rises significantly above the setpoint value of 90%. Thus, we may conclude that the purpose of this level controller is to take over control if the tank level reaches or exceeds the 90% mark.

If neither level controller is selected, the signal that gets passed on to the flow controller as a remote setpoint is the 50% fixed signal entering on the left-hand side of the low-select function. Thus, the flow controller tries to maintain a steady flow rate of 50% in the event neither level controller is selected.

Putting all these pieces together, we may conclude that the purpose of this surge tank control system is to maintain as steady a flow rate as possible out of the tank (and on to some other process), letting the liquid level inside the tank rise and fall significantly before any action is taken to change the flow rate. Only if the level drops below 10% will the flow rate be reduced, and only if the level rises above 90% will the flow rate be increased. Otherwise, the flow rate will hold steady at 50%.

To summarize, the recommended technique for analyzing the purpose of an override control system is as follows:

- First, determine the necessary actions of each controller (assume the selector functions are absent, and each controller gets its turn controlling the process).

- Identify the type of selection (high or low) implemented by each selector function.

- Based on the type of selection and the action of the controller, identify what process condition will cause that controller to become selected. *This is the condition the controller exists to regulate.*

31.9 Review of fundamental principles

Shown here is a partial listing of principles applied in the subject matter of this chapter, given for the purpose of expanding the reader's view of this chapter's concepts and of their general inter-relationships with concepts elsewhere in the book. Your abilities as a problem-solver and as a life-long learner will be greatly enhanced by mastering the applications of these principles to a wide variety of topics, the more varied the better.

- **Negative feedback**: when the output of a system is degeneratively fed back to the input of that same system, the result is decreased (overall) gain and greater stability. Relevant to loop controller action: in order for a control system to be stable, the feedback must be negative.

- **Time constant**: (τ), defined as the amount of time it takes a system to change 63.2% of the way from where it began to where it will eventually stabilize. The system will be within 1% of its final value after 5 time constants' worth of time has passed (5τ). Relevant to process control loops, where natural lags contribute to time constants, usually of multiple order.

References

Austin, George T., *Shreve's Chemical Process Industries*, McGraw-Hill Book Company, New York, NY, 1984.

Dorf, Richard C., *Modern Control Systems*, Fifth Edition, Addison-Wesley Publishing Company, Reading, MA, 1989.

Eckman, Donald P., *Automatic Process Control*, John Wiley & Sons, Inc., New York, NY, 1958.

"Foundation^{TM} Fieldbus Blocks", document 00809-0100-4783, Rev BA, Rosemount, Inc., Chanhassen, MN, 2000.

"Function Blocks Instruction Manual", document FBLOC-FFME, Smar Equipamentos Ind. Ltda., Sertãozinho, Brazil, 2005.

Lavigne, John R., *An Introduction To Paper Industry Instrumentation*, Miller Freeman Publications, Inc., San Francisco, CA, 1972.

Lavigne, John R., *Instrumentation Applications for the Pulp and Paper Industry*, The Foxboro Company, Foxboro, MA, 1979.

Lipták, Béla G. et al., *Instrument Engineers' Handbook – Process Control Volume II*, Third Edition, CRC Press, Boca Raton, FL, 1999.

Mollenkamp, Robert A., *Introduction to Automatic Process Control*, Instrument Society of America, Research Triangle Park, NC, 1984.

Palm, William J., *Control Systems Engineering*, John Wiley & Sons, Inc., New York, NY, 1986.

Shinskey, Francis G., *Energy Conservation through Control*, Academic Press, New York, NY, 1978.

Shinskey, Francis G., *Process-Control Systems – Application / Design / Adjustment*, Second Edition, McGraw-Hill Book Company, New York, NY, 1979.

Chapter 32

Process safety and instrumentation

This chapter discusses instrumentation issues related to industrial process safety. Instrumentation safety may be broadly divided into two categories: how instruments themselves may pose a safety hazard (electrical signals possibly igniting hazardous atmospheres), and how instruments and control systems may be configured to detect unsafe process conditions and automatically shut an unsafe process down.

In either case, the intent of this chapter is to help define and teach how to mitigate hazards encountered in certain instrumented processes. I purposely use the word "mitigate" rather than "eliminate" because the complete elimination of all risk is an impossibility. Despite our best efforts and intentions, no one can absolutely eliminate all dangers from industrial processes[1]. What we can do, though, is *significantly* reduce those risks to the point they begin to approach the low level of "background" risks we all face in daily life, and that is no small achievement.

An important philosophy to follow in the safe design is something called *defense-in-depth*. This is the principle of using multiple layers[2] of protection, in case one or more of those layers fail. Applying defense-in-depth to process design means regarding each and every safety tool and technique as part of a multi-faceted strategy, rather than as a set of mutually-exclusive alternatives.

To give a brief example of defense-in-depth applied to overpressure protection in a fluid processing system, that system might defend against excessive fluid pressure using all of the following techniques:

- A pressure-control system with an operator-adjusted setpoint

- High-pressure alarms to force operator attention

- A safety shutdown system triggered by abnormally high pressure

- Temperature control systems (both regulatory and safety shutdown) to prevent excessive temperature from helping to create excessive fluid pressure

---

[1] For that matter, it is impossible to eliminate all danger from *life in general*. Every thing you do (or don't do) involves some level of risk. The question really should be, "how much risk is there in a given action, and how much risk am I willing to tolerate?" To illustrate, there does exist a non-zero probability that something you will read in this book is so shocking it will cause you to suffer a heart attack. However, the odds of you walking away from this book and never reading it again over concern of epiphany-induced cardiac arrest are just as slim.

[2] Also humorously referred to as the "belt *and* suspenders" school of engineering.

- Pressure-relief valves which automatically open to vent high pressure

- Pressure vessels built with "frangible[3]" tops designed to burst in the safest manner possible

- Locating the process far away from anything (or anyone) that might be harmed by an overpressure event

Any one of these techniques will work to reduce the risk posed by excessive fluid pressure in the system, but all of them used together will provide greater risk reduction than any one used alone.

32.1   Classified areas and electrical safety measures

Any physical location in an industrial facility harboring the potential of explosion due to the presence of flammable process matter suspended in the air is called a *hazardous* or *classified* location. In this context, the label "hazardous" specifically refers to the hazard of explosion, not of other health or safety hazards[4].

---

[3]Frangible roofs are a common design applied to liquid storage tanks harboring the potential for overpressure, such as sulfuric acid storage tanks which may generate accumulations of explosive hydrogen gas. Having the roof seam rupture from overpressure is a far less destructive event than having a side seam or floor seam rupture and consequently spill large volumes of acid. This technique of mitigating overpressure risk does not work to reduce pressure in the system, but it does reduce the risk of damage caused by overpressure in the system.

[4]Chemical corrosiveness, biohazardous substances, poisonous materials, and radiation are all examples of other types of industrial hazards not covered by the label "hazardous" in this context. This is not to understate the danger of these other hazards, but merely to focus our attention on the specific hazard of explosions and how to build instrument systems that will not trigger explosions due to electrical spark.

32.1.1 Classified area taxonomy

In the United States, the National Electrical Code (NEC) published by the National Fire Protection Association (NFPA) defines different categories of "classified" industrial areas and prescribes safe electrical system design practices for those areas. Article 500 of the NEC categorizes classified areas into a system of *Classes* and *Divisions*. Articles 505 and 506[5] of the NEC provide alternative categorizations for classified areas based on *Zones* that is more closely aligned with European safety standards.

The Class and Division taxonomy defines classified areas in terms of hazard type and hazard probability. Each "Class" contains (or may contain) different types of potentially explosive substances: Class I is for gases or vapors, Class II is for combustible dusts, and Class III is for flammable fibers. The three-fold class designation is roughly scaled on the size of the flammable particles, with Class I being the smallest (gas or vapor molecules) and Class III being the largest (fibers of solid matter). Each "Division" ranks a classified area according to the likelihood of explosive gases, dusts, or fibers being present. Division 1 areas are those where explosive concentrations can or do exist under normal operating conditions. Division 2 areas are those where explosive concentrations only exist infrequently or under abnormal conditions[6].

The "Zone" method of area classifications defined in Article 505 of the National Electrical Code applies to Class I (explosive gas or vapor) applications, but the three-fold Zone ranks (0, 1, and 2) are analogous to Divisions in their rating of explosive concentration probabilities. Zone 0 defines areas where explosive concentrations are continually present or normally present for long periods of time. Zone 1 defines areas where those concentrations may be present under normal operating conditions, but not as frequently as Zone 0. Zone 2 defines areas where explosive concentrations are unlikely under normal operating conditions, and when present do not exist for substantial periods of time. This three-fold Zone taxonomy may be thought of as expansion on the two-fold Division system, where Zones 0 and 1 are sub-categories of Division 1 areas, and Zone 2 is nearly equivalent to a Division 2 area[7]. A similar three-zone taxonomy for Class II and Class III applications is defined in Article 506 of the National Electrical Code, the zone ranks for these dust and fiber hazards numbered 20, 21, and 22 (and having analogous meanings to zones 0, 1, and 2 for Class I applications).

An example of a classified area common to most peoples' experience is a vehicle refueling station. Being a (potentially) explosive *vapor*, the hazard in question here is deemed Class I. The Division rating varies with proximity to the fume source. For an upward-discharging vent pipe from an underground gasoline storage tank, the area is rated as Division 1 within 900 millimeters (3 feet) from the vent hole. Between 3 feet and 5 feet away from the vent, the area is rated as Division 2. In relation to an outdoor fuel pump (dispenser), the space internal to the pump enclosure is rated Division 1, and any space up to 18 inches from grade level and up to 20 feet away (horizontally) from the pump is rated Division 2.

---

[5]Article 506 is a new addition to the NEC as of 2008. Prior to that, the only "zone"-based categories were those specified in Article 505.

[6]The final authority on Class and Division definitions is the National Electrical Code itself. The definitions presented here, especially with regard to Divisions, may not be precise enough for many applications. Article 500 of the NEC is quite specific for each Class and Division combination, and should be referred to for detailed information in any particular application.

[7]Once again, the final authority on this is the National Electrical Code, in this case Article 505. My descriptions of Zones and Divisions are for general information only, and may not be specific or detailed enough for many applications.

Within Class I and Class II (but not Class III), the National Electrical Code further sub-divides hazards according to explosive properties called *Groups*. Each group is defined either according to a substance type, or according to specific ignition criteria. Ignition criteria listed in the National Electrical Code (Article 500) include the *maximum experimental safe gap* (MESG) and the *minimum ignition current ratio* (MICR). The MESG is based on a test where two hollow hemispheres separated by a small gap enclose both an explosive air/fuel mixture and an ignition source. Tests are performed with this apparatus to determine the maximum gap width between the hemispheres that will not permit the excursion of flame from an explosion within the hemispheres triggered by the ignition source. The MICR is the ratio of electrical ignition current for an explosive air/fuel mixture compared to an optimum mixture of methane and air. The smaller of either these two values, the more dangerous the explosive substance is.

Class I substances are grouped according to their respective MESG and MICR values, with typical gas types given for each group:

Group	Typical substance	Safe gap	Ignition current
A	Acetylene		
B	Hydrogen	MESG $\leq$ 0.45 mm	MICR $\leq$ 0.40
C	Ethylene	0.45 mm $<$ MESG $\leq$ 0.75 mm	0.40 $<$ MICR $\leq$ 0.80
D	Propane	0.75 mm $<$ MESG	0.80 $<$ MICR

Class II substances are grouped according to material type:

Group	Substances
E	Metal dusts
F	Carbon-based dusts
G	Other dusts (wood, grain, flour, plastic, etc.)

Just to make things confusing, the Class/Zone system described in NEC Article 505 uses a completely different lettering order to describe gas and vapor groups (at the time of this writing there is no grouping of dust or fiber types for the zone system described in Article 506 of the NEC):

Group	Typical substance(s)	Safe gap	Ignition current
IIC	Acetylene, Hydrogen	MESG $\leq$ 0.50 mm	MICR $\leq$ 0.45
IIB	Ethylene	0.50 mm $<$ MESG $\leq$ 0.90 mm	0.45 $<$ MICR $\leq$ 0.80
IIA	Acetone, Propane	0.90 mm $<$ MESG	0.80 $<$ MICR

32.1.2 Explosive limits

In order to have combustion (an explosion being a particularly aggressive form of combustion), certain basic criteria must be satisfied: a proper *oxidizer/fuel ratio*, sufficient *energy* for ignition, and the *potential for a self-sustaining chemical reaction* (i.e. the absence of any chemical inhibitors). We may show these criteria in the form of a *fire triangle*[8], the concept being that removing any of these three critical elements renders a fire (or explosion) impossible:

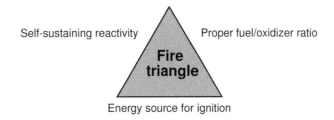

The fire triangle serves as a qualitative guide for *preventing* fires and explosions, but it does not give sufficient information to tell us if the necessary conditions exist to *support* a fire or explosion. In order for a fire or explosion to occur, we need to have an adequate mixture of fuel and oxidizer in the correct proportions, and a source of ignition energy exceeding a certain minimum threshold.

Suppose we had a laboratory test chamber filled with a mixture of acetone vapor (70% by volume) and air at room temperature, with an electrical spark gap providing convenient ignition. No matter how energetic the spark, this mixture would not explode, because there is too *rich* a mixture of acetone (i.e. too much acetone mixed with not enough air). Every time the spark gap discharges, its energy would surely cause some acetone molecules to combust with available oxygen molecules. However, since the air is so dilute in this rich acetone mixture, those scarce oxygen molecules are depleted fast enough that the flame temperature quickly falls off and is no longer hot enough to trigger the remaining oxygen molecules to combust with the plentiful acetone molecules.

The same problem occurs if the acetone/air mixture is too *lean* (not enough acetone and too much air). This is what would happen if we diluted the acetone vapors to a volumetric concentration of only 0.5% inside the test chamber: any spark at the gap would indeed cause some acetone molecules to combust, but there would be too few available to support expansive combustion across the rest of the chamber.

We could also have an acetone/air mixture in the chamber ideal for combustion (about 9.5% acetone by volume) and still not have an explosion if the spark's energy were insufficient. Most combustion reactions require a certain minimum level of *activation energy* to overcome the potential barrier before molecular bonding between fuel atoms and oxidizer atoms occurs. Stated differently, many combustion reactions are not *spontaneous* at room temperature and at atmospheric pressure – they need a bit of "help" to initiate.

---

[8]Traditionally, the three elements of a "fire triangle" were fuel, oxidizer, and ignition source. However, this model fails to account for fuels not requiring oxygen as well as cases where a chemical inhibitor prevents a self-sustaining reaction even in the presence of fuel, oxidizer, and ignition source.

All the necessary conditions for an explosion (assuming no chemical inhibitors are present) may be quantified and plotted as an *ignition curve* for any particular fuel and oxidizer combination. This next graph shows an ignition curve for an hypothetical fuel gas mixed with air:

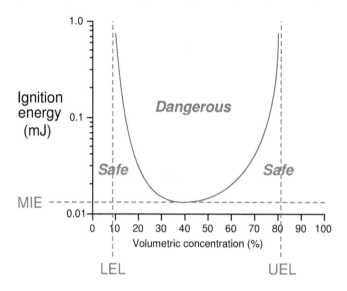

Note how any point in the chart lying *above* the curve is "dangerous," while any point *below* the curve is "safe." The three critical values on this graph are the *Lower Explosive Limit* (LEL), the *Upper Explosive Limit* (UEL), and the *Minimum Ignition Energy* (MIE). These critical values differ for every type of fuel and oxidizer combination, change with ambient temperature and pressure, and may be rendered irrelevant in the presence of a catalyst (a chemical substance that works to promote a reaction without itself being consumed by the reaction). Most ignition curves are published with the assumed conditions of air as the oxidizer, at room temperature and atmospheric pressure, with no catalyst(s) present.

Some substances are so reactive that their minimum ignition energy (MIE) levels are well below the thermal energy of ambient air temperatures. Such fuels will *auto-ignite* the moment they come into contact with air, which effectively means one cannot prevent a fire or explosion by eliminating sources of flame or sparks. When dealing with such substances, the only means for preventing fires and explosions lies with maintaining fuel/air ratios outside of the danger zone (i.e. below the LEL or above the UEL), or by using a chemical inhibitor to prevent a self-sustaining reaction.

The greater the difference in LEL and UEL values, the greater "explosive potential" a fuel gas or vapor presents (all other factors being equal), because it means the fuel may explode over a wider range of mixture conditions. It is instructive to research the LEL and UEL values for many common substances, just to see how "explosive" they are relative to each other:

Substance	LEL (% volume)	UEL (% volume)
Acetylene	2.5%	100%
Acetone	2.5%	12.8%
Butane	1.5%	8.5%
Carbon disulfide	1.3%	50%
Carbon monoxide	12.5%	74%
Ether	1.9%	36%
Ethylene oxide	2.6%	100%
Gasoline	1.4%	7.6%
Kerosene	0.7%	5%
Hydrazine	2.9%	98%
Hydrogen	4.0%	74.2%
Methane	4.4%	17%
Propane	2.1%	9.5%

Note how both acetylene and ethylene oxide have UEL values of 100%. This means it is possible for these gases to explode *even when there is no oxidizer present*. Some other chemical substances exhibit this same property (n-propyl nitrate being another example), where the lack of an oxidizer does not prevent an explosion. With these substances in high concentration, our only practical hope of avoiding explosion is to eliminate the possibility of an ignition source in its presence. Some substances have UEL values so high that the elimination of oxidizers is only an uncertain guard against combustion: hydrazine being one example with a UEL of 98%, and diborane being another example with a UEL of 88%.

32.1.3 Protective measures

Different strategies exist to help prevent electrical devices from triggering fires or explosions in classified areas. These strategies may be broadly divided four ways:

- **Contain the explosion:** enclose the device inside a very strong box that contains any explosion generated by the device so as to not trigger a larger explosion outside the box. This strategy may be viewed as eliminating the "ignition" component of the fire triangle, from the perspective of the atmosphere outside the explosion-proof enclosure (ensuring the explosion inside the enclosure does not ignite a larger explosion outside).

- **Shield the device:** enclose the electrical device inside a suitable box or shelter, then purge that enclosure with clean air (or a pure gas) that prevents an explosive mixture from forming inside the enclosure. This strategy works by eliminating the "proper fuel/oxidizer ratio" component of the fire triangle: by eliminating fuel (if purged by air), or by eliminating oxidizer (if purged by fuel gas), or by eliminating both (if purged by an inert gas).

- **Encapsulated design:** manufacture the device so that it is self-enclosing. In other words, build the device in such a way that any spark-producing elements are sealed air-tight within the device from any explosive atmosphere. This strategy works by eliminating the "ignition" component of the fire triangle (from the perspective of outside the device) or by eliminating the "proper fuel/oxidizer ratio" component (from the perspective of inside the device).

- **Limit total circuit energy:** design the circuit such that there is insufficient energy to trigger an explosion, even in the event of an electrical fault. This strategy works by eliminating the "ignition" component of the fire triangle.

It should be noted that any one of these strategies, correctly and thoroughly applied, is sufficient to mitigate the risk of fire and explosion. For this reason you will seldom see more than one of these strategies simultaneously applied (e.g. an explosion-proof enclosure housing a circuit with insufficient energy to trigger an explosion).

A common example of the first strategy is to use extremely rugged metal *explosion-proof* (NEMA 7 or NEMA 8) enclosures instead of the more common sheet-metal or fiberglass enclosures to house electrical equipment. Two photographs of explosion-proof electrical enclosures reveal their unusually rugged construction:

Note the abundance of bolts securing the covers of these enclosures! This is necessary in order to withstand the enormous forces generated by the pressure of an explosion developing inside the enclosure. Note also how most of the bolts have been removed from the door of the right-hand enclosure. This is an unsafe and very unfortunate occurrence at many industrial facilities, where technicians leave just a few bolts securing the cover of an explosion-proof enclosure because it is so time-consuming to remove all of them to gain access inside the enclosure for maintenance work. Such practices negate the safety of the explosion-proof enclosure, rendering it just as dangerous as a sheet metal enclosure in a classified area.

Explosion-proof enclosures are designed in such a way that high-pressure gases resulting from an explosion within the enclosure must pass through small gaps (either holes in vent devices, and/or the gap formed by a bulging door forced away from the enclosure box) en route to exiting the enclosure. As hot gases pass through these tight metal gaps, they are forced to cool to the point where they will not ignite explosive gases outside the enclosure, thus preventing the original explosion inside the enclosure from triggering a far more violent event. This is the same phenomenon measured in determinations of MESG (Maximum Experimental Safe Gap) for an explosive air/fuel mixture. With an explosion-proof enclosure, all gaps are designed to be less than the MESG for the mixtures in question.

A similar strategy involves the use of a non-flammable *purge gas* pressurizing an ordinary electrical enclosure such that explosive atmospheres are prevented from entering the enclosure. Ordinary compressed air may be used as the purge gas, so long as provisions are made to ensure the air compressor supplying the compressed air is in a non-classified area where explosive gases will never be drawn into the compressed air system.

Devices may be encapsulated in such a way that explosive atmospheres cannot penetrate the device to reach anything generating sufficient spark or heat. *Hermetically sealed* devices are an example of this protective strategy, where the structure of the device has been made completely fluid-tight by fusion joints of its casing. Mercury tilt-switches are good examples of such electrical devices, where a small quantity of liquid mercury is hermetically sealed inside a glass tube. No outside gases, vapors, dusts, or fibers can ever reach the spark generated when the mercury comes into contact (or breaks contact with) the electrodes:

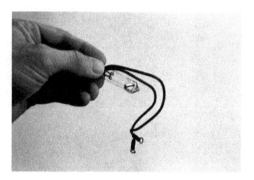 

The ultimate method for ensuring instrument circuit safety in classified areas is to intentionally limit the amount of energy available within a circuit such that it *cannot* generate enough heat or spark to ignite an explosive atmosphere, even in the event of an electrical fault within the circuit. Article 504 of the National Electrical Code specifies standards for this method. Any system meeting these requirements is called an *intrinsically safe* or *I.S.* system. The word "intrinsic" implies that the safety is a natural property of the circuit, since it lacks even the ability to produce an explosion-triggering spark[9].

One way to underscore the meaning of intrinsic safety is to contrast it against a different concept that has the appearance of similarity. Article 500 of the National Electrical Code defines *nonincendive equipment* as devices incapable of igniting a hazardous atmosphere *under normal operating conditions*. However, the standard for nonincendive devices or circuits does not guarantee what will happen under *abnormal* conditions, such as an open- or short-circuit in the wiring. So, a "nonincendive" circuit may very well pose an explosion hazard, whereas an "intrinsically safe" circuit will not because the intrinsically safe circuit simply does not possess enough energy to trigger an explosion under *any* electrical fault condition. As a result, nonincendive circuits are not approved in Class I or Class II Division 1 locations whereas intrinsically safe circuits are approved for all hazardous locations.

---

[9]To illustrate this concept in a different context, consider my own personal history of automobiles. For many years I drove an ugly and inexpensive truck which I joked had "intrinsic theft protection:" it was so ugly, no one would ever want to steal it. Due to this "intrinsic" property of my vehicle, I had no need to invest in an alarm system or any other protective measure to deter theft. Similarly, the components of an intrinsically safe system need not be located in explosion-proof or purged enclosures because the intrinsic energy limitation of the system is protection enough.

Most modern 4 to 20 mA analog signal instruments may be used as part of intrinsically safe circuits so long as they are connected to control equipment through suitable *safety barrier* interfaces, the purpose of which is to limit the amount of voltage and current available at the field device to low enough levels that an explosion-triggering spark is impossible even under fault conditions (e.g. a short-circuit in the field instrument or wiring). A simple intrinsic safety barrier circuit made from passive components is shown in the following diagram[10]:

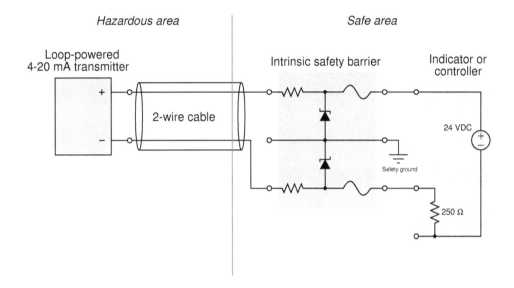

In normal operation, the 4-20 mA field instrument possesses insufficient terminal voltage and insufficient loop current to pose any threat of hazardous atmosphere ignition. The series resistance of the barrier circuit is low enough that the 4-20 mA signal will be unaffected by its presence. As far as the receiving instrument (indicator or controller) is "concerned," the safety barrier might as well not exist.

If a short-circuit develops in the field instrument, the series resistance of the barrier circuit will limit fault current to a value low enough not to pose a threat in the hazardous area. If something fails in the receiving instrument to cause a much greater power supply voltage to develop at its terminals, the zener diode inside the barrier will break down and provide a shunt path for fault current that bypasses the field instrument (and may possibly blow the fuse in the barrier). Thus, the intrinsic safety barrier circuit provides protection against overcurrent *and* overvoltage faults, so that neither type of fault will result in enough electrical energy available at the field device to ignite an explosive atmosphere.

Note that a barrier device such as this *must* be present in the 4-20 mA analog circuit in order for the circuit to be intrinsically safe. The "intrinsic" safety rating of the circuit depends on this barrier, not on the integrity of the field device or of the receiving device. Without this barrier in place, the instrument circuit is not intrinsically safe, even though the *normal* operating voltage and current parameters of the field and receiving devices are well within the parameters of safety for

---

[10]Real passive barriers often used redundant zener diodes connected in parallel to ensure protection against excessive voltage even in the event of a zener diode failing open.

classified areas. It is the barrier and the barrier alone which guarantees those voltage and current levels will remain within safe limits in the event of *abnormal* circuit conditions such as a field wiring short or a faulty loop power supply.

More sophisticated *active* barrier devices are manufactured which provide electrical isolation from ground in the instrument wiring, thus eliminating the need for a safety ground connection at the barrier device.

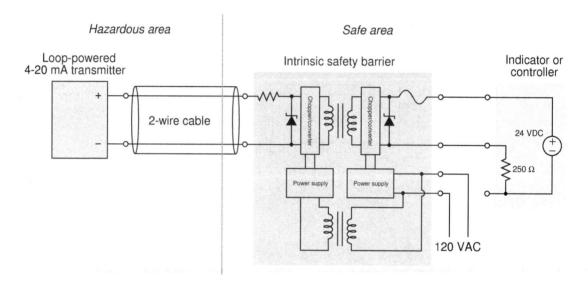

In the example shown here, transformers[11] are used to electrically isolate the analog current signal so that there is no path for DC fault current between the field instrument and the receiving instrument, ground or no ground.

Safety barrier circuits fundamentally limit the amount of power deliverable to a field device from a power supply located in the safe area. Barrier circuits cannot, however, ensure safety for field devices capable of *generating* their own electrical energy. In order for such devices to be considered intrinsically safe, their natural abilities for generating voltage, current, and power must fall below limits defined in NEC article 504. Sensors such as pH electrodes, thermocouples, and photovoltaic light detectors are examples of such field devices, and are called *simple apparatus* by the NEC. The qualifications for a generating device to be a "simple apparatus" is that it cannot generate more than 1.5 volts of voltage, and more than 100 milliamps of current, and more than 25 milliwatts of power. If a device's ability to generate electricity exceeds these limits, the device is not a "simple apparatus" and therefore its circuit is not intrinsically safe.

An example of a generating field device exceeding these limits is a tachogenerator: a small DC generator used to measure the speed of rotating equipment by outputting a DC voltage proportional to speed (typically over a 0-10 volt range). An alternative to a tachogenerator for measuring machine

---

[11]Of course, transformers cannot be used to pass DC signals of any kind, which is why chopper/converter circuits are used before and after the signal transformer to convert each DC current signal into a form of chopped (AC) signal that *can* be fed through the transformer. This way, the *information* carried by each 4-20 mA DC current signal passes through the barrier, but electrical fault current cannot.

speed is an *optical encoder*, using a slotted wheel to chop a light beam (from an LED), generating a pulsed electrical signal of sufficiently low intensity to qualify as a simple apparatus.

Passive (non-generating) field devices may also be classified as "simple apparatus" if they do not dissipate more than 1.3 watts of power. Examples of passive, simple apparatus include switches, LED indicator lamps, and RTD (Resistive Temperature Detector) sensors. Even devices with internal inductance and/or capacitance may be deemed "simple apparatus" if their stored energy capacity is insufficient to pose a hazard.

In addition to the use of barrier devices to create an intrinsically safe circuit, the National Electrical Code (NEC) article 504 specifies certain wiring practices different from normal control circuits. The conductors of an intrinsically safe circuit (i.e. conductors on the "field" side of a barrier) must be separated from the conductors of the non-intrinsically safe circuit (i.e. conductors on the "supply" side of the barrier) by at least 50 millimeters, which is approximately 2 inches. Conductors must be secured prior to terminals in such a way that they cannot come into contact with non-intrinsically safe conductors if the terminal becomes loose. Also, the color *light blue* may be used to identify intrinsically safe conductors, raceways, cable trays, and junction boxes so long as that color is not used for any other wiring in the system.

32.2 Concepts of probability

While the term "probability" may evoke images of imprecision, probability is in fact an exact mathematical concept: *the ratio a specific outcome to total possible outcomes* where 1 (100%) represents certainty and 0 (0%) represents impossibility. A probability value between 1 and 0 describes an outcome that occurs some of the time but not all of the time. Reliability – which is the expression of how likely a device or a system is to function as intended – is based on the mathematics of probability. Therefore, a rudimentary understanding of probability mathematics is necessary to grasp what reliability means.

Before we delve too deeply into discussions of reliability, some definition of terms is in order. We have defined "reliability" to mean the probability of a device or system functioning as designed, which is a good general definition but sometimes not specific enough for our needs. There are usually a variety of different ways in which a device or system can fail, and these different failure modes usually have different probability values. Let's take for example a fire alarm system triggered by a manual pushbutton switch: the intended function of such a system is to activate an alarm whenever the switch is pressed. If we wish to express the reliability of this system, we must first carefully define what we mean by "failure". One way in which this simple fire alarm system could fail is if it remained silent when the pushbutton switch was pressed (i.e. *not* alerting people when it should have). Another, completely different, way this simple system could fail is by accidently sounding the alarm when no one pressed the switch (i.e. alerting people when it had no reason to, otherwise known as a "false alarm"). If we discuss the "reliability" of this fire alarm system, we may need to differentiate between these two different kinds of unreliable system behaviors.

The electrical power industry has an interest in ensuring the safe delivery of electrical power to loads, both ensuring maximum service to customers while simultaneously shutting power off as quickly as possible in the event of dangerous system faults. A complex system of fuses, circuit breakers, and protective relays work together to ensure the flow of power remains uninterrupted as long as safely possible. These protective devices must shut off power when they sense dangerous conditions, but they must also refrain from needlessly shutting off power when there is no danger. Like our fire alarm system which must alert people when needed yet not sound false alarms, electrical protective systems serve two different needs. In order to avoid confusion when quantifying the reliability of electrical protective systems to function as designed, the power industry consistently uses the following terms:

- **Dependability**: The probability a protective system will shut off power when needed

- **Security**: The probability a protective system will allow power to flow when there is no danger

- **Reliability**: A combination of dependability and security

For the sake of clarity I will use these same terms when discussing the reliability of any instrument or control systems. "Dependability" is how reliably a device or system will take appropriate action when it is actively called to do so – in other words, the degree to which we may *depend* on this device or system to do its job when activated. "Security" is how reliably a device or system refrains from taking action when no action should be taken – in other words, the degree to which we may feel *secure* it won't needlessly trigger a system shutdown or generate a false alarm. If there is no need to differentiate, the term "reliability" will be used as a general description of how probable a device or system will do what it was designed to do.

The following matrix should help clarify the meanings of these three terms, defined in terms of what the protective component or system does under various conditions:

	Reliable	Unreliable
Ordinary condition (context of *security*)	No action taken (secure) S	Shut-down (trip)! (unsecure) $\overline{S}$
Emergency condition (context of *dependability*)	Shut-down (trip)! (dependable) D	No action taken (undependable) $\overline{D}$

In summary: a protective function that does not trip when it doesn't need to is *secure*; a protective function that trips when it needs to is *dependable*; a protective system that does both is *reliable*.

The Boolean variables used to symbolize dependability (D), security (S), undependability ($\overline{D}$), and unsecurity ($\overline{S}$) tell us something about the relationships between those four quantities. A bar appearing over a Boolean variable represents the *complement* of that variable. For example, security (S) and unsecurity ($\overline{S}$) are complementary to each other: if we happen to know the probability that a device or system will be secure, then we may calculate with assurance the probability that it is unsecure. A fire alarm system that is 99.3% secure (i.e. 99.3% of the time it generates no false alarms) must generate false alarms the other 0.7% of the time in order to account for all possible system responses 100% of the time no fires occur. If that same fire alarm system is 99.8% dependable (i.e. it alerts people to the presence of a real fire 99.8% of the time), then we may conclude it will fail to report 0.02% of real fire incidents in order to account for all possible responses during 100% of fire incidents.

However, it should be clearly understood that there is no such simple relationship between security (S) and dependability (D) because these two measures refer to completely different conditions and (potentially) different modes of failure. The specific faults causing a fire alarm system to generate a false alarm (an example of an *unsecure* outcome, $\overline{S}$) are quite different from the faults disabling that same fire alarm system in the event of a real fire (an example of an *undependable* outcome, $\overline{D}$). Through the application of redundant components and clever system design we may augment dependability and/or security (sometimes improving one at the expense of the other), but it should be understood that these are really two fundamentally different probability measures and as such are not necessarily related.

32.2.1   Mathematical probability

Probability may be defined as a ratio of specific outcomes to total (possible) outcomes. If you were to flip a coin, there are really only two possibilities[12] for how that coin may land: face-up ("heads") or face-down ("tails"). The probability of a coin falling "tails" is thus one-half ($\frac{1}{2}$), since "tails" is one specific outcome out of two total possibilities. Calculating the probability (P) is a matter of setting up a ratio of outcomes:

$$P(\text{"tails"}) = \frac{\text{"tails"}}{\text{"heads"} + \text{"tails"}} = \frac{1}{2} = 0.5$$

This may be shown graphically by displaying all possible outcomes for the coin's landing ("heads" or "tails"), with the one specific outcome we're interested in ("tails") highlighted for emphasis:

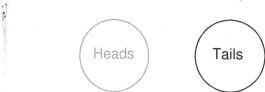

The probability of the coin landing "heads" is of course exactly the same, because "heads" is also *one* specific outcome out of *two* total possibilities.

If we were to roll a six-sided die, the probability of that die landing on any particular side (let's arbitrarily choose the "four" side) is one out of six, because we're looking at one specific outcome out of six total possibilities:

$$P(\text{"four"}) = \frac{\text{"four"}}{\text{"one"} + \text{"two"} + \text{"three"} + \text{"four"} + \text{"five"} + \text{"six"}} = \frac{1}{6} = 0.1\overline{66}$$

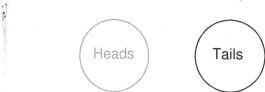

---

[12]To be honest, the coin could also land on its edge, which is a third possibility. However, that third possibility is so remote as to be negligible in the presence of the other two. Strictly speaking, $P(\text{"heads"}) + P(\text{"tails"}) + P(\text{"edge"}) = 1$.

If we were to roll the same six-sided die, the probability of that die landing on an even-numbered side (2, 4, or 6) is three out of six, because we're looking at three specific outcomes out of six total possibilities:

$$P(\text{even}) = \frac{\text{"two"} + \text{"four"} + \text{"six"}}{\text{"one"} + \text{"two"} + \text{"three"} + \text{"four"} + \text{"five"} + \text{"six"}} = \frac{3}{6} = 0.5$$

As a ratio of specific outcomes to total possible outcomes, the probability of any event will always be a number ranging in value from 0 to 1, inclusive. This value may be expressed as a fraction ($\frac{1}{2}$), as a per unit value (0.5), as a percentage (50%), or as a verbal statement (e.g. "three out of six"). A probability value of zero (0) means a specific event is impossible, while a probability of one (1) means a specific event is guaranteed to occur.

Probability values realistically apply only to large samples. A coin tossed ten times may very well fail to land "heads" exactly five times and land "tails" exactly five times. For that matter, it may fail to land on each side exactly 500000 times out of a million tosses. However, so long as the coin and the coin-tossing method are *fair* (i.e. not biased in any way), the experimental results will approach[13] the ideal probability value as the number of trials approaches infinity. Ideal probability values become less and less certain as the number of trials decreases, and are completely useless for singular (non-repeating) events.

A familiar application of probability values is the forecasting of meteorological events such as rainfall. When a weather forecast service provides a rainfall prediction of 65% for a particular day, it means that out of a large number of days sampled in the past having similar measured conditions (cloud cover, barometric pressure, temperature and dew point, etc.), 65% of those days experienced rainfall. This past history gives us some idea of how likely rainfall will be for any present situation, based on similarity of measured conditions.

Like all probability values, forecasts of rainfall are more meaningful with greater samples. If we wish to know how many days with measured conditions similar to those of the forecast day will experience rainfall over the *next ten years*, the forecast probability value of 65% will be quite accurate. However, if we wish to know whether or not rain will fall on any particular (single) day having those same conditions, the value of 65% tells us very little. So it is with all measurements of probability: precise for large samples, ambiguous for small samples, and virtually meaningless for singular conditions[14].

In the field of instrumentation – and more specifically the field of *safety* instrumented systems – probability is useful for the mitigation of hazards based on equipment failures where the probability of failure for specific pieces of equipment is known from mass production of that equipment and years of data gathered describing the reliability of the equipment. If we have data showing the probabilities

---

[13]In his excellent book, *Reliability Theory and Practice*, Igor Bazovsky describes the relationship between true probability (P) calculated from ideal values and estimated probability ($\hat{P}$) calculated from experimental trials as a limit function: $P = \lim_{N \to \infty} \hat{P}$, where N is the number of trials.

[14]Most people can recall instances where a weather forecast proved to be completely false: a prediction for rainfall resulting in a completely dry day, or vice-versa. In such cases, one is tempted to blame the weather service for poor forecasting, but in reality it has more to do with the nature of probability, specifically the meaninglessness of probability calculations in predicting singular events.

of failure for different pieces of equipment, we may use this data to calculate the probability of failure for the system as a whole. Furthermore, we may apply certain mathematical laws of probability to calculate system reliability for different equipment configurations, and therefore minimize the probability of system failure by optimizing those configurations.

As with weather predictions, predictions of system reliability (or conversely, of system failure) become more accurate as the sample size grows larger. Given an accurate probabilistic model of system reliability, a system (or a set of systems) with enough individual components, and a sufficiently long time-frame, an organization may accurately predict the number of system failures and the cost of those failures (or alternatively, the cost of minimizing those failures through preventive maintenance). However, no probabilistic model can accurately predict which component in a large system will fail at any specific point in time.

The ultimate purpose, then, in probability calculations for process systems and automation is to optimize the safety and availability of large systems over many years of time. Calculations of reliability, while useful to the technician in understanding the nature of system failures and how to minimize them, are actually more valuable (more meaningful) at the enterprise level.

32.2.2   Laws of probability

Probability mathematics bears an interesting similarity to Boolean algebra in that probability values (like Boolean values) range between zero (0) and one (1). The difference, of course, is that while Boolean variables may *only* have values equal to zero or one, probability variables range continuously between those limits. Given this similarity, we may apply standard Boolean operations such as NOT, AND, and OR to probabilities. These Boolean operations lead us to our first "laws" of probability for combination events.

The logical "NOT" function

For instance, if we know the probability of rolling a "four" on a six-sided die is $\frac{1}{6}$, then we may safely say the probability of *not* rolling a "four" is $\frac{5}{6}$, the complement of $\frac{1}{6}$. The common "inverter" logic symbol is shown here representing the complementation function, turning a probability of rolling a "four" into the probability of *not* rolling a "four":

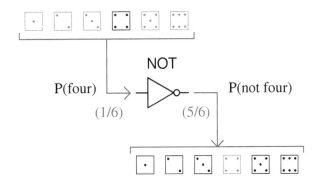

Symbolically, we may express this as a sum of probabilities equal to one:

$$P(\text{total}) = P(\text{``one''}) + P(\text{``two''}) + P(\text{``three''}) + P(\text{``four''}) + P(\text{``five''}) + P(\text{``six''}) = 1$$

$$P(\text{total}) = \frac{1}{6} + \frac{1}{6} + \frac{1}{6} + \frac{1}{6} + \frac{1}{6} + \frac{1}{6} = 1$$

$$P(\text{total}) = P(\text{``four''}) + P(\text{not ``four''}) = \frac{1}{6} + \frac{5}{6} = 1$$

$$P(\text{``four''}) = 1 - P(\text{not ``four''}) = 1 - \frac{5}{6} = \frac{1}{6}$$

We may state this as a general "law" of complementation for any event (A):

$$P(A) = 1 - \overline{P}(A)$$

Complements of probability values find frequent use in reliability engineering. If we know the probability value for the failure of a component (i.e. how likely it is to fail when called upon to function – termed the *Probability of Failure on Demand*, or *PFD* – which is a measure of that component's *undependability*), then we know the *dependability* value (i.e. how likely it is to function on demand) will be the mathematical complement. To illustrate, consider a device with a PFD value of $\frac{1}{100000}$. Such a device could be said to have a dependability value of $\frac{99999}{100000}$, or 99.999%, since $1 - \frac{1}{100000} = \frac{99999}{100000}$.

The logical "AND" function

The AND function regards probabilities of two or more coincidental events (i.e. where the outcome of interest only happens if two or more events happen together, or in a specific sequence). Another example using a die is the probability of rolling a "four" on the first toss, then rolling a "one" on the second toss. It should be intuitively obvious that the probability of rolling this specific combination of values will be less (i.e. less likely) than rolling either of those values in a single toss. The shaded field of possibilities (36 in all) demonstrate the unlikelihood of this sequential combination of values compared to the unlikelihood of either value on either toss:

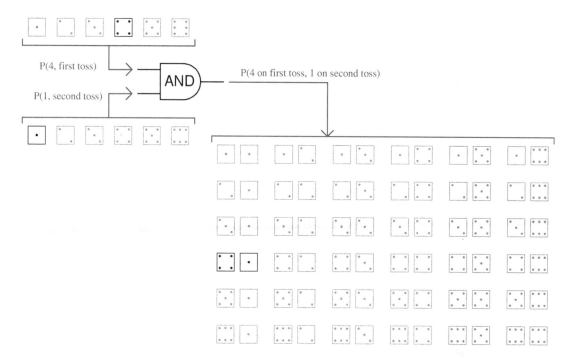

As you can see, there is only one outcome matching the specific criteria out of 36 total possible outcomes. This yields a probability value of one-in-thirty six ($\frac{1}{36}$) for the specified combination, which is the *product* of the individual probabilities. This, then, is our second law of probability:

$$P(\text{A } and \text{ B}) = P(\text{A}) \times P(\text{B})$$

A practical application of this would be the calculation of failure probability for a double-block valve assembly, designed to positively stop the flow of a dangerous process fluid. Double-block valves are used to provide increased assurance of shut-off, since the shutting of *either* block valve is sufficient in itself to stop fluid flow. The probability of failure for a double-block valve assembly – "failure" defined as not being able to stop fluid flow when needed – is the product of each valve's un-dependability (i.e. probability of failing in the open position when commanded to shut off):

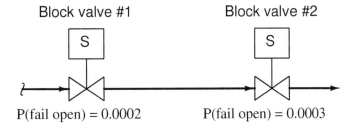

With these two valves in service, the probability of neither valve successfully shutting off flow (i.e. *both* valve 1 *and* valve 2 failing on demand; remaining open when they should shut) is the product of their individual failure probabilities.

$$P(\text{assembly fail}) = P(\text{valve 1 fail open}) \times P(\text{valve 2 fail open})$$

$$P(\text{assembly fail}) = 0.0002 \times 0.0003$$

$$P(\text{assembly fail}) = 0.00000006 = 6 \times 10^{-8}$$

An extremely important assumption in performing such an AND calculation is that the probabilities of failure for each valve are completely unrelated. For instance, if the failure probabilities of both valve 1 and valve 2 were largely based on the possibility of a certain residue accumulating inside the valve mechanism (causing the mechanism to freeze in the open position), and *both* valves were equally susceptible to this residue accumulation, there would be virtually no advantage to having double block valves. If said residue were to accumulate in the piping, it would affect both valves practically the same. Thus, the failure of one valve due to this effect would virtually ensure the failure of the other valve as well. *The probability of simultaneous or sequential events being the product of the individual events' probabilities is true if and only if the events in question are completely independent.*

We may illustrate the same caveat with the sequential rolling of a die. Our previous calculation showed the probability of rolling a "four" on the first toss and a "one" on the second toss to be $\frac{1}{6} \times \frac{1}{6}$, or $\frac{1}{36}$. However, if the person throwing the die is extremely consistent in their throwing technique and the way they orient the die after each throw, such that rolling a "four" on one toss makes it very likely to roll a "one" on the next toss, the sequential events of a "four" followed by a "one" would be far more likely than if the two events were completely random and independent. The probability calculation of $\frac{1}{6} \times \frac{1}{6} = \frac{1}{36}$ holds true only if all the throws' results are completely unrelated to each other.

Another, similar application of the Boolean `AND` function to probability is the calculation of system reliability (R) based on the individual reliability values of components necessary for the system's function. If we know the reliability values for several essential[15] system components, and we also know those reliability values are based on independent (unrelated) failure modes, the overall system reliability will be the product (Boolean `AND`) of those component reliabilities. This mathematical expression is known as *Lusser's product law of reliabilities*:

$$R_{system} = R_1 \times R_2 \times R_3 \times \cdots \times R_n$$

As simple as this law is, it is surprisingly unintuitive. Lusser's Law tells us that any system depending on the performance of several essential components will be *less* reliable than the least-reliable of those components. This is akin to saying that a chain will be *weaker* than its weakest link!

To give an illustrative example, suppose a complex system depended on the reliable operation of six key components in order to function, with the individual reliabilities of those six components being 91%, 92%, 96%, 95%, 93%, and 92%, respectively. Given individual component reliabilities all greater than 90%, one might be inclined to think the overall reliability would be quite good. However, following Lusser's Law we find the reliability of this system (as a whole) is only 65.3% because $0.91 \times 0.92 \times 0.96 \times 0.95 \times 0.93 \times 0.92 = 0.653$.

In his excellent text *Reliability Theory and Practice*, author Igor Bazovsky recounts the German V1 missile project during World War Two, and how early assumptions of system reliability were grossly inaccurate[16]. Once these faulty assumptions of reliability were corrected, development of the V1 missile resulted in greatly increased reliability until a system reliability of 75% (three out of four) was achieved.

---

[15]Here, "essential" means the system will fail if *any* of these identified components fails. Thus, Lusser's Law implies a logical "AND" relationship between the components' reliability values and the overall system reliability.

[16]According to Bazovsky (pp. 275-276), the first reliability principle adopted by the design team was that the system could be no more reliable than its least-reliable (weakest) component. While this is technically true, the mistake was to assume that the system would be *as reliable* as its weakest component (i.e. the "chain" would be exactly as strong as its weakest link). This proved to be too optimistic, as the system would still fail due to the failure of "stronger" components even when the "weaker" components happened to survive. After noting the influence of "stronger" components' unreliabilities on overall system reliability, engineers somehow reached the bizarre conclusion that system reliability was equal to the mathematical *average* of the components' reliabilities. Not surprisingly, this proved even less accurate than the "weakest link" principle. Finally, the designers were assisted by the mathematician Erich Pieruschka, who helped formulate Lusser's Law.

The logical "OR" function

The OR function regards probabilities of two or more redundant events (i.e. where the outcome of interest happens if any one of the events happen). Another example using a die is the probability of rolling a "four" on the first toss *or* on the second toss. It should be intuitively obvious that the probability of rolling a "four" on either toss will be more probable (i.e. more likely) than rolling a "four" on a single toss. The shaded field of possibilities (36 in all) demonstrate the likelihood of this either/or result compared to the likelihood of either value on either toss:

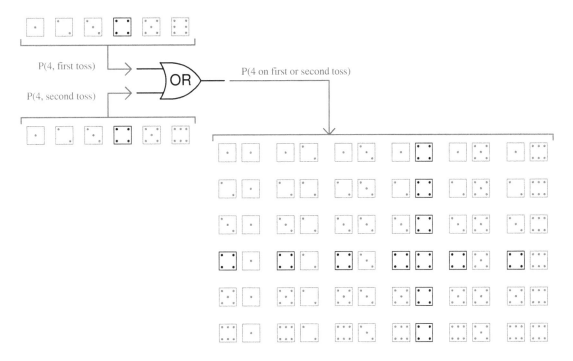

As you can see, there are eleven outcomes matching the specific criteria out of 36 total possible outcomes (the outcome with two "four" rolls counts as a single trial matching the stated criteria, just as all the other trials containing only one "four" roll count as single trials). This yields a probability value of eleven-in-thirty six $\left(\frac{11}{36}\right)$ for the specified combination. This result may defy your intuition, if you assumed the OR function would be the simple *sum* of individual probabilities $\left(\frac{1}{6} + \frac{1}{6} = \frac{2}{6}\right.$ or $\left.\frac{1}{3}\right)$, as opposed to the AND function's *product* of probabilities $\left(\frac{1}{6} \times \frac{1}{6} = \frac{1}{36}\right)$. In truth, there is an application of the OR function where the probability is the simple sum, but that will come later in this presentation.

As with the logical "AND" function, the logical "OR" function assumes the events in question are independent from each other. That is to say, the events lack a common cause, and are not contingent upon one another in any way.

For now, a way to understand why we get a probability value of $\frac{11}{36}$ for our OR function with two $\frac{1}{6}$ input probabilities is to derive the OR function from other functions whose probability laws we already know with certainty. From Boolean algebra, DeMorgan's Theorem tells us an OR function is equivalent to an AND function with all inputs and outputs inverted ($A + B = \overline{\overline{A}\,\overline{B}}$):

(Equivalent logic functions)

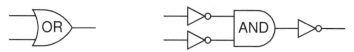

We already know the complement (inversion) of a probability is the value of that probability subtracted from one ($\overline{P} = 1 - P$). This gives us a way to symbolically express the DeMorgan's Theorem definition of an OR function in terms of an AND function with three inversions:

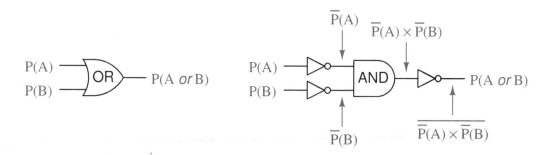

Knowing that $\overline{P}(A) = 1 - P(A)$ and $\overline{P}(B) = 1 - P(B)$, we may substitute these inversions into the triple-inverted AND function to arrive at an expression for the OR function in simple terms of $P(A)$ and $P(B)$:

$$P(A \text{ or } B) = \overline{\overline{P}(A) \times \overline{P}(B)}$$

$$P(A \text{ or } B) = \overline{(1 - P(A))(1 - P(B))}$$

$$P(A \text{ or } B) = 1 - [(1 - P(A))(1 - P(B))]$$

Distributing terms on the right side of the equation:

$$P(A \text{ or } B) = 1 - [1 - P(B) - P(A) + P(A)P(B)]$$

$$P(A \text{ or } B) = P(B) + P(A) - P(A)P(B)$$

This, then, is our third law of probability:

$$P(\text{A } or \text{ B}) = P(B) + P(A) \quad P(A) \times P(B)$$

Inserting our example probabilities of $\frac{1}{6}$ for both $P(A)$ and $P(B)$, we obtain the following probability for the OR function:

$$P(A \text{ or } B) = \frac{1}{6} + \frac{1}{6} - \left(\frac{1}{6}\right)\left(\frac{1}{6}\right)$$

$$P(A \text{ or } B) = \frac{2}{6} - \left(\frac{1}{36}\right)$$

$$P(A \text{ or } B) = \frac{12}{36} - \frac{1}{36}$$

$$P(A \text{ or } B) = \frac{11}{36}$$

This confirms our previous conclusion of there being an $\frac{11}{36}$ probability of rolling a "four" on the first or second rolls of a die.

We may return to our example of a double-block valve assembly for a practical application of OR probability. When illustrating the AND probability function, we focused on the probability of both block valves failing to shut off when needed, since both valve 1 *and* valve 2 would have to fail open in order for the double-block assembly to fail in shutting off flow. Now, we will focus on the probability of *either* block valve failing to open when needed. While the AND scenario was an exploration of the system's un-dependability (i.e. the probability it might fail to stop a dangerous condition), this scenario is an exploration of the system's *un-security* (i.e. the probability it might fail to resume normal operation).

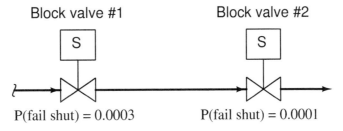

Each block valve is designed to be able to shut off flow independently, so that the flow of (potentially) dangerous process fluid will be halted if *either or both* valves shut off. The probability that process fluid flow may be impeded by the failure of either valve to open is thus a simple (non-exclusive) OR function:

$$P(\text{assembly fail}) = P(\text{valve 1 fail shut}) + P(\text{valve 2 fail shut}) - P(\text{valve 1 fail shut}) \times P(\text{valve 2 fail shut})$$

$$P(\text{assembly fail}) = 0.0003 + 0.0001 - (0.0003 \times 0.0001)$$

$$P(\text{assembly fail}) = 0.0003997 = 3.9997 \times 10^{-4}$$

A similar application of the OR function is seen when we are dealing with *exclusive* events. For instance, we could calculate the probability of rolling either a "three" or a "four" in a single toss of a die. Unlike the previous example where we had two opportunities to roll a "four," and two sequential rolls of "four" counted as a single successful trial, here we know with certainty that the die cannot land on "three" *and* "four" in the same roll. Therefore, the exclusive OR probability (XOR) is much simpler to determine than a regular OR function:

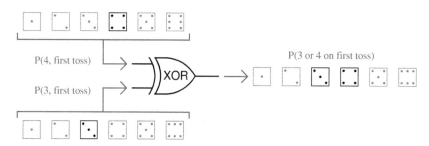

This is the only type of scenario where the function probability is the simple sum of the input probabilities. In cases where the input probabilities are mutually exclusive (i.e. they *cannot* occur simultaneously or in a specific sequence), the probability of one *or* the other happening is the sum of the individual probabilities. This leads us to our fourth probability law:

$$P(\text{A } exclusively \ or \text{ B}) = P(A) + P(B)$$

A practical example of the exclusive-or (XOR) probability function may be found in the failure analysis of a single block valve. If we consider the probability this valve may fail in either condition (stuck open or stuck shut), and we have data on the probabilities of the valve failing open and failing shut, we may use the XOR function to model the system's general unreliability[17]. We know that the exclusive-or function is the appropriate one to use here because the two "input" scenarios (failing open versus failing shut) *absolutely cannot* occur at the same time:

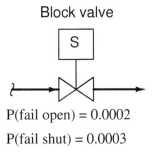

Block valve

P(fail open) = 0.0002

P(fail shut) = 0.0003

$$P(\text{valve fail}) = P(\text{valve fail open}) + P(\text{valve fail shut})$$

$$P(\text{valve fail}) = 0.0002 + 0.0003$$

$$P(\text{valve fail}) = 0.0005 = 5 \times 10^{-4}$$

If the intended safety function of this block valve is to shut off the flow of fluid if a dangerous condition is detected, then the probability of this valve's failure to shut when needed is a measure of its *undependability*. Conversely, the probability of this valve's failure to open under normal (safe) operating conditions is a measure of its *unsecurity*. The XOR of the valve's undependability and its unsecurity therefore represents its *unreliability*. The complement of this value will be the valve's *reliability*: $1 - 0.0005 = 0.9995$. This reliability value tells us we can expect the valve to operate as it's called to 99.95% of the time, and we should expect 5 failures out of every 10,000 calls for action.

---

[17]Here we have an example where dependability and security are lumped together into one "reliability" quantity.

Summary of probability laws

The complement (inversion) of a probability:

$$P(A) = 1 - \overline{P}(A)$$

The probability of coincidental events (where both must happen either simultaneously or in specific sequence) for the result of interest to occur:

$$P(\text{A } and \text{ B}) = P(\text{A}) \times P(\text{B})$$

The probability of redundant events (where either or both may happen) for the result of interest to occur:

$$P(\text{A } or \text{ B}) = P(B) + P(A) - P(A) \times P(B)$$

The probability of exclusively redundant events (where either may happen, but not simultaneously or in specific sequence) for the result of interest to occur:

$$P(\text{A } exclusively \text{ } or \text{ B } exclusively) = P(A) + P(B)$$

32.2.3 Applying probability laws to real systems

The relatively simple concepts of AND and OR Boolean functions become surprisingly complicated when applying them to real-life measures of component reliability, mainly because reliability is measured in multiple ways. As we have already seen, *dependability* (D) and *security* (S) are related concepts in that they both describe the probability of a system or system component functioning properly, but defy simple correlation because they imply different failure modes. "Dependability" for any safety-related system or component is the probability that it will perform its safety function when called upon in an emergency. "Security" by contrast is the probability that the system or component in question will maintain normal operation when there is no emergency.

To illustrate, we will examine the overpressure protection features of a "knock-out drum" used to collect small amounts of liquid entrained in a gas stream. This particular vessel is equipped with two pressure-safety valves (PSV-11 and PSV-12) designed to open and vent gas to atmosphere in the event of an overpressure condition (over 410 PSIG):

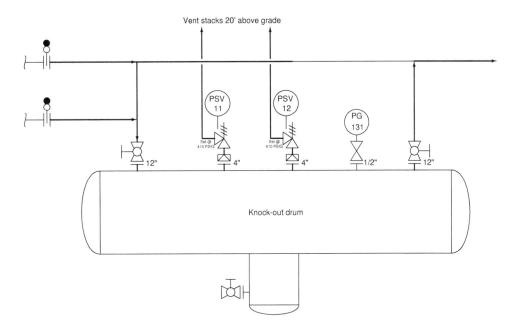

Suppose each of these pressure safety valves has a dependability (D) rating of 0.9992, which means each one has a 99.92% chance of opening up to relieve excess pressure when a high-pressure condition exists. Let us also suppose each of these PSVs has a security (S)[18] rating of 0.995, which means each one has a 99.5% chance of remaining in the shut condition when no overpressure condition exists. Furthermore, assume each of the two pressure safety valves individually has a high enough flow capacity to adequately vent the vessel during an overpressure condition.

---

[18]An easy way to remember what each of these terms mean in the context of a protective system is to associate D (Dependability) with a *dangerous* scenario and S (Security) with a *safe* scenario: D expresses what the system or component will do when a dangerous condition presents itself to the protective system and it needs to act; S expresses what the system or component will do when conditions are safe and there is no need to act.

How might we calculate the overall dependability and security ratings of this dual-PSV overpressure protection system? Clearly, we must use Boolean functions to combine the two valves' D ratings into a D_{system} rating, and likewise with the two valves' S ratings, but which logical function should we use to calculate each measure of reliability? The choice between AND and OR functions may not be obvious at first inspection.

One way to analyze logical functions is in terms of what state (0 or 1) at any input will *guarantee* a certain output state. For an AND function, any 0 state in guarantees a 0 state out. For an OR function, any 1 state in guarantees a 1 state out. These facts are useful when selecting logical functions for a variety of purposes, and they will serve us well in this application of probability values too.

A useful problem-solving technique for this application is called *limiting cases*, where we take some quantity to its extreme limits in an effort to simply the problem at hand. To begin, we will assume that one of the two pressure safety valves in this system has a D rating of 1, which means it is perfectly reliable when called to open by a high-pressure condition. A D rating of 1 is a "limiting case" of the pressure safety valve's dependability: a perfectly dependable PSV. If this were true, would it guarantee the whole overpressure protection system is dependable, or not? Since we know each valve is sized large enough to protect the vessel on its own (without need of the second PSV opening), then the answer to this question is "yes": a single PSV with a D rating of 1 guarantees a D_{system} rating of 1. All we need is for one of these PSVs to vent when it senses a high-pressure condition to protect the vessel from overpressure damage. Therefore, the proper Boolean function to calculate D_{system} from the valves' individual D ratings is the OR function, because given the choices of AND and OR only the OR function guarantees a certain output state with any "1" input. Calculating system dependability using both valves' D ratings:

$$D_{PSV1} = 0.9992 \qquad D_{system} = 0.99999936$$
$$D_{PSV2} = 0.9992$$

The numerical results shown here should make sense: in an overpressure protection system where we only need one of the two valves to vent gas during an overpressure condition, having two valves increases the probability that the vessel will be adequately protected.

Now we will apply this same problem-solving strategy to the system's *security* (S). Taking the high limiting-case value of either PSV's S rating, we ask ourselves the question "Does any one perfectly secure PSV $(S = 1)$ make the system secure?" In other words, if one of these valves was guaranteed not to vent when no overpressure condition exists, would that mean the entire system was guaranteed not to vent when it didn't need to? The answer here is "no", since the presence of *two* pressure safety valves increases the chance of unnecessary leakage. This tells us we cannot use the OR function for security, because a perfectly secure PSV $(S = 1)$ does *not* guarantee a perfectly secure system.

At this point we may conclude that the proper Boolean function for system security in this application is the AND, by process of elimination. However, we may also consider a different limiting-case scenario to verify this conclusion. Let us suppose one of the pressure safety valves failed in such a way that it had *zero* security, meaning there was no chance at all it would remain shut when no overpressure condition existed (i.e. a security rating of $S = 0$ means it is guaranteed to vent when it shouldn't). Would one PSV in this state guarantee a certain system security state? We see here that this is true: any one PSV with an S rating of zero means the system as a whole has a zero S rating as well, because all it takes is one PSV to unnecessarily vent to make the system as a whole unnecessarily vent. Since we know the Boolean AND function guarantees a zero output for any zero input, this is the function we should use when calculating system security. Calculating system security using both valves' S ratings::

$$S_{PSV1} = 0.995$$
$$S_{PSV2} = 0.995$$
$$S_{system} = 0.990025$$

These numerical results should make sense as well: in an overpressure protection system where a leak in one valve is enough to constitute a problem, the presence of multiple valves is a liability and therefore reduces the over-all security.

It is worth noting that a simple change in parameters may strongly impact our reliability calculations. In this scenario we were told each pressure safety valve was sized large enough to adequately vent the vessel on its own, without the help of the other PSV, in the event of an overpressure condition. What if the PSVs were undersized, and *both* of them would be required to vent in order to protect the vessel from overpressure damage? How would this alteration impact our reliability calculations?

It should be obvious that this change will have no effect whatsoever on the system's security, because it still takes just one PSV to leak in order to make the whole system unsecure. However, dependability will definitely be affected by this change because now a single PSV with a $D = 1$ rating is not enough to guarantee a protected system. With undersized PSVs, *both* valves must be dependable in order to guarantee dependable overpressure protection. Conversely, if only one of the PSVs fails in such a way as to be completely undependable ($D = 0$, meaning the valve is guaranteed to fail in the shut condition when faced with high pressure), it makes the whole system undependable because the other valve on its own is not enough to adequately vent the excess gas. From this analysis we can see that the proper Boolean function for dependability will now be AND, because any zero into an AND function guarantees a 0 output. Re-calculating dependability for undersized PSVs:

$$D_{PSV1} = 0.9992$$
$$D_{PSV2} = 0.9992$$
$$D_{system} = 0.99840064$$

32.3 Practical measures of reliability

In reliability engineering, it is important to be able to quantity the reliability (or conversely, the probability of failure) for common components, and for systems comprised of those components. As such, special terms and mathematical models have been developed to describe probability as it applies to component and system reliability.

32.3.1 Failure rate and MTBF

Perhaps the first and most fundamental measure of (un)reliability is the *failure rate* of a component or system of components, symbolized by the Greek letter lambda (λ). The definition of "failure rate" for a group of components undergoing reliability tests is the instantaneous rate of failures per number of surviving components:

$$\lambda = \frac{\frac{dN_f}{dt}}{N_s} \qquad \text{or} \qquad \lambda = \frac{dN_f}{dt}\frac{1}{N_s}$$

Where,
λ = Failure rate
N_f = Number of components failed during testing period
N_s = Number of components surviving during testing period
t = Time

The unit of measurement for failure rate (λ) is inverted time units (e.g. "per hour" or "per year"). An alternative expression for failure rate sometimes seen in reliability literature is the acronym *FIT* ("Failures In Time"), in units of 10^{-9} failures per hour. Using a unit with a built-in multiplier such as 10^{-9} makes it easier for human beings to manage the very small λ values normally associated with high-reliability industrial components and systems.

Failure rate may also be applied to discrete-switching (on/off) components and systems of discrete-switching components on the basis of the number of on/off cycles rather than clock time. In such cases, we define failure rate in terms of cycles (c) instead of in terms of minutes, hours, or any other measure of time (t):

$$\lambda = \frac{\frac{dN_f}{dc}}{N_s} \qquad \text{or} \qquad \lambda = \frac{dN_f}{dc}\frac{1}{N_s}$$

One of the conceptual difficulties inherent to the definition of lambda (λ) is that it is fundamentally a *rate* of failure over time. This is why the calculus notation $\frac{dN_f}{dt}$ is used to define lambda: a "derivative" in calculus always expresses a rate of change. However, a failure *rate* is not the same thing as the number of devices failed in a test, nor is it the same thing as the probability of failure for one or more of those devices. Failure rate (λ) has more in common with the *time constant* of an resistor-capacitor circuit (τ) than anything else.

An illustrative example is helpful here: if we were to test a large batch of identical components for proper operation over some extended period of time with no maintenance or other intervention, the number of failed components in that batch would gradually accumulate while the number of surviving components in the batch would gradually decline. The reason for this is obvious: every component that fails remains failed (with no repair), leaving one fewer surviving component to function. If we limit the duration of this test to a time-span much shorter than the expected lifetime of the components, any failures that occur during the test must be due to random causes ("Acts of God") rather than component wear-out.

This scenario is analogous to another random process: rolling a large set of dice, counting any "1" roll as a "fail" and any other rolled number as a "survive." Imagine rolling the whole batch of dice at once, setting aside any dice landing on "1" aside (counting them as "failed" components in the batch), then only rolling the *remaining* dice the next time. If we maintain this protocol – setting aside "failed" dice after each roll and only continuing to roll "surviving" dice the next time – we will find ourselves rolling fewer and fewer "surviving" dice in each successive roll of the batch. Even though each of the six-sided die has a fixed failure probability of $\frac{1}{6}$, the population of "failed" dice keeps growing over time while the population of "surviving" dice keeps dwindling over time.

Not only does the number of surviving components in such a test dwindle over time, but that number dwindles at an ever-decreasing rate. Likewise with the number of failures: the number of components failing (dice coming up "1") is greatest at first, but then tapers off after the population of surviving components gets smaller and smaller. Plotted over time, the graph looks something like this:

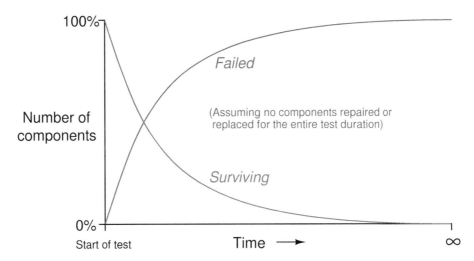

Rapid changes in the failed and surviving component populations occurs at the start of the test when there is the greatest number of functioning components "in play." As components fail due to random events, the smaller and smaller number of surviving components results in a slower approach for both curves, simply because there are fewer surviving components remaining to fail.

These curves are precisely identical to those seen in RC (resistor-capacitor) charging circuits, with voltage and current tracing complementary paths: one climbing to 100% and the other falling to 0%, but both of them doing so at ever-decreasing rates. Despite the asymptotic approach of both curves, however, we can describe their approaches in an RC circuit with a constant value τ, otherwise known as the *time constant* for the RC circuit. Failure rate (λ) plays a similar role in describing the failed/surviving curves of a batch of tested components:

$$N_{surviving} = N_o e^{-\lambda t} \qquad N_{failed} = N_o \left(1 - e^{-\lambda t}\right)$$

Where,

$N_{surviving}$ = Number of components surviving at time t

N_{failed} = Number of components failed at time t

N_o = Total number of components in test batch

e = Euler's constant (≈ 2.71828)

λ = Failure rate (assumed to be a constant during the useful life period)

Following these formulae, we see that 63.2% of the components will fail (36.8% will survive) when $\lambda t = 1$ (i.e. after one "time constant" has elapsed).

Unfortunately, this definition for lambda doesn't make much intuitive sense. There is a way, however, to model failure rate in a way that not only makes more immediate sense, but is also more realistic to industrial applications. Imagine a different testing protocol where we maintain a constant sample quantity of components over the entire testing period by immediately replacing each failed device with a working substitute as soon as it fails. Now, the number of functioning devices under test will remain constant rather than declining as components fail. Imagine counting the number of "fails" (dice falling on a "1") for each batch roll, and then rolling *all* the dice in each successive trial rather than setting aside the "failed" dice and only rolling those remaining. If we did this, we would expect a constant fraction ($\frac{1}{6}$) of the six-sided dice to "fail" with each and every roll. The number of failures per roll divided by the total number of dice would be the failure rate (lambda, λ) for these dice. We do not see a curve over time because we do not let the failed components remain failed, and thus we see a constant number of failures with each period of time (with each group-roll).

We may mathematically express this using a different formula:

$$\lambda = \frac{\frac{N_f}{t}}{N_o} \qquad \text{or} \qquad \lambda = \frac{N_f}{t} \frac{1}{N_o}$$

Where,

λ = Failure rate

N_f = Number of components failed during testing period

N_o = Number of components under test (maintained constant) during testing period by immediate replacement of failed components

t = Time

An alternative way of expressing the failure rate for a component or system is the reciprocal of lambda ($\frac{1}{\lambda}$), otherwise known as *Mean Time Between Failures* (MTBF). If the component or system in question is repairable, the expression *Mean Time To Failure* (MTTF) is often used instead[19]. Whereas failure rate (λ) is measured in reciprocal units of time (e.g. "per hour" or "per year"), MTBF is simply expressed in units of time (e.g. "hours" or "years").

For non-maintained tests where the number of failed components accumulates over time (and the number of survivors dwindles), MTBF is precisely equivalent to "time constant" in an RC circuit: MTBF is the amount of time it will take for 63.2% of the components to fail due to random causes, leaving 36.8% of the component surviving. For maintained tests where the number of functioning components remains constant due to swift repairs or replacement of failed components, MTBF (or MTTF) is the amount of time it will take for the total number of tested components to fail[20].

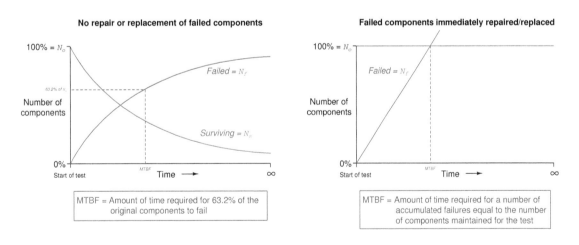

It should be noted that these definitions for lambda and MTBF are idealized, and do not necessarily represent all the complexity we see in real-life applications. The task of calculating

---

[19]Since most high-quality industrial devices and systems are repairable for most faults, MTBF and MTTF are interchangeable terms.

[20]This does not mean the amount of time for *all* components to fail, but rather the amount of time to log a total number of failures equal to the total number of components tested. Some of those failures may be multiple for single components, while some other components in the batch might never fail within the MTBF time.

lambda or MTBF for any real component sample can be quite complex, involving statistical techniques well beyond the scope of instrument technician work.

Simple calculation example: transistor failure rate

Problem: Suppose a semiconductor manufacturer creates a microprocessor "chip" containing 2500000 transistors, each of which is virtually identical to the next in terms of ruggedness and exposure to degrading factors such as heat. The architecture of this microprocessor is such that there is enough redundancy to allow continued operation despite the failure of some of its transistors. This integrated circuit is continuously tested for a period of 1000 days (24000 hours), after which the circuit is examined to count the number of failed transistors. This testing period is well within the useful life of the microprocessor chip, so we know none of the failures will be due to wear-out, but rather to random causes.

Supposing several tests are run on identical chips, with an average of 3.4 transistors failing per 1000-day test. Calculate the failure rate (λ) and the MTBF for these transistors.

Solution: The testing scenario is one where failed components are not replaced, which means both the number of failed transistors and the number of surviving transistors changes over time like voltage and current in an RC charging circuit. Thus, we must calculate lambda by solving for it in the exponential formula.

Using the appropriate formula, relating number of failed components to the total number of components:

$$N_{failed} = N_o \left(1 - e^{-\lambda t}\right)$$

$$3.4 = 2500000 \left(1 - e^{-24000\lambda}\right)$$

$$1.36 \times 10^{-6} = 1 - e^{-24000\lambda}$$

$$e^{-24000\lambda} = 1 - 1.36 \times 10^{-6}$$

$$-24000\lambda = \ln(1 - 1.36 \times 10^{-6})$$

$$-24000\lambda = -1.360000925 \times 10^{-6}$$

$$\lambda = 5.66667 \times 10^{-11} \text{ per hour} = 0.0566667 \text{ FIT}$$

Failure rate may be expressed in units of "per hour," "Failures In Time" (FIT, which means failures per 10^9 hours), or "per year" (pa).

$$\text{MTBF} = \frac{1}{\lambda} = 1.7647 \times 10^{10} \text{ hours} = 2.0145 \times 10^6 \text{ years}$$

Recall that Mean Time Between Failures (MTBF) is essentially the "time constant" for this decaying collection of transistors inside each microprocessor chip.

Simple calculation example: control valve failure rate

Problem: Suppose a control valve manufacturer produces a large number of valves, which are then sold to customers and used in comparable process applications. After a period of 5 years, data is collected on the number of failures these valves experienced. Five years is well within the useful life of these control valves, so we know none of the failures will be due to wear-out, but rather to random causes.

Supposing customers report an average of 15 failures for every 200 control valves in service over the 5-year period, calculate the failure rate (λ) and the MTTF for these control valves.

Solution: The testing scenario is one where failures are repaired in a short amount of time, since these are working valves being maintained in a real process environment. Thus, we may calculate lambda as a simple fraction of failed components to total components.

Using the appropriate formula, relating number of failed components to the total number of components:

$$\lambda = \frac{N_f}{t} \frac{1}{N_o}$$

$$\lambda = \frac{15}{5 \text{ yr}} \frac{1}{200}$$

$$\lambda = \frac{3}{200 \text{ yr}}$$

$$\lambda = 0.015 \text{ per year (pa)} = 1.7123 \times 10^{-6} \text{ per hour}$$

With this value for lambda being so much larger than the microprocessor's transistors, it is not necessary to use a unit such as FIT to conveniently represent it.

$$\text{MTTF} = \frac{1}{\lambda} = 66.667 \text{ years} = 584000 \text{ hours}$$

Recall that Mean Time To Failure (MTTF) is the amount of time it would take[21] to log a number of failures equal to the total number of valves in service, given the observed rate of failure due to random causes. Note that MTTF is largely synonymous with MTBF. The only technical difference between MTBF and MTTF is that MTTF more specifically relates to situations where components are repairable, which is the scenario we have here with well-maintained control valves.

---

[21] The typically large values we see for MTBF and MTTF can be misleading, as they represent a *theoretical* time based on the failure rate seen over relatively short testing times where all components are "young." In reality, the wear-out time of a component will be less than its MTBF. In the case of these control valves, they would likely all "die" of old age and wear long before reaching an age of 66.667 years!

32.3.2   The "bathtub" curve

Failure rate tends to be constant during a component's useful lifespan where the major cause of failure is random events ("Acts of God"). However, lambda does not remain constant over the entire life of the component or system. A common graphical expression of failure rate is the so-called *bathtub curve* showing the typical failure rate profile over time from initial manufacture (brand-new) to wear-out:

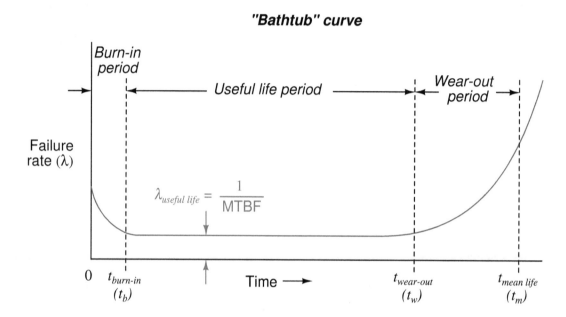

This curve profiles the failure rate of a large sample of components (or a large sample of systems) as they age. Failure rate begins at a relatively high value starting at time zero due to defects in manufacture. Failure rate drops off rapidly during a period of time called the *burn-in period* where defective components experience an early death. After the burn-in period, failure rate remains relatively constant over the useful life of the components, and this is where we typically define and apply the failure rate (λ). Any failures occurring during this "useful life" period are due to random mishaps ("Acts of God"). Toward the end of the components' working lives when the components enter the *wear-out period*, failure rate begins to rise until all components eventually fail. The *mean (average) life* of a component (t_m) is the time required for one-half of the components surviving up until the wear-out time (t_w) to fail, the other half failing after the mean life time.

Several important features are evident in this "bathtub" curve. First, component reliability is greatest between the times of burn-in and wear-out. For this reason, many manufacturers of high-reliability components and systems perform their own burn-in testing prior to sale, so that the customers are purchasing products that have already passed the burn-in phase of their lives. To express this using colloquial terms, we may think of "burnt-in" components as those having already passed through their "growing pains," and are now "mature" enough to face demanding applications.

Another important measure of reliability is the *mean life*. This is an expression of a component's (or system's) operating lifespan. At first this may sound synonymous with MTBF, but it is not.

MTBF – and by extension lambda, since MTBF is the reciprocal of failure rate – is an expression of susceptibility to random ("chance") failures. Both MTBF and λ_{useful} are quite independent of mean life[22]. In practice, values for MTBF often greatly exceed values for mean life.

To cite a practical example, the Rosemount model 3051C differential pressure transmitter has a suggested useful lifetime of 50 years (based on the expected service life of tantalum electrolytic capacitors used in its circuitry), while its demonstrated MTBF is 136 years. The larger value of 136 years is a projection based on the failure rate of large samples of these transmitters when they are all "young," which is why one should never confuse MTBF for service life. In reality, components within the instrument will begin to suffer accelerated failure rates as they reach their end of useful lifetime, as the instrument approaches the right-hand end of the "bathtub" curve.

When determining the length of time any component should be allowed to function in a high-reliability system, the mean life (or even better, the *wear-out* time) should be used as a guide, not the MTBF. This is not to suggest the MTBF is a useless figure – far from it. MTBF simply serves a different purpose, and that is to predict the rate of random failures *during* the useful life span of a large number of components or systems, whereas mean life predicts the service life period where the component's failure rate remains relatively constant.

---

[22]One could even imagine some theoretical component immune to wear-out, but still having finite values for failure rate and MTBF. Remember, λ_{useful} and MTBF refer to *chance* failures, not the normal failures associated with age and extended use.

32.3.3 Reliability

Reliability (R) is the probability a component or system will perform as designed. Like all probability figures, reliability ranges in value from 0 to 1, inclusive. Given the tendency of manufactured devices to fail over time, reliability decreases with time. During the useful life of a component or system, reliability is related to failure rate by a simple exponential function:

$$R = e^{-\lambda t}$$

Where,

R = Reliability as a function of time (sometimes shown as $R(t)$)

e = Euler's constant (≈ 2.71828)

λ = Failure rate (assumed to be a constant during the useful life period)

t = Time

Knowing that failure rate is the mathematical reciprocal of mean time between failures (MTBF), we may re-write this equation in terms of MTBF as a "time constant" (τ) for random failures during the useful life period:

$$R = e^{\frac{-t}{MTBF}} \qquad \text{or} \qquad R = e^{\frac{-t}{\tau}}$$

This inverse-exponential function mathematically explains the scenario described earlier where we tested a large batch of components, counting the number of failed components and the number of surviving components over time. Like the dice experiment where we set aside each "failed" die and then rolled only the remaining "survivors" for the next trial in the test, we end up with a diminishing number of "survivors" as the test proceeds.

The same exponential function for calculating reliability applies to single components as well. Imagine a single component functioning within its useful life period, subject only to random failures. The longer this component is relied upon, the more time it has to succumb to random faults, and therefore the less likely it is to function perfectly over the duration of its test. To illustrate by example, a pressure transmitter installed and used for a period of 1 year has a greater chance of functioning perfectly over that service time than an identical pressure transmitter pressed into service for 5 years, simply because the one operating for 5 years has five times more opportunity to fail. In other words, the reliability of a component over a specified time is a function of time, and not just the failure rate (λ).

Using dice once again to illustrate, it is as if we rolled a single six-sided die over and over, waiting for it to "fail" (roll a "1"). The more times we roll this single die, the more likely it will eventually "fail" (eventually roll a "1"). With each roll, the probability of failure is $\frac{1}{6}$, and the probability of survival is $\frac{5}{6}$. Since survival over multiple rolls necessitates surviving the first roll *and* and next roll *and* the next roll, all the way to the last surviving roll, the probability function we should apply here is the "AND" (multiplication) of survival probability. Therefore, the survival probability after a single roll is $\frac{5}{6}$, while the survival probability for two successive rolls is $\left(\frac{5}{6}\right)^2$, the survival probability for three successive rolls is $\left(\frac{5}{6}\right)^3$, and so on.

The following table shows the probabilities of "failure" and "survival" for this die with an increasing number of rolls:

Number of rolls	Probability of failure (1)	Probability of survival (2, 3, 4, 5, 6)
1	$1 / 6 = 0.16667$	$5 / 6 = 0.83333$
2	$11 / 36 = 0.30556$	$25 / 36 = 0.69444$
3	$91 / 216 = 0.42129$	$125 / 216 = 0.57870$
4	$671 / 1296 = 0.51775$	$625 / 1296 = 0.48225$
n	$1 - \left(\frac{5}{6}\right)^n$	$\left(\frac{5}{6}\right)^n$

A practical example of this equation in use would be the reliability calculation for a Rosemount model 1151 analog differential pressure transmitter (with a demonstrated MTBF value of 226 years as published by Rosemount) over a service life of 5 years following burn-in:

$$R = e^{\frac{-5}{226}}$$

$$R = 0.9781 = 97.81\%$$

Another way to interpret this reliability value is in terms of a large batch of transmitters. If three hundred Rosemount model 1151 transmitters were continuously used for five years following burn-in (assuming no replacement of failed units), we would expect approximately 293 of them to still be working (i.e. 6.564 random-cause failures) during that five-year period:

$$N_{surviving} = N_o e^{\frac{-t}{MTBF}}$$

$$\text{Number of surviving transmitters} = (300) \left(e^{\frac{-5}{226}} \right) = 293.436$$

$$N_{failed} = N_o \left(1 - e^{\frac{-t}{MTBF}} \right)$$

$$\text{Number of failed transmitters} = 300 \left(1 - e^{\frac{-5}{226}} \right) = 6.564$$

It should be noted that the calculation will be linear rather than inverse-exponential if we assume immediate replacement of failed transmitters (maintaining the total number of functioning units at 300). If this is the case, the number of random-cause failures is simply $\frac{1}{226}$ per year, or 0.02212 per transmitter over a 5-year period. For a collection of 300 (maintained) Rosemount model 1151 transmitters, this would equate to 6.637 failed units over the 5-year testing span:

$$\text{Number of failed transmitters} = (300) \left(\frac{5}{226} \right) = 6.637$$

32.3.4 Probability of failure on demand (PFD)

Reliability, as previously defined, is the probability a component or system will perform as designed. Like all probability values, reliability is expressed a number ranging between 0 and 1, inclusive. A reliability value of zero (0) means the component or system is totally unreliable (i.e. it is guaranteed to fail). Conversely, a reliability value of one (1) means the component or system is completely reliable (i.e. guaranteed to properly function). In the context of dependability (i.e. the probability that a safety component or system will function *when called upon to act*), the unreliability of that component or system is referred to as *PFD*, an acronym standing for *Probability of Failure on Demand*. Like dependability, this is also a probability value ranging from 0 to 1, inclusive. A PFD value of zero (0) means there is no probability of failure (i.e. it is 100% dependable – guaranteed to properly perform when needed), while a PFD value of one (1) means it is completely undependable (i.e. guaranteed to fail when activated). Thus:

$$\text{Dependability} + \text{PFD} = 1$$

$$\text{PFD} = 1 - \text{Dependability}$$

$$\text{Dependability} = 1 - \text{PFD}$$

Obviously, a system designed for high dependability should exhibit a small PFD value (very nearly 0). Just how low the PFD needs to be is a function of how critical the component or system is to the fulfillment of our human needs.

The degree to which a system must be dependable in order to fulfill our modern expectations is often surprisingly high. Suppose someone were to tell you the reliability of seatbelts in a particular automobile was 99.9 percent (0.999). This sounds rather good, doesn't it? However, when you actually consider the fact that this degree of probability would mean an average of one failed seatbelt for every 1000 collisions, the results are seen to be rather poor (at least to modern American standards of expectation). If the dependability of seatbelts is 0.999, then the PFD is 0.001:

$$\text{PFD} = 1 - \text{Dependability}$$

$$\text{PFD} = 1 - 0.999$$

$$\text{PFD} = 0.001$$

Let's suppose an automobile manufacturer sets a goal of only 1 failed seatbelt in any of its cars during a 1 million unit production run, assuming each and every one of these cars were to crash. Assuming four seatbelts per car, this equates to a $\frac{1}{4000000}$ PFD. The necessary dependability of this manufacturer's seatbelts must therefore be:

$$\text{Dependability} = 1 - \text{PFD} = 1 - \frac{1}{4000000} = 0.99999975$$

Thus, the dependability of these seatbelts must be 99.999975% in order to fulfill the goal of only 1 (potential) seatbelt failure out of 4 million seatbelts produced.

A common order-of-magnitude expression of desired reliability is the number of "9" digits in the reliability value. A reliability value of 99.9% would be expressed as "three nine's" and a reliability value of 99.99% as "four nine's." Expressed thusly, the seatbelt dependability must be "six nine's" in order to achieve the automobile manufacturer's goal.

32.4 High-reliability systems

As discussed at the beginning of this chapter, instrumentation safety may be broadly divided into two categories: the safety hazards posed by malfunctioning instruments, and special instrument systems designed to reduce safety hazards of industrial processes. This section regards the first category.

All methods of reliability improvement incur some extra cost on the operation, whether it be capital expense (initial purchase/installation cost) or continuing expense (labor or consumables). The choice to improve system reliability is therefore very much an economic one. One of the human challenges associated with reliability improvement is continually justifying this cost over time. Ironically, the more successful a reliability improvement program has been, the less important that program seems. The manager of an operation suffering from reliability problems does not need to be convinced of the economic benefit of reliability improvement as much as the manager of a trouble-free facility. Furthermore, the people most aware of the benefits of reliability improvement are usually those tasked with reliability-improving duties (such as preventive maintenance), while the people least aware of the same benefits are usually those managing budgets. If ever a disagreement erupts between the two camps, pleas for continued financial support of reliability improvement programs may be seen as nothing more than self-interest, further escalating tensions[23].

A variety of methods exist to improve the reliability of systems. The following subsections investigate several of them.

---

[23]Preventive maintenance is not the only example of such a dynamic. Modern society is filled with monetarily expensive programs and institutions existing for the ultimate purpose of avoiding *greater* costs, monetary and otherwise. Public education, health care, and national militaries are just a few that come to my mind. Not only is it a challenge to continue justifying the expense of a well-functioning cost-avoidance program, but it is also a challenge to detect and remove unnecessary expenses (waste) within that program. To extend the preventive maintenance example, an appeal by maintenance personnel to continue (or further) the maintenance budget may happen to be legitimate, but a certain degree of self-interest will always be present in the argument. Just because preventive maintenance is actually necessary to avoid greater expense due to failure, does not mean *all* preventive maintenance demands are economically justified! Proper funding of any such program depends on the financiers being fair in their judgment *and* the executors being honest in their requests. So long as both parties are human, this territory will remain contentious.

32.4.1 Design and selection for reliability

Many workable designs may exist for electronic and mechanical systems alike, but not all are equal in terms of reliability. A major factor in machine reliability, for example, is *balance*. A well-balanced machine will operate with little vibration, whereas an ill-balanced machine will tend to shake itself (and other devices mechanically coupled to it) apart over time[24].

Electronic circuit reliability is strongly influenced by design as well as by component choice. An historical example of reliability-driven design is found in the Foxboro SPEC 200 analog control system. The reliability of the SPEC 200 control system is legendary, with a proven record of minimal failures over many years of industrial use. According to Foxboro technical literature, several design guidelines were developed following application experience with Foxboro electronic field instruments (most notably the "E" and "H" model lines), among them the following:

- All critical switches should spend most of their time in the *closed* state

- Avoid the use of carbon composition resistors – use wirewound or film-type resistors instead

- Avoid the use of plastic-cased semiconductors – use glass-cased or hermetically sealed instead

- Avoid the use of electrolytic capacitors wherever possible – use polycarbonate or tantalum instead

Each of these design guidelines is based on minimization of component failure. Having switches spend most of their lives in the closed state means their contact surfaces will be less exposed to air and therefore less susceptible to corrosion over time (leading to an "open" fault). Wirewound resistors are better able to tolerate vibration and physical abuse than brittle carbon-composition designs. Glass-cased and hermetically-sealed semiconductors are better at sealing out moisture than plastic-cased semiconductors. Electrolytic capacitors are famously unreliable compared to other capacitor types such as polycarbonate, and so their avoidance is wise.

In addition to high-quality component characteristics and excellent design practices, components used in these lines of Foxboro instruments were "burned in" prior to circuit board assembly, thus avoiding many "early failures" due to components "burning in" during actual service.

---

[24]Sustained vibrations can do really strange things to equipment. It is not uncommon to see threaded fasteners undone slowly over time by vibrations, as well as cracks forming in what appear to be extremely strong supporting elements such as beams, pipes, etc. Vibration is almost never good for mechanical (or electrical!) equipment, so it should be eliminated wherever reliability is a concern.

32.4.2   Preventive maintenance

The term *preventive maintenance* refers to the maintenance (repair or replacement) of components prior to their inevitable failure in a system. In order to intelligently schedule the replacement of critical system components, some knowledge of those components' useful lifetimes is necessary. On the standard "bathtub curve," this corresponds with the *wear-out time* or $t_{wear-out}$.

In many industrial operations, preventive maintenance schedules (if they exist at all) are based on past history of component lifetimes, and the operational expenses incurred due to failure of those components. Preventive maintenance represents an up-front cost, paid in exchange for the avoidance of larger costs later in time.

A common example of preventive maintenance and its cost savings is the periodic replacement of lubricating oil and oil filters for automobile engines. Automobile manufacturers provide specifications for the replacement of oil and filters based on testing of their engines, and assumptions made regarding the driving habits of their customers. Some manufacturers even provide dual maintenance schedules, one for "normal" driving and another for "heavy" or "performance" driving to account for accelerated wear. As trivial as an oil change might seem to the average driver, regular maintenance to an automobile's lubrication system is absolutely critical not only to long service life, but also to optimum performance. Certainly, the consequences of not performing this preventive maintenance task on an automobile's engine will be costly[25].

Another example of preventive maintenance for increased system reliability is the regular replacement of light bulbs in traffic signal arrays. For rather obvious reasons, the proper function of traffic signal lights is critical for smooth traffic flow and public safety. It would not be a satisfactory state of affairs to replace traffic signal light bulbs only when they failed, as is common with the replacement of most light bulbs. In order to achieve high reliability, these bulbs must be replaced in advance of their expected wear-out times[26]. The cost of performing this maintenance is undeniable, but then so is the (greater) cost of congested traffic and accidents caused by burned-out traffic light bulbs.

An example of preventive maintenance in industrial instrumentation is the installation and service of *dryer* mechanisms for compressed air, used to power pneumatic instruments and valve actuators. Compressed air is a very useful medium for transferring (and storing) mechanical energy, but problems will develop within pneumatic instruments if water is allowed to collect within air distribution systems. Corrosion, blockages, and hydraulic "locking" are all potential consequences of "wet" instrument air. Consequently, instrument compressed air systems are usually installed separate from utility compressed air systems (used for operating general-purpose pneumatic tools and equipment actuators), using different types of pipe (plastic, copper, or stainless steel rather than black iron or galvanized iron) to avoid corrosion and using *air dryer* mechanisms near the compressor to absorb and expel moisture. These air dryers typically use a beaded *desiccant* material to absorb

---

[25]On an anecdotal note, a friend of mine once destroyed his car's engine, having never performed an oil or filter change on it since the day he purchased it. His poor car expired after only 70000 miles of driving – a mere fraction of its normal service life with regular maintenance. Given the type of car it was, he could have easily expected 200000 miles of service between engine rebuilds had he performed the recommended maintenance on it.

[26]Another friend of mine used to work as a traffic signal technician in a major American city. Since the light bulbs they replaced still had some service life remaining, they decided to donate the bulbs to a charity organization where the used bulbs would be freely given to low-income citizens. Incidentally, this same friend also instructed me on the proper method of inserting a new bulb into a socket: twisting the bulb just enough to maintain some spring tension on the base, rather than twisting the bulb until it will not turn farther (as most people do). Maintaining some natural spring tension on the metal leaf within the socket helps extend the socket's useful life as well!

water vapor from the compressed air, and then this desiccant material is periodically purged of its retained water. After some time of operation, though, the desiccant must be physically removed and replaced with fresh desiccant.

32.4.3 Component de-rating

Some[27] control system components exhibit an inverse relationship between service load (how "hard" the component is used) and service life (how long it will last). In such cases, a way to increase service life is to *de-rate* that component: operate it at a load reduced from its design rating.

For example, a variable-frequency motor drive (VFD) takes AC power at a fixed frequency and voltage and converts it into AC power of varying frequency and voltage to drive an induction motor at different speeds and torques. These electronic devices dissipate some heat owing mostly to the imperfect (slightly resistive) "on" states of power transistors. Temperature is a wear factor for semiconductor devices, with greater temperatures leading to reduced service lives. A VFD operating at high temperature, therefore, will fail sooner than a VFD operating at low temperature, all other factors being equal. One way to reduce the operating temperature of a VFD is to over-size it for the application. If the motor to be driven requires 2 horsepower of electrical power at full load, and increased reliability is demanded of the drive, then perhaps a 5 horsepower VFD (programmed with reduced trip settings appropriate to the smaller motor) could be chosen to drive the motor.

In addition to extending service life, de-rating also has the ability to amplify the mean time between failure (MTBF) of load-sensitive components. Recall that MTBF is the reciprocal of failure rate during the low area of the "bathtub curve," representing failures due to random causes. This is distinct from wear-out, which is an increase in failure rate due to irreversible wear and aging. The main reason a component will exhibit a greater MTBF value as a consequence of de-rating is that the component will be better able to absorb transient overloads, which is a typical cause of failure during the operational life of system components.

Consider the example of a pressure sensor in a process known to exhibit transient pressure surges. A sensor chosen such that the typical process operating pressure spans most of its range will have little overpressure capacity. Perhaps just a few over-pressure events will cause this sensor to fail well before its rated service life. A de-rated pressure sensor (with a pressure-sensing range covering much greater pressures than what are normally encountered in this process), by comparison, will have more pressure capacity to withstand random surges, and therefore exhibit less probability of random failure.

The costs associated with component de-rating include initial investment (usually greater, owing to the greater capacity and more robust construction compared to a "normally" rated component) and reduced sensitivity. The latter factor is an important one to consider if the component is expected to provide high accuracy as well as high reliability. In the example of the de-rated pressure sensor, accuracy will likely suffer because the full pressure range of the sensor is not being used for normal process pressure measurements. If the instrument is digital, resolution will certainly suffer as a result of de-rating the instrument's measurement range. Alternative methods of reliability improvement (including more frequent preventive maintenance) may be a better solution than de-rating in such cases.

---

[27]Many components do not exhibit any relationship between load and lifespan. An electronic PID controller, for example, will last just as long controlling an "easy" self-regulating process as it will controlling a "difficult" unstable ("runaway") process. The same might not be said for the other components of those loops, however! If the control valve in the self-regulating process rarely changes position, but the control valve in the runaway process continually moves in an effort to stabilize it at setpoint, the less active control valve will most likely enjoy a longer service life.

32.4.4 Redundant components

The MTBF of any system dependent upon certain critical components may be extended by duplicating those components in parallel fashion, such that the failure of only one does not compromise the system as a whole. This is called *redundancy*. A common example of component redundancy in instrumentation and control systems is the redundancy offered by distributed control systems (DCSs), where processors, network cables, and even I/O (input/output) channels may be equipped with "hot standby" duplicates ready to assume functionality in the event the primary component fails.

Redundancy tends to extend the MTBF of a system without necessarily extending its service life. A DCS, for example, equipped with redundant microprocessor control modules in its rack, will exhibit a greater MTBF because a random microprocessor fault will be covered by the presence of the spare ("hot standby") microprocessor module. However, given the fact that both microprocessors are continually powered, and therefore tend to "wear" at the same rate, their operating lives will not be additive. In other words, two microprocessors will not function twice as long before wear-out than one microprocessor.

The extension of MTBF resulting from redundancy holds true only if the random failures are truly independent events – that is, not associated by a common cause. To use the example of a DCS rack with redundant microprocessor control modules again, the susceptibility of that rack to a random microprocessor fault will be reduced by the presence of redundant microprocessors *only* if the faults in question are unrelated to each other, affecting the two microprocessors separately. There may exist common-cause fault mechanisms capable of disabling *both* microprocessor modules as easily as it could disable one, in which case the redundancy adds no value at all. Examples of such common-cause faults include power surges (because a surge strong enough to kill one module will likely kill the other at the same time) and a computer virus infection (because a virus able to attack one will be able to attack the other just as easily, and at the same time).

A simple example of component redundancy in an industrial instrumentation system is dual DC power supplies feeding through a diode module. The following photograph shows a typical example, in this case a pair of Allen-Bradley AC-to-DC power supplies for a DeviceNet digital network:

If either of the two AC-to-DC power supplies happens to fail with a low output voltage, the other power supply is able to carry the load by passing its power through the diode redundancy module[28]:

Power supply redundancy module

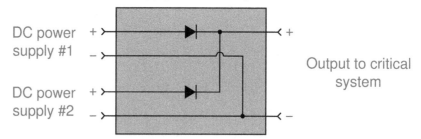

---

[28]This redundancy module has its own MTBF value, and so by including it in the system we are adding one more component that can fail. However, the MTBF rate of a simple diode network greatly exceeds that of an entire AC-to-DC power supply, and so we find ourselves at a greater level of reliability using this diode redundancy module than if we did not (and only had one power supply).

In order for redundant components to actually increase system MTBF, the potential for common-cause failures must be addressed. For example, consider the effects of powering redundant AC-to-DC power supplies from the exact same AC line. Redundant power supplies would increase system reliability in the face of a random power supply failure, but this redundancy would do *nothing at all* to improve system reliability in the event of the common AC power line failing! In order to enjoy the fullest benefit of redundancy in this example, we must source each AC-to-DC power supply from a different (unrelated) AC line.

Another example of redundancy in industrial instrumentation is the use of multiple transmitters to sense the same process variable, the notion being that the critical process variable will still be monitored even in the event of a transmitter failure. Thus, installing redundant transmitters should increase the MTBF of the system's sensing ability.

Here again, we must address common-cause failures in order to reap the full benefits of redundancy. If three liquid level transmitters are installed to measure the exact same liquid level, their combined signals represent an increase in measurement system MTBF *only* for independent faults. A failure mechanism common to all three transmitters will leave the system just as vulnerable to random failure as a single transmitter. In order to achieve optimum MTBF in redundant sensor arrays, the sensors must be immune to common faults.

In this example, three different types of level transmitter monitor the level of liquid inside a vessel, their signals processed by a *selector* function programmed inside a DCS:

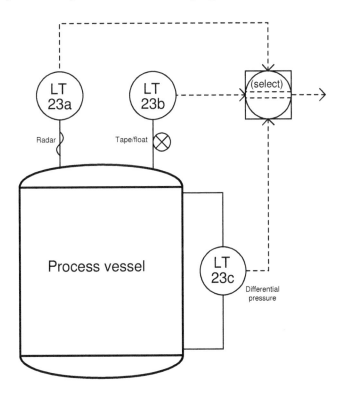

Here, level transmitter 23a is a guided-wave radar (GWR), level transmitter 23b is a tape-and-float, and level transmitter 23c is a differential pressure sensor. All three level transmitters sense liquid level using different technologies, each one with its own strengths and weaknesses. Better redundancy of measurement is obtained this way, since no single process condition or other random event is likely to fault more than one of the transmitters at any given time.

For instance, if the process liquid density happened to suddenly change, it would affect the measurement accuracy of the differential pressure transmitter (LT-23c), but not the radar transmitter nor the tape-and-float transmitter. If the process vapor density were to suddenly change, it might affect the radar transmitter (since vapor density generally affects dielectric constant, and dielectric constant affects the propagation velocity of electromagnetic waves, which in turn will affect the time taken for the radar pulse to strike the liquid surface and return), but this will not affect the float transmitter's accuracy nor will it affect the differential pressure transmitter's accuracy. Surface turbulence of the liquid inside the vessel may severely affect the float transmitter's ability to accurately sense liquid level, but it will have little effect on the differential pressure transmitter's reading nor the radar transmitter's measurement (assuming the radar transmitter is shrouded in a *stilling well*.

If the selector function takes either the median (middle) measurement or an average of the best 2-out-of-3 ("2oo3"), none of these random process occurrences will greatly affect the selected measurement of liquid level inside the vessel. True redundancy is achieved here, since the three level transmitters are not only less likely to (all) fail simultaneously than for any single transmitter to fail, but also because the level is being sensed in three completely different ways.

A crucial requirement for redundancy to be effective is that all redundant components must have precisely the same process function. In the case of redundant DCS components such as processors, I/O cards, and network cables, each of these redundant components must do nothing more than serve as "backup" spares for their primary counterparts. If a particular DCS node were equipped with two processors – one as the primary and another as a secondary (backup) – but yet the backup processor were tasked with some detail specific to it and not to the primary processor (or vice-versa), the two processors would *not* be truly redundant to each other. If one processor were to fail, the other would not perform *exactly* the same function, and so the system's operation would be affected (even if only in a small way) by the processor failure.

Likewise, redundant sensors must perform the exact same process measurement function in order to be truly redundant. A process equipped with triplicate measurement transmitters such as the previous example were a vessel's liquid level was being measured by a guided-wave radar, tape-and-float, and differential pressure based level transmitters, would enjoy the protection of redundancy if and only if all three transmitters sensed the exact same liquid level over the exact same calibrated range. This often represents a challenge, in finding suitable locations on the process vessel for three different instruments to sense the exact same process variable. Quite often, the pipe fittings penetrating the vessel (often called *nozzles*) are not conveniently located to accept multiple instruments at the points necessary to ensure consistency of measurement between them. This is often the case when an existing process vessel is retrofitted with redundant process transmitters. New construction is usually less of a problem, since the necessary nozzles and other accessories may be placed in their proper positions during the design stage[29].

If fluid flow conditions inside a process vessel are excessively turbulent, multiple sensors installed

---

[29]Of course, this assumes good communication and proper planning between all parties involved. It is not uncommon for piping engineers and instrument engineers to mis-communicate during the crucial stages of process vessel design, so that the vessel turns out not to be configured as needed for redundant instruments.

to measure the same variable will sometimes report significant differences. Multiple temperature transmitters located in close proximity to each other on a distillation column, for example, may report significant differences of temperature if their respective sensing elements (thermocouples, RTDs) contact the process liquid or vapor at points where the flow patterns vary. Multiple liquid level sensors, even of the same technology, may report differences in liquid level if the liquid inside the vessel swirls or "funnels" as it enters and exits the vessel.

Not only will substantial measurement differences between redundant transmitters compromise their ability to function as "backup" devices in the event of a failure, such differences may actually "fool" a redundant system into thinking one or more of the transmitters has already failed, thereby causing the deviating measurement to be ignored. To use the triplicate level-sensing array as an example again, suppose the radar-based level transmitter happened to register two inches greater level than the other two transmitters due to the effects[30] of liquid swirl inside the vessel. If the selector function is programmed to ignore such deviating measurements, the system degrades to a duplicate-redundant instead of triplicate-redundant array. In the event of a dangerously low liquid level, for example, only the radar-based and float-based level transmitters will be ready to signal this dangerous process condition to the control system, because the pressure-based level transmitter is registering too high.

---

[30]If a swirling fluid inside the vessel encounters a stationary baffle, it will tend to "pile up" on one side of that baffle, causing the liquid level to actually be greater in that region of the vessel than anywhere else inside the vessel. Any transmitter placed within this region will register a greater level, regardless of the measurement technology used.

32.4.5   Proof tests and self-diagnostics

A reliability enhancing technique related to preventive maintenance of critical instruments and functions, but generally not as expensive as component replacement, is periodic *testing* of component and system function. Regular "proof testing" of critical components enhances the MTBF of a system for two different reasons:

- Early detection of developing problems

- Regular "exercise" of components

First, proof testing may reveal weaknesses developing in components, indicating the need for replacement in the near future. An analogy to this is visiting a doctor to get a comprehensive exam – if this is done regularly, potentially fatal conditions may be detected early and crises averted.

The second way proof testing increases system reliability is by realizing the beneficial effects of regular function. The performance of many component and system types tends to degrade after prolonged periods of inactivity[31]. This tendency is most prevalent in mechanical systems, but holds true for some electrical components and systems as well. Solenoid valves, for instance, may become "stuck" in place if not cycled for long periods of time. Bearings may corrode and seize in place if left immobile. Both primary- and secondary-cell batteries are well known for their tendency to fail after prolonged periods of non-use. Regular cycling of such components actually *enhances* their reliability, decreasing the probability of a "stagnation" related failure well before the rated useful life has elapsed.

An important part of any proof-testing program is to ensure a ready stock of spare components is kept on hand in the event proof-testing reveals a failed component. Proof testing is of little value if the failed component cannot be immediately repaired or replaced, and so these warehoused components should be configured (or be easily configurable) with the exact parameters necessary for immediate installation. A common tendency in business is to focus attention on the engineering and installation of process and control systems, but neglect to invest in the support materials and infrastructure to keep those systems in excellent condition. High-reliability systems have special needs, and this is one of them.

---

[31]The father of a certain friend of mine has operated a used automobile business for many years. One of the tasks given to this friend when he was a young man, growing up helping his father in his business, was to regularly drive some of the cars on the lot which had not been driven for some time. If an automobile is left un-operated for many weeks, there is a marked tendency for batteries to fail and tires to lose their air pressure, among other things. The salespeople at this used car business jokingly referred to this as *lot rot*, and the only preventive measure was to routinely drive the cars so they would not "rot" in stagnation. Machines, like people, suffer if subjected to a lack of physical activity.

Methods of proof testing

The most direct method of testing a critical system is to stimulate it to its range limits and observe its reaction. For a process transmitter, this sort of test usually takes the form of a full-range calibration check. For a controller, proof testing would consist of driving all input signals through their respective ranges in all combinations to check for the appropriate output response(s). For a final control element (such as a control valve), this requires full stroking of the element, coupled with physical leakage tests (or other assessments) to ensure the element is having the intended effect on the process.

An obvious challenge to proof testing is how to perform such comprehensive tests without disrupting the process in which it functions. Proof-testing an out-of-service instrument is a simple matter, but proof-testing an instrument installed in a working system is something else entirely. How can transmitters, controllers, and final control elements be manipulated through their entire operating ranges without actually disturbing (best case) or halting (worst case) the process? Even if all tests may be performed at the required intervals during shut-down periods, the tests are not as realistic as they could be with the process operating at typical pressures and temperatures. Proof-testing components during actual "run" conditions is the most realistic way to assess their readiness.

One way to proof-test critical instruments with minimal impact to the continued operation of a process is to perform the tests on only some components, not all. For instance, it is a relatively simple matter to take a transmitter out of service in an operating process to check its response to stimuli: simply place the controller in manual mode and let a human operator control the process manually while an instrument technician tests the transmitter. While this strategy admittedly is not comprehensive, at least proof-testing some of the instruments is better than proof-testing none of them.

Another method of proof-testing is to "test to shutdown:" choose a time when operations personnel plan on shutting the process down anyway, then use that time as an opportunity to proof-test one or more critical component(s) necessary for the system to run. This method enjoys the greatest degree of realism, while avoiding the inconvenience and expense of an unnecessary process interruption.

Yet another method to perform proof tests on critical instrumentation is to accelerate the speed of the testing stimuli so that the final control elements will not react fully enough to actually disrupt the process, but yet will adequately assess the responsiveness of all (or most) of the components in question. The nuclear power industry sometimes uses this proof-test technique, by applying high-speed pulse signals to safety shutdown sensors in order to test the proper operation of shutdown logic, without actually shutting the reactor down. The test consists of injecting short-duration pulse signals at the sensor level, then monitoring the output of the shutdown logic to ensure consequent pulse signals are sent to the shutdown device(s). Various chemical and petroleum industries apply a similar proof-testing technique to safety valves called *partial stroke testing*, whereby the valve is stroked only part of its travel: enough to ensure the valve is capable of adequate motion without closing (or opening, depending on the valve function) enough to actually disrupt the process.

Redundant systems offer unique benefits and challenges to component proof-testing. The benefit of a redundant system in this regard is that any one redundant component may be removed from service for testing without any special action by operations personnel. Unlike a "simplex" system where removal of an instrument requires a human operator to manually take over control during the duration of the test, the "backup" components of a redundant system should do this automatically, theoretically making the test much easier to conduct. However, the challenge of doing this is the fact that the portion of the system responsible for ensuring seamless transition in the event of a failure is in fact a component liable to failure itself. The only way to test this component is to actually disable one (or more, in highly redundant configurations) of the redundant components to see whether or not the remaining component(s) perform their redundant roles. So, proof-testing a redundant system harbors no danger if all components of the system are good, but risks process disruption if there happens to be an undetected fault.

Let us return to our triplicate level transmitter system once again to explore these concepts. Suppose we wished to perform a proof-test of the pressure-based level transmitter. Being one of three transmitters measuring liquid level in this vessel, we should be able to remove it from service with no preparation (other than notifying operations personnel of the test, and of the potential consequences) since the selector function should automatically de-select the disabled transmitter and continue measuring the process via the remaining two transmitters. If the proof-testing is successful, it proves not only that the transmitter works, but also that the selector function adequately performed its task in "backing up" the tested transmitter while it was removed. However, if the selector function happened to be failed when we disable the one level transmitter for proof-testing, the selected process level signal could register a faulty value instead of switching to the two remaining transmitters' signals. This might disrupt the process, especially if the selected level signal went to a control loop or to an automatic shutdown system. We could, of course, proceed with the utmost caution by having operations personnel place the control system in "manual" mode while we remove that one transmitter from service, just in case the redundancy does not function as designed. Doing so, however, fails to fully test the system's redundancy, since by placing the system in manual mode before the test we do not allow the redundant logic to fully function as it would be expected to in the event of an actual instrument failure.

Regular proof-testing is an essential activity to realize optimum reliability for any critical system. However, in all proof-testing we are faced with a choice: either test the components to their fullest degree, in their normal operating modes, and risk (or perhaps guarantee) a process disruption; or perform a test that is less than comprehensive, but with less (or no) risk of process disruption. In the vast majority of cases, the latter option is chosen simply due to the costs associated with process disruption. Our challenge as instrumentation professionals is to formulate proof tests that are as comprehensive as possible while being the least disruptive to the process we are trying to regulate.

Instrument self-diagnostics

One of the great advantages of digital electronic technology in industrial instrumentation is the inclusion of *self-diagnostic* ability in field instruments. A "smart" instrument containing its own microprocessor may be programmed to detect certain conditions known to indicate sensor failure or other problems, then signal the control system that something is wrong. Though self-diagnostics can never be perfectly effective in that there will inevitably be cases of undetected faults and even false positives (declarations of a fault where none exists), the current state of affairs is considerably better than the days of purely analog technology where instruments possessed little or no self-diagnostic capability.

Digital field instruments have the ability to communicate self-diagnostic error messages to their host systems over the same "fieldbus" networks they use to communicate regular process data. FOUNDATION Fieldbus instruments in particular have extensive error-reporting capability, including a "status" variable associated with every process signal that propagates down through all function blocks responsible for control of the process. Detected faults are efficiently communicated throughout the information chain in the system when instruments have full digital communication ability.

"Smart" instruments with self-diagnostic ability but limited to analog (e.g. 4-20 mA DC) signaling may also convey error information, just not as readily or as comprehensively as a fully digital instrument. The NAMUR recommendations for 4-20 mA signaling (NE-43) provide a means to do this:

Signal level	Fault condition
Output $\leq$ 3.6 mA	Sensing transducer failed low
3.6 mA $<$ Output $<$ 3.8 mA	Sensing transducer failed (detected) low
3.8 mA $\leq$ Output $<$ 4.0 mA	Measurement under-range
21.0 $>$ Output $\geq$ 20.5 mA	Measurement over-range
Output $\geq$ 21.0 mA	Sensing transducer failed high

Proper interpretation of these special current ranges, of course, demands a receiver capable of accurate current measurement outside the standard 4-20 mA range. Many control systems with analog input capability are programmed to recognize the NAMUR error-indicating current levels.

A challenge for any self-diagnostic system is how to check for faults in the "brain" of the unit itself: the microprocessor. If a failure occurs within the microprocessor of a "smart" instrument – the very component responsible for performing logic functions related to self-diagnostic testing – how would it be able to detect a fault in logic? The question is somewhat philosophical, equivalent to determining whether or not a neurologist is able to diagnose his or her own neurological problems.

One simple method of detecting gross faults in a microprocessor system is known as a *watchdog timer*. The principle works like this: the microprocessor is programmed to output continuous a low-frequency pulse signal, with an external circuit "watching" that pulse signal for any interruptions or freezing. If the microprocessor fails in any significant way, the pulse signal will either skip pulses or "freeze" in either the high or low state, thus indicating a microprocessor failure to the "watchdog" circuit.

One may construct a watchdog timer circuit using a pair of solid-state timing relays connected to the pulse output channel of the microprocessor device:

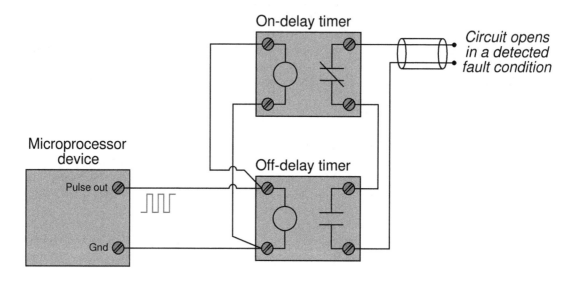

Both the on-delay and off-delay timers receive the same pulse signal from the microprocessor, their inputs connected directly in parallel with the microprocessor's pulse output. The off-delay timer immediately actuates upon receiving a "high" signal, and begins to time when the pulse signal goes "low." The on-delay timer begins to time during a "high" signal, but immediately de-actuates whenever the pulse signal goes "low." So long as the time settings for the on-delay and off-delay timer relays are greater than the "high" and "low" durations of the watchdog pulse signal, respectively, neither relay contact will open as long as the pulse signal continues in its regular pattern.

When the microprocessor is behaving normally, outputting a regular watchdog pulse signal, the off-delay timer's contact will hold in a closed state because it keeps getting energized with each "high" signal and never has enough time to drop out during each "low" signal. Likewise, the on-delay timer's contact will remain in its normally closed state because it never has enough time to pick up during each "high" signal before being de-actuated with each "low" signal. Both timing relay contacts will be in a closed state when all is well.

However, if the microprocessor's pulse output signal happens to freeze in the "low" state (or skip a "high" pulse), the off-delay timer will de-actuate, opening its contact and signaling a fault. Conversely, if the microprocessor's pulse signal happens to freeze in the "high" state (or skip a "low" pulse), the on-delay timer will actuate, opening its contact and signaling a fault. Either timing relay opening its contact signals an interruption or cessation of the watchdog pulse signal, indicating a serious microprocessor fault.

32.5 Overpressure protection devices

Fluid pressure exerts force on any surface area it contacts, as described by the formula $F = PA$. One practical consequence of this fact is that process vessels and pipelines may catastrophically burst if subjected to excessive fluid pressure. If subjected to excessive vacuum, some vessels and piping may implode (collapse in on themselves). Not only do these potential failures pose operational problems, but they may also pose severe safety and environmental hazards, especially if the process fluid in question is toxic, flammable, or both.

Special safety devices exist to help prevent such unfortunately events from occurring, among them being *rupture disks*, *relief valves*, and *safety valves*. The following subsections describe each of these protective devices and their intended operation. In a P&ID, rupture disks and relief valves are represented by the following symbols:

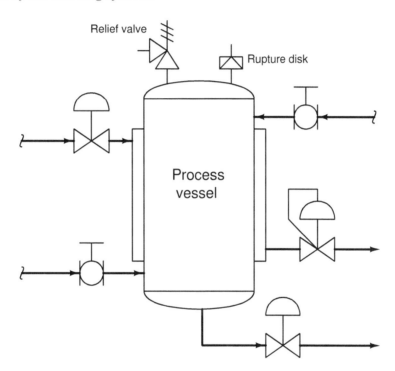

A rupture disk acts like an electrical fuse for overpressure protection: when the burst pressure is exceeded, the disk ruptures to let fluids escape through it. Safety and relief valves work like self-resetting circuit breakers: they open to relieve pressure, then re-close to seal the process system once more.

Two common causes of process overpressure are *piping blockages* and overheating caused by *fires*. Although it may sound ridiculous, a number of fatal industrial accidents have been caused by something as simple as shut block valves that should have been left open. When fluid cannot escape a process vessel, the pumping forces may exceed the burst rating of the vessel, causing catastrophic failure. Fires may also cause overpressure conditions, owing to the expansion of process fluids inside sealed vessels. Overpressure protection devices play a crucial role in such scenarios, venting process fluid so as to avoid bursting the vessel. It should be mentioned that these two causes of overpressure may have vastly differing protection requirements: the required flow rate of exiting fluid to safely limit pressure may be far greater in a "fire case" than it is for a "blockage case," which means overpressure protection devices sized for the latter may be insufficient to protect against the former.

Overpressure protection device selection is a task restricted to the domain of process safety engineers.   Instrument technicians may be involved in the installation and maintenance of overpressure protection devices, but only a qualified and licensed engineer should decide which specific device(s) to use for a particular process system.

32.5.1 Rupture disks

One of the simplest forms of overpressure protection for process lines and vessels is a device known as a *rupture disk*. This is nothing more than a thin sheet of material (usually alloy steel) designed to rupture in the event of an overpressure condition. The amount of force applied to this thin metal sheet is given by the formula $F = PA$ (force equals pressure times area). The thin metal sheet is designed to rupture at a certain threshold of force equivalent to the burst pressure ($P = \frac{F}{A}$). Once the disk ruptures, the fluid vents through new-formed path, thus relieving pressure. Like an electrical fuse, a rupture disk is a one-time device which must be replaced after it has "blown."

A photograph of a small rupture disk (prior to being placed in service) appears here:

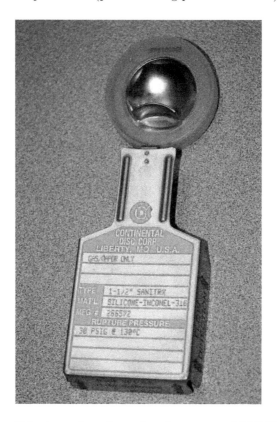

The particular rupture disk shown here has a burst pressure of 30 PSI at a temperature of 130 ^{o}C. Temperature is an important factor in the rating of a rupture disk, as the physical strength of the thin metal rupture element changes with temperature. This metal disk is usually quite thin, usually in the order of 0.002 to 0.060 inches in thickness.

Some modern rupture disks use a *graphite* rupture element instead of metal. Not only does graphite exhibit better corrosion resistance to process fluids than metal, but it also does not fatigue in the same way that metal will over time. Burst pressure for a graphite rupture disk, therefore, may be more consistent than with a metal disk. A significant disadvantage of graphite rupture disks, however, is their tendency to shatter upon bursting. Metal rupture disks merely tear, but a graphite rupture disk tends to break into small pieces which are then carried away by the exiting fluid. These graphite shards may exit as shrapnel, or even lodge inside the mechanism of a valve if one is installed downstream of the rupture disk.

32.5.2   Direct-actuated safety and relief valves

Pressure Relief Valves (PRVs) and *Pressure Safety Valves* (PSVs) are special types of valves designed to open up in order to relieve excess pressure from inside a process vessel or piping system. These valves are normally shut, opening only when sufficient fluid pressure develops across them to relieve that process fluid pressure and thereby protect the pipes and vessels upstream. Unlike regular control valves, PRVs and PSVs are actuated by the process fluid pressure itself rather than by some external pressure or force (e.g. pneumatic signal pressure, electrical motor or solenoid coil).

While the terms "Relief Valve" and "Safety Valve" are sometimes interchanged, there is a distinct difference in operation between them. A *relief valve* opens in direct proportion to the amount of overpressure it experiences in the process piping. That is, a PRV will open slightly for slight overpressures, and open more for greater overpressures. Pressure Relief Valves are commonly used in liquid services. By contrast, a *safety valve* opens fully with a "snap action" whenever it experiences a sufficient overpressure condition, not closing until the process fluid pressure falls significantly below that "lift" pressure value. In other words, a PSV's action is *hysteretic*[32]. Pressure Safety Valves are commonly used in gas and vapor services, such as compressed air systems and steam systems.

Safety valves typically have two pressure ratings: the pressure value required to initially open ("lift") the valve, and the pressure value required to reseat (close) the valve. The difference between these two pressure is called the *blowdown* pressure. A safety valve's lift pressure will always exceed its reseat pressure, giving the valve a hysteretic behavior.

---

[32]A simple "memory trick" I use to correctly distinguish between relief and safety valves is to remember that a safety valve has snap action (both words beginning with the letter "s").

This photograph shows a Varec pressure relief valve on an industrial hot water system, designed to release pressure to atmosphere if necessary to prevent damage to process pipes and vessels in the system:

The vertical pipe is the atmospheric vent line, while the bottom flange of this PRV connects to the pressurized hot water line. A large spring inside the relief valve establishes the lift pressure.

A miniature pressure relief valve manufactured by Nupro, cut away to show its internal components, appears in this next photograph. The pipe fittings on this valve are 1/4 inch NPT, to give a sense of scale:

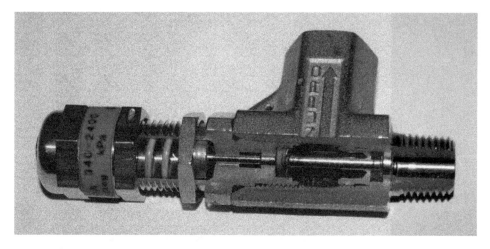

A close-up photograph shows the plug and seat inside this PRV, pointed to by the tip of a ball-point pen:

A simple tension-adjusting mechanism on a spring establishes this valve's lift pressure. The spring exerts a force on the stem to the right, pressing the plug against the face of the seat. A knob allows manual adjustment of spring tension, relating directly to lift pressure:

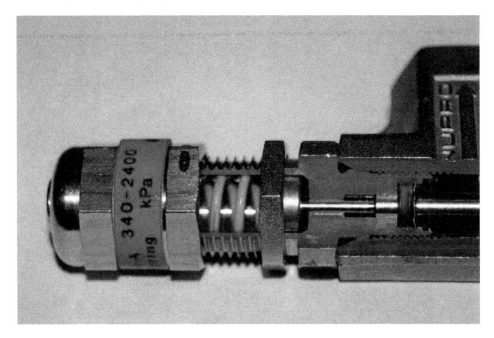

The operation of this relief valve mechanism is quite simple: process fluid pressure entering the right-hand side fitting exerts force against the plug, which normally blocks passage of the fluid through to the side fitting. The area of the plug serves as a piston for the fluid pressure to push against, the amount of force predicted by the familiar force-pressure-area formula $F = PA$. If the fluid pressure exerts enough force on the plug's end to lift it off the seat against the restraining force of the spring (on the left-hand side of the valve mechanism), the plug lifts and vents fluid pressure through the side port.

It is worthy to note that most relief valve mechanisms work on the exact same principle of actuation: *the valve's plug serves as its own actuator*. The pressure difference across this plug provides all the motive force necessary to actuate the valve. This simplicity translates to a high degree of reliability, a desirable quality in any safety-related system component.

Another style of overpressure valve appears in this next photograph. Manufactured by the Groth corporation, this is a combination pressure/vacuum safety valve assembly for an underground tank, designed to vent excess pressure to atmosphere *or* introduce air to the tank in the event of excess vacuum forming inside:

Even when buried, the threat of damage to the tank from overpressure is quite real. The extremely large surface area of the tank's interior walls represents an incredible amount of force potential even with low gas pressures[33]. By limiting the amount of differential gas pressure which may exist between the inside and outside of the tank, the amount of stress applied to the tank walls by gas pressure or vacuum is correspondingly limited.

Large storage tanks – whether above-ground or subterranean – are typically thin-wall for reasons of economics, and cannot withstand significant pressures or vacuums. An improperly vented storage tank may burst with only slight pressure inside, or collapse inwardly with only a slight vacuum inside. Combination pressure/vacuum safety valves such as this Groth model 1208 unit reduce the chances of either failure from happening.

Of course, an alternative solution to this problem is to continuously vent the tank with an open vent pipe at the top. If the tank is always vented to atmosphere, it cannot build up either a pressure or a vacuum inside. However, continuous venting means vapors could escape from the tank if the liquid stored inside is volatile. Escaping vapors may constitute product loss and/or negative environmental impact, being a form of *fugitive emission*. In such cases it is prudent to vent the tank

---

[33]To illustrate, consider a (vertical) cylindrical storage tank 15 feet tall and 20 feet in diameter, with an internal gas pressure of 8 inches water column. The total force exerted radially on the walls of this tank from this very modest internal pressure would be in excess of *39000 pounds!* The force exerted by the same pressure on the tank's circular lid would exceed *13000 pounds* (6.5 tons)!

with an automatic valve such as this only when needed to prevent pressure-induced stress on the tank walls.

An illustration shows the interior construction of this safety valve:

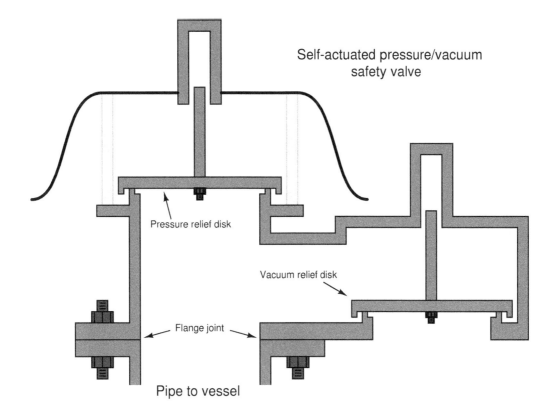

Like the miniature Nupro relief valve previously shown, the trim of this Groth safety valve acts as its own actuator: process gas pressure directly forces the vent plug off its seat, while process gas vacuum forces the vacuum plug off its seat. The lift pressure and vacuum ratings of the Groth valve are quite low, and so no spring is used to provide restraining force to the plugs. Rather, the *weight* of the plugs themselves holds them down on their seats against the force of the process gas.

This set of illustrations shows a pressure/vacuum safety valve in both modes of operation:

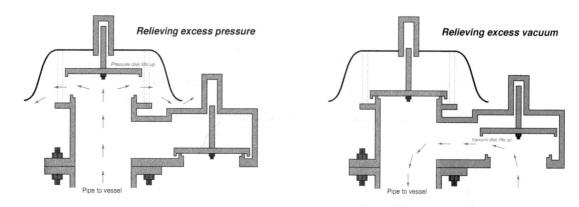

In each mode, the respective disk lifts up against the force of its own weight to allow gases to flow through the valve. If a greater lift pressure (or lift *vacuum*) rating is desired, precise weights may be fixed to the top of either disk. Greater weights equate to greater pressures, following the familiar equation $P = \frac{F}{A}$, where F is the force of gravity acting on the disk and weight(s) and A is the area of the disk.

For example, suppose the disk in one of these safety valves weighs 8 pounds and has a diameter of 9 inches. The surface area for a circular disk nine inches in diameter is 63.62 square inches ($A = \pi r^2$), making the lift pressure equal to 0.126 PSI ($P = \frac{F}{A}$). Such low pressures are often expressed in units other than PSI in order to make the numbers more manageable. The lift pressure of 0.126 PSI for this safety valve might alternatively be described as 3.48 inches water column or 0.867 kPa.

A close inspection of this valve design also provides clues as to why it is technically a *safety* valve rather than a *relief* valve. Recall that the distinction between these two types of overpressure-protection valves was that a relief valve opens proportionally to the experienced overpressure, while a safety valve behaves in a "snap" action manner[34], opening at the lift pressure and not closing again until a (lower) re-seating pressure is achieved.

The "secret" to achieving this snap-action behavior characteristic of safety valves is to design the valve's plug in such a way that it presents a larger surface area for the escaping process fluid to act upon once open than it does when closed. This way, less pressure is needed to hold the valve open than to initially lift it from a closed condition.

---

[34]Think: a safety valve has snap action!

Examining the pressure-relief mechanism of the Groth valve design closer, we see how the plug's diameter exceeds that of the seating area, with a "lip" extending down. This wide plug, combined with the lip forms an effective surface area when the plug is lifted that is larger than that exposed to the process pressure when the plug is seated. Thus, the process fluid finds it "easier" to hold the plug open than to initially lift it off the seat. This translates into a reseating pressure that is less than the lift pressure, and a corresponding "snap action" when the valve initially lifts off the seat.

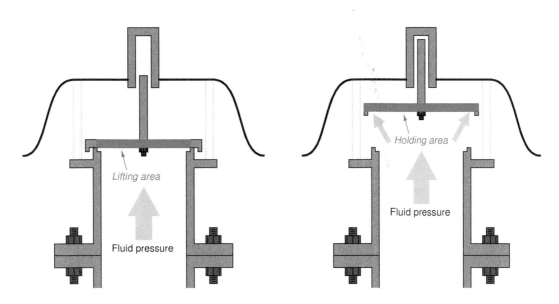

The extra area on the plug's lower surface enclosed by the lip (i.e. the holding area minus the lifting area) is sometimes referred to as a *huddling chamber*. The size of this "huddling chamber" and the length of the lip establishes the degree of hysteresis (blowdown) in the safety valve's behavior.

A certain class of overpressure valve called a *safety relief valve* is designed with an adjustable "blowdown ring" to allow variations in the huddling chamber's geometry. Essentially, the blowdown ring acts as an inner lip on the valve seat to complement the outer lip on the plug. Adjusting this inner lip farther away from the plug allows more process fluid to escape laterally without touching the plug, thereby minimizing the effect of the huddling chamber and making the valve behave as a simple relief valve with no snap-action. Adjusting the blowdown ring closer to the plug forces the escaping fluid to travel toward the plug's face before reversing direction past the outer lip, making the huddling chamber more effective and therefore providing snap-action behavior. This adjustability allows the safety relief valve to act as a simple relief valve (i.e. opening proportional to overpressure) or as a safety valve (snap action) with varying amounts of blowdown ($P_{blowdown} = P_{lift} - P_{reseat}$) as determined by the user. This blowdown ring's position is typically locked into place with a seal to discourage tampering once the valve is installed in the process.

This next photograph shows a cutaway of a safety relief valve manufactured by Crosby, mounted on a cart for instructional use at Bellingham Technical College:

The adjusting bolt marked by the letter "A" at the top of the valve determines the lift pressure setting, by adjusting the amount of pre-load on the spring. Like the Nupro and Groth valves shown previously, the Crosby valve's plug serves as its own actuator, the actuating force being a function of differential pressure across the valve and plug/seat area ($F = PA$).

The toothed gear-like component directly left of the letter "J" is called a *guide ring*, and it functions as a blowdown adjustment. This ring forms a "lip" around the valve seat's edge much like the lip shown in the Groth valve diagrams. If the guide ring is turned to set it at a lower position (extending further past the seat), the volume of the huddling chamber increases, thereby increasing the blowdown value (i.e. keeping the valve open longer than it would be otherwise as the pressure falls).

An interesting combination of overpressure-protection technologies sometimes seen in industry are rupture disks combined with safety valves. Placing a rupture disk before a safety valve provides the benefits of ensuring zero leakage during normal operation as well as isolating the safety valve from potentially corrosive effects of the process fluid:

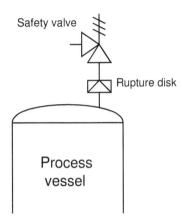

Potential problems with this strategy include the possibility of accumulating vapor pressure between the rupture disk and the safety valve (thereby increasing the effective burst pressure of the disk), and also the possibility of rupture disk shards becoming lodged in the safety valve mechanism, restricting flow and/or preventing re-closure.

32.5.3   Pilot-operated safety and relief valves

While many safety and relief valves actuate by the direct action of the process fluid forcing against the valve plug mechanism, others are more sophisticated in design, relying on a secondary pressure-sensing mechanism to trigger and direct fluid pressure to the main valve assembly to actuate it. This pressure-sensing mechanism is called a *pilot*, and usually features a widely-adjustable range to give the overall valve assembly a larger variety of applications.

In a pilot-operated overpressure-protection valve, the "lift" pressure value is established by a spring adjustment in the pilot mechanism rather than by an adjustment made to the main valve mechanism. A photograph[35] of a pilot-operated pressure relief valve used on a liquid petroleum pipeline appears here:

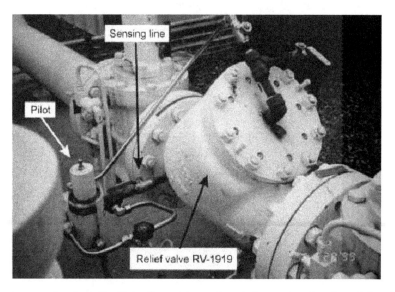

The relief valve mechanism itself is the white-painted flanged valve found in the center-right region of the photograph (RV-1919). This particular relief valve happens to be a Fisher model 760 with 8-inch, ANSI 300# flanges. The actuating pilot mechanism is the small unit connected to the relief valve body via stainless-steel tubing. When this pilot senses fluid pressure in the pipeline exceeding the lift pressure, it switches fluid pressure to the piston actuating mechanism of the main relief valve, opening it to relieve fluid pressure from the pipeline. Thus, the lift pressure value for the relief valve is set within the pilot rather than within the main valve mechanism. Altering this lift pressure setting is a matter of adjusting spring tension within the pilot mechanism, and/or replacing components within the pilot mechanism.

---

[35]This photograph courtesy of the National Transportation Safety Board's report of the 1999 petroleum pipeline rupture in Bellingham, Washington. Improper setting of this relief valve pilot played a role in the pipeline rupture, the result of which was nearly a quarter-million gallons of gasoline spilling into a creek and subsequently igniting. One of the lessons to take from this event is the importance of proper instrument maintenance and configuration, and how such technical details concerning industrial components may have consequences reaching far beyond the industrial facility where those components are located.

32.6 Safety Instrumented Functions and Systems

A *Safety Instrumented Function*, or *SIF*, is one or more components designed to execute a specific safety-related task in the event of a specific dangerous condition. The over-temperature shutdown switch inside a clothes dryer or an electric water heater is a simple, domestic example of an SIF, shutting off the source of energy to the appliance in the event of a detected over-temperature condition. Safety Instrumented Functions are alternatively referred to as *Instrument Protective Functions*, or *IPF*s.

A *Safety Instrumented System*, or *SIS*, is a collection of SIFs designed to bring an industrial process to a safe condition in the event of any dangerous detected conditions. Also known as *Emergency Shutdown* (ESD) or *Protective Instrument Systems* (PIS), these systems serve as an additional "layer" of protection against process equipment damage, adverse environmental impact, and/or human injury beyond the protection normally offered by a properly operating regulatory control system. Like all automatic control systems, an SIS consists of three basic sections: (1) Sensor(s) to detect a dangerous condition, (2) Controller to decide when to shut down the process, and (3) Final control element(s) to actually perform the shutdown action necessary to bring the process to a safe condition. Sensors may consist of process switches and/or transmitters separate from the regulatory control system. The controller for an SIS is usually called a *logic solver*, and is also separate from the regular control system. The final control elements for an SIS may be special on/off valves (often called "chopper" valves) or override solenoids used to force the normal control valve into a shutdown state.

Some industries, such as chemical processing and nuclear power, have extensively employed safety instrumented systems for many decades. Likewise, automatic shutdown controls have been standard on steam boilers and combustion furnaces for years. The increasing capability of modern instrumentation, coupled with the realization of enormous costs (both social and fiscal) resulting from industrial disasters has pushed safety instrumentation to new levels of sophistication and new breadths of application. It is the purpose of this section to explore some common safety instrumented system concepts as well as some specific industrial applications.

One of the challenges inherent to safety instrumented system design is to balance the goal of maximum safety against the goal of maximum economy. If an industrial manufacturing facility is equipped with enough sensors and layered safety shutdown systems to virtually ensure no unsafe condition will ever prevail, that same facility will be plagued by "false alarm" and "spurious trip" events[36] where the safety systems malfunction in a manner detrimental to the profitable operation of the facility. In other words, a process system designed with an emphasis on automatic shut-down will probably shut down more frequently than it actually needs to. While the avoidance of unsafe process conditions is obviously a noble goal, it cannot come at the expense of economically practical operation or else there will be no reason for the facility to exist at all[37]. A safety system must fulfill

---

[36]Many synonyms exist to describe the action of a safety system needlessly shutting down a process. The term "nuisance trip" is often (aptly) used to describe such events. Another (more charitable) label is "fail-to-safe," meaning the failure brings the process to a safe condition, as opposed to a dangerous condition.

[37]Of course, there do exist industrial facilities operating at a financial loss for the greater public benefit (e.g. certain waste processing operations), but these are the exception rather than the rule. It is obviously the point of a *business* to turn a profit, and so the vast majority of industries simply cannot sustain a philosophy of safety at *any* cost. One could argue that a "paranoid" safety system even at a waste processing plant is unsustainable, because too many "false trips" result in inefficient processing of the waste, posing a greater public health threat the longer it remains unprocessed.

its intended protective function, but not at the expense of compromising the intended purpose of the facility.

This tension is understood well within the electric power generation and distribution industries. Faults in high-voltage electrical lines can be very dangerous, as well as destructive to electrical equipment. For this reason, special protective devices are placed within power systems to monitor conditions and halt the flow of electricity if those conditions become threatening. However, the very presence of these devices means it is possible for power to accidently shut off, causing unnecessary power outages for customers. In the electrical industry, the word "dependability" refers to the probability that the protective systems will cut power when required. By contrast, the word "security" is used in the electrical industry to refer to the avoidance of unnecessary outages. We will apply these terms to general process systems.

To illustrate the tension between dependability and security in a fluid process system, we may analyze a double-block shutoff valve[38] system for a petroleum pipeline:

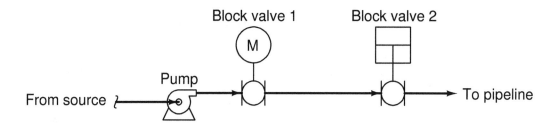

The safety function of these block valves is, of course, to shut off flow from the petroleum source to the distribution pipeline in the event that the pipeline suffers a leak or rupture. Having two block valves in "series" adds an additional layer of safety, in that only one of the block valves need shut to fulfill the safety (dependability) function. Note the use of two different valve actuator technologies: one electric (motor) and the other a piston (either pneumatic or hydraulically actuated). This diversity of actuator technologies helps avoid common-cause failures, helping to ensure both valves will not simultaneously fail due to a single cause.

However, the typical operation of the pipeline demands both block valves be open in order for petroleum to flow through it. The presence of redundant (dual) block valves, while increasing safety, decreases security for the pipeline. If *either* of the two block valves happened to fail shut when there was no need to shut off the pipeline, flow through the pipeline would needlessly halt. Having two series-plumbed block valves instead of one block valve increases the probability of unnecessary pipeline shutdowns.

---

[38] As drawn, these valves happen to be ball-design, the first actuated by an electric motor and the second actuated by a pneumatic piston. As is often the case with redundant instruments, an effort is made to diversify the technology applied to the redundant elements in order to minimize the probability of common-cause failures. If both block valves were electrically actuated, a failure of the electric power supply would disable both valves. If both block valves were pneumatically actuated, a failure of the compressed air supply would disable both valves. The use of one electric valve and one pneumatic valve grants greater independence of operation to the double-block valve system.

A precise notation useful for specifying dependability and security in redundant systems compares the number of redundant elements necessary to achieve the desired result compared to the total number of redundant elements. If the desired result for our double-block valve array is to shut down the pipeline in the event of a detected leak or rupture, we would say the system is *one out of two* (1oo2) redundant for dependability. In other words, only one out of the two redundant valves needs to function properly (shut off) in order to bring the pipeline to a safe condition. If the desired result is to allow flow through the pipeline when the pipeline is leak-free, we would say the system is *two out of two* (2oo2) redundant for security. This means *both* of the two block valves need to function properly (open up) in order to allow petroleum to flow through the pipeline.

This numerical notation showing the number of essential elements versus number of total elements is often referred to as *MooN* ("*M* out of *N*") notation, or sometimes as *NooM* ("*N* out of *M*") notation[39]. When discussing safety instrumented systems, the ISA standard 84 defines redundancy in terms of the number of agreeing channels necessary to perform the safety (shutdown) function – in other words, the ISA's usage of "MooN" notation implies dependability, rather than security.

A complementary method of quantifying dependability and security for redundant systems is to label in terms of how many element failures the system may sustain while still achieving the desired result. For this series set of double block valves, the safety (shutdown) function has a *fault tolerance* of one (1), since one of the valves may fail to shut when called upon but the other valve remains sufficient in itself to shut off the flow of petroleum to the pipeline. The normal operation of the system, however, has a fault tolerance of zero (0). Both block valves must open up when called upon in order to establish flow through the pipeline.

---

[39]For what it's worth, the ISA safety standard 84 defines this notation as "MooN," but I have seen sufficient examples of the contrary ("NooM") to question the authority of either label.

It should be clearly evident that a series set of block valves emphasizes dependability (the ability to shut off flow through the pipeline when needed) at the expense of security (the ability to allow normal flow through the pipeline when there is no leak). We may now analyze a parallel block valve scheme to compare its redundant characteristics:

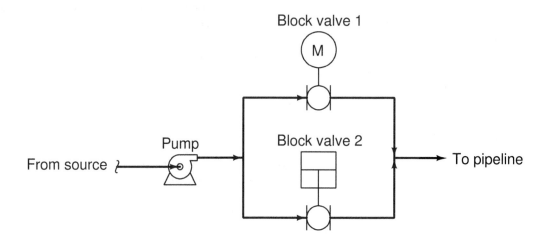

In this system, the safety (dependability) redundancy function is 2oo2, since *both* block valves would have to shut off in order to bring the pipeline to a safe condition in the event of a detected pipeline leak. However, security would be 1oo2, since only one of the two valves would have to open up in order to establish flow through the pipeline. Thus, a parallel block valve array emphasizes production (the ability to allow flow through the pipeline) over safety (the ability to shut off flow through the pipeline).

Another way to express the redundant behavior of the parallel block valve array is to say that the safety function has a fault tolerance of zero (0), while the production function has a fault tolerance of one (1).

One way to avoid compromises between dependability and security is to increase the number of redundant components, forming arrays of greater complexity. Consider this quadruple block valve array, designed to serve the same function on a petroleum pipeline:

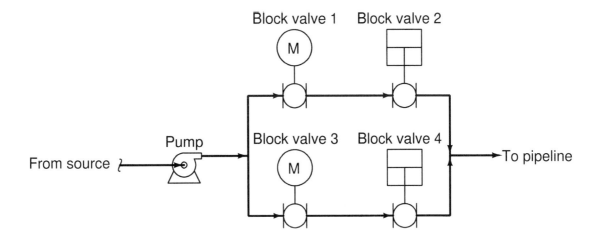

In order to fulfill its safety function of shutting off the flow of petroleum to the pipeline, both parallel pipe "branches" must be shut off. At first, this might seem to indicate a two-out-of-four (2oo4) dependability, because all we would need is for one valve in each branch (two valves total) out of the four valves to shut off in order to shut off flow to the pipeline. We must remember, however, that we do not have the luxury of assuming idealized faults. If only two of the four valves function properly in shutting off, they just might happen to be two valves *in the same branch*, in which case two valves properly functioning is not enough to guarantee a safe pipeline condition. Thus, this redundant system actually exhibits *three*-out-of-four (3oo4) dependability (i.e. it has a safety fault tolerance of one), because we need three out of the four block valves to properly shut off in order to *guarantee* a safe pipeline condition.

Analyzing this quadruple block valve array for security, we see that three out of the four valves need to function properly (open up) in order to guarantee flow to the pipeline. Once again, it may appear at first as though all we need are two of the four valves to open up in order to establish flow to the pipeline, but this will not be enough if those two valves happen to be in different parallel branches. So, this system exhibits three-out-of-four (3oo4) security (i.e. it has an production fault tolerance of one).

32.6.1   SIS sensors

Perhaps the simplest form of sensor providing process information for a safety instrumented function is a *process switch*. Examples of process switches include temperature switches, pressure switches, level switches, and flow switches[40]. SIS sensors must be properly calibrated and configured to indicate the presence of a dangerous condition. They must be separate and distinct from the sensors used for regulatory control, in order to ensure a level of safety protection beyond that of the basic process control system.

Referring to the clothes dryer and domestic water heater over-temperature shutdown switches, these high-temperature shutdown sensors are distinctly separate from the regulatory (temperature-controlling) sensors used to maintain the appliance's temperature at setpoint. As such, they should only ever spring into action in the event of a high-temperature *failure* of the basic control system. That is, the over-temperature safety switch on a clothes dryer or a water heater should only ever reach its high-temperature limit if the normal temperature control system of the appliance fails to do its job of regulating temperature to normal levels.

Industrial Safety Instrumented Systems (SIS) always use dedicated transmitters and/or process switches to detect abnormal process conditions. As a rule, one should always use independent sensors for safety shutdown, and never rely on the regulatory control sensor(s) for safety functions. In the electric power industry we see this same segregation of functions: separate instrument transformers (PTs and CTs) are used to sense line voltage and line current for metering and control (regulatory) versus for protective relay (safety shutdown) equipment. It would be foolish to depend on one sensor for both functions. We see this general rule applied even in home appliances such as electric water heaters: the safety shutdown temperature switch is a separate component from the thermostat switch used to regulate water temperature. This way, a failure in the regulatory sensor does not compromise the integrity of the safety function.

A modern trend in safety instrumented systems is to use continuous process transmitters rather than discrete process switches to detect dangerous process conditions. Any process transmitter – analog or digital – may be used as a safety shutdown sensor if its signal is compared against a "trip" limit value by a comparator relay or function block. This comparator function provides an on-or-off (discrete) output based on the transmitter's signal value relative to the trip point.

---

[40]For a general introduction to process switches, refer to chapter 9 beginning on page 649.

A simplified example of a continuous transmitter used as a discrete alarm and trip device is shown here, where analog comparators generate discrete "trip" and "alarm" signals based on the measured value of liquid in a vessel. Note the necessity of *two* level switches on the other side of the vessel to perform the same dual alarm and trip functions:

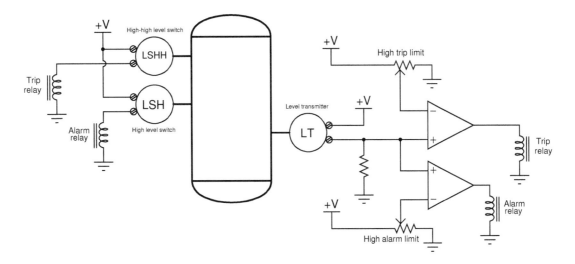

Benefits to using a continuous transmitter instead of discrete switches include the ability to easily change the alarm or trip value, and better diagnostic capability. The latter point is not as obvious as the former, and deserves more explanation. A transmitter continuously measuring liquid level will produce an output signal that varies over time with the measured process variable. A "healthy" transmitter should therefore exhibit a continuously changing output signal, proportional to the degree of change in the process. Discrete process switches, in contrast to transmitters, provide no indication of "healthy" operation. The only time a process switch should ever change states is when its trip limit is reached, which in the case of a safety shutdown sensor indicates a dangerous (rare) condition. A process switch showing a "normal" process variable may indeed be functional and indicating properly, but it might also be failed and incapable of registering a dangerous condition should one arise – there is no way to tell by monitoring its un-changing status. The continuously varying output of a process transmitter therefore serves as an indicator[41] of proper function.

---

[41]Of course, the presence of some variation in a transmitter's output over time is no guarantee of proper operation. Some failures may cause a transmitter to output a randomly "walking" signal when in fact it is not registering the process at all. However, being able to measure the continuous output of a process transmitter provides the instrument technician with far more data than is available with a discrete process switch. A safety transmitter's output signal may be correlated against the output signal of another transmitter measuring the same process variable, perhaps even the transmitter used in the regulatory control loop. If two transmitters measuring the same process variable agree closely with one another over time, chances are extremely good are both functioning properly.

In applications where Safety Instrumented Function (SIF) reliability is paramount, *redundant* transmitters may be installed to yield additional reliability. The following photograph shows triple-redundant transmitters measuring liquid flow by sensing differential pressure dropped across an orifice plate:

A single orifice plate develops the pressure drop, with the three differential pressure transmitters "tubed" in parallel with each other, all the "high" side ports connected together through common[42] impulse tubing and all the "low" side ports connected together through common impulse tubing. These particular transmitters happen to be FOUNDATION Fieldbus rather than 4-20 mA analog electronic. The yellow instrument tray cable (ITC) used to connect each transmitter to a segment coupling device may be clearly seen in this photograph.

---

[42]It should be noted that the use of a single orifice plate and of common (parallel-connected) impulse lines represents a point of common-cause failure. A blockage at one or more of the orifice plate ports, or a closure of a manual block valve, would disable all three transmitters. As such, this might not be the best method of achieving high flow-measurement reliability.

The "trick" to using redundant transmitters is to have the system self-determine what the actual process value is in the event one or more of the redundant transmitters disagree with each other. *Voting* is the name given to this important function, and it often takes the form of signal selector functions:

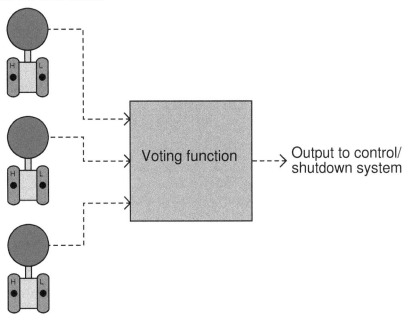

Multiple selection criteria are typically offered by "voting" modules, including *high, low, average,* and *median.* A "high" select voter would be suitable for applications where the dangerous condition is a large measured value, the voting module selecting the highest-valued transmitter signal in an effort to err on the side of safety. This would represent a 1oo3 safety redundancy (since only one transmitter out of the three would have to register beyond the high trip level in order to initiate the shutdown). A "low" select voter would, of course, be suitable for any application where the dangerous condition is a small measured value (once again providing a 1oo3 safety redundancy).

The "average" selection function merely calculates and outputs the mathematical average of all transmitter signals – a strategy prone to problems if one of the redundant transmitters happens to fail in the "safe" direction (thus skewing the average value away from the "dangerous" direction and thereby possibly causing the system to respond to an actual dangerous condition later than it should).

The *median select* criterion is very useful in safety systems because it effectively ignores any measurements deviating substantially from the others. Median selector functions may be constructed of high- and low-select function blocks in either of the following[43] manners:

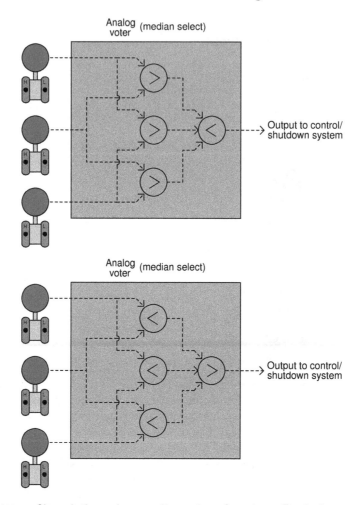

Three transmitters filtered through a median select function effectively provide a 2oo3 safety redundancy, since just a single transmitter registering a value beyond the safety trip point would be ignored by the voting function. *Two* or more transmitters would have to register values past the trip point in order to initiate a shutdown.

It should be stressed that redundant transmitter strategies are only effective if the transmitters all sense the exact same process variable, and if their failure modes are independent (i.e. no common-cause failure modes exist). If, for example, a set of redundant transmitters are attached to the

---

[43]The best way to prove to yourself the median-selecting abilities of both function block networks is to perform a series of "thought experiments" where you declare three arbitrary transmitter signal values, then follow through the selection functions until you reach the output. For any three signal values you might choose, the result should always be the same: the *median* signal value is the one chosen by the voter.

process at different points such that they may legitimately sense different measurement values, the effectiveness of their redundancy will be compromised. Similarly, if a set of redundant transmitters are susceptible to failure from a shared condition (e.g. multiple liquid level transmitters that may be fooled by changes in process fluid density), then reliability will suffer.

32.6.2   SIS controllers (logic solvers)

Control hardware for safety instrumented functions should be separate from the control hardware used to regulate the process, if only for the simple reason that the SIF exists to bring the process to a safe state in the event of any unsafe condition arising, including dangerous failure of the basic regulatory controls. If a single piece of control hardware served the dual purposes of regulation *and* shutdown, a failure within that hardware resulting in loss of regulation (normal control) would not be protected because the safety function would be disabled by the same fault.

Safety controls are usually discrete with regard to their output signals. When a process needs to be shut down for safety reasons, the steps to implement the shutdown often take the form of opening and closing certain valves fully rather than partially. This sort of all-or-nothing control action is most easily implemented in the form of discrete signals triggering solenoid valves or electric motor actuators. A digital controller specially designed for and tasked with the execution of safety instrumented functions is usually called a *logic solver*, or sometimes a *safety PLC*, in recognition of this discrete-output nature.

A photograph of a "safety PLC" used as an SIS in an oil refinery processing unit is shown here, the controller being a Siemens "Quadlog" model:

Some logic solvers such as the Siemens Quadlog are adaptations of standard control systems (in the case of the Quadlog, its standard counterpart is called APACS). In the United States, where Rockwell's Allen-Bradley line of programmable logic controllers holds the dominant share of the PLC market, a version of the ControlLogix 5000 series called *GuardLogix* is manufactured specifically for safety system applications. Not only are there differences in hardware between standard and safety controllers (e.g. redundant processors), but some of the programming instructions are unique to these safety-oriented controllers as well.

An example of a safety-specific programming instruction is the GuardLogix DCSRT instruction, which compares two redundant input channels for agreement before activating a "start" bit which may be used to start some equipment function such as an electric motor:

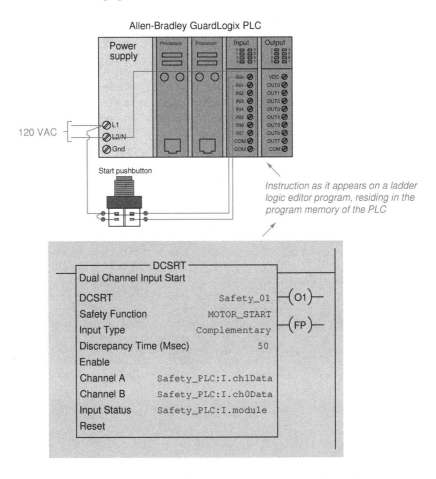

In this case, the DCSRT instruction looks for two discrete inputs to be in the correct complementary states (Channel A = 1 and Channel B = 0) before allowing a motor to start. These states must not conflict for a time-span longer than 50 milliseconds, or else the DCSRT instruction will set a "Fault Present" (FP) bit. As you can see, the form-C pushbutton contacts are wired to two discrete inputs on the GuardLogix PLC, giving the PLC dual (complementary) indication of the switch status.

For specialized and highly critical applications, dedicated safety controllers exist which share no legacy with standard control platforms. Triconex and ICS-Triplex are two such manufacturers, producing *triple-modular redundant* (TMR) control systems implementing 2oo3 voting at the hardware level, with redundant signal conditioning I/O circuits, redundant processors, and redundant communication channels between all components. The nuclear power industry boasts a wide array of application-specific digital control systems, with triple (or greater!) component redundancy for extreme reliability. An example of this is Toshiba's TOSMAP system for boiling-water nuclear power reactors, the digital controller and electro-hydraulic steam turbine valve actuator subsystem having a stated MTBF[44] of over 1000 years!

32.6.3   SIS final control elements

When a dangerous condition in a volatile process is sensed by process transmitters (or process switches), triggering a shutdown response from the logic solver, the final control elements must move with decisive and swift action. Such positive response may be obtained from a standard regulatory control valve (such as a globe-type throttling valve), but for more critical applications a rotary ball or plug valve may be more suitable. If the valve in question is used for safety shutdown purposes only and not regulation, it is often referred to as a *chopper* valve for its ability to "chop" (shut off quickly and securely) the process fluid flow. A more formal term for this is an *Emergency Isolation Valve*, or *EIV*.

Some process applications may tolerate the over-loading of both control and safety functions in a single valve, using the valve to regulate fluid flow during normal operation and fully stroke (either open or closed depending on the application) during a shutdown condition. A common method of achieving this dual functionality is to install a solenoid valve in-line with the actuating air pressure line, such that the valve's normal pneumatic signal may be interrupted at any moment, immediately driving the valve to a fail-safe position at the command of a discrete "trip" signal.

---

[44]*MTBF* stands for *Mean Time Between Failure*, and represents the reliability of a large collection of components or systems. For any large batch of identical components or systems constantly subjected to ordinary stresses, MTBF is the theoretical length of time it will take for 63.2% of them to fail based on ordinary failure rates within the lifetime of those components or systems. Thus, MTBF may be thought of as the "time constant" (τ) for failure within a batch of identical components or systems.

Such a "trip" solenoid (sometimes referred to as a *dump* solenoid, because it "dumps" all air pressure stored in the actuating mechanism) is shown here, connected to a fail-closed (air-to-open) control valve:

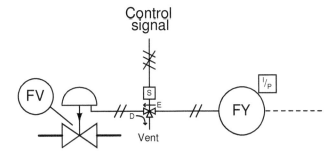

Compressed air passes through the solenoid valve from the I/P transducer to the valve's pneumatic diaphragm actuator when energized, the letter "E" and arrow showing this path in the diagram. When de-energized, the solenoid valve blocks air pressure coming from the I/P and vents all air pressure from the valve's actuating diaphragm as shown by the letter "D" and arrow. Venting all actuating air pressure from a fail-closed valve will cause the valve to fail closed, obviously.

If we wished to have the valve fail open on demand, we could use the exact same solenoid and instrument air plumbing, but swap the fail-closed control valve for a fail-open control valve. When energized (regular operation), the solenoid would pass variable air pressure from the I/P transducer to the valve actuator so it could serve its regulating purpose. When de-energized, the solenoid would force the valve to the fully-open position by "dumping" all air pressure from the actuator.

For applications where it is safer to lock the control valve in its last position than to have it fail either fully closed or fully open, we might elect to use a solenoid valve in a different manner:

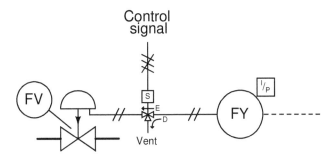

Here, de-energization of the solenoid valve causes the I/P transducer's air pressure output to vent, while trapping and holding all air pressure inside the actuator at the trip time. Regardless of the valve's "natural" fail-safe state, this system forces the valve to lock position[45] until the solenoid is re-energized.

---

[45]This is assuming, of course, that there are no air leaks anywhere in the actuator, tubing, or solenoid which would cause the trapped pressure to decrease over time.

An example of a trip solenoid installed on a control valve appears in the following photograph. This valve also happens to have a *hand jack* wheel installed in the actuating mechanism, allowing a human operator to manually override the valve position by forcing it closed (or open) when the hand wheel is turned sufficiently:

Of all the components of a Safety Instrumented System (SIS), the final control elements (valves) are generally the least reliable, contributing most towards the system's probability of failure on demand (PFD). Sensors generally come in at second place in their contribution toward unreliability, and logic solvers a distant third place. Redundancy may be applied to control elements by creating valve networks where the failure of a single valve does not cause the system as a whole to fail. Unfortunately, this approach is extremely expensive, as valves have both high capital and high maintenance costs compared to SIS sensors and logic solvers.

A less expensive approach than redundancy to increasing safety valve reliability is to perform regular proof tests of their operation. This is commonly referred to in the industry as *partial stroke testing*. Rather than proof-test each safety valve to its full travel, which would interrupt normal process operations, the valve is commanded to move only part of its full travel. If the valve responds well to this "partial stroke" test, there is a high probability that it is able to move all the way, thus fulfilling the basic requirements of a proof test without actually shutting the process down[46].

---

[46]Of course, if there is opportunity to fully stroke the safety valve to the point of process shutdown without undue interruption to production, this is the superior way of performing valve proof tests. Such "test-to-shutdown" proof testing may be scheduled at a time convenient to operations personnel, such as at the beginning of a planned process shutdown.

32.6.4 Safety Integrity Levels

A common way of ranking the dependability of a Safety Instrumented Function (SIF) is to use a simple numerical scale from one to four, with four being extremely dependable and one being only moderately dependable:

SIL number	Required Safety Availability (RSA)	Probability of Failure on Demand (PFD)
1	90% to 99%	0.1 to 0.01
2	99% to 99.9%	0.01 to 0.001
3	99.9% to 99.99%	0.001 to 0.0001
4	99.99% to 99.999%	0.0001 to 0.00001

The Required Safety Availability (RSA) value is synonymous with *dependability*: the probability[47] that a Safety Instrumented Function will perform its duty when faced with a dangerous process condition. Conversely, the Probability of Failure on Demand (PFD) is synonymous with *undependability*: the mathematical complement of RSA (PFD = 1 − RSA), expressing the probability that the SIF will fail to perform as needed, when needed.

Conveniently, the SIL number matches the minimum number of "nines" in the Required Safety Availability (RSA) value. For instance, a safety instrumented function with a Probability of Failure on Demand (PFD) of 0.00073, will have an RSA value of 99.927%, which equates to a SIL 3 rating.

It is important to understand what SIL is, and what SIL is not. The SIL rating refers to the reliability of a safety *function*, not to individual components of a system nor to the entire process itself. An overpressure protection system on a chemical reactor process with a SIL rating of 2, for example, has a Probability of Failure on Demand between 0.01 and 0.001 *for the specific shutdown function as a whole*. This PFD value incorporates failure probabilities of the sensor(s), logic solver, final control element(s), and the process piping including the reactor vessel itself plus any relief valves and other auxiliary equipment. If there arises a need to improve the PFD of this reactor's overpressure protection, safety engineers have a variety of options at their disposal for doing so. The safety instruments themselves might be upgraded, a different redundancy strategy implemented, preventive maintenance schedules increased in frequency, or even process equipment changed to make an overpressure event less likely.

SIL ratings do not apply to an entire process. It is quite possible that the chemical reactor mentioned in the previous paragraph with an overpressure protection system SIL rating of 3 might have an over*temperature* protection system SIL rating of only 2, due to differences in how the two different safety systems function.

Adding to this confusion is the fact that many instrument manufacturers rate their products as approved for use in certain SIL-rated applications. It is easy to misunderstand these claims, thinking that a safety instrumented function will be rated at some SIL value simply because instruments rated for that SIL value are used to implement it. In reality, the SIL value of any safety function is a much more complex determination. It is possible, for instance, to purchase and install a pressure transmitter rated for use in SIL 2 applications, and have the safety function as a whole be less than

---

[47]*Probability* is a quantitative measure of a particular outcome's likelihood. A probability value of 1, or 100%, means the outcome in question is certain to happen. A probability value of 0 (0%) means the outcome is impossible. A probability value of 0.3 (30%) means it will happen an average of three times out of ten.

99% reliable (PFD greater than 0.01, or a SIL level no greater than 1) due to the effect of *Lusser's Law*[48].

As with so many other complex calculations in instrumentation engineering, there exist software packages with all the necessary formulae pre-programmed for engineers and technicians alike to use for calculating SIL ratings of safety instrumented functions. These software tools not only factor in the inherent reliability ratings of different system components, but also correct for preventive maintenance schedules and proof testing intervals so the user may determine the proper maintenance attention required to achieve a given SIL rating.

---

[48]Lusser's Law of Reliability states that the total reliability of a system dependent on the function of several independent components is the mathematical product of those components' individual reliabilities. For example, a system with three essential components, each of those components having an individual reliability value of 70%, will exhibit a reliability of only 34.3% because $0.7 \times 0.7 \times 0.7 = 0.343$. This is why a safety function may utilize a pressure transmitter rated for use in SIL-3 applications, but exhibit a much lower total SIL rating due to the use of an ordinary final control element.

32.6.5 SIS example: burner management systems

One "classic" example of an industrial automatic shutdown system is a *Burner Management System* (or *BMS*) designed to monitor the operation of a combustion burner and shut off the fuel supply in the event of a dangerous condition. Sometimes referred to as *flame safety systems*, these systems watch for such potentially dangerous conditions as *low fuel pressure*, *high fuel pressure*, and *loss of flame*. Other dangerous conditions related to the process being heated (such as *low water level* for a steam boiler) may be included as additional trip conditions.

The safety shutdown action of a burner management system is to halt the flow of fuel to the burner in the event of any hazardous detected condition. The final control element is therefore one or more shutoff valves (and sometimes a vent valve in addition) to positively stop fuel flow to the burner.

A typical ultraviolet flame sensor appears in this photograph:

This flame sensor is sensitive to ultraviolet light only, not to visible or infrared light. The reason for this specific sensitivity is to ensure the sensor will not be "fooled" by the visible or infrared glow of hot surfaces inside the firebox if ever the flame goes out unexpectedly. Since ultraviolet light is emitted *only* by an active gas-fueled flame, the sensor acts as a true flame detector, and not a heat detector.

One of the more popular models of fuel gas safety shutoff valve used in the United States for burner management systems is shown here, manufactured by Maxon:

This particular model of shutoff valve has a viewing window on it where a metal tag linked to the valve mechanism marked "Open" (in red) or "Shut" (in black) positively indicates the valve's mechanical status. Like most safety shutoff valves on burner systems, this valve is electrically actuated, and will automatically close by spring tension in the event of a power loss.

Another safety shutoff valve, this one manufactured by ITT, is shown here:

Close inspection of the nameplate on this ITT safety valve reveals several important details. Like the Maxon safety valve, it is electrically actuated, with a "holding" current indicated as 0.14 amps at 120 volts AC. Inside the valve is an "auxiliary" switch designed to actuate when the valve has mechanically reached the full "open" position. An additional switch, labeled *valve seal overtravel interlock*, indicates when the valve has securely reached the full "shut" position. This "valve seal" switch generates a *proof of closure* signal used in burner management systems to verify a safe shutdown condition of the fuel line. Both switches are rated to carry 15 amps of current at 120

VAC, which is important when designing the electrical details of the system to ensure the switch will not be tasked with too much current.

A simple P&ID for a gas-fired combustion burner system is shown here. The piping and valving shown is typical for a single burner. Multiple-burner systems are often equipped with individual shutoff valve manifolds and individual fuel pressure limit switches. Each burner, if multiple exist in the same furnace, *must* be equipped with its own flame sensor:

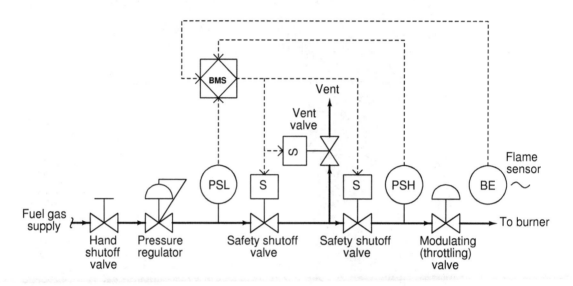

Note the use of double-block and bleed shutdown valves to positively isolate the fuel gas supply from the burner in the event of an emergency shutdown. The two block valves are specially designed for the purpose (such as the Maxon and ITT safety valves previously shown), while the bleed valve is often nothing more than an ordinary electric solenoid valve.

Most burner management systems are charged with a dual role: both to manage the safe shutdown of a burner in the event of a hazardous condition, *and* the safe start-up of a burner in normal conditions. Start-up of a large industrial burner system usually includes a lengthy *purge time* prior to ignition where the combustion air damper is left wide-open and the blower running for several minutes to positively purge the firebox of any residual fuel vapors. After the purge time, the burner management system will ignite the burner (or sometimes ignite a smaller burner called the *pilot*, which in turn will light the main burner). A burner management system executes all these pre-ignition and timing functions to ensure the burners will ignite safely and without incident.

While many industrial burners are managed by electromechanical relay or analog electronic control systems, the modern trend is toward microprocessor-based digital electronic controls. One popular system is the Honeywell 7800 series burner control system, an example of which is shown in this photograph:

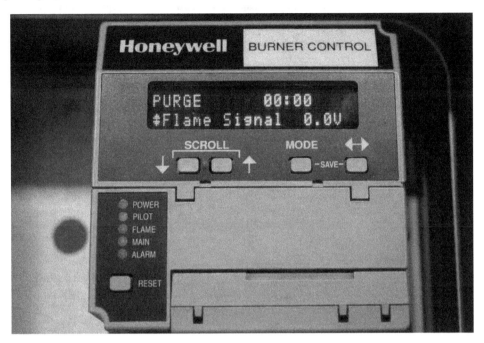

Microprocessor controls provide numerous advantages over relay-based and analog electronic burner management systems. Timing of purge cycles is far more accurate with microprocessor control, and the requisite purge time is more difficult to override[49]. Microprocessor-based burner controls usually have digital networking capability as well, allowing the connection of multiple controls to a single computer for remote monitoring.

---

[49]Yes, maintenance and operations personnel alike are often tempted to bypass the purge time of a burner management system out of impatience and a desire to resume production. I have personally witnessed this in action, performed by an electrician with a screwdriver and a "jumper" wire, overriding the timing function of a flame safety system during a troubleshooting exercise simply to get the job done faster. The electrician's rationale was that since the burner system was having problems lighting, and had been repeatedly purged in prior attempts, the purge cycle did not have to be full-length in subsequent attempts. I asked him if he would feel comfortable repeating those same words in court as part of the investigation of why the furnace exploded. He didn't think this was funny.

The Honeywell 7800 series additionally offers local "annunciator" modules to visually indicate the status of permissive (interlock) contacts, showing maintenance personnel which switches are closed and what state the burner control system is in:

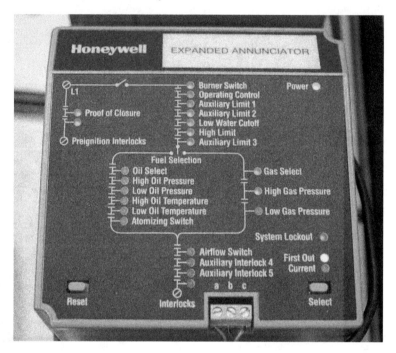

The entire "gas train" piping system for a dual-fuel boiler at a wastewater treatment facility appears in the following photograph. Note the use of double-block and bleed valves on both "trains" (one for utility-supplied natural gas and the other for "sludge gas" produced by the facility's anaerobic digesters), the block valves for each train happening to be of different manufacture. A Honeywell 7800 flame safety control system is located in the blue enclosure:

32.6.6   SIS example: water treatment oxygen purge system

One of the processes of municipal wastewater treatment is the aerobic digestion of organic matter by bacteria. This process emulates one of many waste-decomposition processes in nature, performed on an accelerated time frame for the needs of large wastewater volumes in cities. The process consists of supplying naturally occurring bacteria within the wastewater with enough oxygen to metabolize the organic waste matter, which to the bacteria is food. In some treatment facilities, this aeration is performed with ambient air. In other facilities, it is performed with nearly pure oxygen.

Aerobic decomposition is usually part of a larger process called *activated sludge*, whereby the effluent from the decomposition process is separated into solids (sludge) and liquid (supernatant), with a large fraction of the sludge recycled back to the aerobic chamber to sustain a healthy culture of bacteria and also ensure adequate retention time for decomposition to occur. Separating liquids from solids and recycling the solids ensures a short retention time for the liquid (allowing high processing rates) and a long retention time for the solids (ensuring thorough digestion of organic matter by the bacteria).

A simplified P&ID of an activated sludge water treatment system is shown here, showing how both the oxygen flow into the aeration chamber and the sludge recycle flow back to the aeration chamber are controlled as a function of influent wastewater flow:

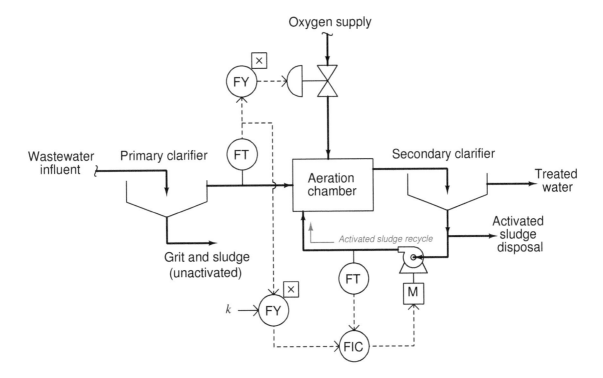

Aerobic decomposition performed with ambient air as the oxidizer is a very simple and safe process. Pure oxygen may be chosen instead of ambient air because it accelerates the metabolism of the bacteria, allowing more processing flow capacity in less physical space. For the same reason that pure oxygen accelerates bacterial metabolism, it also accelerates combustion of any flammable substances. This means if ever a flammable vapor or liquid were to enter the aeration chamber, there would be a risk of explosion.

Although flammable liquids are not a normal component of municipal wastewater, it is possible for flammable liquids to find their way to the wastewater treatment plant. One possibility is the event of a fuel carrier vehicle spilling its cargo, with gasoline or some other volatile fuel draining into a sewer system tunnel through holes in a grate. Such an occurrence is not normal, but certainly possible. Furthermore, it may occur without warning for the operations personnel to take preemptive action at the wastewater treatment plant.

To decrease this safety hazard, *Low Explosive Limit* (LEL) sensors installed on the aeration chamber detect and signal the presence of flammable gases or vapors inside the chamber. If any of the sensors register the presence of flammable substances, a safety shutdown system purges the chamber of pure oxygen by taking the following steps:

- Stop the flow of pure oxygen into the aeration chamber

- Open large vent valves to atmosphere

- Start air blowers to purge the chamber of residual pure oxygen

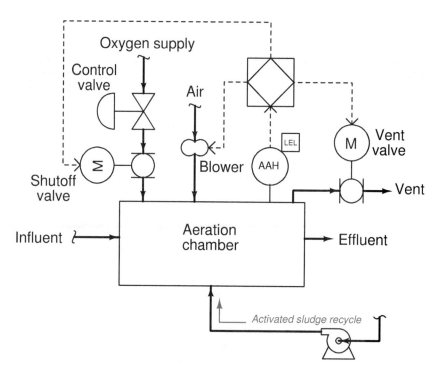

As with the P&ID, this diagram is a simplified representation of the real safety shutdown system. In a real system, multiple analytical high-alarm (LEL) sensors work to detect the presence of flammable gases or vapors, and the oxygen block valve arrangement would most likely be a double block and bleed rather than a single block valve.

The following photograph shows an LEL sensor mounted inside an insulated enclosure for protection from cold weather conditions at a wastewater treatment facility:

In this photograph, we see a purge air blower used to sweep the aeration chamber of pure oxygen (replacing it with ambient air) during an emergency shutdown condition:

Since this is a centrifugal blower, providing no seal against air flow through it when stopped, an automatic purge valve located downstream (not to be confused with the manually-actuated vent valve seen in this photograph) is installed to block off the blower from the oxygen-filled chamber. This purge valve remains shut during normal operation, and opens only after the blower has started to initiate a purge.

32.6.7 SIS example: nuclear reactor scram controls

Nuclear fission is a process by which the nuclei of specific types of atoms (most notably uranium-235 and plutonium-239) undergo spontaneous disintegration upon the absorption of an extra neutron, with the release of significant thermal energy and additional neutrons. A quantity of fissile material subjected to a source of neutron particle radiation will begin to fission, releasing massive quantities of heat which may then be used to boil water into steam and drive steam turbine engines to generate electricity. The "chain reaction" of neutrons splitting fissile atoms, which then eject more neutrons to split more fissile atoms, is inherently exponential in nature, but may be regulated by natural and artificial feedback loops.

A simplified diagram of a pressurized[50] water reactor (PWR) appears here:

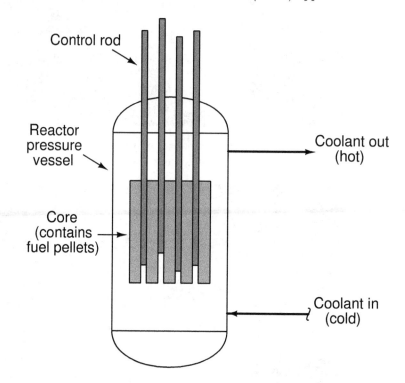

---

[50]Boiling-water reactors (BWR), the other major design type in the United States, output saturated steam at the top rather than heated water. Control rods enter a BWR from the bottom of the pressure vessel, rather than from the top as is standard for PWRs.

In the United States of America, nuclear reactors are designed to exhibit what is called a *negative temperature coefficient*, which means the chain reaction naturally slows as the temperature of the coolant increases. This physical tendency, engineered by the configuration of the reactor core and the design of the coolant system, adds a measure of self-stabilization to what would otherwise be an inherently unstable ("runaway") process. This is an example of a "natural" negative-feedback loop in action: a process by which the very laws of physics conspire to regulate the activity of the fission reaction.

Additional regulation ability comes from the insertion of special *control rods* into the reactor core, designed to absorb neutrons and prevent them from "splitting" more atoms. With enough control rods inserted into a reactor core, a chain reaction cannot self-sustain. With enough control rods withdrawn from a freshly-fueled reactor core, the chain reaction will grow to an intensity strong enough to damage the reactor. Control rod position thus constitutes the primary method of power control for a fission reactor, and also the first[51] means of emergency shutdown. These control rods are inserted and withdrawn in order to exert demand-control over the fission reaction. If the reaction rate is too low to meet demand, either a human operator or an automatic control system may withdraw the rods until the desired reactivity is reached. If the reaction rate becomes excessive, the rods may be inserted until the rate falls down to the desired level. Control rods are therefore the final control element (FCE) of an "artificial" negative-feedback loop designed to regulate reaction rate at a level matching power demand.

Due to the intense radiation flux near an operating power reactor, these control rods must be manipulated remotely rather than by direct human actuation. Nuclear reactor control rod actuators are typically special electric motors developed for this critical application.

---

[51]Other means of reactor shutdown exist, such as the purposeful injection of "neutron poisons" into the coolant system which act as neutron-absorbing control rods on a molecular level. The insertion of "scram" rods into the reactor, though, is by far the *fastest* method for quenching the chain-reaction.

A photograph[52] showing the control rod array at the top of the ill-fated reactor at Three-Mile Island nuclear power plant appears here, with a mass of control cables connecting the rod actuators to the reactor control system:

Rapid insertion of control rods into a reactor core for emergency shutdown purposes is called a *scram.* Accounts vary as to the origin of this term, whether it has meaning as a technical acronym or as a colloquial expression to evacuate an area. Regardless of its etymology, a "scram" is an event to be avoided if possible. Like all industrial processes, a nuclear reactor fulfills its intended purpose only when operating. Shutdowns represent not only loss of revenue for the operating company, but also loss of power to local utilities and possible disruption of critical public services (heating, cooling, water pumping, fire protection, traffic control, etc.). An emergency shutdown system at a nuclear power plant must fulfill the opposing roles of dependability and security, with an extremely high degree of instrument reliability.

The electric motor actuators intended for normal operation of control rods are generally too slow to use for scram purposes. Hydraulic actuators capable of overriding the electric motor actuation may be used for scram insertion. Some early pressurized-water reactor scram system designs used a simple mechanical latch, disengaging the control rods from their motor actuators and letting gravity draw the rods fully into the reactor core.

---

[52]This appears courtesy of the Nuclear Regulatory Commission's special inquiry group report following the accident at Three Mile Island, on page 159.

A partial list of criteria sufficient to initiate a reactor scram is shown here:

- Detected earthquake

- Reactor pressure high

- Reactor pressure low

- Reactor water level low (BWR only)

- Reactor differential temperature high

- Main steam isolation valve shut

- Detected high radioactivity in coolant loop

- Detected high radioactivity in containment building

- Manual shutdown switch(es)

- Control system power loss

- Core neutron flux high

- Core neutron flux rate-of-change (period) high

The last two criteria bear further explanation. Since each fission event (the "splitting" of one fuel atom's nucleus by an absorbed neutron) results in a definite amount of thermal energy release and also a definite number of additional neutrons released, the number of neutrons detected in the reactor core at any given moment is an approximate indication of the core's thermal power as well as its reactivity. Neutron radiation flux measurement is therefore a fundamental process variable for fission reactor control, and also for safety shutdown. If sensors detect an excessive neutron flux, the reactor should be "scrammed" to avoid damage due to overheating. Likewise, if sensors detect a neutron flux level that is *rising* at an excessive *rate*, it indicates the possibility of a runaway chain-reaction which should also initiate a reactor "scram."

In keeping with the high level of reliability and emphasis on safety for nuclear reactor shutdown controls, a common redundant strategy for sensors and logic is *two-out-of-four*, or *2oo4*. A contact logic diagram showing a 2oo4 configuration appears here:

2oo4 redundant logic for reactor scram systems

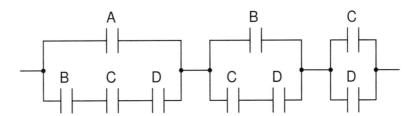

*Any two contacts (A, B, C, or D) opening
will interrupt power flow and "scram" the reactor*

32.7   Review of fundamental principles

Shown here is a partial listing of principles applied in the subject matter of this chapter, given for the purpose of expanding the reader's view of this chapter's concepts and of their general inter-relationships with concepts elsewhere in the book. Your abilities as a problem-solver and as a life-long learner will be greatly enhanced by mastering the applications of these principles to a wide variety of topics, the more varied the better.

- **Activation energy**: the amount of energy necessary to initiate a chemical reaction. Relevant to minimum ignition energy (MIE) for explosive mixtures of fuel and oxidizer, and also to intrinsic safety (preventing enough energy from reaching the process energy to possibly ignited a hazardous atmosphere).

- **Common-cause failures**: when multiple functions in a system depend on a single element, failure of that element will cause all dependent functions to fail. Relevant to design of safety shutdown systems and reliability calculations.

- **Defense-in-Depth**: a design philosophy relying on multiple layers of protection, the goal being to maintain some degree of protection in the event of one or more other layers failing.

References

Adamski, Robert S., *Design Critical Control or Emergency Shut Down Systems for Safety AND Reliability*, Revision 2, Premier Consulting Services, Irvine, CA.

Andrew, William G., *Applied Instrumentation in the Process Industries*, Volume I, Second Edition, Gulf Publishing Company, Houston, TX, 1979.

ANSI/ISA-84.00.01-2004 Part 1 (IEC 61151-1 Mod), "Functional Safety: Safety Instrumented Systems for the Process Industry Sector – Part 1: Framework, Definitions, System, Hardware and Software Requirements", ISA, Research Triangle Park, NC, 2004.

ANSI/ISA-84.00.01-2004 Part 2 (IEC 61151-2 Mod), "Functional Safety: Safety Instrumented Systems for the Process Industry Sector – Part 2: Guidelines for the Application of ANSI/ISA-84.00.01-2004 Part 1 (IEC 61151-1 Mod)", ISA, Research Triangle Park, NC, 2004.

Bazovsky, Igor, *Reliability Theory and Practice*, Prentice-Hall, Inc., Englewood Cliffs, NJ, 1961.

da Silva Cardoso, Gabriel; de Lima, Marcelo Lopes; dos Santos da Rocha, Maria Celia; Ferreira Lemos, Solange Soares, "Safety Instrumented Systems standardization for Fluid Catalytic Cracking Units at PETROBRAS", ISA, presented at ISA EXPO 2005, Chicago, IL, 2005.

"Engineer's Guide", Pepperl+Fuchs.

"Failure Mode / Mechanism Distributions" (FMD-97), Reliability Analysis Center, Rome, NY, 1997.

Grebe, John and Goble, William, *Failure Modes, Effects and Diagnostic Analysis; Project: 3051C Pressure Transmitter*, Report number Ros 03/10-11 R100, exida.com L.L.C., 2003.

"GuardLogix Safety Application Instruction Set", Publication 1756-RM095D-EN-P, Rockwell Automation, Inc., Milwaukee, WI, 2009.

Hattwig, Martin, and Steen, Henrikus, *Handbook of Explosion Prevention and Protection*, Wiley-VCH Verlag GmbH & Co. KGaA, Weinheim, Germany, 2004.

Hellemans, Marc *The Safety Relief Valve Handbook, Design and Use of Process Safety Valves to ASME and International Codes and Standards*, Elsevier Ltd, Oxford, UK, 2009.

Hicks, Tyler G., *Standard Handbook of Engineering Calculations*, McGraw-Hill Book Company, New York, NY, 1972.

"Identification and Description of Instrumentation, Control, Safety, and Information Systems and Components Implemented in Nuclear Power Plants", EPRI, Palo Alto, CA: 2001. 1001503.

"IEC 61508 Frequently Asked Questions", Rosemount website http://mw4rosemount.usinternet.com/solution/faq61508.html, updated December 1, 2003.

IEEE PSRC, WG I 19, "Redundancy Considerations for Protective Relaying Schemes", version 1.0, IEEE, 2007.

Lipták, Béla G. et al., *Instrument Engineers' Handbook – Process Measurement and Analysis Volume I*, Fourth Edition, CRC Press, New York, NY, 2003.

Lipták, Béla G. et al., *Instrument Engineers' Handbook – Process Control Volume II*, Third Edition, CRC Press, Boca Raton, FL, 1999.

Lipták, Béla G. et al., *Instrument Engineers' Handbook – Process Software and Digital Networks*, Third Edition, CRC Press, New York, NY, 2002.

"Modern Instrumentation and Control for Nuclear Power Plants: A Guidebook", Technical Reports Series No. 387, International Atomic Energy Agency (IAEA), Vienna, 2009.

Newnham, Roger and Chau, Paul, *Safety Controls and Burner Management Systems (BMS) on Direct-Fired Multiple Burner Heaters*, Born Heaters Canada Ltd.

"NFPA 70", National Electrical Code, 2008 Edition, National Fire Protection Association.

"NIOSH Pocket Guide to Chemical Hazards", DHHS (NIOSH) publication # 2005-149, Department of Health and Human Services (DHHS), Centers for Disease Control and Prevention (CDC), National Institute for Occupational Safety and Health (NIOSH), Cincinnati, OH, September 2005.

Perrow, Charles, *Normal Accidents: living with high-risk technologies*, Princeton University Press, Princeton, NJ, 1999.

Rogovin, Mitchell and Frampton, George T. Jr., *Three Mile Island Volume I, A Report to the Commissioners and to the Public*, Nuclear Regulatory Commission Special Inquiry Group, Washington DC, 1980.

Schultz, M. A., *Control of Nuclear Reactors and Power Plants*, McGraw-Hill Book Company, New York, NY, 1955.

Showers, Glenn M., "Preventive Maintenance for Burner-Management Systems", *HPAC – Heating/Piping/Air Conditioning Engineering*, February 2000.

Svacina, Bob, and Larson, Brad, *Understanding Hazardous Area Sensing*, TURCK, Inc., Minneapolis, MN, 2001.

"The SPEC 200 Concept", Technical Information document TI 200-100, The Foxboro Company, Foxboro, MA, 1972.

Ward, S.; Dahlin, T.; Higinbotham, W.; "Improving Reliability for Power System Protection", paper presented before the 58th annual Protective Relay Conference in Atlanta, GA, April 28-30, 2004.

Wehrs, Dave, "Detection of Plugged Impulse Lines Using Statistical Process Monitoring Technology", Emerson Process Management, Rosemount Inc., Chanhassen, MN, December 2006.

Chapter 33

Instrumentation cyber-security

As digital technology finds greater application in industrial measurement and control systems, these systems become subject to digital vulnerabilities. Cyber-security, which used to be strictly limited to information technology (IT) systems such as those used in office and research environments (e.g. desktop computers, printers, internet routers), is now a pressing concern for industrial measurement and control systems.

There exist many points of commonality between digital IT and digital control systems, and it is at these points where mature protection concepts may be borrowed from the world of IT for use protecting industrial control systems. However, digital measurement and control systems have many unique features, and it is here we must develop protection strategies crafted specifically for industrial applications.

The chief difference between industrial controls and IT systems is, of course, the fact that industrial controls directly manage real physical processes. The purpose of an IT system, in constrast, is to manage *information*. While information can be dangerous in the wrong hands, physical processes such as chemical plants, nuclear power stations, water treatment facilities, hazardous waste treatment facilities, can be even more so.

This chapter will primarily focus on digital security as it applies to industrial measurement and control systems. The opening section is a case study on what has become a famous example of an industrial-scale cyber-attack: the so-called *Stuxnet* virus.

It should be noted that cyber-security is a very complex topic, and that this chapter of the book is quite unfinished at the time of this writing (2016). Later versions of the book will likely have much more information on this important topic.

33.1　Stuxnet

In November of 2007 a new computer virus was submitted to a virus scanning service. The purpose of this new virus was not understood at the time, but it was later determined to be an early version of the so-called *Stuxnet* virus which was designed to infiltrate and attack programmable logic controllers (PLCs) installed at the uranium enrichment facility in Iran, a critical part of that country's nuclear program located in the city of Natanz. Stuxnet stands as the world's first known computer virus ever designed to specifically attack an industrial control platform, in this case Siemens model S7 PLCs.

Later forensic analysis revealed the complexity and scope of Stuxnet for what it was: a digital weapon, directed against the Iranian nuclear program for the purpose of delaying that program's production of enriched uranium. Although the origins of Stuxnet are rather unique as viruses go, the lessons learned from Stuxnet help us as industrial control professionals to fortify our own control systems against similarly-styled digital attacks. The next such attack may not come from a nation-state like Stuxnet did, but you can be sure whoever attacks next will have gained from the lessons Stuxnet taught the world.

Since the Stuxnet attack was directed against a nuclear facility, it is worthwhile to know a little about what that facility did and how it functioned. The next subsection will delve into some of the details of modern uranium enrichment processes, while further subsections will outline how Stuxnet attacked those physical processes through the PLC control system.

The sections following this one on Stuxnet will broaden the scope of the conversation to vulnerabilities and fortification strategies common to many industrial control networks and systems.

33.1.1 A primer on uranium enrichment

Uranium is a naturally occurring metal with interesting properties lending themselves to applications of nuclear power and nuclear weaponry. Uranium is extremely dense, and also (mildly) radioactive. Of greater importance, though, is that some of the naturally occurring isotopes[1] of uranium are *fissile*, which means those atoms may be easily "split" by neutron particle bombardment, releasing huge amounts of energy as well as more neutrons which may then go on to split more uranium atoms in what is called a *chain reaction*. Such a chain-reaction, when controlled, constitutes the energy source of a fission reactor. Nuclear weapons employ violently uncontrolled chain reactions.

The most fissile isotope of uranium is uranium 235, that number being the total count of protons and neutrons within the nucleus of each atom. Unfortunately (or fortunately, depending on your view of nuclear fission), ^{235}U constitutes only 0.7% of all uranium found in the earth's crust. The vast majority of naturally occurring uranium is the isotope ^{238}U which has all the same chemical properties of ^{235}U but is non-fissile (i.e. an atom of ^{238}U will not be "split" by neutron particle bombardment[2]).

Naturally-occurring uranium at a concentration of only 0.7% ^{235}U is too "dilute" for most[3] nuclear reactors to use as fuel, and certainly is not concentrated enough to construct a nuclear weapon. Most power reactors require uranium fuel at a ^{235}U concentration of at least 3% for practical operation, and a concentration of at least 20% is considered the low threshold for use in constructing a uranium-based nuclear weapon. Mildly concentrated uranium useful for reactor fuel is commonly referred to "low-enriched uranium" or *LEU*, while uranium concentrated enough to build a nuclear weapon is referred to as "highly enriched uranium" or *HEU*. Modern uranium-based nuclear bombs rely on the uranium being concentrated to at least 90% ^{235}U, as do military power reactors such as the extremely compact designs used to power nuclear submarines. All of this means that an industrial-scale process for concentrating (enriching) ^{235}U is a necessary condition for building and sustaining a nuclear program of any kind, whether its purpose be civilian (power generation, research) or military (weapons, nuclear-powered vehicles).

Different technologies currently exist for uranium enrichment, and more are being developed. The technical details of uranium enrichment set the background for the Stuxnet story, the site of this cyber-attack being the Natanz uranium enrichment facility located in the middle-eastern nation of Iran.

---

[1]The term *isotope* refers to differences in atomic mass for any chemical element. For example, the most common isotope of the element *carbon* (C) has six neutrons and six protons within each carbon atom nucleus, giving that isotope an atomic mass of twelve (^{12}C). A carbon atom having two more neutrons in its nucleus would be an example of the isotope ^{14}C, which just happens to be radioactive: the nucleus is unstable, and will over time *decay*, emitting energy and particles and in the process change into another element.

[2]It is noteworthy that ^{238}U can be converted into a different, fissile element called plutonium through the process of neutron bombardment. Likewise, naturally-occurring thorium 232 (^{232}Th) may be converted into ^{233}U which is fissile. However, converting non-fissile uranium into fissile plutonium, or converting non-fissile thorium into fissile uranium, requires intense neutron bombardment at a scale only seen within the core of a nuclear reactor running on some other fuel such as ^{235}U, which makes ^{235}U the critical ingredient for any independent nuclear program.

[3]Power reactors using "heavy" water as the moderator (such as the Canadian "CANDU" design) are in fact able to use uranium at natural ^{235}U concentration levels as fuel, but most of the power reactors in the world do not employ this design.

Like all 2-phase separation processes, uranium enrichment breaks a single input "feed" stream into two out-going streams of differing composition. Since in the case of uranium enrichment only one stream is of strategic interest, the stream containing concentrated ^{235}U is called the *product*. The other stream coming exiting the separation process, having been largely depleted of valuable ^{235}U, is called the *tails*:

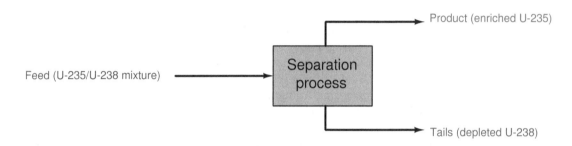

During the United States' Manhattan Project of World War Two, the main process chosen to enrich uranium for the first atomic weapons and industrial-scale reactors was *gaseous diffusion*. In this process, the uranium metal is first chemically converted into *uranium hexafluoride* (UF$_6$) gas so that it may be compressed, transported through pipes, processed in vessels, and controlled with valves. Then, the UF$_6$ gas is run through a long series of diffusion membranes (similar to fine-pore filters). At each membrane, those UF$_6$ molecules containing ^{235}U atoms will preferentially cross through the membranes because they are slightly less massive than the UF$_6$ molecules containing ^{238}U atoms. The mass difference between the two isotopes of uranium is so slight, though, that this membrane diffusion process must be repeated thousands of time in order to achieve any significant degree of enrichment. Gaseous diffusion is therefore an extremely inefficient process, but nevertheless one which may be scaled up to industrial size and used to enrich uranium at a pace sufficient for a military nuclear program. At the time of its construction, the world's first gaseous diffusion enrichment plant (built in Oak Ridge, Tennessee) also happened to be the world's largest industrial building.

An alternative uranium enrichment technology considered but later abandoned by the Manhattan Project scientists was *gas centrifuge* separation. A gas centrifuge is a machine with a hollow rotor spun at extremely high speed. Gas is introduced into the interior of the rotor, where centrifugal force causes the heavier molecules to migrate toward the walls of the rotor while keeping the lighter molecules toward the center. Centrifuges are commonly used for separating a variety of different liquids and solids dissolved in liquid (e.g. separating cells from plasma in blood, separating water from cream in milk), but gas centrifuges face a much more challenging task because the difference in density between various gas molecules is typically far less than the density differential in most liquid mixtures. This is especially true when the gas in question is uranium hexafluoride (UF$_6$), and the only difference in mass between the UF$_6$ molecules is that caused by the miniscule[4] difference in mass between the uranium isotopes ^{235}U and ^{238}U.

---

[4]The formula weight for UF$_6$ containing fissile ^{235}U is 349 grams per mole, while the formula weight for UF$_6$ containing non-fissile ^{238}U is only slightly higher: 352 grams per mole. Thus, the difference in mass between the two molecules is less than one percent.

Gas centrifuge development was continued in Germany, and then later within the Soviet Union. The head of the Soviet gas centrifuge effort – a captured Austrian scientist named Gernot Zippe – was eventually brought to the United States where he shared the refined centrifuge design with American scientists and engineers. As complex as this technology is, it is far[5] more energy-efficient than gas diffusion, making it the uranium enrichment technology of choice at the time of this writing (2016).

An illustration of Gernot Zippe's design is shown below. The unenriched UF_6 feed gas is introduced into the middle of the spinning rotor where it circulates in "counter-current" fashion both directions parallel to the rotor's axis. Lighter (^{235}U) gas tends to stay near the center of the rotor and is collected at the bottom by a stationary "scoop" tube where the inner gas current turns outward. Heavier (^{238}U) gas tends to stay near the rotor wall and is collected at the top by another stationary "scoop" where the outer current turns inward:

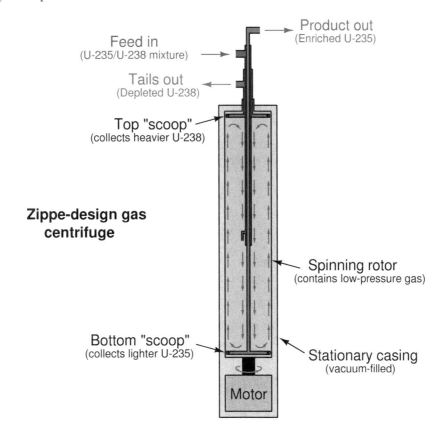

Like the separation membranes used in gaseous diffusion processes, each gas centrifuge is only able to enrich the UF^6 gas by a very slight amount. The modest enrichment factor of each centrifuge necessitates many be connected in series, with each successive centrifuge taking in the out-flow of the previous centrifuge in order to achieve any practical degree of enrichment. Furthermore, gas

---

[5]By some estimates, gas centrifuge enrichment is 40 to 50 *times* more energy efficient than gaseous diffusion enrichment.

centrifuges are by their very nature rather limited in their flow capacity[6]. This low "throughput" necessitates parallel-connected gas centrifuges in order to achieve practical production rates for a national-scale nuclear program. A set of centrifuges connected in parallel for higher flow rates is called a *stage*, while a set of centrifuge stages connected in series for greater enrichment levels is called a *cascade*.

A gas centrifuge *stage* is very simple to understand, as each centrifuge's feed, product, and tails lines are simply paralleled for additional throughput:

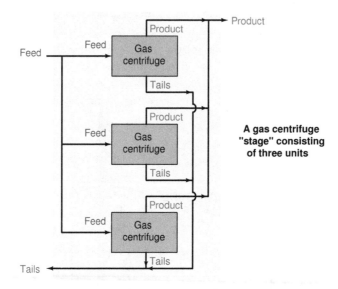

A gas centrifuge *cascade* is a bit more complex to grasp, as each centrifuge's product gets sent to the feed inlet of the next stage for further enrichment, and the tails gets sent to the feed inlet of the previous stage for further depletion. The main feed line enters the cascade at one of the middle stages, with the main product line located at one far end and the main tails line located at the other far end:

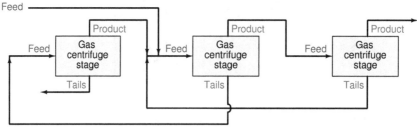

---

[6]A typical gas centrifuge's mass flow rating is on the order of milligrams per second. At their very low (vacuum) operating pressures, a typical centrifuge rotor will hold only a few grams of gas at any moment in time.

This US Department of Energy (DOE) photograph shows an array of 1980's-era American gas centrifuges located in Piketon, Ohio. Each of the tall cylinders is a single gas centrifuge machine, with the feed, product and tails tubing seen connecting to the spinning rotor at the top of the stationary casing:

The size of each stage in a gas centrifuge cascade is proportional to its feed flow rate. The stage processing the highest feed rate must be the largest (i.e. contain the most centrifuges), while the stages at the far ends of the cascade contain the least centrifuges. A cascade similar to the one at the Natanz enrichment facility in Iran – the target of the Stuxnet cyber-attack – is shown here without piping for simplicity, consisting of 164 individual gas centrifuges arranged in 15 stages. The main feed enters in the middle of the cascade at the largest stage, while enriched product exits at the right-hand end and depleted tails at the left-hand end:

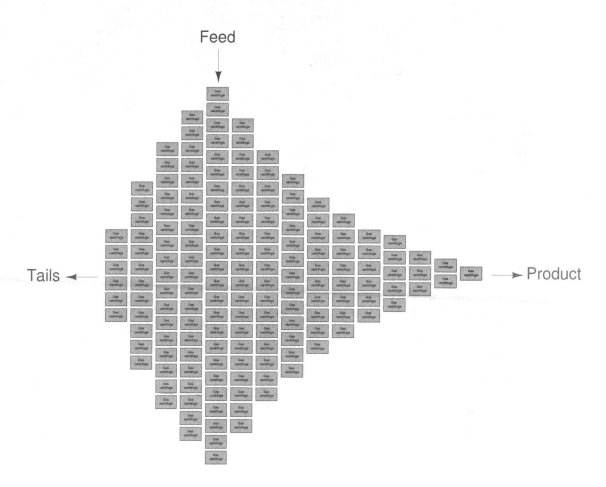

The sheer number of gas centrifuges employed at a large-scale uranium enrichment facility is quite staggering. At the Natanz facility, where just one cascade contained 164 centrifuges, cascades were paralleled together in *sub-units* of six cascades each (984 centrifuges per sub-unit), and three of these sub-units made one cascade *unit* (2952 centrifuges total).

33.1.2 Gas centrifuge vulnerabilities

It would be an understatement to say that a gas centrifuge is a delicate machine. In order to perform their task efficiently[7], gas centrifuge rotors must be long and made to rotate at extremely high rates of speed. Maintaining any rotating machine in a state of near-perfect balance is difficult, much more so when the rotating element is very long[8]. Furthermore, since the gas pressure inside each centrifuge rotor is sub-atmospheric, leak-free seals must be maintained between the spinning rotor and the stationary components (the casing and internal tubing). The extremely high rotational speeds of modern gas centrifuges (many tens of thousands of revolutions per minute!) necessitate advanced materials be used in rotor construction, optimizing light weight and high strength so that the rotors will not be torn to pieces by their own centrifugal force.

A peculiar problem faced by any high-speed rotating machine is a phenomenon called *critical speed*. Any object possessing both mass and resilience is capable of oscillating, which of course includes any and every rotating machine component. If the rotating component of a machine happens to spin at a rate equal to its own natural oscillating frequency, a condition of *mechanical resonance* occurs. *Any* amount of imbalance in the rotating component while spinning at this speed, however slight, will generate a force driving the assembly into continuous oscillation. The speed at which this resonance occurs is called the "critical speed" of the machine, and it should be avoided whenever possible.

Destructive resonance will be avoided so long as the machine is maintained at any speed significantly below or above its critical speed. Most modern gas centrifuges are classified as *supercritical* machines, because they are designed to operate at rotational speeds exceeding their critical speeds. The only time resonance becomes a problem in a supercritical machine is during start-up and shut-down, when the speed must momentarily pass through the critical value. So long as this moment is brief, however, oscillations will not have enough time to grow to destructive levels.

In addition to the problems faced by all high-speed rotating machines, a problem unique to gas centrifuges is gas pressure control. Since the rotor of a gas centrifuge spins inside of an evacuated[9] stationary casing, the existence of any gas pressure inside the rotor creates additional stress acting in the same outward direction as the rotor's own centrifugal force. This means rotor gas pressure must be maintained at a very low level in order to minimize rotor stress. Furthermore, if pressure and temperature conditions are not carefully controlled in a gas centrifuge, the gas may actually *sublimate* into a solid state which will deposit material on the inside wall of the rotor and surely throw it out of balance.

---

[7]Three major factors influence the efficiency of a gas centrifuge: rotor wall speed, rotor length, and gas temperature. Of these, rotor wall speed is the most influential. Higher speeds separate isotopes more effectively, because higher wall speeds result in greater amounts of radial acceleration, which increases the amount of centrifugal force experienced by the gas molecules. Longer rotors also separate isotopes more effectively because they provide more opportunity for the counter-flowing gas streams to separate lighter molecules toward the center and heavier molecules toward the wall. Higher temperatures reduce separation efficiency, because gas molecules at higher temperatures are more mobile and therefore diffuse (i.e. mix together) at higher rates. Therefore, the optimum gas centrifuge design will be long, spin as fast as possible, and operate as cool as possible.

[8]To give you an idea of just how long some gas centrifuge rotors are, the units built for the US Department of Energy facility in Ohio used rotors *40 feet in length*!

[9]This means the hollow casing exists in a state of vacuum, with no air or other gases present. This is done in order to help thermally insulate the rotor from ambient conditions, as well as avoid generating heat from air friction against the rotor's outside surface. Remember, elevated temperatures cause the gas to diffuse at a faster rate, which in turn causes the gas to randomly mix and therefore not separate into light and heavy isotopes as intended.

One could argue that the temperamental nature of gas centrifuges is a good thing, because it makes the manufacture of enriched uranium difficult to achieve, which in turn complicates the development of nuclear weapons. This fragility also makes gas centrifuges an ideal target for anyone interested in halting or delaying nuclear weapons development, which was precisely the aim of the Stuxnet computer virus.

33.1.3 The Natanz uranium enrichment facility

Iran used an obsolete gas centrifuge design, perhaps the best they could obtain at the time, as the uranium enrichment platform of choice for their Natanz facility. By modern standards, this design was inefficient and troublesome, but the Iranians were able to coax serviceable performance from this centrifuge design by means of extensive instrumentation and controls.

Simply put, the Iranian strategy was to manufacture centrifuges faster than they would break and equip the centrifuge cascades with enough piping and supervisory instrumentation that they could detect and isolate failed centrifuges without stopping production, rather than wait until they had perfected the design of the centrifuges themselves. The extensive network of sensors, valves, piping, and PLCs (Programmable Logic Controllers) installed at the Natanz facility facilitated this fault-tolerant design.

The key to the Natanz system's fault tolerance was a set of isolation ("block") valves installed at each gas centrifuge. Each machine was also equipped with a sufficient array of sensors to detect malfunctions. If a centrifuge experienced trouble, such as excessive vibration, the PLC control system would automatically shut all the isolation valves for that failed centrifuge and turn off its drive motor. Since most stages in each cascade contained multiple centrifuges in parallel, the isolation of a single centrifuge within a stage would not shut down the entire cascade. Instead, maintenance personnel could repair the failed centrifuge while production continued, and return it to service when ready.

One undesired consequence of shutting isolation valves on operating centrifuges, though, was increased gas pressure in portions of the cascade. With fewer centrifuges left to handle a constant feed flow, the pressure drop across that stage increases. All upstream stages therefore experience more gas pressure, which as described earlier increases the stress imparted on the spinning centrifuge rotors. In answer to this problem was another innovation at the Natanz facility: using the "dump system" (a standard feature in any gas centrifuge cascade, for evacuating gas from the centrifuges in the event of an emergency shut-down event) as a pressure relief in the event of overpressure resulting from too many isolated centrifuges. Of course, engaging this "dump" system as a means of pressure control would reduce production rates, but it was a better outcome for the system operators than a complete shut-down of the cascade.

In summary, the instrumentation employed in the Natanz facility would automatically detect problems in each centrifuge, isolate any failed centrifuges from the running cascade, and open dump valves as necessary to reduce gas pressure on the remaining centrifuges. This so-called Cascade Protection System was implemented by Siemens model S7-417 PLCs, one per sub-unit (six cascades, each sub-unit containing 984 individual gas centrifuges). All-digital *Profibus* technology was used to communicate process data over network cables between the field instruments and the PLCs, as a means of reducing what would have otherwise been a huge amount of analog and discrete signal wiring.

Additional Siemens PLCs were used at the Natanz facility to control the gas centrifuges, notably the model S7-315 employed to issue commands to variable-frequency drive units sending power to the rotor drive motors. Like the larger S7-417 PLC units, one S7-315 PLC was used to control the motor drives of each cascade sub-unit (six cascades, 984 centrifuges). As subsequent portions of this chapter will detail, both of these Siemens PLC platforms were targets of the Stuxnet virus.

33.1.4   Stuxnet version 0.5

Multiple versions of the Stuxnet virus were aimed at the Natanz facility, at least two significantly different "major" versions which are publicly known at the time of this writing (2016). The first major Stuxnet version, developed as early as November of 2005 and labeled as version 0.5 by the Symantec Corporation, differed from later versions both in its means of delivery (the *dropper* portion of the virus code) and its means of attack (the *payload* portion of the virus code). Later versions of Stuxnet (compiled in 2009-2010 and dubbed versions 1.x by Symantec) employed a much more sophisticated "dropper" and a payload designed to affect a completely different portion of the Iranian centrifuge control system.

A summary of Stuxnet version 0.5 appears here:

- **Infection point**: The infection begins with files written to a removable drive (e.g. USB flash drive), automatically run by the Windows operating system upon connection to a personal computer.

- **Dropper vector**: Stuxnet searches for and infects any Siemens Step 7 PLC project archives found on the personal computer.

- **Payload target**: Siemens S7-417 programmable logic controllers (PLCs) implementing the Cascade Protection System for isolation and overpressure control of centrifuges.

- **Payload vector**: Install a DLL (Dynamically Linked Library) file in the Siemens Step 7 software library collection designed to alter any Step 7 programming code downloaded to a PLC, inserting attack code in the infected PLCs.

- **Payload task**: Shut off isolation valves and mis-calibrate the pressure sensors to cause mild over-pressuring of the centrifuges.

- **Goal**: Increase stress on operating centrifuges, leading to premature failure. Avoid catastrophic cascade failure, which would raise suspicion.

- **Stop date**: July 4, 2009.

The "dropper" portion of Stuxnet version 0.5 exploited a vulnerability in the Siemens "Step 7" PLC programming software which runs on Windows-based personal computers, but did not exploit any vulnerabilities within the Windows operating system itself. In fact, this early version of Stuxnet lacked the ability to self-propagate over the internet, and had to be installed on a personal computer running the Siemens Step 7 software. The most popular hypothesis to date is that the infection happened via a USB flash drive, or "memory stick" used to store digital data.

The "payload" portion of Stuxnet version 0.5 was incredibly sophisticated by comparison.

33.1.5 Stuxnet version 1.x

Subsequent versions of Stuxnet have been labeled as version 1.x and are treated here as one major release. A summary of Stuxnet versions 1.x appears here:

- **Infection point**: The infection begins with files written to a removable drive (e.g. USB flash drive), automatically run by the Windows operating system upon connection to a personal computer. The infection is then able to spread from one Windows PC to another over networks using multiple Windows vulnerabilities.

- **Dropper vector**: Exploit multiple "zero day[10]" vulnerabilities in Windows XP and Vista operating systems to aggressively propagate the virus over computer networks, then infect any Siemens Step 7 project files found on those computers.

- **Payload target**: Siemens S7-315 programmable logic controllers (PLCs) regulating centrifuge rotor speeds.

- **Payload vector**: Install a DLL (Dynamically Linked Library) file in the Siemens Step 7 software library collection designed to alter any Step 7 programming code downloaded to a PLC, inserting attack code in the infected PLCs.

- **Payload task**: Change rotor speeds over time so as to make them pass through their "critical speed" range.

- **Goal**: Increase stress on operating centrifuges, leading to premature failure. Again, avoid catastrophic cascade failure which would raise suspicion.

- **Stop date**: June 24, 2012.

33.2 Motives

There are multiple motives for compromising the security of an industrial control system, some of which overlap motives for attacking IT systems, and some of which are unique to the industrial world. This section details some of the reasons why people might wish to attack an industrial control system.

---

[10]The term *zero-day* in the digital security world refers to vulnerabilities that are unknown to the manufacturer of the software, as opposed to known vulnerabilities that have been on record with the manufacturer for some time. The fact that Stuxnet 1.x employed no less than four zero-day Windows exploits strongly suggests it was developed by an agency with highly sophisticated resources. In other words, Stuxnet 1.x wasn't made by amateurs. This is literally world-class hacking in action!

33.2.1 Technical challenge

Computer experts tend to be a demographic of people motivated by technical challenges and problem-solving. To this type of person, the challenge of breaking in to a computer system designed to foil intruders may be too tempting to resist.

To the person interested in compromising a digital system just for the sake of seeing whether it can be done, the reward is in achieving access, not necessarily inflicting any damage. These people are generally not a direct threat, but may pose an indirect threat if they share their expertise with others harboring sinister motives.

Other individuals motivated by the technical challenge of accessing a digital system are interested in seeing just how much havoc they can wreak once they gain access. Such individuals are analogous to *digital arsonists*, interested in starting the biggest fire that they can simply for the sake of the fire's size.

33.2.2 Profit

The major motive driving IT cyber-attacks today is *profit*: the theft of credit card and other sensitive digital information which may be sold on the black market. Criminal organizations benefit from this style of digital attack, with many attackers becoming millionaires by way of their digital exploits.

Another form of profit-driven attack is commonly called *ransomware*, where an attacker inserts malicious software on the victim's computer(s) preventing access to the system or encrypting files such that they become unusable. This malware then presents a message to the victim asking for monetary payment in exchange for normal system access.

Neither of these attacks is novel to industrial systems, and in fact are commonplace in the IT world. What is novel in industrial systems is the severity of the repercussions. One might imagine the response from an oil drilling rig's management team to ransomware preventing startup-up of a new oil well, where downtime may be in the range of millions of US dollars per day of production. Not only is the imperative to get back online stronger than it would be for a private individual whose home computer was being held ransom, but the ability for an oil company to immediately pay the attacker is much greater than any private individual.

Another potential application of the profit motive in industrial system attacks is *commodities trading*. Traders who profit from the purchase and sale of commodities produced by industrial manufacturers might stand to gain by knowing the day-to-day operational status of those manufacturers. If such people were to access the production and inventory logs residing in a facility's digital control system, for example, they may be able to make more profitable trading decisions based on this privileged information. Eavesdropping on industrial control system data therefore poses another mode of *insider trading*.

33.2.3 Espionage

Aside from gathering data from industrial systems for the direct purpose of profit, less direct motives for attacking industrial control systems exist. One such motive is the theft of proprietary process data, for example recipes and formulae for producing chemical products such as craft foods and drinks, as well as pharmaceuticals.

Special control strategies and process designs critical to the manufacture of certain products are valuable to competing organizations as well. A chemical company eager to discover how to control a temperamental new chemical reaction process might wish to sample the controller algorithms and instrument configurations used by a successful competitor. Even if these design details were not stolen outright, the attacker may gather valuable test data and learn from the developmental mistakes of their competitor, thereby saving time and money pursuing their own design.

Militaries also stand to gain from espionage of industrial measurement and control systems, since the military capabilities of other nations are founded on industrial-scale operations. A country interested in tracking the development of an adversary's nuclear weapons potential, for example, would have a motive to perform digital espionage via the control systems of those foreign nuclear facilities.

33.2.4   Sabotage

Here, at least in my view, is where cyber-security as it relates to industrial control systems becomes really interesting. The major factor distinguishing digital control system security from IT system security is the former's supervision of a real physical process. This means a control system cyber-attack has far more *direct potential for harm* than any IT cyber-attack.

Corporations and nation-states both have an interest in industrial sabotage if it means they may diminish the economic productivity of a competitor. A country, for example, whose export market is dominated by a single product may be tempted to launch cyber-attacks against facilities producing that same product in other countries, as a means to either maintain or elevate their power in the world economy. Corporations have the exact same interest, just at a different level within the global economy.

Certain activists may also have an interest in sabotaging an industrial facility. Shutting down production of a facility they deem dangerous or unethical, or perhaps just causing the company financial loss through poor product quality and/or non-compliance, are potential motivators for activists to target specific industrial processes.

Military interest in industrial sabotage is practically a "given" assumption, as such a cyber-attack merely constitutes a new type of weapon to add to their existing arsenals. Unlike conventional weapons, cyber-weapons are relatively inexpensive.

Another category of sabotage relevant to cyber-attacks is that perpetrated by *malicious insiders*. This last category is especially troubling, as it involves personnel with in-depth knowledge of the digital systems in question. This simple fact makes defense against such attacks extremely challenging, because these are people normally authorized to access the system and therefore are able to bypass most (if not all) security measures. A few notable examples of internal sabotage are listed here:

- Secret agents of foreign nations

- Recently discharged (former) employees

- Disgruntled employees within a corporation

The destructive potential of a government operative with access to critical systems needs no further explanation. Employees, however, do. An employee who gets laid off or fired may still have access to their former employer's critical systems if their system account is not promptly closed. The same is true if the company maintains a lax password policy, such as multiple people sharing a common user account. Even current employees may be motivated to sabotage their employer's systems, especially where there might be an economic advantage[11] to doing so.

---

[11]Consider what forms of sabotage *striking* employees might be willing to do in order to gain leverage at the bargaining table.

33.2.5 Terrorism

This last motive is especially troubling when one considers the proliferation of digital technology and the disconcerting rise of terror-related attacks around the world. The goal of terrorists is quite simply to instill terror as a means of manipulating and/or punishing perceived enemies. Driven by ideology, terrorists tend not to discriminate when selecting their targets. Like arsonists previously mentioned, success is measured by the magnitude of terror and carnage instilled by the event. Common concerns of ethics are trumped by the dictates of the ideology.

The attacks of September 11, 2001 taught the world how ordinary technologies and systems (in that case, fully-fueled jet passenger aircraft) may be exploited as weapons capable of killing and injuring thousands of people. Industrial process designers would do well to think in similar terms, examining their systems not just from the perspective of their intended purpose but also as potential weapons wielded by terrorists.

33.3 Lexicon of cyber-security terms

Cyber-security seems to have its own vocabulary, ranging from unwieldy technical acronyms to slang terms borrowed from amateur computer enthusiasts. What follows is a partial listing of some common terms and their definitions. This list is not only useful as a definitional reference when encountering such terms in cyber-security literature, but it also serves to outline a number of common attack strategies:

- **Active attack**: an attack involving data written to a network or to device. See *passive attack* for contrast.

- **Backdoor**: an easy-to-access pathway into a system, typically used by system developers for convenience in their work. There is nothing wrong with a backdoor during development, but backdoors are very dangerous when left in place on commissioned systems.

- **Blacklist**: a database of prohibited messages or users or software applications.

- **Broadcast**: a form of network where all transmissions are heard by all connected devices, even those devices the data is not intended for. Any communication network sharing a common physical channel is a broadcast network.

- **Brute-force attack**: attempting every combination of characters in an effort to forge a working password.

- **Comsec**: shorthand for "communications security".

- **Crypto**: shorthand for "cryptography", which is the purposeful scrambling of data to render it unintelligible to all but the intended recipient.

- **Data diode**: a device permitting only one-way (simplex) data communication. Data diodes eliminate the possibility of active attacks, because they make writing data to the protected system impossible.

- **Denial-of-Service**: a form of attack where the intended function of the system is either downgraded or entirely faulted.

- **Dictionary attack**: attempting common words and character combinations in an effort to forge a working password.

- **Firewall**: a software or hardware application intended to limit connectivity between networked devices.

- **Flooding**: an attack technique consisting of overloading a digital system with data or requests for data, generally the point of which being to achieve denial of service (DoS) when the target system becomes overloaded.

- **FTP**: an acronym standing for File Transfer Protocol, a protocol used for reading and writing files on one computer remotely from another computer. FTP is a predecessor to *SFTP* which includes public-private key encryption for much better security.

- **HTTP**: an acronym standing for Hyper Text Transfer Protocol, the method used for computers to exchange web page data (encoded in HTML files). HTTP is not encrypted.

- **HTTPS**: an acronym standing for Hyper Text Transfer Protocol Secure, the method used for computers to exchange web page data (encoded in HTML files) using encryption.

- **IP**: an acronym standing for Internet Protocol, the packaging of data into "packets" which may be routed independently of each other across a large network.

- **IT**: an acronym standing for Information Technology, used to broadly describe general-purpose digital data systems and communications.

- **Key**: a relatively small segment of digital data that serves to either encrypt or decrypt other digital data. The imagery here is that of a key used to engage or disengage a physical lock.

- **LAN**: an acronym standing for Local Area Network, a network connecting multiple devices over a limited distance, such as the span of an office building or campus. See *WAN* for contrast.

- **Malware**: software written with malicious intent.

- **Man-in-the-Middle**: an attack where the attacker is positioned directly in between sender and receiver, in such a way as to be able to modify messages sent over the network without either sender or receiver being aware.

- **Operating system**: software installed on a computer for the purpose of directly managing that computer's hardware resources, functioning as an intermediate layer between the application and the hardware itself. The existence of operating system software vastly simplifies the design and development of application software. Popular consumer-grade operating systems at the time of this writing (2016) include Microsoft Windows, Apple OS X, Linux, and BSD.

- **Packet sniffing**: the act of passively monitoring data transmitted over a broadcast network.

- **Passive attack**: an attack only involving the reading of data from a network or device. See *active attack* for contrast.

- **Passphrase**: an easily-memorized sentence which may be used to generate complex passwords. For example, one could take the first letter of every word in the passphrase "What we have here is a failure to communicate" to generate the password `wwhhiaftc`. Passphrases are useful because they make complex passwords easy to remember, and in fact may be used to generate multiple passwords from the same phrase (e.g. using the *last* letter of each word instead of the first, to create the password `teeesaeoe` from the same passphrase used previously).

- **Ping**: a simple network utility used on IP networks to test connectivity. The ping message is sent from one computer to another, with the receiving computer replying to declare successful receipt of the ping message.

- **Private key**: a cryptographic key useful for decrypting encrypted data. "Private" refers to the fact that this key must be held in confidence by authorized parties only, since it has the ability to unlock coded messages.

- **Public key**: a cryptographic key useful only for encrypting data. "Public" refers to the fact that this key may be shared openly, as it cannot be used to unlock a coded message, but instead is only useful for encoding messages sent to a party holding a *private key* which can decode the message.

- **Replay attack**: a form of attack where a message is intercepted, recorded, and later broadcast to the network in order to inflict damage. An interesting feature of replay attacks is that they may work on encrypted messages, and even when the purpose of the message is unknown to the attacker!

- **SFTP**: an acronym standing for Secure File Transfer Protocol, a protocol used for reading and writing files on one computer remotely from another computer. SFTP is a successor to *FTP* which lacked encryption.

- **Sniffing**: inspecting network communications for important data. So-called "packet sniffers" monitor data traffic on a broadcast network for certain information such as network addresses and data content.

- **Spoofing**: presenting a false identification to the receiver of digital data. This commonly takes the form of presenting fake network address information, to trick the receiver into thinking the source is from a legitimate location or device.

- **SSH**: an acronym standing for Secure Shell, a remote-access utility commonly used in Unix operating systems allowing users to log into a computer from another computer connected to the same network. SSH is a successor to *telnet*, which lacked encryption.

- **TCP**: an acronym standing for Terminal Control Protocol, the protocol used to ensure segments of data make it to their intended destinations after being routed by IP (see *Internet Protocol*).

- **Telnet**: a remote-access utility commonly used in Unix operating systems allowing users to log into a computer from another computer connected to the same network. Telnet is a predecessor to *SSH* which includes public-private key encryption for much greater security.

- **Trusted**: a component or section of a digital system that is assumed to be safe from intrusion.

- **VPN**: an acronym standing for Virtual Private Network, which encrypts every aspect of a transaction between two computers connected on a network. The effect is to form a "virtual network" between the machines, the privacy of which being ensured by the encryption used to scramble the data.

- **Vuln**: shorthand for "vulnerability" or weakness in a system.

- **Walled garden**: a term used to describe an area of a digital system assumed to be safe from intrusion. See *trusted*.

- **WAN**: an acronym standing for Wide Area Network, a network connecting multiple devices over a long range, such as the span of a city. See *LAN* for contrast.

- **War dialing**: the exploratory practice of dialing random phone numbers in search of telephone modem connections, which may connect to computer systems.

- **Whitelist**: a database of permitted messages or users or software applications.

33.4 Fortifying strategies

As control system professionals, it is in our interest to ensure our measurement and control systems are secure from unauthorized access. It is helpful to regard system *security* similarly to how we regard system *safety* or *reliability*, as these concerns share many common properties:

- Just as accidents and faults are inevitable, so are unauthorized access attempts

- Just as 100% perfect safety and 100% perfect reliability is unattainable, so is 100% security

- Digital security needs to be an important criterion in the selection and setup of industrial instrumentation equipment, just as safety and reliability are important criteria

- Maximizing security requires a security-savvy culture within the organization, just as maximizing safety requires a safety-savvy culture and maximizing reliability requires a reliability-centric design philosophy

Also similar to safety and reliability is the philosophy of *defense-in-depth*, which is simply the idea of having multiple layers of protection in case one or more fail. Applied to digital security, defense-in-depth means not relying on a single mode of protection (e.g. passwords only) to protect a system from attack.

33.4.1 Policy-based strategies

These strategies focus on human behavior rather than system design or component selection. In some ways these strategies are the simplest to implement, as they generally require little in the way of technical expertise. This is not to suggest, however, that policy-based security strategies are therefore the *easiest* to implement. On the contrary, changing human behavior is usually a very difficult feat. Policy-based strategies are not necessarily cheap, either: although little capital is generally required, operational costs will likely rise as a result of these policies. This may take the form of monetary costs, additional staffing costs, and/or simply costs associated with impeding normal work flow (e.g. pulling personnel away from their routine tasks to do training, requiring personnel to spend more time doing things like inventing and tracking new passwords, slowing the pace of work by limiting authorization).

Foster awareness

Ensure all personnel tasked with using and maintaining the system are fully aware of security threats, and of best practices to mitigate those threats. Given the ever-evolving nature of cyber-attacks, this process of educating personnel must be continuous.

A prime mechanism of cyber-vulnerability is the casual sharing of information between employees, and with people outside the organization. Information such as passwords and network design should be considered "privileged" and should only be shared on a need-to-know basis. Critical security information such as passwords should never be communicated to others or stored electronically in plain ("cleartext") format. When necessary to communicate or store such information electronically, it should be encrypted so that only authorized personnel may access it.

In addition to the ongoing education of technical personnel, it is important to keep management personnel aware of cyber threat and threat potentials, so that the necessary resources will be granted toward cyber-security efforts.

Employ security personnel

For any organization managing important processes and services, "important" being defined here as *threatening* if compromised by the right type of cyber-attack, it is imperative to employ qualified and diligent personnel tasked with the ongoing maintenance of digital security. These personnel must be capable of securing the control systems themselves and not just general data systems.

One of the routine tasks for these personnel should be evaluations of risks and vulnerabilities. This may take the form of security audits or even simulated attacks whereby the security of the system is tested with available tools.

Utilize effective authentication

Simply put, it is imperative to correctly identify all users accessing a system. This is what "authentication" means: correctly identifying the person (or device) attempting to use the digital system. *Passwords* are perhaps the most common authentication technique.

The first and foremost precaution to take with regard to authentication is to never use default (manufacturer) passwords, since these are public information. This precautionary measure may seem so obvious as to not require any elaboration, but sadly it remains a fact that too many password-protected devices and systems are found operating in industry with default passwords.

Another important precaution to take with passwords is to not use the same password for all systems. The reasoning behind this precaution is rather obvious: once a malicious party gains knowledge of that one password, they have access to all systems protected by it. The scenario is analogous to using the exact same key to unlock every door in the facility: all it takes now is one copied key and suddenly intruders have access to every room.

Passwords must also be changed on a regular basis. This provides some measure of protection even after a password becomes compromised, because the old password(s) no longer function.

Passwords chosen by system users should be "strong," meaning difficult for anyone else to guess. When attackers attempt to guess passwords, they do so in two different ways:

- Try using common words or phrases that are easy to memorize

- Try every possible combination of characters until one is found that works

The first style of password attack is called a *dictionary attack*, because it relies on a database of common words and phrases. The second style of password attack is called a *brute force attack* because it relies on a simple and tireless ("brute") algorithm, practical only if executed by a computer.

A password resistant to dictionary-style attacks is one not based on a common word or phrase. Ideally, that password will appear to be nonsense, not resembling any discernible word or simple pattern. The only way to "crack" such a password, since a database of common words will be useless against it, will be to attempt every possible character combination (i.e. a brute-force attack).

A password resistant to brute-force-style attacks is one belonging to a huge set of possible passwords. In other words, there must be a very large number of possible passwords limited to the same alphabet and number of characters. Calculating the brute-force strength of a password is a matter of applying a simple exponential function:

$$S = C^n$$

Where,
S = Password strength (i.e. the number of unique password combinations possible)
C = Number of available characters (i.e. the size of the alphabet)
n = Number of characters in the password

For example, a password consisting of four characters, each character being a letter of the English alphabet where lower- and upper-case characters are treated identically, would give the following strength:

$$S = 26^4 = 456976 \text{ possible password combinations}$$

If we allowed case-sensitivity (i.e. lower- and upper-case letters treated differently), this would double the value of C and yield more possible passwords:

$$S = 52^4 = 7311616 \text{ possible password combinations}$$

Obviously, then, passwords using larger alphabets are stronger than passwords with smaller alphabets.

Cautiously grant authorization

While *authentication* is the process of correctly identifying the user, *authorization* is the process of assigning rights to each user. The two concepts are obviously related, but not identical. Under any robust security policy, users are given only as much access as they need to perform their jobs efficiently. Too much access not only increases the probability of an attacker being able to cause maximum harm, but also increases the probability that benevolent users may accidently cause harm.

Perhaps the most basic implementation of this policy is for users to log in to their respective computers using the lowest-privilege account needed for the known task(s), rather than to log in at the highest level of privilege they *might* need. This is a good policy for all people to adopt when they use personal computers to do any sort of task, be it work- or leisure-related. Logging in with full ("administrator") privileges is certainly convenient because it allows you to do anything on the system (e.g. install new software, reconfigure any service, etc.) but it also means any malware accidently engaged[12] under that account now has the same unrestricted level of access to the system. Habitually logging in to a computer system with a low-privilege account helps mitigate this risk, for any accidental execution of malware will be similarly limited in its power to do harm.

Another implementation of this policy is called *application whitelisting*, where only trusted software applications are allowed to be executed on any computer system. This stands in contrast to "blacklisting" which is the philosophy behind anti-virus software: maintaining a list of software applications known to be harmful (malware) and prohibiting the execution of those pre-identified applications. Blacklisting (anti-virus) only protects against malware that has been identified and notified to that computer. Blacklisting cannot protect against "zero-day" malware known by no one except the attacker. In a whitelisting system, each computer is pre-loaded with a list of acceptable applications, and no other applications – benign or malicious – will be able to run on that machine.

---

[12]Consider the very realistic scenario of logging in as administrator (or "root" in Unix systems) and then opening an email message which happens to carry an attached file infected with malware. Any file executed by a user is by default run at that user's level of privilege because the operating system assumes that is the user's intent.

Maintain good documentation

While this is important for effective maintenance in general, thorough and accurate documentation is especially important for digital security because it helps identify vulnerabilities. Details to document include:

- Network diagrams

- Software version numbers

- Device addresses

Close unnecessary access pathways

All access points to the critical system must be limited to those necessary for system function. This means all other potential access points in the critical system must be closed so as to minimize the total number of access points available to attackers. Examples of access points which should be inventoried and minimized:

- Hardware communication ports (e.g. USB serial ports, Ethernet ports, wireless radio cards)

- Software TCP ports

- Shared network file storage ("network drives")

- "Back-door" accounts used for system development

That last category deserves some further explanation. When engineers are working to develop a new system, otherwise ordinary and sensible authentication/authorizations measures become a major nuisance. The process of software development always requires repeated logins, shutdowns, and tests forcing the user to re-authenticate themselves and negotiate security controls. It is therefore understandable when engineers create simpler, easier access routes to the system under development, to expedite their work and minimize frustration.

Such "back-door" access points become a problem when those same engineers forget (or simply neglect) to remove them after the developed system is released for others to use. An interesting example of this very point was the so-called *basisk* vulnerability discovered in some Siemens S7 PLC products. A security researcher named Dillon Beresford working for NSS Labs discovered a telnet[13] service running on certain models of Siemens S7 PLCs with a user account named "basisk" (the password for this account being the same as the user name). All one needed to do in order to gain privileged access to the PLC's operating system was connect to the PLC using a telnet client and enter "basisk" for the user name and "basisk" for the password! Clearly, this was a back-door account used by Siemens engineers during development of that PLC product line, but it was not closed prior to releasing the PLC for general use.

---

[13]Telnet is a legacy software utility used to remotely access command-line computer operating systems. Inherently unsecure, telnet exchanges login credentials (user name and password) unencrypted over the network connection. A modern replacement for telnet is SSH (Secure SHell).

Maintain operating system software

All operating system software manufacturers periodically release "patches" designed to improve the performance of their products. This includes patches for discovered security flaws. Therefore, it is essential for all computers belonging to a critical system to be regularly "patched" to ensure maximum resistance to attack.

This is a significant problem within industry because so much industrial control system software is built to run on consumer-grade operating systems such as Microsoft Windows. Popular operating systems are built with maximum convenience in mind, not maximum security or even maximum reliability. New features added to an operating system for the purpose of convenient access and/or new functionality often present new vulnerabilities[14].

Another facet to the consumer-grade operating system problem is that these operating systems have relatively short lifespans. Driven by consumer demand for more features, software manufacturers develop new operating systems and abandon older products at a much faster rate than industrial users upgrade their control systems. Upgrading the operating systems on computers used for an industrial control system is no small feat, because it usually means disruption of that system's function, not only in terms of the time required to install the new software but also (potentially) re-training required for employees. Upgrading may even be impossible in cases where the new operating system no longer supports features necessary for that control system[15]. This would not be a problem if operating system manufacturers provided the same long-term (multi-decade) support for their products as industrial hardware manufacturers typically do, but this is not the case for consumer-grade products such as Microsoft Windows[16].

---

[14]I am reminded of an example from the world of "smart" mobile telephones, commonly equipped with *accelerometer* sensors for detecting physical orientation. Accelerometers detect the force of acceleration and of gravity, and are useful for a variety of convenient "apps" having nothing to do with telephony. Smart phone manufacturers include such sensors in their mobile devices and link those sensors to the phone's operating system because doing so permits innovative applications, which in turn makes the product more desirable to application developers and ultimately consumers. It was discovered, though, that the signals generated by these accelerometers could be used to detect "keystrokes" made by the user, the sensors picking up vibrations made as the user taps their finger against the glass touch-screen of the smart phone. With the right signal processing, the accelerometers' signals could be combined in such a way to identify which characters the user was tapping on the virtual keyboard, and thereby eavesdrop on their text-based communications!

[15]An example of this is where a piece of obsolete industrial software runs on the computer's operating system, for example a data acquisition program or data-analysis program made by a company that no longer exists. If this specialized software was written to run on a particular operating system, and no others, future versions of that operating system might not permit proper function of that specialized software. I have seen such cases in industry, where industrial facilities continue to run obsolete (unsupported) operating systems in order to keep running some specialized industrial software (e.g. PLC programming editors), which is needed to operate or maintain some specialized piece of control hardware which itself is obsolete but still functions adequately for the task. In order to upgrade to a modern operating system on that computer (e.g. an obsolete version of Microsoft Windows), one must upgrade the specialized software (e.g. the PLC programming editor software), which in turn would mean upgrading the control hardware (e.g. the PLCs themselves). All of this requires time and money, much more than just what is required to upgrade the operating system software itself.

[16]As a case in point, there are still a great many industrial computers running Microsoft Windows XP at the time of this writing (2016), even though this operating system is no longer supported by Microsoft. This means no more Service Pack upgrades from Microsoft, security patches, or even research on vulnerabilities for this obsolete operating system. All users of Windows XP are "on their own" with regard to cyber-attacks.

Routinely archive critical data

The data input into and generated by digital control systems is a valuable commodity, and must be treated as such. Unlike material commodities, data is easily replicated, and this fact provides some measure of protection against loss from a cyber-attack. Routine "back-ups" of critical data, therefore, is an essential part of any cyber-security program. It should be noted that this includes not just operational data collected by the control system during operation, but also data such as:

- PID tuning parameters

- Control algorithms (e.g. function block programs, configuration data, etc.)

- Network configuration parameters

- Software installation files

- Software license (authorization) files

- Software drivers

- Firmware files

- User authentication files

- All system documentation (e.g. network cable diagrams, loop diagrams)

This archived data should be stored in a medium immune to cyber-attacks, such as read-only optical disks. It would be foolish, for example, to store this sort of critical data only as files on the operating drives of computers susceptible to attack along with the rest of the control system.

Create response plans

Just as no industrial facility would be safe without incident response plans to mitigate physical crises, no industrial facility using digital control systems is secure without response plans for cyber-attacks. This includes such details as:

- A chain of command for leading the response

- Instructions on how to restore critical data and system functions

- Work-arounds for minimal operation while critical systems are still unavailable

Limit mobile device access

Mobile digital devices such as cell phones and even portable storage media (e.g. USB "flash" drives) pose digital security risks because they may be exploited as an attack vector bypassing air gaps and firewalls. It should be noted that version 0.5 of Stuxnet was likely inserted into the Iranian control system in this manner, through an infected USB flash drive.

A robust digital security policy will limit or entirely prohibit personal electronic devices into areas where they might connect to the facility's networks or equipment. Where mobile devices are essential for job functions, those devices should be owned by the organization and registered in such a way as to authenticate their use. Computers should be configured to automatically reject non-registered devices such as removable flash-memory storage drives. Portable computers not owned and controlled by the organization should be completely off-limits[17] from the process control system.

Above all, one should never underestimate the potential harm allowing uncontrolled devices to connect to critical, trusted portions of an industrial control system. The degree to which any portion of a digital system may be considered "trusted" is a function of *every* component of that system. Allowing connection to untrusted devices violates the confidence of that system.

Secure all toolkits

A special security consideration for industrial control systems is the existence of software designed to create and edit controller algorithms and configurations. The type of software used to write and edit Ladder Diagram (LD) code inside of programmable logic controllers (PLCs) is a good example of this, such as the Step7 software used to program Siemens PLCs in Iran's Natanz uranium enrichment facility. Instrumentation professionals use such software on a regular basis to do their work, and as such it is an essential tool of the trade. However, this very same software is a weapon in the hands of an attacker, or when hijacked by malicious code.

A common practice in industry is to leave computers equipped with such "toolkit" software connected to the control network for convenience. This is a poor policy, and one that is easily remedied by simply disconnecting the programming computer from the control network immediately after downloading the edited control code. An even more secure policy is to never connect such "toolkit" computers to a network at all, but only to controllers directly, so that the toolkit software cannot be hijacked.

Another layer of defense is to utilize robust password protection on the programmable control devices when available, rather than leaving password fields blank which then permits any user of the toolkit software full access to the controller's programming.

---

[17]This raises a potential problem from the perspective of outside technical support, since such support often entails contracted or manufacturer-employed personnel entering the site and using their work computers to perform system configuration tasks. For any organization implementing a strong security access policy, this point will need to be negotiated into every service contract to ensure all the necessary pieces of hardware and software exist "in-house" for the service personnel to use while on the job.

Close abandoned accounts

Given the fact that disgruntled technical employees constitute a significant security threat to organizations, it stands to reason that the user accounts of terminated employees should be closed as quickly as possible. Not only do terminated employees possess authentication knowledge in the form of user names and passwords, but they may also possess extensive knowledge of system design and vulnerabilities.

33.4.2 Design-based strategies

A *design-based* security strategy is one rooted in technical details of system architecture and functionality. Some of these strategies are quite simple (e.g. air gaps) while others are quite complex (e.g. encryption). In either case, they are strategies ideally addressed at the inception of a new system, and at every point of system alteration or expansion.

Advanced authentication

The authentication security provided by passwords, which is the most basic and popular form of authentication at the time of this writing, may be greatly enhanced if the system is designed to not just reject incorrect passwords, but to actively inconvenience the user for entering wrong passwords.

Password timeout systems introduce a mandatory waiting period for the user if they enter an incorrect password, typically after a couple of attempts so as to allow for innocent entry errors. *Password lockout* systems completely lock a user out of their digital account if they enter multiple incorrect passwords. The user's account must then be reset by another user on that system possessing high-level privileges.

The concept behind both password timeouts and password lockouts is to greatly increase the amount of time required for any dictionary-style or brute-force password attack to be successful, and therefore deter these attacks. Unfortunately timeouts and lockouts also present another form of system vulnerability to a *denial of service* attack. Someone wishing to deny access to a particular system user need only attempt to sign in as that user, using incorrect passwords. The timeout or lockout system will then delay (or outright deny) access to the legitimate user.

Authentication based on the user's knowledge (e.g. passwords) is but one form of identification, though. Other forms of authentication exist which are based on the possession of physical items called *tokens*, as well as identification based on unique features of the user's body (e.g. retinal patterns, fingerprints) called *biometric* authentication.

Token-based authentication requires all users to carry tokens on their person. This form of authentication so long as the token does not become stolen or copied by a malicious party.

Biometric authentication enjoys the advantage of being extremely difficult to replicate and nearly[18] impossible to lose. The hardware required to scan fingerprints is relatively simple and inexpensive. Retinal scanners are more complex, but not beyond the reach of organizations possessing expensive digital assets. Presumably, there will even be DNA-based authentication technology available in the future.

---

[18]Before you laugh at the idea of losing one's own body, consider something as plausible as a fingerprint scanner programmed to accept the image of just one finger, and then that user suffering an injury to that same finger either obscuring the fingerprint or destroying the finger entirely.

Air gaps

An *air gap* is precisely what the name implies: a physical separation between the critical system and any network, preventing communication. Although it seems so simple that it ought to be obvious, an important design question to ask is whether or not the system in question really needs to have connectivity at all. Certainly, the more networked the system is, the easier it will be to access useful information and perform useful operational functions. However, connectivity is also a liability: that same convenience makes it easier for attackers to gain access.

While it may seem as though air gaps are the ultimate solution to digital security, this is not entirely true. A control system that *never* connects to a network is still vulnerable to cyber-attack, and that is through detachable programming and data-storage devices. For example, a PLC without a permanent network connection may become compromised by way of an infected portable computer used to edit the PLC's code. A DCS completely isolated from the facility's local area network (LAN) may become compromised if someone plugs in an infected data storage device such as a USB "flash" memory module.

In order for air gaps to be completely effective, they must be permanent and include portable devices as well as network connections. This is where effective security policy comes into play, ensuring portable devices are not allowed into areas where they might connect (intentionally or otherwise) to critical systems. Effective air-gapping of critical networks also necessitates physical security of the network media: ensuring attackers cannot gain access to the network cables themselves, so as to "tap" into those cables and thereby gain access. This requires careful planning of cable routes and possibly extra infrastructure (e.g. separate cable trays, conduits, and access-controlled equipment rooms) to implement.

Wireless (radio) data networks pose a special problem for the "air gap" strategy, because the very purpose of radio communication is to bridge physical air gaps. A partial measure applicable to some wireless systems is to use *directional antennas* to link separated points together, as opposed to *omnidirectional* antennas which transmit and receive radio energy equally in all directions. This complicates the task of "breaking in" to the data communication channel, although it is not 100 percent effective since no directional antenna has a perfectly focused radiation pattern, nor do directional antennas preclude the possibility of an attacker intercepting communications directly between the two antennae. Like all security measures, the purpose of using directional antennas is to make an attack *less probable*.

Network segregation

This simply means to divide digital networks into separate entities (or at least into layers) in order to reduce the exposure of digital control systems to any sources of harm. Air gaps constitute an elementary form of network segregation, but are not practical when at least some data must be communicated between networks.

At the opposite end of the network segregation spectrum is a scenario where all digital devices, control systems and office computers alike, connect to the facility's common Local Area Network (LAN). This is almost always a bad policy, as it invites a host of problems not limited to cyber-attacks but extending well beyond that to innocent mistakes and routine faults which may compromise system integrity. At the very least, control systems deserve their own dedicated network(s) on which to communicate, free of traffic from general information technology (IT) office systems.

In facilities where control system data absolutely must be shared on the general LAN, a *firewall* should be used to connect those two networks. Firewalls are either software or hardware entities designed to filter data passed through based on pre-set rules. In essence, each network on either side of a firewall is a "zone" of communication, while the firewall is a "conduit" between zones allowing only certain types of messages through. A rudimentary firewall might be configured to "blacklist" any data packets carrying hyper-text transfer protocol (HTTP) messages, as a way to prevent web-based access to the system. Alternatively, a firewall might be configured to "whitelist" only data packets carrying Modbus messages for a control system and block everything else.

Some specialized firewalls are manufactured specifically for industrial control systems. One such firewall at the time of this writing (2016) is manufactured by the Tofino, and has the capability to screen data packets based on rules specific to industrial control system platforms such as popular PLC models. Industrial firewalls differ from general-purpose data firewalls in their ability to recognize control-specific data.

In systems where different communication zones must have different levels of access to the outside world, multiple firewalls may be set up in such a way as to create a so-called *demilitarized zone* (DMZ). A DMZ is a network existing between a pair of firewalls, one firewall filtering data to and from a protected network, and the other firewall filtering data to and from an unprotected (or less-protected) network. Any devices connected to the DMZ will have access to either network, through different firewall rule sets. Any data exchanged between the protected and unprotected networks, though, must pass through both firewalls.

Encryption

Encryption refers to the intentional scrambling of data by means of a designated code called a *key*, a similar (or in some cases identical) key being used to un-scramble (decrypt) that data on the receiving end. The purpose of encryption, of course, is to foil passive attacks by making the data unintelligible to anyone but the intended recipient, and also to foil active attacks by making it impossible for an attacker's transmitted message to be successfully received.

A popular form of encrypted communication used for general networking is a *Virtual Private Network* or VPN. This is where two computer systems use VPN software (or two VPN hardware devices) to encrypt messages sent to each other over an unsecure medium. Since every single packet of data exchanged between the two computers is encrypted, the communication will be unintelligible to anyone else who might be "listening" on that unsecure network. In essence, VPNs create a secure "tunnel" for data to travel between points on an otherwise unprotected network.

Encryption may also be applied to dedicated, non-broadcast networks such as telephone channels and serial data communication lines. Special cryptographic modems and serial data translators are manufactured specifically for this purpose, and may be applied to legacy SCADA and control networks based on telephony or serial communication cables.

Important files stored on computer drives may also be encrypted, such that only users possessing the proper key(s) may decrypt and use the files.

It should be noted that encryption does not necessarily protect against so-called *replay* attacks, where the attacker records a communicated message and later re-transmits that same message to the network. For example, if a control system uses an encrypted message to command a remotely-located valve to shut, an attacker might simply re-play that same message at any time in the future to force the valve to shut without having to decrypt the message.

An interesting form of encryption applicable to certain wireless (radio) data networks is *spread-spectrum* communication. This is where all the data is not communicated over the same radio carrier frequency, but rather is divided or "spread" among a range of frequencies. Various techniques exist for spreading digital data across a spectrum of radio frequencies, but they all comprise a form of data encryption because the spreading of that data is orchestrated by means of a cryptographic key. Perhaps the simplest spread-spectrum method to understand is *frequency-hopping* or *channel-hopping*, where the transmitters and receivers both switch frequencies on a keyed schedule. Any receiver uninformed by the same key will not "know" which channels will be used, or in what order, and therefore will be unable to intercept anything but isolated pieces of the communicated data. Spread-spectrum technology was invented during the second World War as a means for Allied forces to encrypt their radio transmissions such that Axis forces could not interpret them.

Spread-spectrum capability is built into several wireless data communication standards, including Bluetooth and *Wireless*HART.

Read-only system access

One way to thwart so-called "active" attacks (where the attacker inserts or modifies data in a digital system to achieve malicious ends) is to engineer the system in such a way that all communicated data is *read-only* and therefore cannot be written or edited by anyone. This, of course, by itself will do nothing to guard against "passive" (read-only) attacks such as eavesdropping, but passive attacks are definitely the lesser of the two evils with regard to industrial control systems.

In systems where the digital data is communicated serially using protocols such as EIA/TIA-232, read-only access may be ensured by simply disconnecting one of the wires in the EIA/TIA-232 cable. By disconnecting the wire leading to the "receive data" pin of the critical system's EIA/TIA-232 serial port, that system cannot receive external data but may only transmit data. The same is true for EIA/TIA-485 serial communications where "transmit" and "receive" connection pairs are separate.

Certain serial communication schemes are inherently simplex (i.e. one-way communication) such as EIA/TIA-422. If this is an option supported by the digital system in question, the use of that option will be an easy way to ensure remote read-only access.

For communication standards such as Ethernet which are inherently duplex (bi-directional), devices called *data diodes* may be installed to ensure read-only access. The term "data diode" invokes the functionality of a semiconductor rectifying diode, which allows the passage of electric current in one direction only. Instead of blocking reverse current flow, however, a "data diode" blocks reverse *information* flow.

The principle of read-only protection applies to computing systems as well as communication networks. Some digital systems do not strictly require on-board data collection or modification of operating parameters, and in such cases it is possible to replace read/write magnetic data drives with read-only (e.g. optical disk) drives in order to create a system that cannot be compromised. Admittedly, applications of this strategy are limited, as there are few control systems which never store operational data nor require any editing of parameters. However, this strategy should be considered where it applies[19].

Many digital devices offer *write-protection* features in the form of a physical switch or key-lock preventing data editing. Just as some types of removable data drives have a "write-protect" tab or switch located on them, some "smart" field instruments also have write-protect switches inside their enclosures which may be toggled only by personnel with direct physical access to the device. Programmable Logic Controllers (PLCs) often have a front-panel write-protect switch allowing protection of the running program.

---

[19]An example of this strategy in action is an internet-connected personal computer system I once commissioned, running the Linux operating system from a DVD-ROM optical disk rather than a magnetic hard drive. The system would access the optical disk upon start-up to load the operating system kernel into its RAM memory, and then access the disk as needed for application executable files, shared library files, and other data. The principal use of this system was web browsing, and my intent was to make the computer as "hacker-proof" as I possibly could. Since the operating system files were stored on a read-only optical disk, it was impossible for an attacker to modify that data without having physical access to the machine. In order to thwart attacks on the data stored in the machine's RAM memory, I configured the system to automatically shut down and re-start every day at an hour when no one would be using it. Every time the computer re-booted, its memory would be a *tabula rasa* ("clean slate"). Of course, this meant no one could permanently store downloaded files or other data on this machine from the internet, but from a security perspective that was the very point.

Not only do write-protect switches guard against malicious attacks, but they also help prevent innocent mistakes from causing major problems in control systems. Consider the example of a PLC network where each PLC connected to a common data network has its own hardware write-protect switch. If a technician or engineer desires to edit the program in one of these PLCs from their remotely-located personal computer, that person must first go to the location of that PLC and disable its write protection. While this may be seen as an inconvenience, it ensures that the PLC programmer will not mistakenly access the wrong PLC from their office-located personal computer, which is especially easy to do if the PLCs are similarly labeled on the network.

Making regular use of such features is a policy measure, but ensuring the exclusive use of equipment with this feature is a system design measure.

Control platform diversity

In control and safety systems utilizing redundant controller platforms, an additional measure of security is to use different models of controller in the redundant array. For example, a redundant control or safety system using two-out-of-three voting (2oo3) between three controllers might use controllers manufactured by three different vendors, each of those controllers running different operating systems and programmed using different editing software. This mitigates against device-specific attacks, since no two controllers in the array should have the exact same vulnerabilities.

A less-robust approach to process control security through diverse platforms is simply the use of effective Safety Instrumented Systems (SIS) applied to critical processes, which always employ controls different from the base-layer control system. An SIS system is designed to bring the process to a safe (shut down) condition in the event that the regular control system is unable to maintain normal operating conditions. In order to avoid common-cause failures, the SIS must be implemented on a control platform independent from the regular control system. The SIS might even employ analog control technology (and/or discrete relay-based control technology) in order to give it complete immunity from digital attacks.

In either case, improving security through the use of multiple, diverse control systems is another example of the *defense in depth* philosophy in action: building the system in such a way that no essential function depends on a single layer or single element, but rather multiple layers exist to ensure that essential function.

33.5 Review of fundamental principles

Shown here is a partial listing of principles applied in the subject matter of this chapter, given for the purpose of expanding the reader's view of this chapter's concepts and of their general inter-relationships with concepts elsewhere in the book. Your abilities as a problem-solver and as a life-long learner will be greatly enhanced by mastering the applications of these principles to a wide variety of topics, the more varied the better.

- **Blacklisting**: the concept of flagging certain users, software applications, etc. as "forbidden' from accessing a system.

- **Chemical isotopes**: variants of chemical elements differing fundamentally in atomic mass. Relevant to the subject of uranium enrichment for nuclear reactors and nuclear weapons, where one particular isotope must be separated from ("enriched") another isotope in order to be useful.

- **Defense-in-Depth**: a design philosophy relying on multiple layers of protection, the goal being to maintain some degree of protection in the event of one or more other layers failing.

- **Reliability**: a statistical measure of the probability that a system will perform its design function. Relevant here with regard to control systems, in that proper control system design can significantly enhance the reliability of a large system if the controls are able to isolate faulted redundant elements within that system. This is the strategy used by designers of the Iranian uranium enrichment facility, using PLC controls to monitor the health of many gas centrifuges used to enrich uranium, and taking failed centrifuges off-line while maintaining continuous production.

- **Whitelisting**: the concept of only permitting certain users, software applications, etc. to access a system.

References

"21 Steps to Improve Cyber Security of SCADA Networks", Department of Energy, USA, May 2011.

Bartman, Tom and Carson, Kevin, "Securing Communications for SCADA and Critical Industrial Systems", Technical Paper 6678-01, Schweitzer Engineering Laboratories, Inc., Pullman, WA, January 22, 2015.

Beresford, Dillon, "Siemens Simatic S7 PLC Exploitation", technical presentation at Black Hat USA conference, 2011.

Byres, Eric, "Building Intrinsically Secure Control and Safety Systems – Using ANSI/ISA-99 Security Standards for Improved Security and Reliability", Byres Security Inc., May 2009.

"Common Cybersecurity Vulnerabilities in Industrial Control Systems", Department of Homeland Security, Control Systems Security Program, National Cyber Security Division, USA, May 2011.

Falliere, Nicolas; Murchu, Liam O.; Chien, Eric; "W32.Stuxnet Dossier", version 1.4, Symantec Corporation, Mountain View, CA, February 11, 2011.

Fischer, Ted, "Private and Public Key Cryptography and Ransomware", Center for Internet Security, Inc., Pullman, WA, December 2014.

Grennan, Mark, "Firewall and Proxy Server HOWTO", version 0.8, February 26, 2000.

Horak, Ray, *Webster's New World Telecom Dictionary*, Wiley Publishing, Inc., Indianapolis, IN, 2008.

Kemp, R. Scott, "Gas Centrifuge Theory and Development: A Review of US Programs", Program on Science and Global Security, Princeton University, Princeton, NJ, Taylor & Francis Group, LLC, 2009.

Langner, Ralph, "To Kill A Centrifuge – A Technical Analysis of What Stuxnet's Creators Tried to Achieve", The Langner Group, Arlington, MA, November 2013.

Lee, Jin-Shyan; Su, Yu-Wei; Shen, Chung-Chou, "A Comparative Study of Wireless Protocols: Bluetooth, UWB, ZigBee, and Wi-Fi", Industrial Technology Research Institute, Hsinchu, Taiwan, November 2007.

Leidigh, Christopher, "Fundamental Principles of Network Security", White Paper #101, American Power Conversion (APC), 2005.

Leischner, Garrett and Whitehead, David, "A View Through the Hacker's Looking Glass", Technical Paper 6237-01, Schweitzer Engineering Laboratories, Inc., Pullman, WA, April 2006.

Makhijani, Arjun Ph.D.; Chalmers, Lois; Smith, Brice Ph.D.; "Uranium Enrichment – Just Plain Facts to Fuel an Informed Debate on Nuclear Proliferation and Nuclear Power", Institute for Energy and Environmental Research, October 15, 2004.

McDonald, Geoff; Murchu, Liam O.; Doherty, Stephen; Chien, Eric; "Stuxnet 0.5: The Missing Link", version 1.0, Symantec Corporation, Mountain View, CA, February 26, 2013.

Oman, Paul W.; Risley, Allen D.; Roberts, Jeff; Schweitzer, Edmund O. III, "Attack and Defend Tools for Remotely Accessible Control and Protection Equipment in Electric Power Systems", Schweitzer Engineering Laboratories, Inc., Pullman, WA, March 12, 2002.

Postel, John, *Internet Protocol – DARPA Internet Program Protocol Specification*, RFC 791, Information Sciences Institute, University of Southern California, Marina Del Ray, CA, September 1981.

Rescorla, E. and Korver, B.; "Guidelines for Writing RFC Text on Security Considerations" (RFC 3552), The Internet Society, July 2003.

Risley, Allen; Marlow, Chad; Oman, Paul; Dolezilek, Dave, "Securing SEL Ethernet Products With VPN Technology", Application Guide 2002-05, Schweitzer Engineering Laboratories, Inc., Pullman, WA, July 11, 2002.

"Seven Strategies to Effectively Defend Industrial Control Systems", National Cybersecurity and Communications Integration Center (NCCIC), Department of Homeland Security (DHS), USA.

"W32.DuQu – The Precursor to the next Stuxnet", version 1.4, Symantec Corporation, Mountain View, CA, November 23, 2011.

Whitehead, David and Smith, Rhett, "Cryptography: A Tutorial for Power Engineers", Technical Paper 6345-01, Schweitzer Engineering Laboratories, Inc., Pullman, WA, October 20, 2008.

Zippe, Gernot, "A Progress Report: Development of Short Bowl Centrifuges", Department of Physics, University of Virginia, July 1, 1959.

Chapter 34

Problem-solving and diagnostic strategies

The ability to solve complex problems is the most valuable technical skill an instrumentation professional can cultivate. A great many tasks associated with instrumentation work may be broken down into simple step-by-step instructions that any marginally qualified person may perform, but effective problem-solving is different. Problem-solving requires creativity, attention to detail, and the ability to approach a problem from multiple mental perspectives.

"Problem-solving" often refers to the solution of abstract problems, such as "word" problems in a mathematics class. However, in the field of industrial instrumentation it most often finds application in the form of "troubleshooting:" the diagnosis and correction of problems in instrumented systems. Troubleshooting is really just a form of problem-solving, applied to real physical systems rather than abstract scenarios. As such, many of the techniques developed to solve abstract problems work well in diagnosing real system problems. As we will see in this chapter, problem-solving in general and troubleshooting in particular are closely related to *scientific method*, where hypotheses are proposed, tested, and modified in the quest to discern cause and effect.

Like all skills, problem-solving may be improved with practice and persistence. The goal of this chapter is to outline several problem-solving tools and techniques.

34.1 Learn principles, not procedures

"As to methods there may be a million and then some, but principles are few. The man who grasps principles can successfully select his own methods. The man who tries methods, ignoring principles, is sure to have trouble." – **Ralph Waldo Emerson**

Effective problem-solvers always reason from fundamental principles, rather than follow memorized procedures. Following this logic, the wise strategy for any student learning how to solve any type of problem is to internalize as many general principles as possible, and to connect common threads of principle to different applications and scenarios.

A good way for any student to self-check that their focus is indeed on principles rather than procedures is to continually ask themselves if they are able to explain *why* what they have learned is true. It is one thing to be able to explain what you have learned, but it is quite another to explain *why* your new knowledge is valid. "Why" may be the most important question you ever ask, and you need to ask it often.

34.2 Active reading

Learning from reading printed text is a kind of problem-solving activity it is own right. The problem is how to acquire new information from the pages of a book, and the solution requires active engagement of your mind as you read. This is often more difficult that it may seem at first, especially when the subject matter is complex and/or the source text is poorly written. Given the fact that much more technical information is available in text form than in any other format, and also the fact that continuous learning is absolutely essential in the field of instrumentation, active reading is an indispensable tool for student success as well as for continued professional growth in the field of instrumentation.

34.2.1 Don't limit yourself to one text!

A very common mistake made by struggling students is to limit themselves to the reading assigned by their instructor(s), when better books might be available. If homework assignments are given from a particular assigned text, it is understandable why a student might think this is the *only* appropriate text to use. However, textbook selection is an imperfect process, often influenced by factors other than optimal learning (e.g. instructor bias, publisher influence, etc.). Sadly, textbooks are sometimes chosen not on their merits as a learning resource but rather by whether or not the textbook comes complete with pre-made exercises and examinations to be used in class (thus relieving the instructors of much work creating these on their own). As a student you must take responsibility for your own learning and seek the best books available for learning, even if this means only using the assigned text as a source of homework problems!

Thanks to the internet, searching for high-quality books is very easy. Not only may you peruse titles and reviews, but many booksellers also allow limited access to these texts online so you can see firsthand how the books are written and decide whether or not each book would suit your needs.

If you are a student and you approach your instructor asking about other texts to read on the subject, you may find your instructor doesn't like the assigned text any better than you, and has alternatives ready to suggest. This is often the case in courses where the text has been selected by committee.

If you examine the personal library of any highly competent technician or engineer, you will likely find multiple books covering the same topics. Building your own collection of useful texts for learning is a sound strategy. Not only will you find yourself referencing these books throughout your career, but you will find their multiple explanations of common concepts an easier way to learn than by limiting yourself to just one point of view.

34.2.2   Marking versus outlining a text

A practical and common method to increase engagement while reading is to "mark" the paper pages of a book with notations, the idea being to note points of interest and thereby stimulate thinking as you read. For a brief primer on this subject I recommend Mortimer Adler's essay *How to Mark a Book*.

We will begin our exploration of active reading with an example taken from page 101 of the classic text *The Measurement of High Temperatures* written by George Burgess and Henry Louis Le Châtelier in 1912. One reason for choosing such an old text is that the style of writing adds another challenge to the task of reading. For those already familiar with the subject of temperature measurement, the archaic writing style will help give you the perspective of a new student, encountering something unfamiliar for the first time:

CHAPTER IV.

THERMOELECTRIC PYROMETER

Principle. -- The junction of two metals heated to a given temperature is the seat of an electromotive force which is a function of the temperature only, at least under certain conditions which we shall define further on. In a circuit including several different junctions at different temperatures, the total electromotive force is equal to their algebraic sum. In a closed circuit there is produced a current equal to the quotient of this resultant electromotive force and the total resistance.

Experiments of Becquerel, Pouillet, and Regnault. -- It was Becquerel who first had the idea to profit from the discovery of Seebeck to measure high temperature (1830). He used a platinum-palladium couple, and estimated the temperature of the flame of an alcohol lamp, finding it equal to 135°. In reality the temperature of a wire heated in a flame is not that of the gases in combustion; it is inferior to this.
The method was studied and used for the first time in a systematic manner by Pouillet; he employed an iron-platinum couple which he compared with the air thermometer previously described (page 61). In order to protect the platinum from the action of the furnace gases, he inclosed it in an iron gun barrel which constituted the second metal of the junction. Pouillet does not seem to have made applications of this method, which must have given him very discordant results.
Edm. Becquerel resumed the study of his father's couple (platinum-palladium). He was the first to remark the great importance of using in these measurements a galvanometer of high resistance. It is the electromotive force which is a function

101

This is the original text as it appears on page 101 of the book. Next, we will explore different ways of "marking" the text, as if we were a new reader to this subject.

The most rudimentary method of marking a book consists of underlining and/or highlighting with a felt-tipped pen:

CHAPTER IV.

THERMOELECTRIC PYROMETER

Principle. -- The junction of two metals heated to a given temperature is the seat of an electromotive force which is a function of the temperature only, at least under certain conditions which we shall define further on. In a circuit including several different junctions at different temperatures, the total electromotive force is equal to their algebraic sum. In a closed circuit there is produced a current equal to the quotient of this resultant electromotive force and the total resistance.

Experiments of Becquerel, Pouillet, and Regnault. -- It was Becquerel who first had the idea to profit from the discovery of Seebeck to measure high temperature (1830). He used a platinum-palladium couple, and estimated the temperature of the flame of an alcohol lamp, finding it equal to 135°. In reality the temperature of a wire heated in a flame is not that of the gases in combustion; it is inferior to this.
   The method was studied and used for the first time in a systematic manner by Pouillet; he employed an iron-platinum couple which he compared with the air thermometer previously described (page 61). In order to protect the platinum from the action of the furnace gases, he inclosed it in an iron gun barrel which constituted the second metal of the junction. Pouillet does not seem to have made applications of this method, which must have given him very discordant results.
   Edm. Becquerel resumed the study of his father's couple (platinum-palladium). He was the first to remark the great importance of using in these measurements a galvanometer of high resistance. It is the electromotive force which is a function

101

Here you see how the reader has highlighted the paragraph introductory words in yellow ink, underlined names and dates with green lines, and underlined concepts with red lines. While such exercises may have helped the reader remain awake as well as generate cues for later cramming, it is doubtful they assisted the reader in understanding the concepts. A large degree of blame for this rather shallow and unproductive practice may be laid on instructional curricula emphasizing memorization and execution of procedures over conceptual understanding. Simply put, *this is what you get when students expect to be quizzed on isolated bits of data:* you get poor study habits such as this. Unfortunately, not only will this approach fail to yield deep understanding of the concepts, but it also reduces the act of reading to drudgery. When such practices are so common, it's no wonder a great many students loathe academic reading.

The choice of highlighting versus underlining, or of one color over another, is relatively unimportant. Any process of "marking" a book merely by drawing attention to certain words within it suffers the same weakness: all you are doing is emphasizing specific words and phrases, not incorporating your own thoughts into the text. In order to be *actively engaged* in your reading, you must expose your own thoughts and reflections on what you read, not just emphasize specific statements made by the author. As Adler points out:

> "Understanding is a two-way operation; learning doesn't consist in being an empty receptacle. The learner has to question himself and question the teacher. He even has to argue with the teacher, once he understands what the teacher is saying. And marking a book is literally an expression of differences, or agreements of opinion, with the author."

The thrust of Adler's essay is that the reader gains the greatest understanding of a text by expressing their own thoughts about what they are reading: articulating questions, drawing conclusions, linking ideas, and otherwise being an active participant in the reading process rather than a passive observer.

With this in mind, let's re-approach the text. Instead of simply emphasizing words and phrases, we will now show how a reader may articulate some of their own thoughts on the same page:

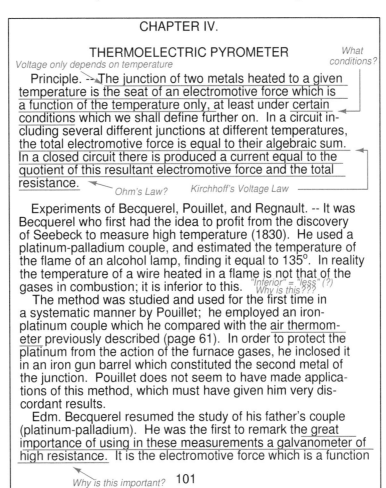

Here you can see the focus has completely shifted away from facts and figures, and toward concepts. Note the questions raised by the reader, either doubting their own understanding of the book or the author's assertions. These questions need not be fully resolved at the conclusion of reading. Indeed, these may be excellent topics to raise in class once the student returns to school and has the instructor's attention. Questions should be taken as a positive sign for active reading and not as an indication of trouble: if you read a large body of prose and have no questions of your own, you probably weren't thinking deeply enough!

If this manner of marking the text seems messy and cluttered, one may opt to make all the notations on a separate piece of paper, or even typed into a computer for later printing and retrieval.

Even deeper engagement may be achieved if one takes the time to write an *outline* of ideas in the text. This is an exercise best done on a separate sheet of paper, or using a computer. An example is shown here, side-by-side with the original text. Note how an outline may include graphical sketches as well as words:

CHAPTER IV.

THERMOELECTRIC PYROMETER

Principle. -- The junction of two metals heated to a given temperature is the seat of an electromotive force which is a function of the temperature only, at least under certain conditions which we shall define further on. In a circuit including several different junctions at different temperatures, the total electromotive force is equal to their algebraic sum. In a closed circuit there is produced a current equal to the quotient of this resultant electromotive force and the total resistance.

Experiments of Becquerel, Pouillet, and Regnault. -- It was Becquerel who first had the idea to profit from the discovery of Seebeck to measure high temperature (1830). He used a platinum-palladium couple, and estimated the temperature of the flame of an alcohol lamp, finding it equal to 135°. In reality the temperature of a wire heated in a flame is not that of the gases in combustion; it is inferior to this.
The method was studied and used for the first time in a systematic manner by Pouillet; he employed an iron-platinum couple which he compared with the air thermometer previously described (page 61). In order to protect the platinum from the action of the furnace gases, he inclosed it in an iron gun barrel which constituted the second metal of the junction. Pouillet does not seem to have made applications of this method, which must have given him very discordant results.
Edm. Becquerel resumed the study of his father's couple (platinum-palladium). He was the first to remark the great importance of using in these measurements a galvanometer of high resistance. It is the electromotive force which is a function

101

Joining two different metals together and heating that junction produces a voltage dependent on temperature. Other factors are involved which I'll learn about later.

The voltages of multiple thermoelectric junctions adds in series, with current = V/R.

Becquerel used a platinum-palladium junction to try to measure a candle flame, but he got a result that was too low. (I need to find out why an object placed in a flame could be cooler than the actual flame!)

Pouillet used a platinum-iron junction using a hollow iron tube as the second conductor. This tube protected the platinum. (Not sure exactly what "action" the platinum needed protection from -- maybe chemical corrosion?)

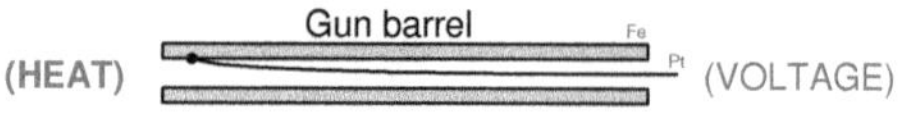

(HEAT)   Gun barrel   Fe   Pt   (VOLTAGE)

Becquerel's son discovered circuit resistance is an important factor. You need a meter with high resistance . . .

What you see here are notes for just one page of Burgess' and Le Châtelier's text. A complete outline of thermoelectric pyrometers would of course cover multiple pages of the source text, and not just the one page shown. Note how the outline includes cues for future learning – hints that there is more to this subject than what is immediately presented on this page of the text.

Writing your own outline of a text is especially useful when the text in question is densely packed with ideas and/or difficult to understand, because the act of outlining serves as a self-check for your own comprehension. With highlighting and underlining it is all too easy to lazily read the words and make these marks without really understanding what the text is saying. While outlining you know quite well when you haven't understood the text because you simply won't be able to express it in your own words. Instead of continuing to highlight in a state of blissful unawareness, you will find yourself stuck and unable to continue outlining. This prompts you to go back and scrutinize the text, going over it more carefully than before, until you find yourself able to continue outlining again.

Perhaps the most common objection to outlining text as you read is that the process is slow. This raises a very important point, namely that *active reading should be slow*. Facts and figures may be skimmed, but complex ideas take time to penetrate into your mind. Not only will outlining force you to slow down when you need to, but it will also serve as a gauge for later study when you review your own notes to see if you still agree with them. In the course of studying some topic, you will often find that your understanding of that topic changes from your first impression. Seeing this change for yourself allows you to better understand how you learn, and thereby gives you practical insight into the workings of your own mind when it comes time to learn something new.

34.3 General problem-solving techniques

A variety of problem-solving techniques have been presented for students over the years which are all helpful in tackling problems both in the classroom and in the real world. Several of these techniques are presented here in this section.

34.3.1 Identifying and classifying all "known" conditions

An important step in solving certain types of problems, especially quantitative problems where calculations are necessary to obtain precise answers, it is often useful to list all the known quantities available to us relevant to the problem. Similarly, taking the time to list all relevant (and possibly relevant) mathematical formulae we might apply to the solution is a helpful step.

One way to save time applying the latter suggestion in a classroom setting is to keep a concise reference card or file filled with formulae you've been learning within that course. This reference may be referred to as often as necessary, without having to re-write the equations for each and every problem, thus eliminating unnecessary effort.

34.3.2 Re-cast the problem in a different format

Many people find it easier to grasp the nature of a problem – and by extension, that problem's solution – if they can look at an illustration of the problem. Therefore, a helpful step in solving problems described to you in words is to translate those words into a picture to look at.

If you are one of those people for whom drawing is a challenge, take heart in the fact that this is a skill you can build. Practice is the key to honing this skill. With this in mind, make it a habit to sketch some kind of illustration for every problem you are asked to solve. If you are working in teams to solve a problem, a collaborative sketch goes a long way toward coordinating problem-solving efforts and ensuring everyone on the team has the same view of the problem.

For some people, describing a problem verbally is helpful in solving it. If your brain tends to work like this – understanding concepts and situations better when they are cast into clear prose – then you may find it helpful to first draft an explanatory paragraph of the problem in your own words. This is also an exercise lending itself well to team-based problem solving, as the entire team can help each other describe the nature of the problem.

34.3.3   Working backwards from a known solution

Sometimes we may gain insight into the solution of a problem by assuming we already know the answer to a similar problem, then working "backward" to find the problem from that assumed solution.

An application of this problem-solving strategy is found learning how to decode binary bits that have been encoded using the Manchester standard. With Manchester encoding, binary bits are represented by the rising and falling *edges* of square-shaped waveforms rather than high and low states themselves. For example:

Manchester-encoded waveform for binary 011110100011

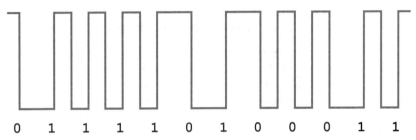

Seeing this example, we note how each binary "0" bit is represented by a *falling* edge, while each "1" bit is represented by a *rising* edge.

Where most students encounter trouble is in situations where they have been given a Manchester encoded waveform and must decode it into its representative bit stream. Take this for example:

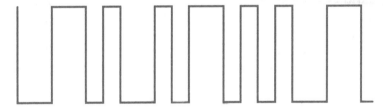

Most students' first inclination is to ask their instructor or their classmates for an algorithm to decode the waveform. "What steps should I take to figure out where the data bits are?" they will ask. This sort of "give me the answer" mind-set should always be discouraged, because it is the polar opposite of true problem-solving technique, where the student methodically searches for patterns and develops algorithms on their own.

A better approach is to encourage the strategy of working the problem backwards: begin with a known series of binary bits, and then develop a Manchester waveform from that. The act of *encoding* a binary string provides insight that will be useful in *decoding* the next Manchester waveform they encounter.

For example, let's begin with the binary string 100011101:

1   0   0   0   1   1   1   0   1

We may begin the process of encoding this into Manchester format by sketching the rising- and falling-edges we know we will need for each bit:

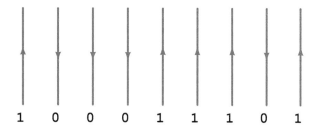

Next, we can try connecting the tops and bottoms of these pulse edges to form a complete waveform. Soon, however, we will find that this is only possible where opposite bit states are adjacent to each other. Where identical bits follow in sequence, we are faced with sequential rising edges or sequential falling edges, which we cannot simply bridge at the tops or bottoms to make a full pulse:

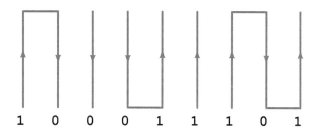

This observation leads to the realization of why we need *reversals* in a Manchester waveform. The only way to connect repeating bits' edges together is if the waveform goes through another rising or falling edge in order to be properly set up for the next edge we need to represent a bit:

Reversals needed to "set up" for next real bit transitions

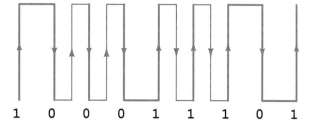

Here we see the power and utility of working a problem "backwards": it reveals to us the *reason why* things are the way they are. Without this understanding, problem-solving is nothing more than rote recall of algorithms, and limited in application. Any problem becomes simpler to solve once we fully understand its rationale.

Once we realize the purpose for reversals in a Manchester waveform, it becomes obvious to see that these reversals always fall *between* the bit transitions, and thus are always *out of step* with the frequency of the bits. Those edge transitions representing real data bits must always fall along a regular timed interval, with reversals being "half-steps" in between those intervals. We need only to look for the widest-spaced intervals in a Manchester waveform to distinguish those pulse edges representing real data bits, and then we know to ignore any pulse edges out of step with them.

Returning to our sample problem, where we were given a Manchester waveform and asked to decode it:

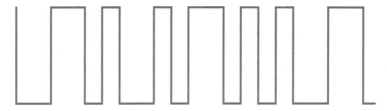

First, we identify the real data bit edges by widest spacing:

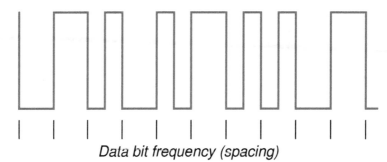

Data bit frequency (spacing)

Now that we know which pulse edges represent bits, we may ignore those that do not (the reversals), and decode the waveform:

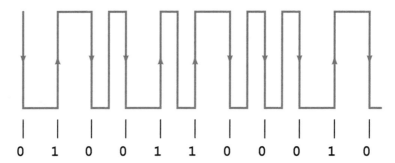

0   1   0   0   1   1   0   0   0   1   0

34.3.4 Using thought experiments

One of the most powerful problem-solving techniques available for general use is something called a *thought experiment*. Scientists use experiments to confirm or refute hypotheses, testing their explanations by seeing whether or not they can successfully predict the outcome of a certain situation by comparing their predictions against real outcomes. While this technique is extremely useful, it might not always be practical or expedient. A useful alternative to real experiments is to mentally "model" the system and then imagine changing certain elements or variables within that model to deduce the effects.

Albert Einstein famously applied "thought experiments" to the formulation of his Theory of Relativity, for the very simple reason that he lacked the resources and technology to actually test his ideas in real life. Working as a patent clerk, he would imagine what might happen if an observer were to travel at or near the speed of light. One particular example of this is the anecdote of Einstein observing a clock tower as he rode a trolley traveling away from the tower. "What would an observer see," he wondered, "as he viewed the clock's face while traveling away from it at the speed of light?" Concluding that the clock's face would appear to be frozen in time was one of the surprising "experimental" results leading Einstein to a more rigorous examination of physical effects at extremely high velocities.

"Thought experiments" are useful in solving a wide variety of problems, because they allow us to test our understanding of a system's behavior. By imagining certain conditions or variables changing in a system and then asking ourselves what the effects will be, we probe our own understanding of that system, often times with the result being that we are able to predict its behavior under conditions that baffle us at first.

You will find "thought experiments" scattered throughout this book, used both as illustrations of problem-solving strategies and also as a tool to explain how certain technologies function. An example of this is the section explaining non-dispersive analyzers, which are instruments employing the absorption of light by certain species of chemicals in order to detect the presence and measure the quantities of those chemicals. Beginning in section 23.6 on page 1807, a series of "thought experiments" are used to explore the principles used to identify chemicals by light absorption. This series of virtual experiments becomes most valuable when this section explores the analyzer's ability to *selectively* measure the presence of one light-absorbing chemical to the exclusion of other light-absorbing chemicals within the same mixture.

34.3.5   Explicitly annotating your thoughts

Suppose you were asked to solve this multiplication problem, without the use of a calculating machine of any kind, but with access to paper and a writing tool:

$$
\begin{array}{r}
3418 \\
\times\ 572 \\
\hline
\end{array}
$$

Your primary school education should have prepared you to solve elementary arithmetic problems of this kind, by a process of digit-by-digit multiplication and addition, to arrive at an answer of 1,955,096. The procedure, while tedious, is rather simple: manually multiply the top numeral three times over by successive digits of the bottom numeral, noting any "carried" quantities as you do so, then sum those three subtotals together (padded with zeros to represent the place of the bottom numeral's digit) to arrive at the final product.

Now suppose you were asked to solve the exact same multiplication problem, but this time doing the same digit-by-digit arithmetic all in your mind, without the use of a writing tool to annotate your work. Suddenly this elementary task becomes nearly impossible for anyone who isn't a mathematical savant. What made the difference between this problem as an elementary exercise and this same problem as a nearly impossible feat? The answer to this question is *short-term memory*: most people do not possess a good enough short-term memory to mentally manage all the intermediate calculations necessary to complete the calculation. This is why people learn to annotate their work when performing manual multiplication, so they don't have to rely on their limited short-term memories. The freedom to write your steps on paper converts what would otherwise be a Herculean feat of arithmetic into a rather trivial exercise.

Annotating your intermediate steps as you solve a problem is actually an excellent general problem-solving strategy, applicable to far more than just arithmetic. Some examples of annotating intermediate steps are listed here:

- **Reading a complex document**: *annotating your thoughts, questions, and epiphanies as you read the text allows you to derive a better understanding of the text as a whole.*

- **Learning a new computer application**: *noting how features are accessed and identifying the necessary conditions for each feature to work helps you navigate the software more efficiently.*

- **Following a route on a map**: *marking where you started, where your destination is, and where you have traveled thus far helps you see how far you still need to go, and which alternative routes are open to you.*

- **Analyzing an electric circuit**: *annotating all calculated voltages, currents, and impedances on the diagram helps you keep track of what you know about the circuit and where to go next in your analysis.*

- **Troubleshooting a system fault**: *noting all your diagnostic steps and conclusions along the way helps you confirm or disprove hypotheses.*

Sadly, many students attempt to solve new types of problems analogously to performing multiplication without paper and pencil: they attempt to mentally manage all their intermediate steps, not writing anything down that would help them later. As a result, students tend to get "lost" when trying to solve new problems simply because they cannot readily reference of all their thoughts along the way. Most people simply give up when they begin to feel "lost" in solving a problem, thinking that if they cannot mentally picture the solution in its entirety then they have no hope of attaining it. Let's face it: how soon would you give up on multiplying 3418×572 without a calculator if you believed the only alternative was to manage all the arithmetic in your head?

One reason why students default to the "mental-only" approach when approaching new problems is that their educational experience has only presented annotation for specific types of problems. Thus, marking all the carry digits and subtotals is something they "only do" when performing multiplication by hand; marking calculated voltages and currents on a schematic diagram is something they "only do" when solving DC resistor circuits; taking notes when reading is something they "only do" when completing a book report. In other words, *students see annotation only in very specific contexts, and so they may fail to see just how widely applicable annotation is as a problem-solving strategy.* What teachers should do is model and encourage annotation as a problem-solving technique for *all* types of problems, not just for *some* types of problems.

To illustrate how this might be done in the context of control system analysis, let us suppose we were asked to determine the effect of flow transmitter FT-24 failing with a low (no-flow) signal in this ratio control system, part of a process for manufacturing ammonium nitrate fertilizer:

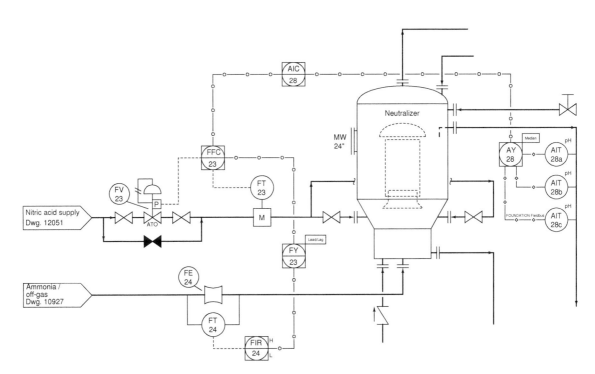

Before it is possible to analyze the effects of a transmitter failure, we must first determine what the system ought to do in normal operation. Natural questions to ask might include the following:

- Where do the instrument signals come from and where do they go to?

- What does each instrument signal represent?

- What is the direction of action for each controller in the system?

With just a basic understanding of ratio control systems, we may answer all of these questions by close examination of the P&ID segment, and also annotate those thoughts and conclusions on the diagram in order to help us analyze the system's response. Starting with the first two questions of where signals originate and terminate and what each signal represents, we may annotate this with arrows and text (shown in red):

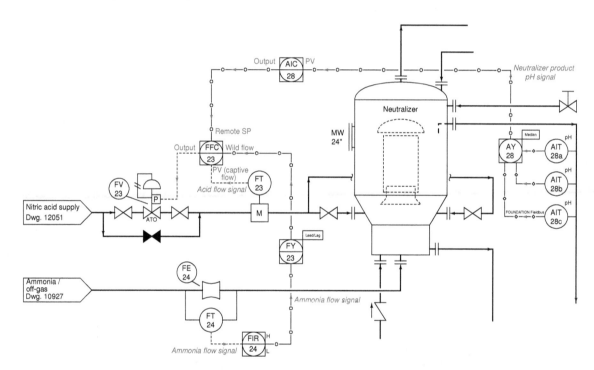

We know that all transmitters *output* data, and so all signal arrows should point *away* from all transmitters and *toward* all controllers. We know that all valves *receive* data, which means arrows must point toward the control valve. The first tag letter of each transmitter (AIT, FT) tells us its measurement function: chemical pH and flow, respectively. The fact that FT-23 is mounted on the same pipe as FV-23 tells us FT-23 must send controller FFC-23 its process variable (captive flow), making the other flow signal (from FT-24) the "wild" flow in this ratio control scheme. AIC-28's task is to control pH exiting the neutralizer, so we know its output signal must call for a neutralizing reagent, in this case nitric acid. This tells us the signal between AIC-28 and FFC-23 must be a cascade output-setpoint link, with AIC-28 as the master controller and FFC-23 as the slave controller.

Now we turn to the question of controller action, since we know the direction of each controller's action (e.g. direct or reverse) is significant to how each controller will react to any given change in signal. Here, 'thought experiments" are useful as we imagine the process variable changing due to some load condition, and then determine how the controller must respond to bring that process variable back to setpoint.

When we annotate the action of each controller, it is best to use symbols more descriptive than the words "direct" and "reverse," especially due to the confusion this often causes when distinguishing the effects of a changing PV signal versus a changing SP signal. In this case, we will write a short formula next to each controller denoting its action according to how the error is calculated ($e = \text{PV} - \text{SP}$ for direct action and $e = \text{SP} - \text{PV}$ for reverse action). We may also write "+" and "−" symbols next to each input on each controller to further reinforce the direction of each signal's influence:

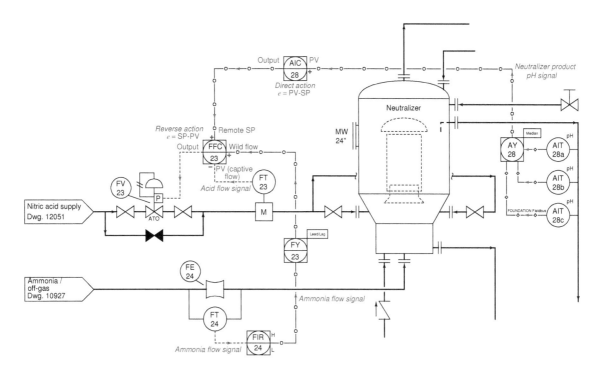

FFC-23 is the best controller to start with, since it is the slave controller (in the "inner-most" control loop of this cascade/ratio system). Here, we see that FFC-23 must be reverse-acting, for if FT-23 reports a higher flow we will want FV-23 to close down. This means the remote SP input must have a non-inverting effect on the output: a greater signal from AIC-28 will increase nitric acid flow into the neutralizer. Following this reasoning, we see that AIC-28 should be direct-acting, calling for more nitric acid flow into the neutralizer as product pH becomes more alkaline (pH increases).

The purpose of the ratio control strategy is to balance the "wild" flow of ammonia into the neutralizer with a proportional flow of nitric acid. This is in keeping with principles of chemical reactions (stoichiometry) and mass balance. Therefore, we would expect an increase in ammonia flow to call for a proportionate increase in nitric acid flow, giving the wild flow signal a non-inverting effect on FFC-23.

Only at this point in time are we fully ready to analyze the effects of FT-24 failing with a low-flow signal. Once again, we may annotate the failure on the diagram as well, arbitrarily electing to use blue "up" and "down" arrows and bold text to indicate the directions of change for each signal immediately following the failure of FT-24:

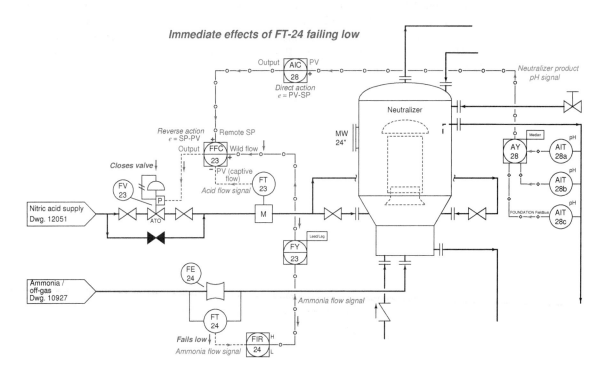

As FT-24's signal fails low, the "wild" flow signal to FFC-23 goes low as well. Since we have already determined that input has a non-inverting effect on the ratio controller, we may conclude control valve FV-23 will close as a result, decreasing the flow of nitric acid into the neutralizer. This analysis becomes trivial after we've done the work of annotating the diagram with our own notes showing how the instruments are supposed to function. Without this set-up, the task of analyzing the effects of FT-24 failing low would be much more difficult.

34.4 Mathematical problem-solving techniques

Some problem-solving techniques are unique to quantitative problems, involving mathematical calculations. In this section we will explore some useful tips to help you solve such problems.

34.4.1   Manipulating algebraic equations

One of the most useful problem-solving techniques in all of algebra is the art of manipulating, or re-writing, equations to solve for a particular variable. The key to this technique is the fact that we may subject an equation to any mathematical operation we desire, so long as we apply that same operation to both sides of the equation.

Let us begin with an illustration showing why this is true. We will represent the equation $x + 9 = y$ as a balance-beam scale, showing the quantity $x + 9$ in one pan of the scale and y in the other pan:

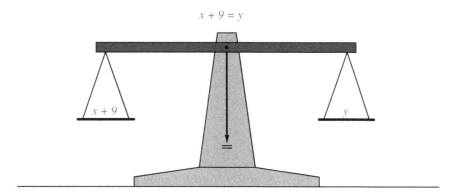

The balance-beam analogy merely represents the fact that the two expressions $x + 9$ and y are equal; that is to say, they possess the same mathematical value as described by the equation $x + 9 = y$. It should be intuitively obvious that this equality will remain unaltered if we were to add some equal quantity to both sides of the equation, such as the number 3:

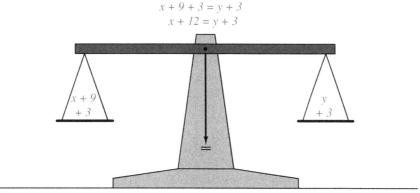

Adding 3 equally to both sides does not affect the balance

If the left-hand quantity is now larger by 3 units, and the right-hand quantity is also larger by 3 units, and those two quantities began as equal to each other, than those two quantities *must still remain equal to one another*. We have not altered the "truth" of the equation by adding 3 to both sides, any more than we would have upset the balance of a real scale by adding 3 units of mass to both pans.

We could similarly multiply each of these quantities by the same factor (say, 4) and still remain equal. The validity of the equation $x + 9 = y$ is not harmed by multiplying both sides by 4 to get $4(x + 9) = 4y$. Similarly, we could take the square-root of both sides to get $\sqrt{x + 9} = \sqrt{y}$ and still have an equality. We could reciprocate both sides to get $\frac{1}{x+9} = \frac{1}{y}$ and still have an equality. We could take the logarithm of both sides to get $\log(x + 9) = \log y$ and still have an equality. We could raise both sides to the power z to get $(x + 9)^z = y^z$ and still have an equality. The lesson here should be perfectly clear: if we start with an equation (two equal expressions), and apply the same mathematical operation to both sides of that equation, we'll still have an equation (two equal expressions).

This fact – that we may apply any mathematical operation equally to both sides of an equation without harming the validity of that equation – might seem at first to be pointless. However, it turns out to be a powerful tool for isolating variables within an equation. If we are creative in the operation(s) we choose to apply to an equation, we may do so in such a way that isolates one variable by itself, placing all other portions of the equation on the other side of the "equals" sign.

Returning to our expression $x + 9 = y$, suppose we were tasked with solving for x. In other words, we need to manipulate this equation to get x by itself on one side of the equals sign, with everything else on the other side of the equals sign. Clearly, the problem here is how to get rid of the "9" that's being added to x on the left-hand side of the equation. One way to do this is to subtract 9 from both sides of the equation, thus canceling the "9" on the left-hand side and shifting it over to the right-hand side. This results in x being left all by itself on the left-hand side of the equation, just like we want:

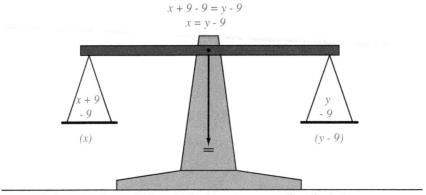

Subtracting 9 equally from both sides does not affect the balance

Although we have the mathematical freedom to do *anything* we wish to both sides of the equation so long as we do it equally, we can see here that only one operation will work to isolate x by itself. This, therefore, is the challenge of algebraic manipulation: *how to determine which operation(s) we should apply to both sides in order to end up with the equation re-written the way we want?*

A general principle to help us answer this challenge is that of *inverse functions*: that is, pairs of mathematical operations known to "un-do" each other. The inverse function to addition is subtraction, which is why we *subtracted* 9 from each side of the equation $x + 9 = y$ in order to "un-do" the *addition* of 9 to x on the left-hand side. In other words, we determine the mathematical operation being applied to the variable we wish to isolate, and then we intentionally apply the opposite mathematical operation to both sides of the equation to "strip away" that interfering operation.

Here is a table showing a few inverse mathematical functions:

Function $f(x)$	Inverse function $f^{-1}(x)$
Addition ($+$)	Subtraction ($-$)
Multiplication ($\times$)	Division ($\div$)
Power (x^n)	Root ($\sqrt[n]{x}$)
Exponent (n^x)	Logarithm ($\log_n x$)

It should be noted that sometimes we can use the same function to "un-do" itself if we apply the quantity creatively. For example, instead of subtracting 9 from each side of the equation $x + 9 = y$ to solve for x, we could have just as easily *added* -9 to both sides, which is mathematically equivalent. Likewise, instead of using division to "un-do" multiplication, we may simply multiply by the reciprocal, as in the following example where we wish to solve for x in the equation $\frac{5}{3}x = y$. Since we can see that the fraction $\frac{5}{3}$ is being *multiplied* by x, we may get rid of the fraction by multiplying both sides by its reciprocal ($\frac{3}{5}$), leaving x multiplied by a factor of 1, which is the same as x all by itself:

$$\frac{5}{3}x = y$$

$$\left(\frac{3}{5}\right)\frac{5}{3}x = \left(\frac{3}{5}\right)y$$

$$\frac{1}{1}x = \left(\frac{3}{5}\right)y$$

$$x = \frac{3}{5}y$$

Had we chosen to *divide* both sides of this equation by the fraction $\frac{5}{3}$, we still would have arrived at a mathematically correct result, but it would have taken the form of a *compound fraction*, which is not as "elegant" a presentation:

$$\frac{5}{3}x = y$$

$$\frac{\frac{5}{3}x}{\frac{5}{3}} = \frac{y}{\frac{5}{3}}$$

$$x = \frac{y}{\frac{5}{3}}$$

The task of solving for a variable becomes more complicated when there are multiple operations needing to be stripped away to isolate a particular variable. Take for example the equation $5x - 4 = y$, where we wish to solve for x. Clearly, we must find a way to "un-do" the subtraction of 4, as well as the multiplication by 5, but which inverse operation should we apply first? The key here is to properly recognize the *order of operations* expressed in the original equation.

If we were to evaluate the equation $5x - 4 = y$ to arrive at a value for y given a value for x, our proper order of operations would be to first multiply x by 5, and then subtract 4. Multiplication/division precedes addition/subtraction, if there are no other influences such as parentheses. If our goal is to "un-do" each of these operations in order to arrive at x by itself, we must do so in *reverse order of operations*. This means first un-doing the subtraction of 4, and then un-doing the multiplication by 5. The following steps show how this is done:

Step	Equation	Explanation
1	$5x - 4 = y$	Original equation
2	$5x - 4 + 4 = y + 4$	Adding 4 to both sides
3	$5x = y + 4$	Simplifying
4	$\frac{5x}{5} = \frac{y+4}{5}$	Dividing both sides by 5
5	$x = \frac{y+4}{5}$	Simplifying

Note how these steps would have been different if the original equation were written with a different order of operations, such as $5(x - 4) = y$. With the parentheses forcing the order of operations such that the subtraction occurs before the multiplication, our steps for isolating x must reverse as well – we must first divide by 5, then add 4:

Step	Equation	Explanation
1	$5(x - 4) = y$	Original equation
2	$\frac{5(x-4)}{5} = \frac{y}{5}$	Dividing both sides by 5
3	$x - 4 = \frac{y}{5}$	Simplifying
4	$x - 4 + 4 = \frac{y}{5} + 4$	Adding 4 to both sides
5	$x = \frac{y}{5} + 4$	Simplifying

34.4.2   Linking formulae to solve mathematical problems

Most practical problems with mathematical solutions do not come to us with the proper formula pre-packaged for our use. Instead, we must identify relevant formulae relating the given variables together, and then use those formulae in combination to solve for the quantity we seek. The task of linking formulae to solve mathematical problems is one many students find quite challenging, and so it is worth our time to explore this in some detail.

Example: sensing valve position

For our first example we will consider a scenario where we wish to have a computer sense the stem position of a control valve. A linear potentiometer coupled to the valve stem will provide us with a suitable sensor to convert the valve stem's position into an electrical signal the computer can measure with appropriate analog I/O hardware:

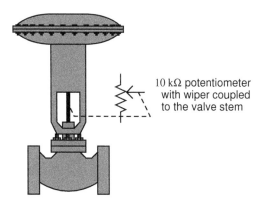

This potentiometer may be used as one-half of a voltage divider network, with a constant-voltage power supply for the source, and a fixed resistor across which the computer's analog I/O can read a voltage drop. As valve stem position changes, the potentiometer's resistance in this circuit will change, thereby changing the voltage seen by the analog input module connected to the computer:

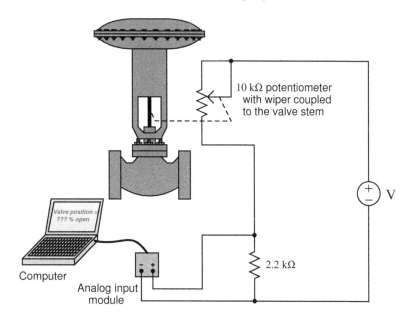

The problem we are now faced with is this: how do we program the computer to display the valve stem's position in *percent* (on a 0% to 100% range), when all it directly senses from the circuit is a varying DC voltage? It certainly would not be helpful to have the computer display the raw signal in units of volts. What we need is a mathematical formula to translate that sensed voltage drop across the 2200 ohm resistor into a percentage value representing valve stem position.

Our first step needs to be identifying all the relevant mathematical formulae in this problem. Since the computer is sensing the voltage drop across one of two resistors in a series circuit, it would appear the voltage divider formula is relevant here (relating source voltage and circuit resistances to voltage drops):

$$V = V_{source} \left(\frac{R}{R_{total}} \right)$$

We also need to somehow relate the valve's stem position (we'll use x to represent the per unit value of valve opening) to the resistance of the potentiometer. If 0% stem position (i.e. $x = 0$, valve fully closed) drives the potentiometer wiper fully down and 100% stem position (i.e. $x = 1$, valve fully open) drives the wiper fully up, the potentiometer's resistance in this circuit should be a simple proportion of its full 10000 ohm value:

$$R_{pot} = 10000x$$

Now that we have a formula relating valve stem position to electrical resistance, and another formula relating electrical resistance to voltage drop, we are ready to link these two formulae together and derive a function expressing voltage drop as a function of valve stem position.

A useful strategy for identifying how multiple formulae link together is to "mark up" those formulae on paper to show how their variables relate to the given information as well as to each other. A standard I find easy to remember and apply is to draw a *circle* around the variable I'm need to solve the value of, and to draw a *square* or a *rectangle* around any variables whose values are given. To begin this process, I will write both the voltage divider and potentiometer proportion formulae near each other, then circle x (valve stem position, per unit) as the variable I need to solve for:

$$R_{pot} = 10000\textcircled{x}$$

Of course, a basic rule of algebra is that for any *one* formula it is only possible to solve for the value of *one* variable. This means *all* other quantities in a formula must be known in order to solve for that one variable. In the R_{pot} formula we have x which we're trying to solve the value of, the full-scale resistance value of 10000 ohms which is given to us in the problem, and the potentiometer's manifested resistance value in the circuit (R_{pot}). Since 10000 is a constant, the only other piece of information we need in this formula to solve for x is the resistance value R_{pot}. The fact that the variable R_{pot} remains unmarked makes this point clear: R_{pot} *is the missing piece of the puzzle to solve for x.*

At this juncture we look to the other formula on hand to see if any of its variables will provide us this missing information. Here, it should be clear we need to find out where R_{pot} might fit in the voltage divider formula. The R variable in the numerator of the fraction in this formula refers to the resistance across which we are measuring the voltage drop V. In this case, R must be the 2200 ohm fixed resistor, so we will enclose R in a square to represent the fact we know its value. Since the power supply's voltage is constant, we may also enclose V_{source} in a rectangle to show that value will be available to us:

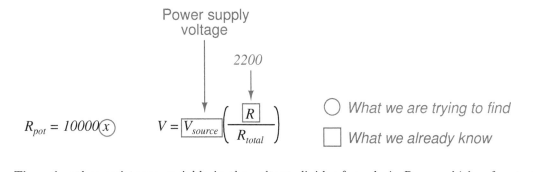

The only other resistance variable in the voltage divider formula is R_{total}, which refers to the total resistance of the series-connected resistors in a voltage divider circuit. This particular circuit has two resistors: the 2200 ohm fixed resistor and the potentiometer. We know that total resistance in a two-resistor series circuit is the sum of those two resistors' individual values ($R_{total} = R_1 + R_2$), we we will write this as a third formula in our collection, with R_1 being the 2200 ohm fixed resistor and R_2 being the potentiometer's resistance:

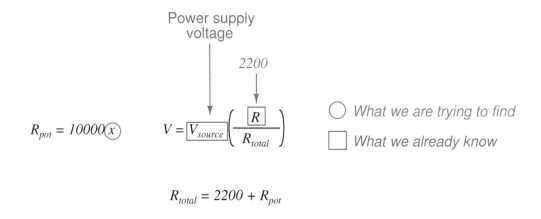

$$R_{total} = 2200 + R_{pot}$$

Now we are ready to link these formulae together. Recall how we needed to find the value of R_{pot} in the potentiometer's formula before we could calculate the value of x (valve stem position). Note how the R_{total} formula relates that potentiometer resistance value to the other resistances in the circuit. This means the total resistance formula can provide us the value of R_{pot} we need to solve for x. We will draw a circle around R_{pot} in the total resistance formula reminding us we need that value, then show a link between this and the potentiometer formula by drawing an arrow extending from one to the other:

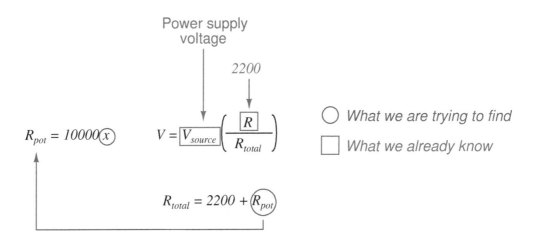

So far, this markup tells us R_{pot} is the missing puzzle piece to calculate x, and that R_{total} is the missing puzzle piece to calculate R_{pot}.

Looking at our three formulae, we see that the voltage divider formula is able to provide us with the value of R_{total} which we need to calculate R_{pot} which we need to calculate x. We will circle R_{total} in the voltage divider formula and draw another arrow showing the link:

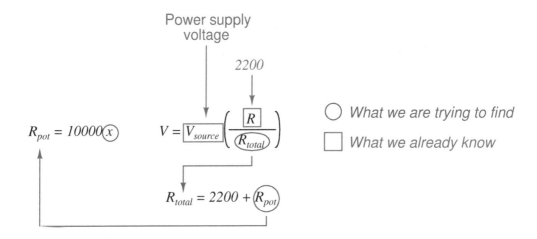

The only variable unmarked and unlinked now is V, which is the voltage sensed by the computer's analog input module. This is now the one independent variable which will tell us the position of the valve stem (x).

Each arrow linking formulae together shows us where one formula will be *substituted* for a variable in another formula. The only thing we must do now prior to this substitution is algebraically manipulate each formula to solve for the one circled variable within it. I will skip these algebraic steps for brevity, and simply re-write the three formulae in manipulated form:

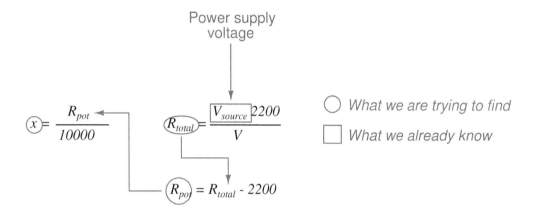

The logic chain of dependency linking these three formulae together is now crystal-clear: we begin with a measured voltage drop value of V to give us the value of R_{total}, which then plugs into the total resistance formula to give us the potentiometer's value R_{pot}, which then plugs into the potentiometer formula to give us valve stem position (x) as a per unit value. The algebraic substitutions are shown here:

Substituting $R_{total} - 2200$ for R_{pot} in the $x = \frac{R_{pot}}{10000}$ formula:

$$x = \frac{R_{total} - 2200}{10000}$$

Substituting $\frac{V_{source} 2200}{V}$ for R_{total} in the $x = \frac{R_{total} - 2200}{10000}$ formula:

$$x = \frac{\frac{V_{source} 2200}{V} - 2200}{10000}$$

This final formula can now be programmed into the computer, telling the computer how to calculate x as a function of V.

To summarize this problem-solving strategy:

1. Begin by writing every formula you can think of relevant to the problem at hand.

2. Identify the final value you're trying to solve for, and circle it.

3. Identify all given values, and show them by drawing squares or rectangles around those variables.

4. Identify any and all "missing puzzle pieces" in the formula with the circled variable.

5. If there is another formula containing a "missing puzzle piece," circle that variable and draw an arrow linking it to the previous formula, then identify any and all "missing puzzle pieces" needed to solve for that variable. There should only be one circled variable per formula.

6. Repeat the previous step as often as needed until there are no missing pieces left.

7. Algebraically manipulate all formulae to solve for their circled variables.

8. Algebraically substitute all variables as shown by the arrows.

Example: gas pressure inside a cylinder

Let's apply this technique to a practical problem related to engine and compressor mechanisms – how to determine the amount of gas pressure generated inside a cylinder when a piston is moved to compress the gas, given the distance of the piston's motion (x):

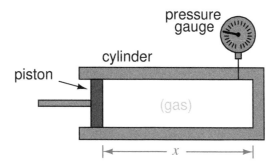

Compression is nothing more than a forced reduction in volume, and so a good first step for us to take is to recall any mathematical formulae relating *pressure* and *volume* for a confined gas. In this case, the Ideal Gas Law is our clear choice, describing the relationship between gas pressure, volume, temperature, and the molar gas quantity inside any enclosed volume:

$$PV = nRT$$

Where,

P = Absolute pressure (atmospheres)
V = Volume (liters)
n = Gas quantity (moles)
R = Universal gas constant (0.0821 L · atm / mol · K)
T = Absolute temperature (K)

However, the Ideal Gas Law formula contains no variable x to represent piston position. Somehow, we must find a way to relate x to the Ideal Gas Law formula in order to progress to a solution for this problem.

The first step in applying this formula is to identify which variables are given to us in the problem, and which we need to solve for. In order to solve for the value of any one variable in an equation, there cannot be any other unknown quantities – this is a basic law of algebra.

In this case, the variable we're ultimately trying to solve for is *pressure* (P). The only other quantity found in the Ideal Gas Law formula that we know the value of at this point is R, which is a natural constant. V, n, and T are all unknown to us at this point in time. Once we do know the values of these three variables, we may algebraically manipulate the Ideal Gas Law formula to solve for P.

Once again we will begin by "marking up" the formula by drawing a circle around the variable we wish to solve for, and drawing squares around any variables or constants for which we already know the values. Any variables left unmarked (i.e. not enclosed in a circle or a square) are unknown quantities which we *must* determine before we can solve for the circled variable:

$$\boxed{\textbf{(P)}\ V\ =\ n\ \fbox{R}\ T}$$

◯ *What we are trying to find*

☐ *What we already know*

This first step of identifying known and unknown quantities is critically important to the process of problem-solving, because it directs us to what we need to do next. An obstacle so many students and professionals alike experience is paralysis in problem-solving: they get part-way into solving a problem and encounter a point where they have no idea what to do or where to go next. This is why having a strategy to identify the next step is so vitally important.

After identifying these unknown variables of V, n, and T, we now know we need to either find other formulae to solve for those values, or investigate the real mechanism to see if we may directly measure their values.

It should be clear that temperature (T) is one of those variables we should be able to directly measure in the mechanism. We know in order to solve for pressure (P) given piston position (x), one of the real-world parameters we must first measure is temperature.

It should also be clear that volume (V) is a variable related to piston position (x), and as such we should be able to identify a formula relating the two, which we may then combine with the Ideal Gas Law formula to arrive at one step closer to our solution. Since the interior volume of the machine's cylinder is, well, *cylindrical*, we know that the formula for calculating the volume of a cylinder is what we need to incorporate next:

$$V = \pi r^2 l$$

Where,
    V = Volume of cylinder
    $\pi \approx 3.1415927$
    r = Radius of cylinder
    l = Length of cylinder

As we did with the Ideal Gas Law formula, our next step is to identify in the cylinder volume formula those quantities we currently know versus those we don't (and therefore those we need to find). Marking up the formula with circles and squares is helpful, as is drawing an arrow between the variable we're trying to solve for with the cylinder formula and where it goes in the Ideal Gas Law formula. This arrow visually links the two formulae together, showing us where we will eventually need to perform algebraic substitution:

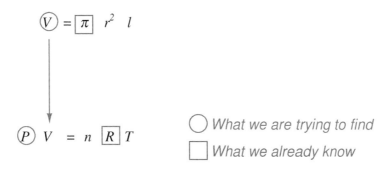

Pi (π) of course is a natural constant, just like R, so we enclose it in a square to show that we know its value. r is the radius of the cylinder, which in the mechanism's case is the same as the radius of the piston. As with temperature, this is a quantity we will need to determine in order to solve this problem. Length (l) is exactly the same as the piston position given to us in the problem: x, meaning we may re-write the cylinder volume formula as $V = \pi r^2 x$. Representing all of these relationships graphically:

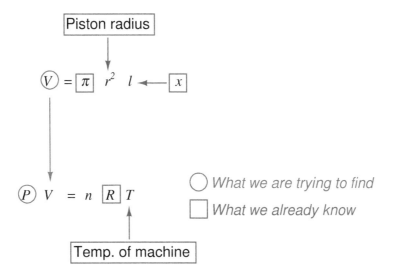

The only unknown still existing is n, the number of moles of gas held inside the cylinder. This is not something one would be able to directly measure, like temperature, piston radius, or piston position. Either this is a quantity that would have to be provided to us as a given condition in the problem, or we must know something else about the problem in order to find the value of n. We will set aside this issue for the time being and concentrate for now on how to combine the cylinder and Ideal Gas Law formulae.

As our graphical markup of the two formulae show us, volume (V) calculated by the cylinder formula is what gets "plugged in" to the Ideal Gas Law formula (V as well). In algebra, this operation is called *substitution*: replacing a single variable in an equation with another equation.

Substituting $V = \pi r^2 x$ for V in $PV = nRT$ . . .

$$PV = nRT$$

$$P(\pi r^2 x) = nRT$$

$$P\pi r^2 x = nRT$$

Now, solving for P

$$P = \frac{nRT}{\pi r^2 x}$$

We now have a formula for P written as a function of x. Provided we know the temperature (T), the piston's dimensions, and the molecular quantity of gas inside the cylinder (n), we may plug any arbitrary value for piston position (x) and very easily calculate pressure (P).

Returning to the question of knowing the value of molecular gas quantity (n), we might ask ourselves, "What if no one provided us with a value for n? How could we possibly solve for pressure then?" Recall once more the basic law of algebra telling us we can only solve for one unknown quantity in a single equation at one time. What would we need to know in our custom formula in order to solve for n? A graphic markup of the formula is helpful once again:

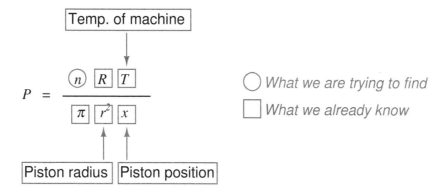

From this we can see it would be possible to calculate n, if only we knew P. Clearly, this is a Catch-22: our goal is to calculate the value of P from n, yet the only algebraic means we have of determining n is to first know the value of P.

However, there is a solution to this conundrum. Most engine and compressor mechanisms work by moving the piston back and forth in coordination with valves to vent and direct the flow of gas into and out of the cylinder. This usually means there will be *other* piston positions where vent valve(s) are open to ensure atmospheric pressure inside the cylinder. In other words, there will be other values of x for which the pressure P will be known. In such a case where both x and P are known, we may solve for n, and then be confident that value of n will be the same for other piston positions because the vent valves close off before compression begins, trapping all gas molecules inside the cylinder.

Representing two such conditions, using subscripts to distinguish pressures and piston positions in each of the two conditions. Here, we will assume that the gas temperature is the same in both conditions, to simplify the problem:

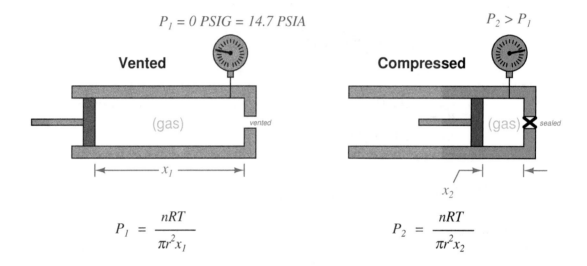

Now that we have a means to solve for n, we may use substitution again to replace n in the "compressed" formula (condition 2) with the n from the "vented" formula (condition 1). A key assumption in making this substitution is that we know n will actually be the same value for those two conditions, which is a safe assumption because compression machines seal off the cylinder (preventing gas entry or escape) prior to the compression stroke of the piston.

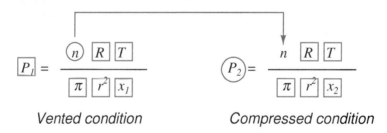

To do this substitution, we must first manipulate the first equation to solve for n, then replace n in the second equation with our manipulated first equation:

$$P_1 = \frac{nRT}{\pi r^2 x_1}$$

$$n = \frac{\pi r^2 x_1 P_1}{RT}$$

Substituting . . . n

$$P_2 = \frac{\left(\frac{\pi r^2 x_1 P_1}{RT}\right) RT}{\pi r^2 x_2}$$

Here we see something very interesting: following the substitution, many of the variables and constants cancel each other out, leaving us with a much simpler formula solving for P_2 in terms of x_1 and x_2. First, we cancel RT out of the numerator:

$$P_2 = \frac{\left(\frac{\pi r^2 x_1 P_1}{RT}\right) RT}{\pi r^2 x_2}$$

$$P_2 = \frac{\pi r^2 x_1 P_1}{\pi r^2 x_2}$$

Next, we cancel out πr^2 from numerator and denominator:

$$P_2 = \frac{x_1 P_1}{x_2}$$

$$P_2 = \frac{x_1}{x_2} P_1$$

Thus, the machine's gas pressure in the compressed condition is a simple ratio of its two piston positions multiplied by the absolute pressure in the vented condition.

34.4.3   Double-checking calculations

When performing calculations to arrive at an answer to some problem, it is important to check your work. Even if all the algebraic work you've done is perfectly correct, it is still possible to commit simple "keystroke" errors while entering numbers or executing operations on your calculator, and/or to make simple mental-math calculation errors. For this reason, teachers from time immemorial have encouraged their students to *double-check* their work.

What many students tend to do, unfortunately, is check their work by following all the same steps one more time and seeing if they get the same answer as before. While this might make sense to do at first, it actually invites the exact same errors, because our short-term memory for calculator keystrokes and mental math operations tends to be quite good. If we made a keystroke error the first time, we are very likely to make the exact same keystroke error the second time, simply following muscle memory.

A good way to avoid this error is to check your mathematical work *backwards*, beginning with the answer you previously calculated, working backwards through the equations to see if you can arrive at one of the values given to you at the start of the problem. This technique forces you to approach the problem differently, using different keystrokes in different orders.

For example, suppose you were tasked with calculating the pressure generated by a vertical column of liquid inside a process vessel, in order to properly set the lower- and upper-range values (LRV and URV) for the hydrostatic level transmitter:

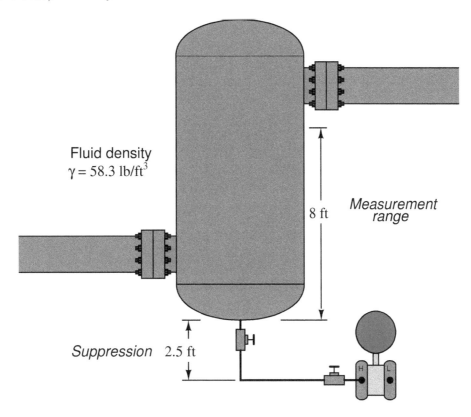

The pressure seen by the transmitter at a 0% level condition (empty vessel) should be only that produced by the 2.5 foot height of the impulse tubing filled with process fluid:

$$P = \gamma h$$

$$P = (58.3 \text{ lb/ft}^3)(2.5 \text{ ft})$$

$$P = 145.75 \text{ lb/ft}^2 = 1.012 \text{ PSI}$$

Likewise, the pressure seen by the transmitter at a 100% level condition should be that produced by the impulse tubing height *and* the vessel's internal fill of 8 feet:

$$P = \gamma h$$

$$P = (58.3 \text{ lb/ft}^3)(2.5 \text{ ft} + 8 \text{ ft})$$

$$P = (58.3 \text{ lb/ft}^3)(10.5 \text{ ft})$$

$$P = 612.15 \text{ lb/ft}^2 = 4.251 \text{ PSI}$$

If we wished to check our final answer of 4.25 PSI, we could work backwards from this result to try to calculate the total fluid height of 10.5 feet, or work backwards to calculate the fluid density of 58.3 lb/ft^3. Let's do the latter and see what we get:

$$\gamma = \frac{P}{h}$$

$$\gamma = \frac{612.15 \text{ lb/ft}^2}{10.5 \text{ ft}}$$

$$\gamma = 58.3 \text{lb/ft}^3$$

Although this technique may seem obvious, it nevertheless avoids the pitfalls of repeating keystroke errors, which is an error plaguing the work of many students!

A variation on this theme is to calculate a quantity not given or reached in any of the initial problems. Here, for example, we were tasked with determining the transmitter's lower- and upper-range values (pressures at 0% fill and 100% fill). One way to check our work is to see if a different fill condition such as 50% gives us a pressure value lying exactly half-way between the LRV of 1.012 PSI and the URV of 4.251 PSI.

If we imagine the vessel half-full, our total liquid height seen by the transmitter will be 4 feet plus the suppression value of 2.5 feet, or 6.5 feet total. Calculating hydrostatic pressure at this liquid height:

$$P = \gamma h$$

$$P = (58.3 \text{ lb/ft}^3)(2.5 \text{ ft} + 4 \text{ ft})$$

$$P = (58.3 \text{ lb/ft}^3)(6.5 \text{ ft})$$

$$P = 378.95 \text{ lb/ft}^2 = 2.632 \text{ PSI}$$

A simple way to check that 2.632 lies half-way in between 1.012 and 4.251 is to calculate the average value of 1.012 and 4.251:

$$\frac{1.012 + 4.251}{2} = 2.632$$

Sure enough, it checks out to be correct. This is good validation of our initial work in calculating the transmitter's LRV of 1.012 PSI and URV of 4.251 PSI.

Yet another variation on this same theme of checking your work different ways is to approach the problem differently altogether. We know we may calculate the LRV and URV pressure values for this process vessel and fluid if we convert the fluid's density into a specific gravity value (58.3 lb/ft^3 = 0.9339) and then calculate hydrostatic pressure as though we were dealing with inches water column, corrected for the specific gravity of this process fluid.

A height of 2.5 feet (0% level) is 30 inches, which when multiplied by the specific gravity value of 0.9339 yields 28.016 inches WC, which is equivalent to our first calculated pressure value of 1.012 PSI.

A height of 10.5 feet (100% level) is 126 inches, which when multiplied by the specific gravity value of 0.9339 yields 117.67 inches WC, which is equivalent to our first calculated pressure value of 4.251 PSI.

As you can see, solving for LRV and URV pressures using an entirely different technique yields the same answers as before, which is good validation of our original work.

It is important to note that double-checking your work in this manner merely *helps confirm* your original work, but it does not *conclusively prove* your original answers are correct. For example, one of the numbers we keep using in the URV calculations is the total liquid height of 10.5 feet. Suppose, though, that this height value was something we incorrectly calculated mentally by adding the suppression and the range heights. For example, suppose our actual given heights in the problem were a 2.5 feet suppression height and a 7 foot measurement range, and we mistakenly added 2.5 feet and 7 feet in our heads to get 10.5 feet. If that were the case, both the original URV pressure calculation and the subsequent double-checks might still agree with one another, but could still be wrong because they all depend on that same 10.5 foot height value.

The most realistic attitude to maintain in all problem-solving is *scientific skepticism*: there is no such thing as *absolute proof* so long as the potential for error exists. Given that you are a human being – liable to all manner of fallacies and mistakes – absolute proof will forever lie outside your grasp. The best we can do is incrementally minimize the potential for errors and mistakes, and accept the inevitable uncertainty.

34.4.4 Maintaining relevance of intermediate calculations

A very common attitude I've noticed with students is something I like to call "any procedure, so long as it works." When learning to solve particular types of quantitative problems, the natural tendency is to identify procedural techniques for calculating the correct answer(s), and then practice those procedures until they can be performed without flaw. Unfortunately, it is all too easy to lose focus on principles when the learning emphasis is on procedure. The allure of a procedure guaranteed to yield correct answers overshadows the greater need to master and apply fundamental principles, leading to poor problem-solving ability masked by an illusion of competence.

This is a significant obstacle to deep and significant learning in the sciences, and there is no one solution to it. However, there is a way to identify and self-correct this behavior in some contexts, and it relies on a habit of identifying the real-world relevance of *all* intermediate calculations within a quantitative problem.

To illustrate, let us consider a simple voltage divider circuit comprised of two resistors, where some amount of supply voltage is divided into a smaller proportion to become the output voltage signal. The problem at hand is calculating the output voltage of this divider circuit, knowing the values of supply voltage and resistor resistances:

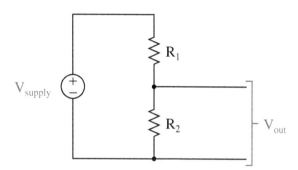

One of the basic formulae all beginning electronics students learn is the appropriately named *voltage divider formula*, shown here:

$$V_{out} = V_{supply} \left(\frac{R_2}{R_1 + R_2} \right)$$

Calculating V_{out} is a simple matter of substituting known values of voltage and resistance into this formula, then performing the necessary arithmetic. However, the procedure by which a student evaluates this formula – particularly in regard to the understanding of fundamental concepts – matters greatly to that student's mastery of circuit analysis.

Suppose a student evaluates the voltage divider formula according to the order of operations enforced by the parentheses. Note the sequence of steps in this procedure, and how each step yields a value relevant to the voltage divider circuit:

Step	Calculation	Unit of measurement	Meaning
1	$R_1 + R_2$	Ohms	Total circuit resistance (R_{total})
2	$R_2 \div R_{total}$	(unitless)	Voltage division ratio
3	$V_{supply} \times$ Ratio	Volts	Output voltage (V_{out})

Now let us suppose a student calculates the exact same output voltage for this divider circuit using a modified version of the voltage divider formula. This formula may be derived from the original version by applying the commutative property of multiplication, simply swapping the positions of R_2 and V_{supply}:

$$V_{out} = R_2 \left(\frac{V_{supply}}{R_1 + R_2} \right)$$

The proper order of operations for this modified formula will be different from the original version, but the final result (V_{out}) will be identical. Each step of the evaluation still yields a value relevant and applicable to the voltage divider circuit:

Step	Calculation	Unit of measurement	Meaning
1	$R_1 + R_2$	Ohms	Total circuit resistance (R_{total})
2	$V_{supply} \div R_{total}$	Amps	Circuit current (I)
3	$R_2 \times I$	Volts	Output voltage (V_{out})

Finally, let us suppose a student calculates the exact same output voltage for this divider circuit using another modified version of the voltage divider formula. This formula may be derived from the original version by applying the associative property of multiplication, grouping R_2 and V_{supply} together into the denominator of the fraction:

$$V_{out} = \frac{R_2 V_{supply}}{R_1 + R_2}$$

The proper order of operations for this modified formula will be different from the original version, but the final result (V_{out}) will be identical. Note, however, the irrelevance of the result in step #2:

Step	Calculation	Unit of measurement	Meaning
1	$R_1 + R_2$	Ohms	Total circuit resistance (R_{total})
2	$R_2 \times V_{supply}$	Volt-ohms	???
3	$R_{total} \times$ previous result	Volts	Output voltage (V_{out})

Despite the mathematical equivalence of this last formula to the prior two, step #2 makes no *conceptual* sense whatsoever. The product of resistance and voltage, while mathematically useful in solving for V_{out}, has no practical meaning within this or any other circuit.

Conceptual problem-solving is an important skill, because it is only by mastering fundamental concepts can one become proficient in solving any arbitrary problem involving those concepts. Procedural problem-solving is useful only when applied to the specific type of problem the procedure was developed for, and useless when faced with any other type of problem. A student who understands the meaning of each step as they evaluate a voltage divider circuit will have little problem solving for quantities in other electrical circuits. A student who has merely memorized a step-by-step procedure for evaluating a voltage divider will struggle trying to solve for quantities in other circuits, unless they happen to have memorized step-by-step procedures for all those other circuits as well.

Troubleshooting is also closely linked with conceptual understanding. In my years as an instructor of electronics and instrumentation, I have seen an almost perfect correlation between conceptual understanding and diagnostic proficiency: procedural thinkers are invariably poor troubleshooters, and poor troubleshooters are invariably procedural thinkers.

The practical lesson we may draw from this example of voltage divider circuit evaluation is the importance of identifying the meaning of every intermediate result. If you ever find yourself performing calculations, unable to explain the practical significance of every step you take, it means *you are thinking procedurally rather than conceptually*. Instructors may apply this standard to their students' work by asking students to explain the meaning of each and every calculation they perform.

34.5 Problem-solving by simplification

A whole class of problem-solving techniques focuses on *altering* the given problem into a simpler form that is easier to analyze. Once a solution is found to the simplified problem, fresh ideas for attacking the original problem often become clear. This section will highlight multiple techniques for problem-simplification, as well as other useful techniques for problem-solving.

The first step, however, to problem simplification is to "give yourself the right" to alter the problem into a different form! Many students tend to avoid this, for fear of "getting off track" and losing sight of the original problem. What is needed is a spirit of adventure: a willingness and a confidence to explore the possibilities. Do not think you *must* solve exactly the problem that is given to you at first. Modify the problem, solve the simpler version of that problem, then apply the lessons and patterns obtained from that solution to the original (more complex) problem!

34.5.1   Limiting cases

A powerful method for analyzing the effects of a change in some system is to consider the effects of "extreme" changes, which are often easier to visualize than subtle changes. Such "extreme" changes are examples of what is generally known in science as a *limiting case*: a special case of a more general rule or trend, possessing fewer possibilities. By virtue of the fact that limiting cases have fewer possibilities, applying a limiting case to a given problem generally simplifies that problem.

Consider, for example, this Wheatstone bridge circuit, where changes in the thermistor's resistance (with temperature) affect the output voltage of the bridge:

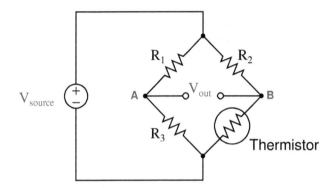

A realistic question to ask of this circuit is, "what will happen to V_{out} when the thermistor's resistance increases?" If our only goal is to arrive at a qualitative answer (e.g. increase/decrease, positive/negative), we may simplify the problem by considering the effects of the thermistor failing completely open, because an "open" fault is nothing more than an extreme example (a *limiting case*) of a resistance increase.

If we perform this "thought experiment" on the bridge circuit, the circuit becomes simpler because we have eliminated one resistor (the thermistor):

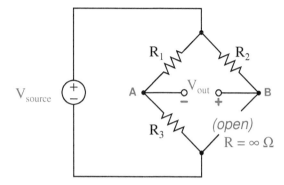

*Limiting case of **increased** thermistor resistance*

With the thermistor eliminated from the circuit, we see that test point **B** has lost its connection to the negative terminal of the voltage source. This can only mean one thing for the potential at test point **B**: it will become more positive (less negative). If the bridge circuit happened to be balanced prior to the thermistor fault, V_{out} will now be such that **B** is positive and **A** is negative by comparison.

Analyzing the results of this limiting case even further, we can see that resistor R_2 now carries zero current (thanks to the thermistor now being failed open), which means R_2 will now drop zero voltage. If R_2 drops no voltage at all, test point **B** must now be at the exact same potential as the positive terminal of the voltage source. This being the case, measuring V_{out} between test points **A** and **B** will be equivalent[1] to measuring voltage across R_1. Thus, the limiting case of V_{out} for an increase in thermistor resistance is V_{R1}, with **B** positive and **A** negative.

---

[1]With R_2 dropping zero voltage, test point **B** is now essentially common to the node at the top of the bridge circuit. With test point **A** already common with the lower terminal of R_1 and now test point **B** common to the upper terminal of R_1, V_{out} is exactly the same as V_{R1}.

Another realistic question to ask of this circuit is, "what will happen to V_{out} when the thermistor's resistance decreases?" Once again, the problem-solving technique of limiting cases helps us by transforming the four-resistor bridge circuit into a three-resistor bridge circuit. The limiting case of a resistance decrease would be a condition of no resistance: a *shorted* thermistor:

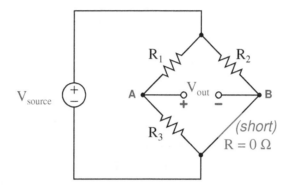

*Limiting case of **decreased** thermistor resistance*

With the thermistor shorted in this "thought experiment," we see that test point **B** now becomes electrically common with the negative terminal of the voltage source. This, of course, has the effect of making test point **B** as negative as it can possibly be. More specifically, by making test point **B** electrically common with the bottom node of the bridge, it makes V_{out} equal[2] to the voltage drop across R_3. Thus, the limiting case of V_{out} for a decrease in thermistor resistance is V_{R3}, with **A** positive and **B** negative.

---

[2]As before, the limiting case of a thermistor fault causes test points **A** and **B** to become synonymous with the terminals of one of the remaining resistors, in this case R_3. Since point **A** is already common with the upper terminal of R_3 and the shorted fault has now made point **B** common with the lower terminal of R_3, V_{out} must be exactly the same as V_{R3}.

Let us consider another application of this problem-solving technique, this time to the analysis of a passive filter circuit:

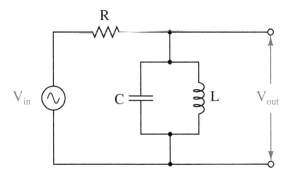

If the type of filter circuit shown here were unknown (i.e. the student could not identify it as a low-pass, high-pass, band-pass, or band-stop filter circuit at first sight), the technique of limiting cases could be applied to determine its behavior. In this case, the limit to apply is one of frequency: we may perform "thought experiments" whereby we imagine the input frequency being extremely low, versus being extremely high.

We know that the reactance of an inductor is directly proportional to frequency ($X_L = 2\pi f L$) and that the reactance of a capacitor is inversely proportional to frequency ($X_C = \frac{1}{2\pi f C}$). Therefore, at an extremely low frequency ($f = 0$ Hz), the inductor will act like a short while the capacitor acts like an open:

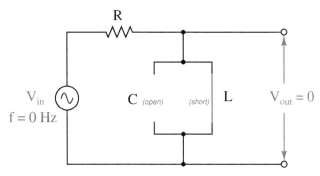

*Limiting case of **decreased** frequency*

Likewise, at extremely high frequencies ($f = \infty$), the capacitor will act like a short while the inductor acts like an open:

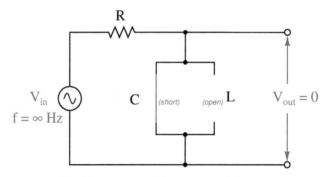

*Limiting case of **increased** frequency*

From these two limiting-case "thought experiments" we may conclude that the filter circuit is neither a low-pass nor a high-pass, because it neither passes low-frequency signals nor high-frequency signals. We may also conclude that it is not a band-stop filter, because that would pass both low-frequency and high-frequency signals. This means it must be a band-pass filter, by eliminating the other three alternatives.

If we would wish to confirm the band-pass nature of this filter by a positive experimental result rather than merely by eliminating what it is *not*, we could perform one more limiting-case "thought experiment:" a condition where the signal frequency exactly equals the resonant frequency of the LC network ($f = \frac{1}{2\pi\sqrt{LC}}$). Here, we must recall the principle that a parallel LC network has infinite impedance at its resonant frequency:

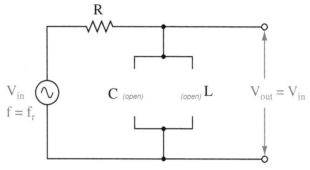

*Limiting case of **resonant** frequency*

In this "thought experiment" we see that the LC network will be completely "open" and allow 100% of the input signal to appear at the output terminals. Thus, it becomes clear that this passive circuit functions as a *band-pass* filter.

As with the Wheatstone bridge circuit, the value of limiting-case analysis is that it acts to simplify the system by effectively eliminating components (replacing them with "shorts" or "opens"). Even in non-electrical problems, limiting cases works the same by simplifying a system's behavior so that it becomes easier to apprehend, and from these simplified cases we may usually determine behavioral trends of the system (e.g. which way it tends to respond as some variable increases or decreases).

34.6 Scientific system diagnosis

At the root of successful system diagnosis is a rigorous adherence to *scientific reasoning*. There exists no single algorithmic approach to solving problems, but rather a singular *mind-set* characterized by the following traits:

- Curiosity

- Persistence

- Attention to detail

- Diligence in checking conclusions

- Regular checking of assumptions

- A willingness to abandon ideas based on contrary evidence

Science is, at its heart, a methodology useful to identify causes and effects. Thus, it is well-suited to the problem of system diagnosis, where our goal is to quickly and accurately identify the cause(s) behind improper operation (effects).

34.6.1   Scientific method

Although no one technique seems to be universally recognized as "the scientific method," the following steps are commonly applied in science to determine causes and effects:

- Observe effects, and then create *hypotheses* (explanations accounting for those observations)

- Design a test for one or more of those hypotheses

- Perform the test (experiment), and collect data from it

- Validate or invalidate the hypotheses based on the data

- Repeat

Perhaps the most challenging step in this method is designing a good test for the hypotheses. By "test" I mean a trial that really challenges each hypothesis, and doesn't just collect more data to support it. A good way to help yourself devise a rigorous test of any hypothesis is to keep these two questions in mind:

"If this hypothesis is true, what other effects should we see if we look for them?"

. . . and . . .

"If this hypothesis is false, what other effects should we *not* see if we look for them?"

An ideal test (experiment) is one that answers both of these questions at once, providing both positive and negative evidence.

In contrast to scientific diagnosis is a technique a colleague of mine refers to as "Easter-egging," where the troubleshooter tries to find the problem by individually checking every component or possible fault they can think of, in serial fashion. The term "Easter-egging" invokes the image of children hunting for hidden eggs on Easter morning, randomly searching in every place they can think of where an egg might be hidden. There is no logical reasoning to "Easter-egging" and so it is a very inefficient method of solving system problems.

34.6.2   Occam's Razor

A very helpful principle in scientific testing is something called *Occam's Razor*, a rule stating that the simplest explanation for any observed effects is usually the most accurate. While not infallible, Occam's Razor is nevertheless a valid "gambling strategy" based on simple probability. In system troubleshooting, it means that a single fault is more likely to account for the symptoms than a set of coincidental faults. For this reason, it is generally good practice to enter a troubleshooting scenario with the assumption that only one thing is wrong, unless the data conclusively points otherwise.

34.6.3 Diagnosing intermittent problems

Intermittent faults are some of the most challenging to diagnose, for the simple reason that the relevant symptoms come and go. A persistent fault is easier to solve because the data is continuously there for inspection.

The key to troubleshooting intermittent faults is to set up test equipment to capture events that occur when you are not directly observing them. Some suggested methods include:

- Using the "Min/Max" capture mode on a digital multimeter (DMM)

- Using a data recorder or event logger to capture signal history

- Looking for evidence left by certain intermittent faults (e.g. if the suspected fault is high temperature at a certain location, looking for evidence such as charring or discoloration that would be caused by high temperature at some past time)

- Using videorecording equipment to capture events

Perhaps one of the most useful features of modern digital multimeters is the ability to capture minimum and maximum signal levels. Many times I have used this feature on my own meter to monitor the highs and lows of some signal in order to capture evidence of an intermittent fault. This is also useful for monitoring signal changes that happen too fast to see on the display of a meter (e.g. detecting the peak pulse amplitude of a fast signal). While limited to the sample rate of the digital meter, it remains a powerful tool in the hands of a knowledgeable technician.

A colleague of mine once diagnosed a complex, intermittent problem on a natural gas compressor unit by setting up a video camera to film the control panel gauges on the compressor, then reviewing the video recording frame-by-frame after the camera had recorded a "trip" event. This kind of creativity is often key to diagnosing intermittent problems.

34.6.4   Strategy: tracing data paths

A method often useful for tracing the location of faults in complex systems is to identify where data is coming from (source), where it is going (destination), and all paths taken by the data in between. If we then plot those paths on a one-line diagram of the system, the intersection of paths often tells us where the problem lies.

For example, consider this system of networked devices in a data acquisition system:

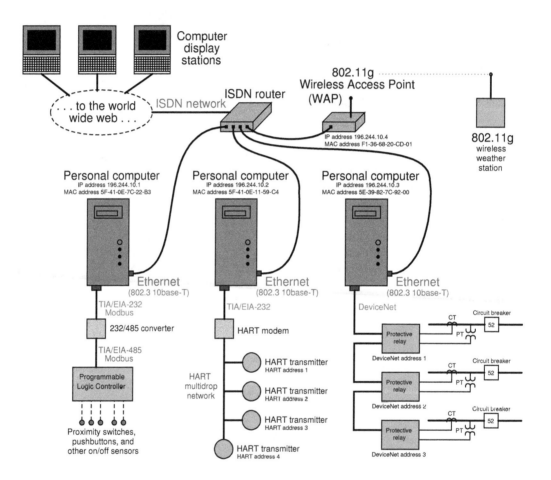

Suppose operations personnel noticed they could no longer access any protective relay data from the left-most display station connected to the world-wide web (Internet), but they could still access live weather station data from that same display station. Applying the technique of tracing data paths may be helpful to us in locating the fault in this complex system, and also devising a good test to pinpoint the location.

Here we see the same system with green and red lines overlaid showing good data paths and failed data paths:

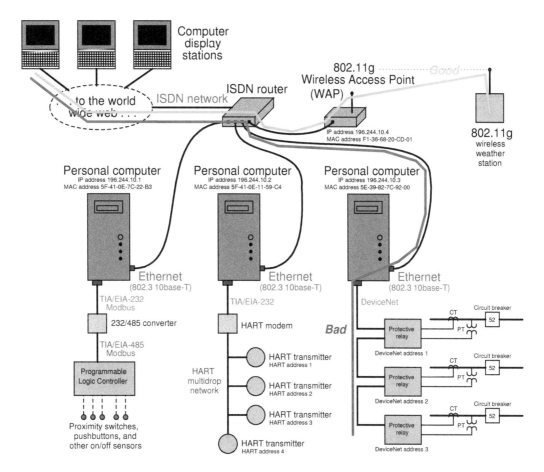

Note how these two paths overlap in the display station, the world-wide-web, and the ISDN router. Since data from the weather station is getting through this part of the path just fine, yet protective relay data is not, the most likely location of the fault is in a part of the system where these two data paths are *not* common. In other words, it is unlikely that the problem lies within the display station, the Internet, or the ISDN router, because all those component are proven to work just fine for the weather station data. A more probable location of the fault is in an area of the "bad" data path that is *not* common to the "good" data path. In this particular case, it points to a problem from the ISDN router to the DeviceNet network.

A good test to do at this point is to try "pinging" the right-most personal computer from one of the other two personal computers connected directly to the ISDN router. This would be testing a data path from one PC to the other, thereby testing the integrity of the right-most PC and the cable connecting it to the ISDN router. If this test is successful, the problem likely lies farther beyond the PC (e.g. in the DeviceNet network) ; if this test is unsuccessful, the problem likely lies within that PC or within the cabling connecting it to the router.

Note the careful use of the words "likely" and "probable" in the hypotheses. Hypotheses are uncertain by their very nature, and are never "proven" in any final sense. Even though our initial approach of sketching and comparing data pathways suggests a problem between the ISDN router and the DeviceNet network connecting the protective relays together, it is still possible for the problem to lie somewhere closer to the display station. If, for example, the display station relied on dedicated software "drivers" to properly poll and interpret data from the protective relay network that were not used to do the same for weather station data, the problem could actually lie within the display station! A corrupted relay driver would prevent that station from properly displaying protective relay data and yet permit the display of weather station data. In order for our data path tracing procedure to encompass this possibility, we would need to show the pathways splitting within the display station, so they would no longer be common to each other:

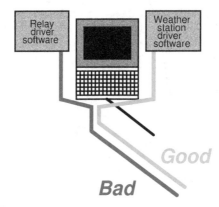

As always, the assumptions we embed into our hypotheses can skew their validity. If we assume completely overlapping data paths for protective relay and weather data within the display station, we would not recognize the possibility of a driver problem.

34.7   Common diagnostic mistakes

Volumes could be written about poor diagnostic technique. The following mistakes are not intended to comprise a comprehensive list, but are merely warnings against errors that are all too common among students and within the profession.

34.7.1 Failing to gather data

Perhaps the most common mistake made by technicians attempting to diagnose a system problem is failing to gather data (i.e. taking measurements and performing simple system tests) during the troubleshooting process. Even a small amount of data gathered from a system may profoundly accelerate the process of diagnosis.

A colleague of mine has a very descriptive term for the poor habit of looking for faults before gathering data: *Easter-Egging*. The idea is that a technician goes about finding the problem in a system the same way they might go about searching for eggs hidden on Easter morning: randomly. With Easter egg hunting, the eggs could literally be hidden *anywhere*, and so there is no rational way to proceed on a search. In like manner, a technician who lacks information about the nature or source of a system problem is likely to hunt in random fashion for its source. Not only will this likely require significant time and effort, but it may very well fail entirely.

A much more efficient way to proceed is to gather new data with each and every step in the troubleshooting process. By "gathering data," I mean the following:

- Taking measurements with test equipment (multimeter, pressure gauges, etc.)

- Observing equipment indicator lights

- Stimulating the system and observing its response(s)

- Using your other senses (smell, hearing, touch) to gather clues

- Documenting new data in a notepad to help track and analyze the results of your measurements and tests

34.7.2 Failing to use relevant documentation

Diagrams are indispensable "maps" for solving problems in complex systems. A critical first step in diagnosing any system problem is to obtain correct diagrams of the system, so that you may see the pathways of power and signals in the system. Attempting to diagnose a system problem without consulting the relevant diagrams is like trying to find your way around a city without a map.

A corollary to the rule of obtaining relevant diagrams is to *use them* when reasoning through fault scenarios. All too often I see students locate the diagram for a system, glance at it, then set it aside and proceed to stumble through the rest of the diagnosis because they try to trace all the signal paths in the real world as they assess fault possibilities and devise tests. Diagrams are laid out in a clean and logical format for a reason: it is much easier to follow the flow of signals and power in a diagram than it is to follow the same flow through the convoluted paths of real-world wires and cables. If you reject the diagram in favor of tracing all pathways by looking at the real-world system, you are needlessly adding complexity to the problem: not only do you have to reason through the fault hypotheses and diagnostic tests, but you also must mentally "un-tangle" the signal paths as they are laid out in the real world (which can be a daunting task in itself!). *Do all your diagnostic thinking while looking at the diagram, and refer to the real-life system only when the time comes to execute a diagnostic test.* The result will be a much faster and less frustrating experience than if you try to trace everything in real life.

When diagnosing a problem in a system where one or more of the key components are unfamiliar to you, it is important to consult the relevant technical literature on those components. This is especially true if the component in question has been identified as suspect by your diagnostic test(s) and it is quite complex (e.g. loop controller, PLC, motor drive, data acquisition module, etc.). Just a few minutes' worth of reading the manual may save you hours of fruitless diagnosis.

This point also underscores the necessity of technical reading as a skill to be practiced and honed at every opportunity. Being able to quickly locate pertinent information in a dense technical document is key to fast and efficient troubleshooting!

34.7.3 Fixating on the first hypothesis

When diagnosing a faulted system, an efficient strategy is to brainstorm *multiple* hypotheses accounting for the symptoms, then devise tests to support or discredit those hypotheses in the fewest steps.

Let me illustrate by example. Suppose a pump motor is controlled by a PLC, the PLC sending a command signal to a variable-frequency motor drive (VFD) to command the motor to start and stop:

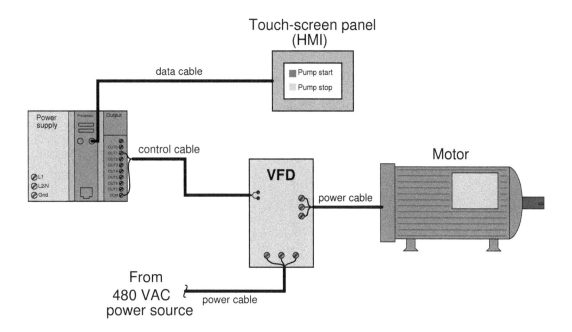

If the motor refuses to start when the "Pump start" icon is pressed on the HMI screen, a competent troubleshooter will begin to mentally list a range of problems that could prevent the motor from starting:

- Motor faulted

- VFD lacking power

- VFD not configured properly to receive signal from PLC

- PLC output defective, not sending signal to VFD

- PLC program halted or faulty

- HMI not sending signal to PLC

- . . . etc.

After brainstorming such a list, a competent troubleshooter will then devise a simple test to "divide the problem space in half." One such test[3] in this system would be to use a multimeter to measure the electrical signal from the PLC's output card to the VFD input terminals. If a signal appears when the "Pump start" icon on the HMI is pressed, it means everything in the HMI and PLC is working as it should, and that the problem must lie with the VFD or beyond. If no signal appears, it means the problem lies with the VFD, HMI, or associated cabling. Again, the wise strategy is to brainstorm multiple hypotheses explain why the motor won't start, then execute simple tests to eliminate most of those hypotheses so you may focus on those that are most likely.

By contrast, a novice might only think of one possibility – such as the VFD being improperly configured – and then immediately fixate on that hypothesis by inspecting the drive parameters looking for one that is improperly set. If the fault lies elsewhere, the novice could spend all day reviewing VFD parameters and never find the problem. Given the wide range of possible faults, fixating on any one fault from the start is very likely a waste of time. I have watched technicians and students alike waste hours of time trying to find a fault that was not where they were looking, simply because that was the first area they thought of to look toward. Only after squandering valuable time on one failed hypothesis will the novice then consider other possibilities and other tests.

---

[3]Other possible tests include inspecting the LED status light on that PLC output card channel (a light indicates the HMI and PLC program are working correctly, and that the problem could lie within the output card or beyond to the motor) or measuring voltage at the drive output (voltage there indicates the problem must lie with the motor or the cable to the motor rather than further back).

34.7.4 Failing to build and test a new system in stages

Technicians must sometimes assemble new systems from components. A very common mistake is to assemble the system completely before attempting to test it for proper operation. This is almost always a grievous mistake.

The number of potential mistakes one can make when assembling a brand-new system is quite large. Given this large set of potential mistakes, the probability of making multiple mistakes when assembling the system is very high. Since diagnosis of a system with multiple faults is always more complicated than diagnosing a system with one fault, waiting for the entire system to be assembled before checking it invites multi-fault scenarios.

To illustrate, consider this pressure-control system, where an electronic pressure transmitter sends a 4-20 mA signal to a loop controller, which in turn drives a control valve with another 4-20 mA signal:

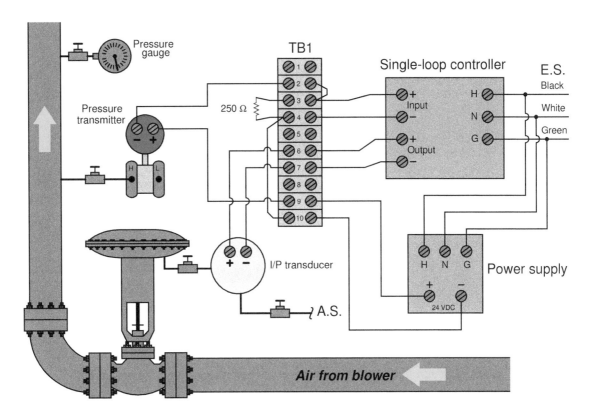

Imagine building this system, placing each component in the proper location, connecting all wires together, and testing it for proper operation. If one were to wait until the entire system were assembled before testing, the probability of having to diagnose multiple faults would be great.

A better strategy would be to assemble and test the system in stages. Consider this sequence of steps as a more practical alternative:

1. Install the I/P transducer, connecting air tubes to supply and valve diaphragm.

2. *Test the I/P and control valve operation using a loop calibrator in "source" mode to drive a 4-20 mA signal to the I/P.*

3. Install and wire power to the loop controller, ensuring it powers up properly.

4. Connect cabling between the I/P and the loop controller's output.

5. *Test the controller's ability in manual mode to "stroke" the control valve throughout its entire range.*

6. Connect wires between the loop controller's input and the 250 ohm resistor on the terminal block.

7. *Test the controller's ability to properly read an input signal by using a loop calibrator to drive 4-20 mA through the 250 ohm resistor.*

8. Install the pressure transmitter, connecting impulse line between it and the process line.

9. *Power the transmitter with a portable DC power supply (or loop calibrator set to the appropriate mode) and check its calibration by applying known pressures to the input tube.*

10. Connect wires between the permanent DC power supply, the transmitter, and the controller's input.

11. *Apply pressure to the transmitter input and check to see that it reads properly on the controller's digital display.*

12. *Test the controller's ability to monitor and control process pressure in manual mode.*

13. Perform manual-mode (open-loop) tests to verify process characteristics and obtain data needed for loop tuning (e.g. lag time, dead time, etc.).

14. Enter preliminary PID tuning parameter values.

15. *Test the controller's ability to monitor and control process pressure in automatic mode.*

16. Modify PID tuning parameter values and re-test in automatic mode until robust control is obtained.

Note how the pressure control instrumentation is constructed and then immediately tested as a series of sub-systems, rather than assembling the entire thing and testing only at the very end. Although the *built-test-build* sequence shown here may appear to be more time-intensive at first blush, it will actually save a lot of time and confusion over the *build-everything-then-test-last* method favored by novices.

34.8 Helpful "tricks" using a digital multimeter (DMM)

The digital multimeter (DMM) is quite possibly the most useful tool in the instrument technician's collection[4]. This one piece of test equipment, properly wielded, yields valuable insight into the status and operation of many electrical and electronic systems. Not only is a good-quality multimeter capable of precisely indicating electrical voltage, current, and resistance, but it is also useful for more advanced tests. The subject of this section is how to use a digital multimeter for some of these advanced tests[5].

For all these tests, I suggest the use of a top-quality field multimeter. I am personally a great fan of *Fluke* brand meters, having used this particular brand for nearly my whole professional career. The ability of these multimeters to accurately measure true RMS amplitude, discriminate between AC and DC signals, measure AC signals over a wide frequency range, and survive abuse both mechanical and electrical, is outstanding.

34.8.1 Recording unattended measurements

Many modern multimeters have a feature that records the highest and lowest measurements sensed during the duration of a test. On Fluke brand multimeters, this is called the *Min/Max* function. This feature is extremely useful when diagnosing intermittent problems, where the relevant voltages or currents indicating or causing the problem are not persistent, but rather come and go. Many times I have used this feature to monitor a signal with an intermittent "glitch," while I attended to other tasks.

The most basic high-low capture function on a multimeter only tells you what the highest and lowest measured readings were during the test interval (and that only within the meter's scan time – it is possible for a very brief transient signal to go undetected by the meter if its duration is less than the meter's scan time). More advanced multimeters actually log the *time* when an event occurs, which is obviously a more useful feature. If your tool budget can support a digital multimeter with "logging" capability, spend the extra money and take the time to learn how this feature works!

---

[4]As a child, I often watched episodes of the American science-fiction television show *Star Trek*, in which the characters made frequent use of a diagnostic tool called a *tricorder*. Week after week the protagonists of this show would avoid trouble and solve problems using this nifty device. The *sonic screwdriver* was a similar tool in the British science-fiction television show *Doctor Who*. Little did I realize while growing up that my career would make just as frequent use of another diagnostic tool: the electrical multimeter.

[5]I honestly considered naming this section "Stupid Multimeter Tricks," but changed my mind when I realized how confusing this could be for some of my readers not familiar with colloquial American English.

34.8.2   Avoiding "phantom" voltage readings

My first "trick" is not a feature of a high-quality DMM so much as it is a solution to a common problem *caused* by the use of a high-quality DMM. Most digital multimeters exhibit very high input impedance in their voltage-measuring modes. This is commendable, as an ideal voltmeter should have infinite input impedance (so as to not "load" the voltage signal it measures). However, in industrial applications, this high input impedance may cause the meter to register the presence of voltage where none should rightfully appear.

Consider the case of testing for the absence of AC voltage on an isolated power conductor that happens to lie near other (energized) AC power conductors within a long run of conduit:

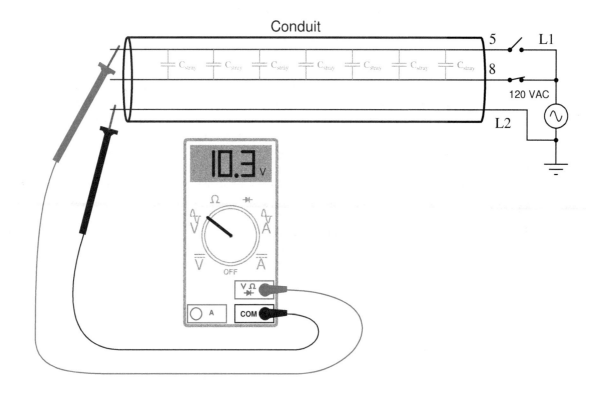

With the power switch feeding wire 5 in the open state, there should be no AC voltage measured between wire 5 and neutral (L2), yet the voltmeter registers slightly over 10 volts AC. This "phantom voltage" is due to capacitive coupling between wire 5 and wire 8 (still energized) throughout the length of their mutual paths within the conduit.

Such phantom voltages may be very misleading if the technician encounters them while troubleshooting a faulty electrical system. Phantom voltages give the impression of connection (or at least high-resistance connection) where no continuity actually exists. The example shown, where the phantom voltage is 10.3 volts compared to the source voltage value of 120 volts, is actually quite modest. With increased stray capacitance between the conductors (longer wire runs in close proximity, and/or more than one energized "neighboring" wire), phantom voltage magnitude begins to approach that of the source voltage[6].

The equivalent circuit is shown here, with the DMM modeled as a 10 MΩ resistance:

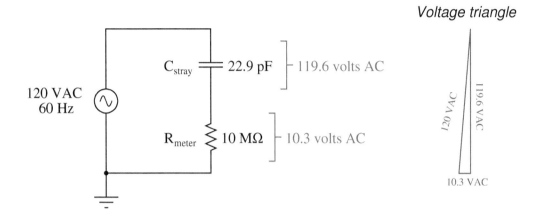

An analog voltmeter would never have registered 10.3 volts under the same conditions, due to its substantially lower input impedance. Thus, "phantom voltage" readings are a product of modern test equipment more than anything else.

---

[6]I have personally measured "phantom" voltages in excess of 100 volts AC, in systems where the source voltage was 120 volts AC.

The obvious solution to this problem is to use a different voltmeter – one with a much lesser input impedance. But what is a technician to do if their only voltmeter is a high-impedance DMM? Connect a modest resistance in parallel with the meter input terminals, of course! Fluke happens to market just this type of accessory[7], the SV225 "Stray Voltage Adapter" for the purpose of eliminating stray voltage readings on a high-impedance DMM:

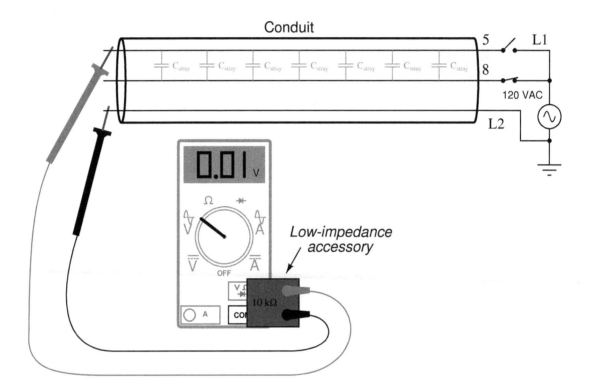

With the voltmeter's input impedance artificially decreased by the application of this accessory, the capacitive coupling is insufficient to produce any substantial voltage dropped across the voltmeter's input terminals, thus eliminating. The technician may now proceed to test for the presence of AC control signal (or power) voltages with confidence.

---

[7]Before there was such an accessory available, I used a 20 kΩ high-power resistor network connected in parallel with my DMM's input terminals, which I fabricated myself. It was ugly and cumbersome, but it worked well. When I made this, I took great care in selecting resistors with power ratings high enough that accidental contact with a truly "live" AC power source (up to 600 volts) would not cause damage to them. A pre-manufactured device such as the Fluke SV225, however, is a much better option.

34.8.3 Non-contact AC voltage detection

While the last multimeter "trick" was the elimination of a parasitic effect, this trick is the exploitation of that same effect: "phantom voltage" readings obtained through capacitive coupling of a high-impedance voltmeter to a conductor energized with AC voltage (with respect to ground). You may use a high-impedance AC voltmeter to perform qualitative measurements of ground-referenced AC power voltage by setting the meter to the most sensitive AC range possible, grounding one test lead, and simply touching the other test lead to the insulation of the conductor under test. The presence of voltage (usually in the range of millivolts AC) upon close proximity to the energized conductor will indicate the energization of that conductor.

This trick is useful for determining whether or not particular AC power or control wires are energized in a location where the only access you have to those wires is their insulating sheaths. An example of where you might encounter this situation is where you have removed the cover from a conduit elbow or other fitting to gain access to a wire bundle, and you find those wires labeled for easy identification, but the wires do not terminate to any exposed metal terminals for you to contact with your multimeter's probe tips. In this case, you may firmly connect one probe to the metal conduit fitting body, while individually touching the other probe tip to the desired conductors (one at a time), watching the meter's indication in AC millivolts.

Several significant caveats limit the utility of this "trick:"

- The impossibility of quantitative measurement

- The potential for "false negative" readings (failure to detect a voltage that is present)

- The potential for "false positive" readings (detection of a "phantom voltage" from an adjacent conductor)

- The exclusive applicability to AC voltages of significant magnitude (≥ 100 VAC)

Being a qualitative test only, the millivoltage indication displayed by the high-impedance voltmeter tells you nothing about the actual magnitude of AC voltage between the conductor and ground. Although the meter's input impedance is quite constant, the parasitic capacitance formed by the surface area of the test probe tip and the thickness (and dielectric constant) of the conductor insulation is quite variable. However, in conditions where the validity of the measurement may be established (e.g. cases where you can touch the probe tip to a conductor known to be energized in order to establish a "baseline" millivoltage signal), the technique is useful for quickly checking the energization status of conductors where ohmic (metal-to-metal) contact is impossible.

For the same reason of wildly variable parasitic capacitance, this technique should *never* be used to establish the de-energization of a conductor for safety purposes. The only time you should trust a voltmeter's non-indication of line voltage is when that same meter is validated against a known source of similar voltage in close proximity, and when the test is performed with direct metal-to-metal (probe tip to wire) contact. A non-indicating voltmeter *may* indicate the absence of dangerous voltage, or it may indicate an insensitive meter.

34.8.4   Detecting AC power harmonics

The presence of *harmonic* voltages[8] in an AC power system may cause many elusive problems. Power-quality instruments exist for the purpose of measuring harmonic content in a power system, but a surprisingly good qualitative check for harmonics may be performed using a multimeter with a frequency-measuring function.

Setting a multimeter to read AC voltage (or AC current, if that is the quantity of interest) and then activating the "frequency" measurement function should produce a measurement of exactly 60.0 Hz in a properly functioning power system (50.0 Hz in Europe and some other parts of the world). The only way the meter should ever read anything significantly different from the base frequency is if there is significant harmonic content in the circuit. For example, if you set your multimeter to read frequency of AC voltage, then obtained a measurement of 60 Hz that intermittently jumped up to some higher value (say 78 Hz) and then back down to 60 Hz, it would suggest your meter was detecting harmonic voltages of sufficient amplitude to make it difficult for your meter to "lock on" to the fundamental frequency.

It is very important to note that this is a crude test of power system harmonics, and that measurements of "solid" base frequency do not guarantee the absence of harmonics. Certainly, if your multimeter produces unstable readings when set to measure frequency, it suggests the presence of strong harmonics in the circuit. However, the absence of such instability does not necessarily mean the circuit is free of harmonics. In other words, a stable reading for frequency is *inconclusive*: the circuit might be harmonic-free, or the harmonics may be weak enough that your multimeter ignores them and only displays the fundamental circuit frequency.

---

[8]These are AC voltages having frequencies that are integer-multiples of the fundamental powerline frequency. In the United States, where 60 Hz is standard, harmonic frequencies would be whole-number multiples of 60: 120 Hz, 180 Hz, 240 Hz, 300 Hz, etc.

34.8.5 Identifying noise in DC signal paths

An aggravating source of trouble in analog electronic circuits is the presence of AC "noise" voltage superimposed on DC signals. Such "noise" is immediately evident when the signal is displayed on an oscilloscope screen, but how many technicians carry a portable oscilloscope with them for troubleshooting?

A high-quality multimeter exhibiting good discrimination between AC and DC voltage measurement is very useful as a qualitative noise-detection instrument. Setting the multimeter to read AC voltage, and connecting it to an signal source where pure (unchanging) DC voltage is expected, should yield a reading of nearly zero millivolts. If noise is superimposed on this DC signal, it will reveal itself as an AC voltage, which your meter will display.

Not only is the AC voltage capability of a high-quality (discriminating) multimeter useful in detecting the presence of "noise" voltage superimposed on analog DC signals, it may also give clues as to the source of the noise. By activating the frequency-measuring function of the multimeter while measuring AC voltage (or AC millivoltage), you will be able to track the frequency of the noise to see its value and stability.

Once on a job I was diagnosing a problem in an analog power control system, where the control device was acting strangely. Suspecting that noise on the measurement signal line might be causing the problem, I set my Fluke multimeter to measure AC volts, and read a noise voltage of several tenths of a volt (superimposed on a DC signal a few volts in magnitude). This told me the noise *was* indeed a significant problem. Pressing the "Hz" button on my multimeter, I measured a noise frequency of 360 Hz, which happens to be the "ripple" frequency of a six-pulse (three-phase) AC-to-DC rectifier operating on a base frequency of 60 Hz. This told me where the likely source of the noise was, which led me to the physical location of the problem (a bad shield on a cable run near the rectified power output wiring).

34.8.6   Generating test voltages

Modern digital multimeters are fantastically capable measurement tools, but did you know they are also capable of *generating* simple test signals? Although this is not the design purpose of the resistance and diode-check functions of a multimeter, the meter does output a low DC voltage in each of these settings.

This is useful when qualitatively testing certain instruments such as electronic indicators, recorders, controllers, data acquisition modules, and alarm relays, all designed to input a DC voltage signal from a 250 ohm resistor conducting the 4-20 mA electronic transmitter signal. By setting a multimeter to either the resistance (Ω) or diode check function and then connecting the test leads to the input terminals of the instrument, the instrument's response may be noted.

Of course, this is a *qualitative* test only, since multimeters are not designed to output any precise amount of voltage in either the resistance or diode-check modes. However, for testing the basic response of a process indicator, recorder, controller, data acquisition channel, DCS input, or any other DC-signal-receiving devices, it is convenient and useful. In every multimeter I have ever tried this with, the diode-check function outputs *more* voltage than the resistance measurement function[9]. This gives you two levels of "test signal" generation: a low level (resistance) and a high level (diode check). If you are interested in using your multimeter to generate test voltages, I recommend you take the time to connect your multimeter to a high-impedance voltmeter (such as another digital multimeter set to measure DC volts) and note just how much voltage your meter outputs in each mode. Knowing this will allow you to perform tests that are more quantitative than qualitative.

---

[9]There is a design reason for this. Most digital multimeters are designed to be used on semiconductor circuits, where the minimum "turn-on" voltage of a silicon PN junction is approximately 500 to 700 millivolts. The diode-check function must output more than that, in order to force a PN junction into forward conduction. However, it is useful to be able to check ohmic resistance in a circuit *without* activating any PN junctions, and so the resistance measurement function typically uses test voltages *less than 500 millivolts.*

34.8.7 Using the meter as a temporary jumper

Often in the course of diagnosing problems in electrical and electronic systems, there is a need to temporarily connect two or more points in a circuit together to force a response. This is called "jumpering," and the wires used to make these temporary connections are called *jumper wires*.

More than once I have found myself in a position where I needed to make a temporary "jumper" connection between two points in a circuit, but I did not have any wires with me to make that connection. In such cases, I learned that I could use my multimeter test leads while plugged into the *current-sensing* jacks of the meter. Most digital multimeters have a separate jack for the red test lead, internally connected to a low-resistance *shunt* leading to the common (black) test lead jack. With the red test lead plugged into this jack, the two test leads are effectively common to one another, and act as a single length of wire.

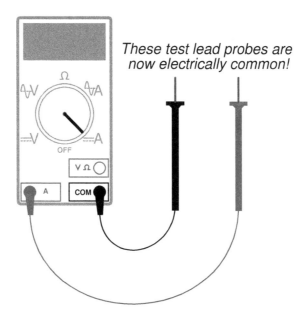

Touching the meter's test leads to two points in a circuit will now "jumper" those two points together, any current flowing through the shunt resistance of the multimeter. If desired, the meter may be turned on to monitor how much current goes through the "jumper" if this is diagnostically relevant.

An additional benefit to using a multimeter in the current-measuring mode as a test jumper is that this setting is usually current-protected by a fuse inside the meter. Applying jumper wires to a live circuit may harbor some danger if significant potential and current-sourcing capability exist between those two points: the moment a jumper wire bridges those points, a dangerous current may develop within the wire. Using the multimeter in this manner gives you a *fused* jumper wire: an added degree of safety in your diagnostic procedure.

References

Adler, Mortimer, "How to Mark a Book", *The McGraw-Hill Reader*, McGraw-Hill Book Company, New York, NY, 1982.

Appendix A

Flip-book animations

This appendix demonstrates certain principles through the use of "flip-book" animation. Each page of these appendix sections forms one frame of the "animation," viewed by rapidly flipping pages (if the book is printed), or rapidly clicking the "next page" button (if the book is viewed on a computer). While crude, this animation technique enjoys the benefits of low technology (it even works in paper form!) and convenient pausing at critical frames.

Enjoy!

A.1   Polyphase light bulbs animated

The key to understanding how three-phase electric motors work is to have an accurate mental picture of the *rotating magnetic field* created by the stator windings of a polyphase motor. One of the best ways to visualize this phenomenon is to observe a string of "chaser" lights blinking in a polyphase sequence. Just in case you don't happen to have a string of polyphase lights at your viewing convenience, I have provided a simulation here to demonstrate how the illusion of motion is created by the sequential energization of light bulbs.

Note how the lights appear to "move" from left to right as the energization sequences moves from A to C. After a few cycles of ABC, two of the wires are crossed to reverse phase sequence. This has the result of reversing the apparent direction of motion! It matters not which two phases are reversed. In this animation, I reverse phases A and B, but I could have just as well swapped phases B and C, or phases A and C, and created the exact same effect.

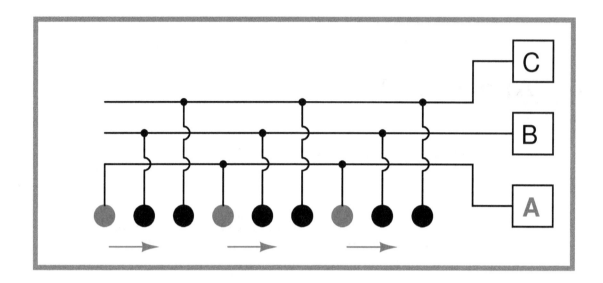

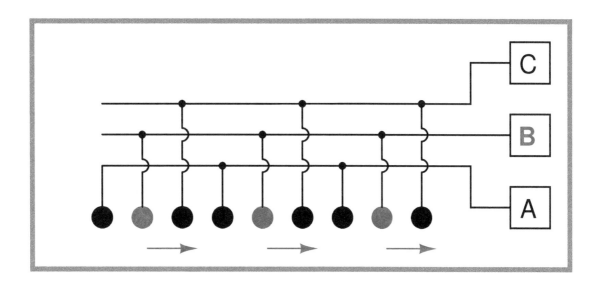

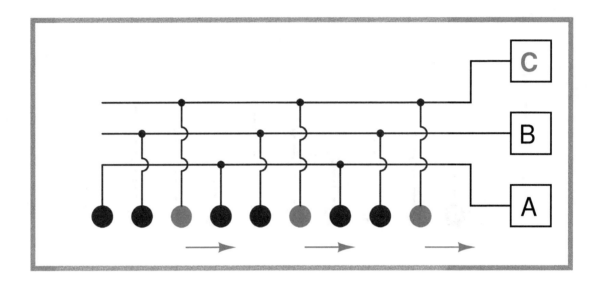

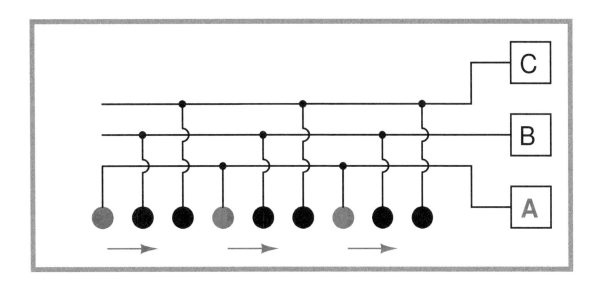

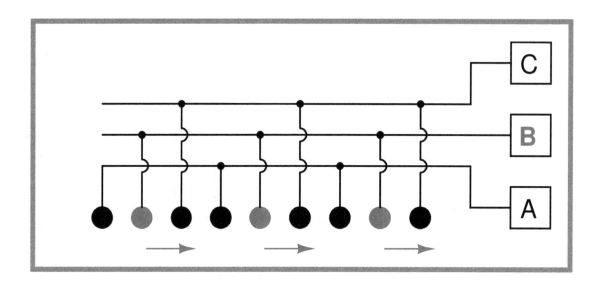

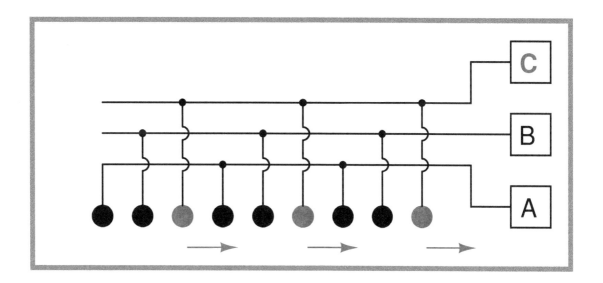

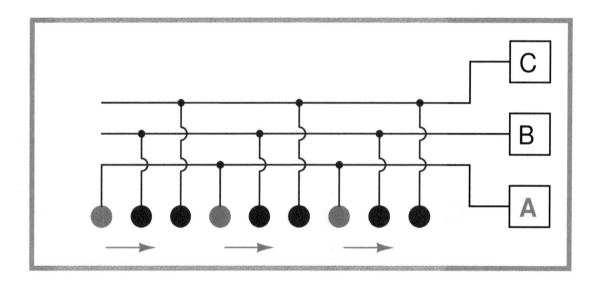

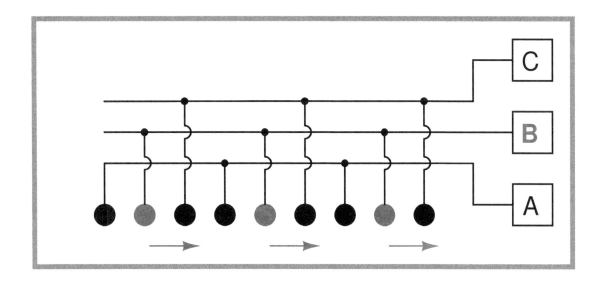

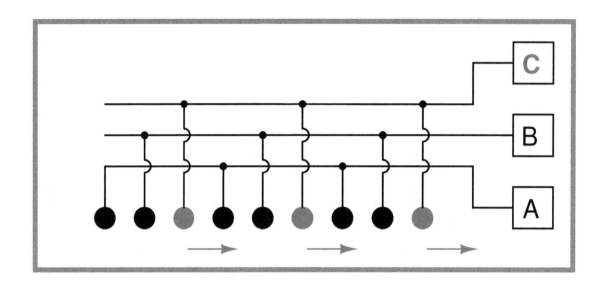

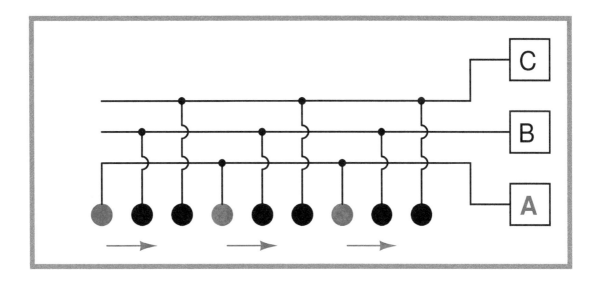

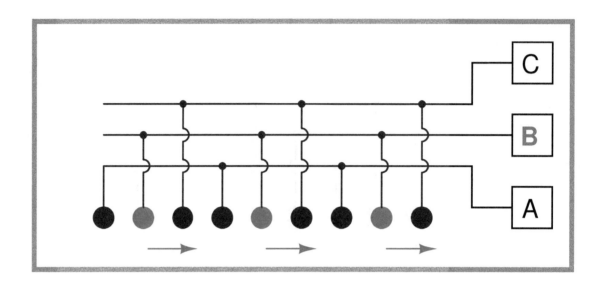

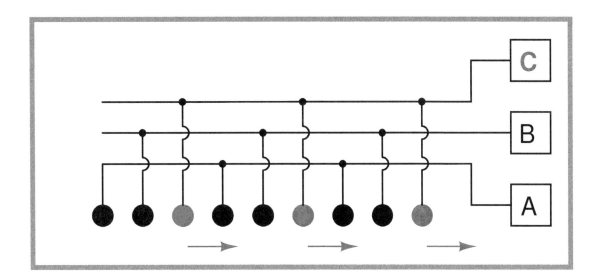

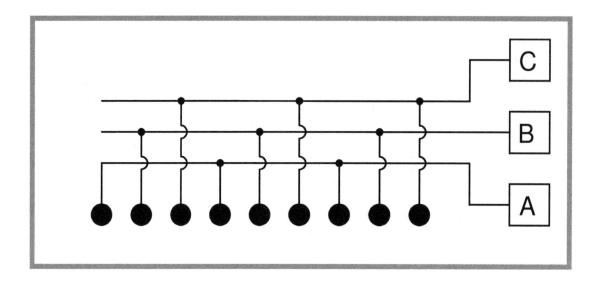

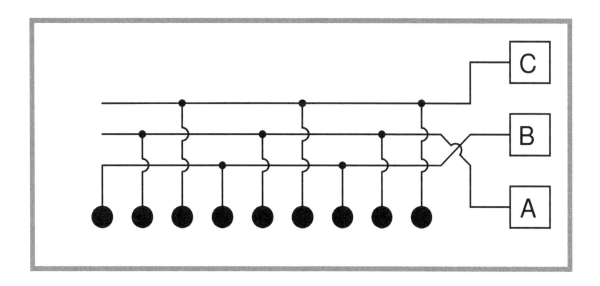

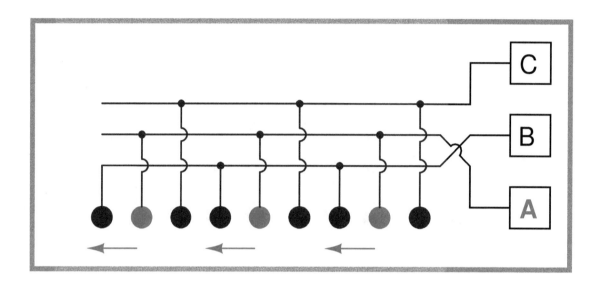

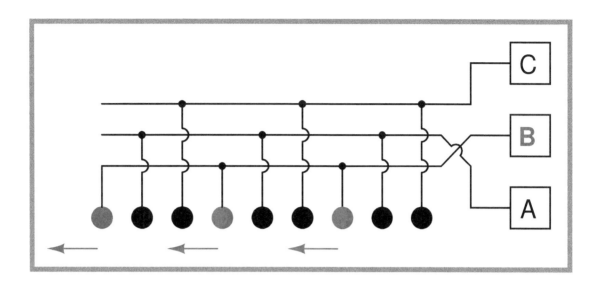

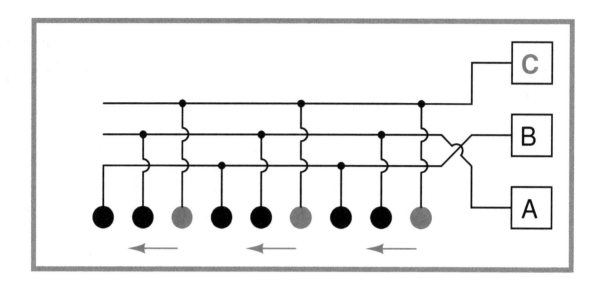

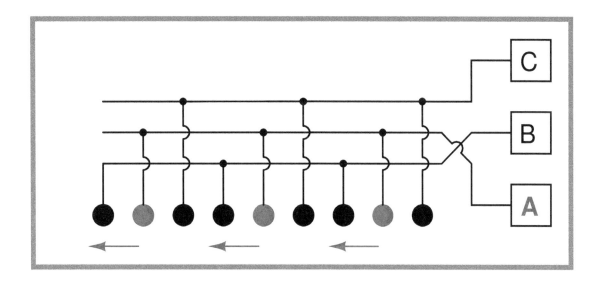

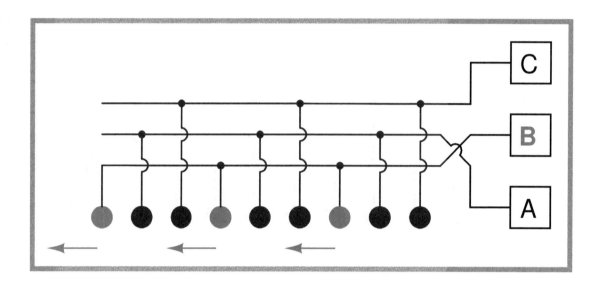

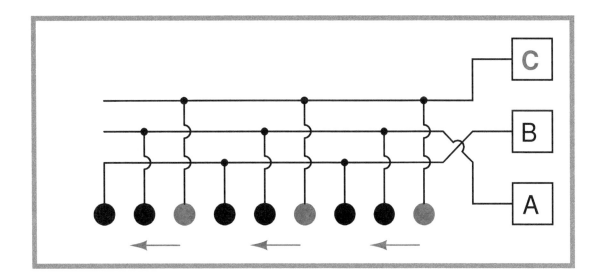

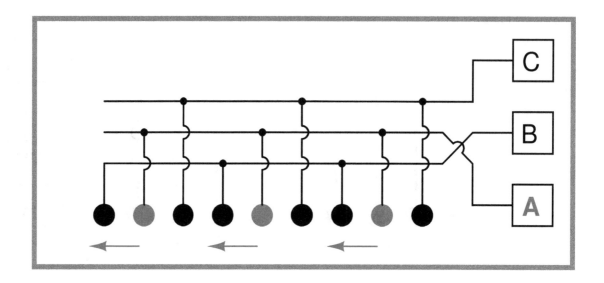

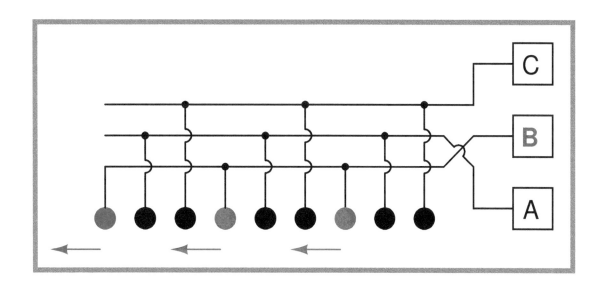

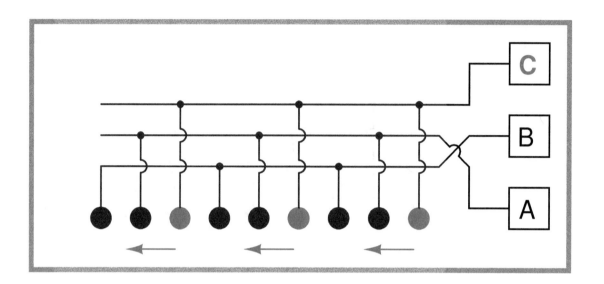

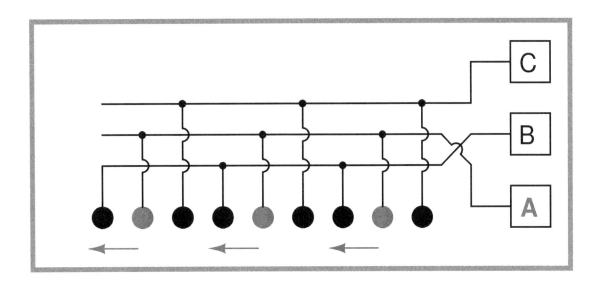

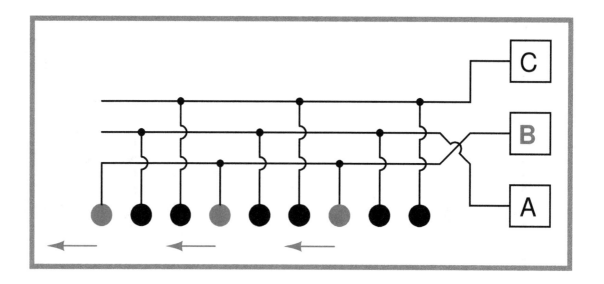

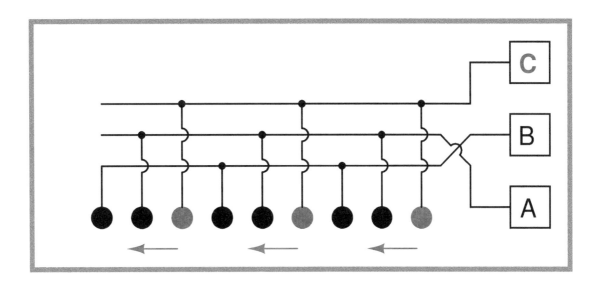

A.2   Polyphase induction motor animated

The following animation shows how the "rotating" magnetic field of a three-phase AC induction motor is produced by the interaction of three stator winding sets energized with different phases (A, B, and C) of a three-phase AC power source. A red arrow shows the direction of the *resultant* magnetic field created by the interaction of the three winding sets.

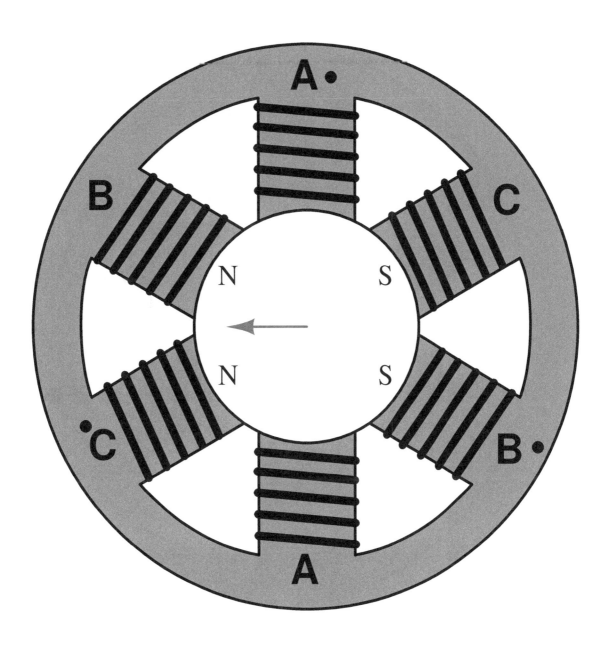

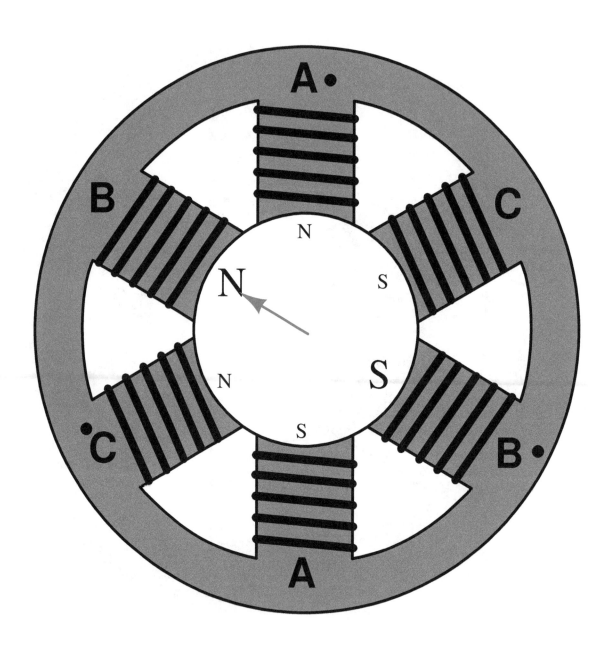

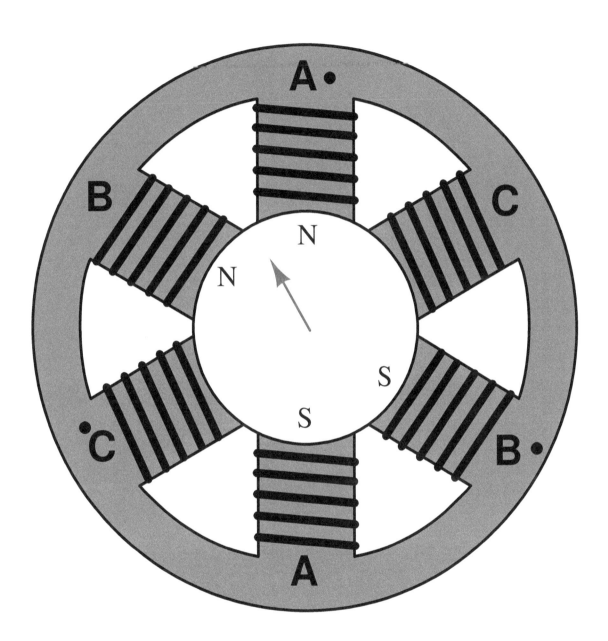

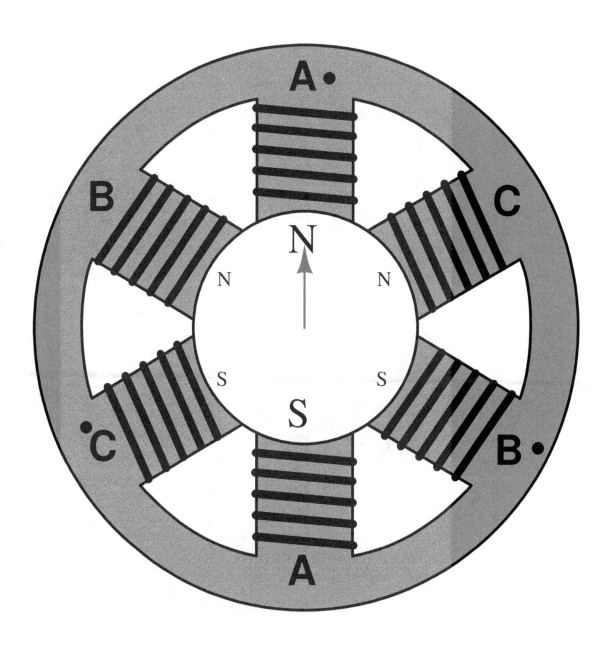

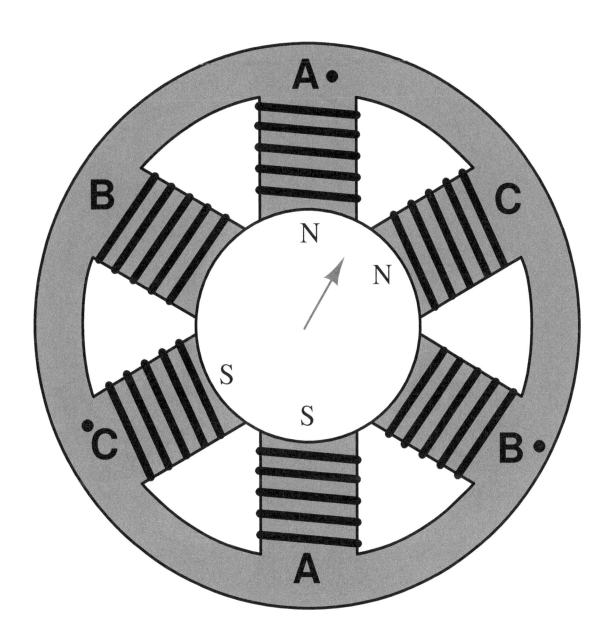

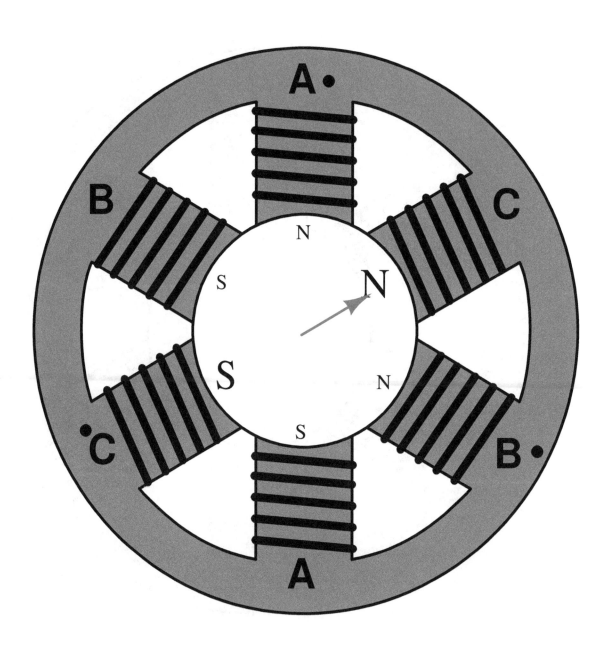

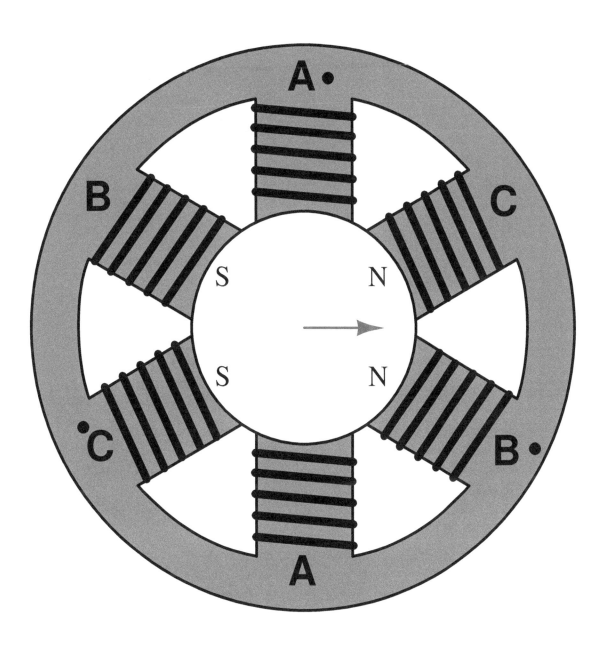

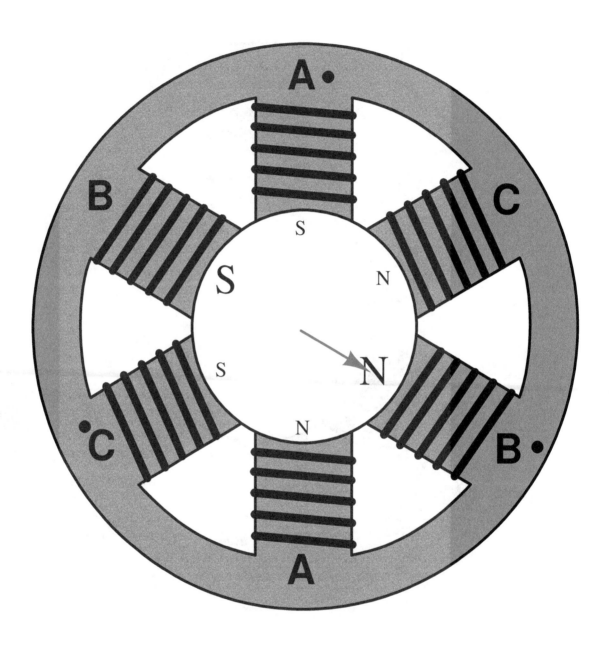

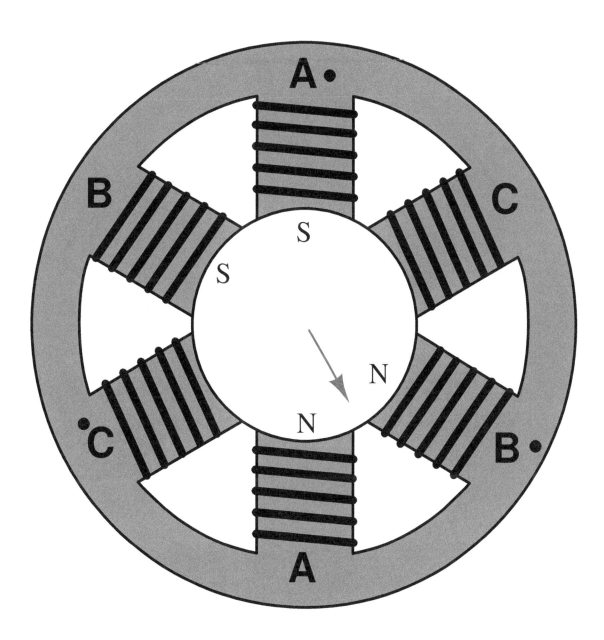

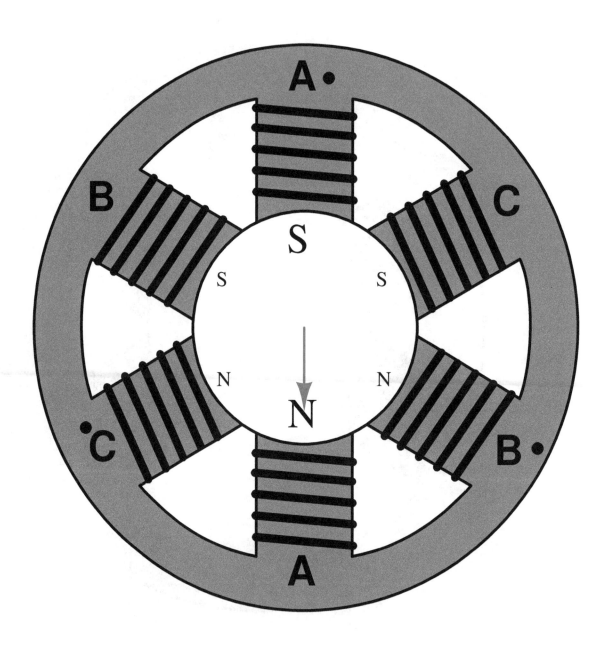

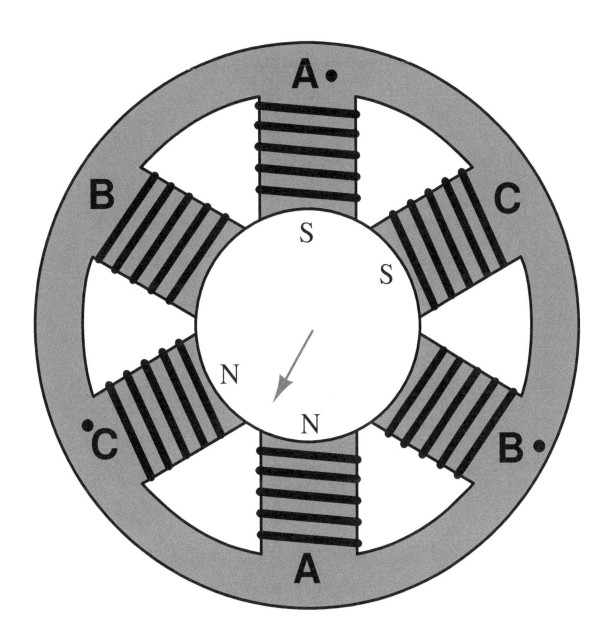

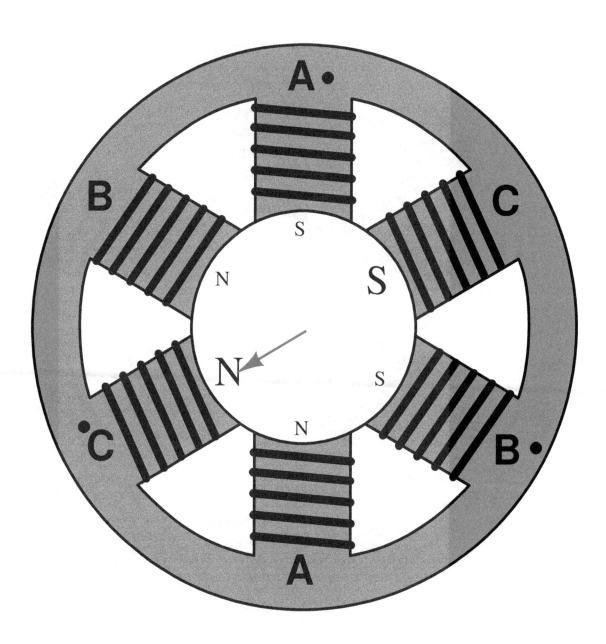

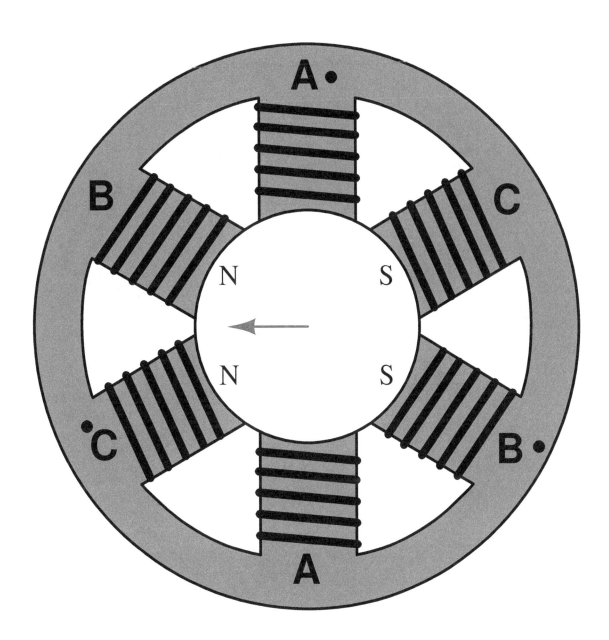

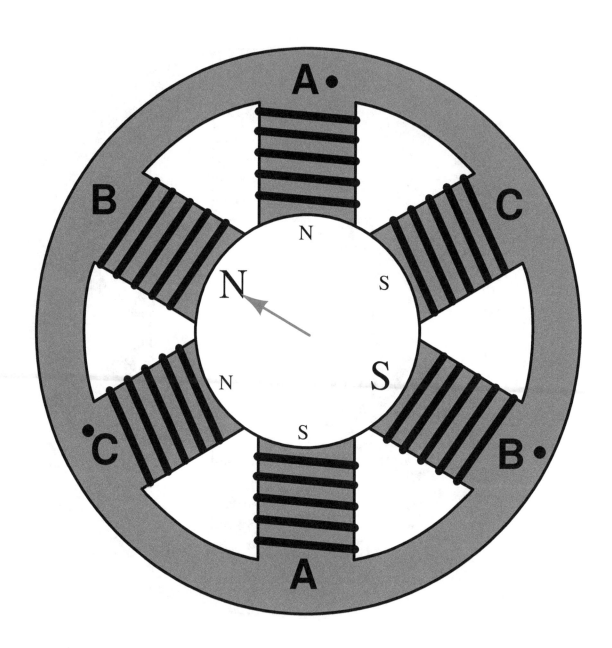

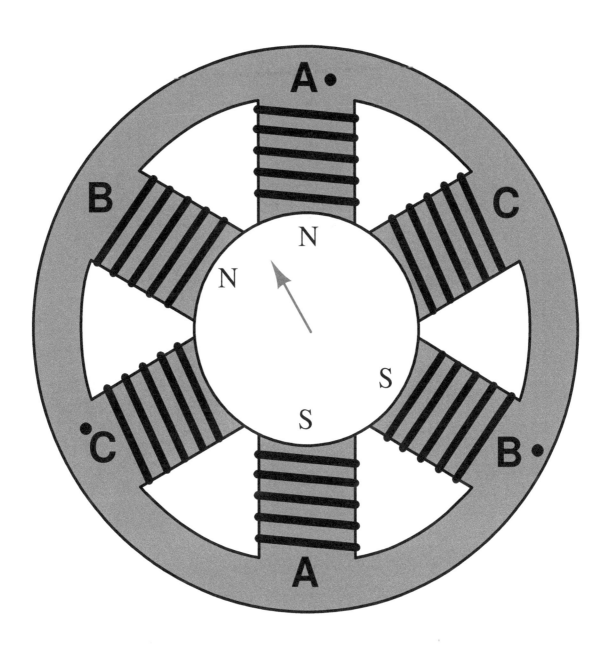

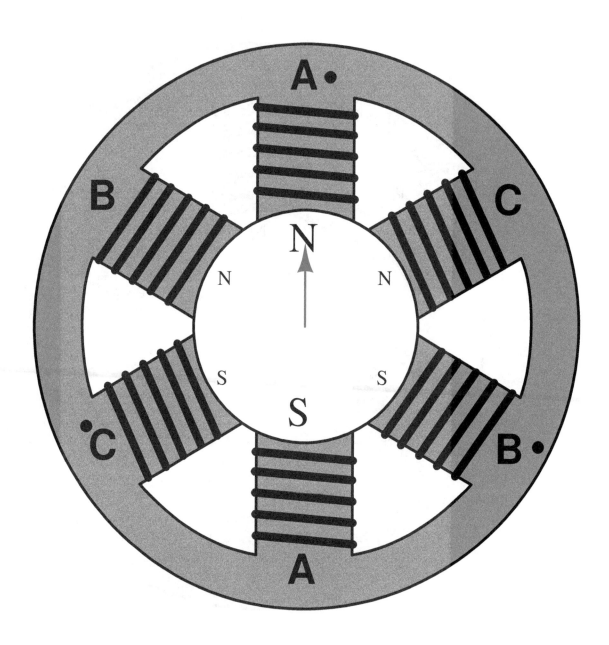

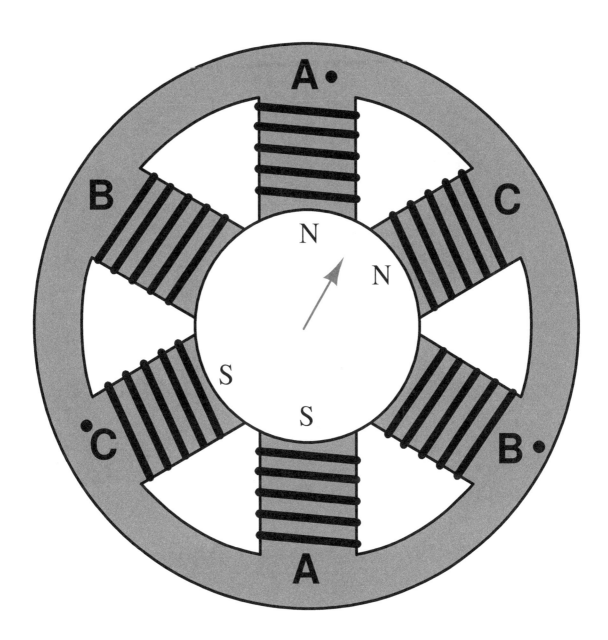

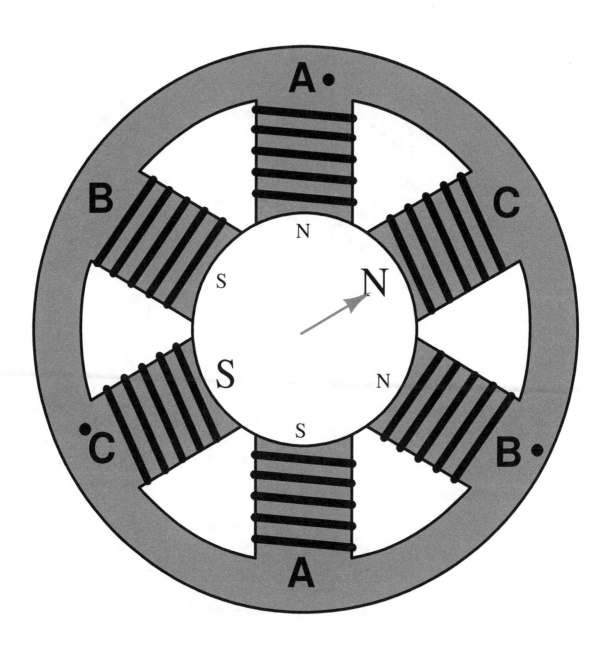

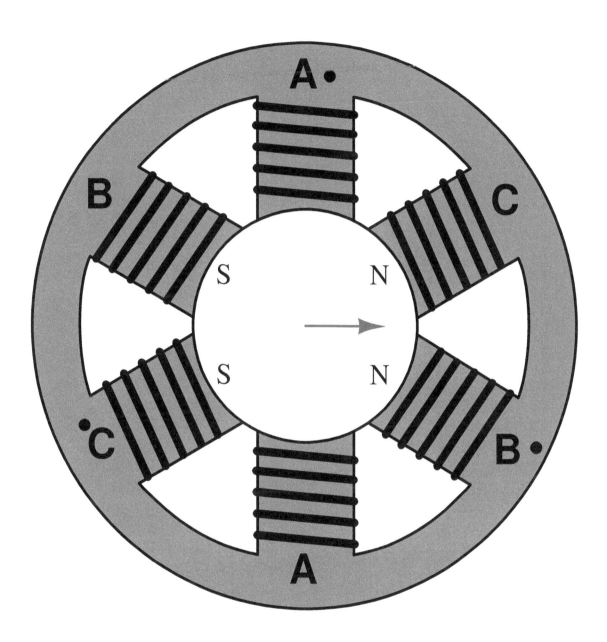

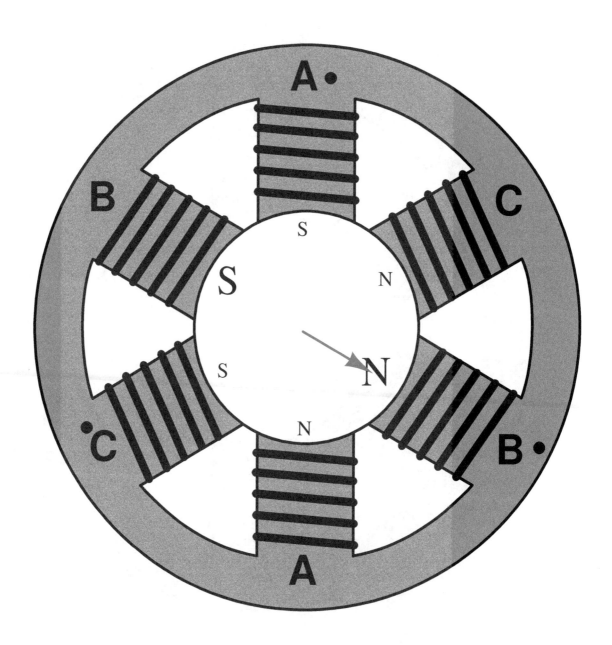

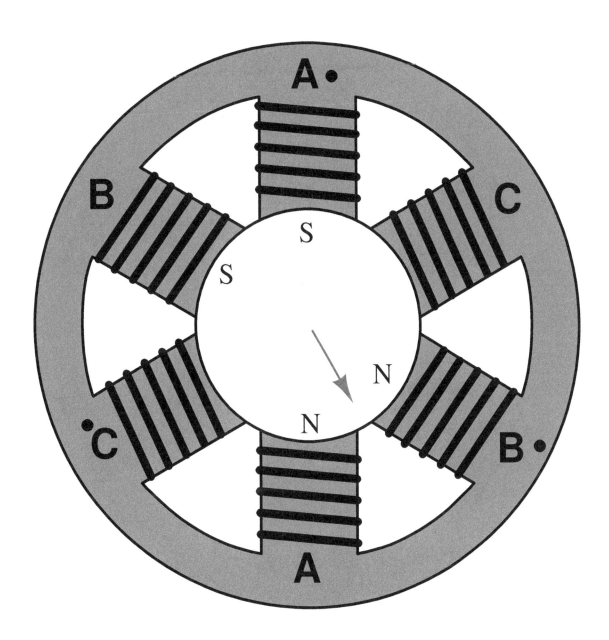

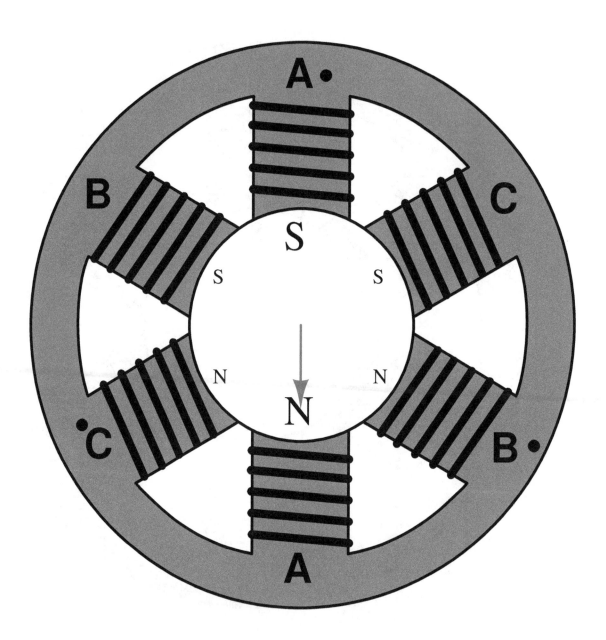

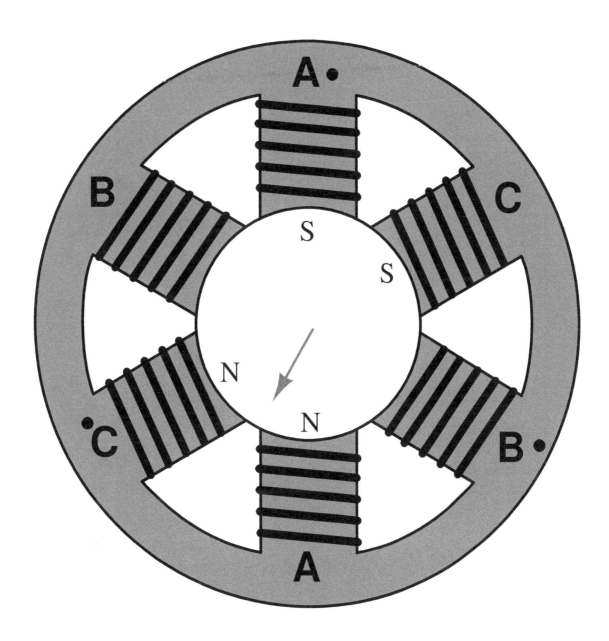

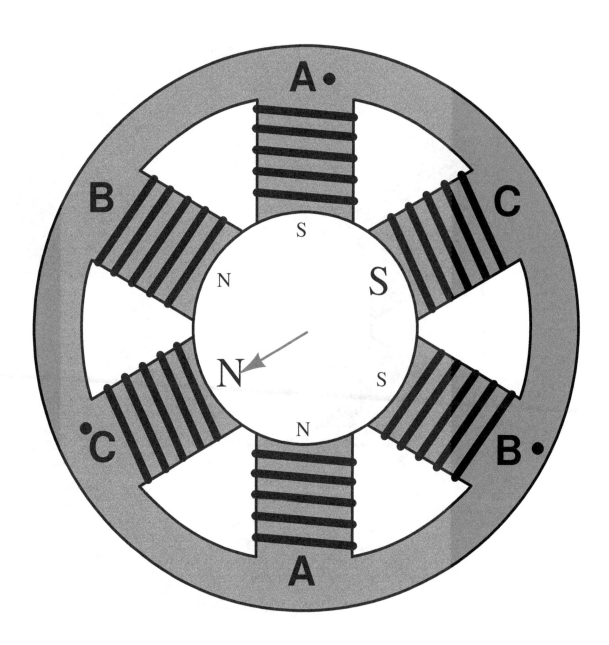

A.3   Rotating phasor animated

The following animation shows a rotating phasor in three-dimensional form. The phasor rotates in a complex plane (with real and imaginary axes), but travels linearly along the time axis. In doing so it traces a path that looks like a circle when viewed along the centerline (time axis) but looks like a sinusoidal wave when viewed from above or along the side.

Euler's Relation describes the phasor's position in the complex plane:

$$e^{j\omega t} = \cos \omega t + j \sin \omega t$$

Where,

e = Euler's number (approximately equal to 2.718281828)

ω = Angular velocity, in radians per second

t = Time, in seconds

$\cos \omega t$ = Horizontal projection of phasor (along a real number line) at time t

$j \sin \omega t$ = Vertical projection of phasor (along an imaginary number line) at time t

If you imagine the phasor's length either growing or decaying exponentially over time, the result will be a spiral that either widens like a horn or shrinks like a funnel. This would be a visualization of a *complex* exponential, where the s variable defines both the rate of growth/decay (the envelope of the spiral) and the angular velocity (the pitch of the spiral):

$$e^{st} = e^{(\sigma + j\omega)t} = e^{\sigma t} e^{j\omega t}$$

Where,

s = Complex growth/decay rate and frequency (sec^{-1})

$\sigma = \frac{1}{\tau}$ = Real growth/decay rate (time constants per second, or sec^{-1})

$j\omega$ = Imaginary frequency (radians per second, or sec^{-1})

t = Time (seconds)

The example of a unit-length rotating phasor is nothing more than a special case of the complex exponential, where $\sigma = 0$ (i.e. there is no growth or decay over time):

$$e^{st} = e^{(0 + j\omega)t} = e^0 e^{j\omega t} = e^{j\omega t}$$

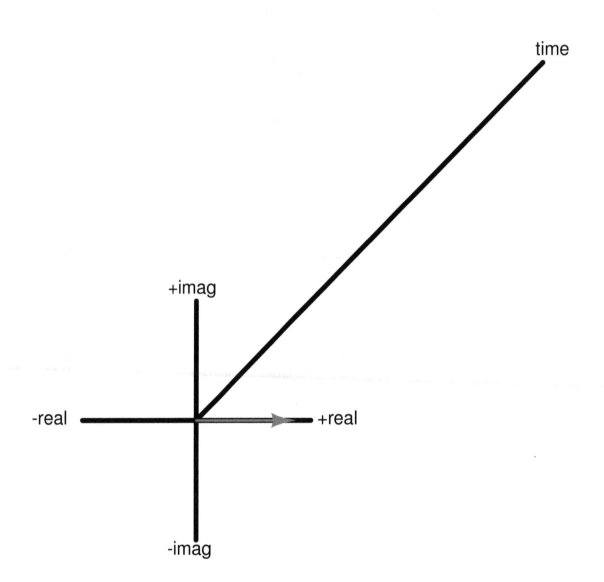

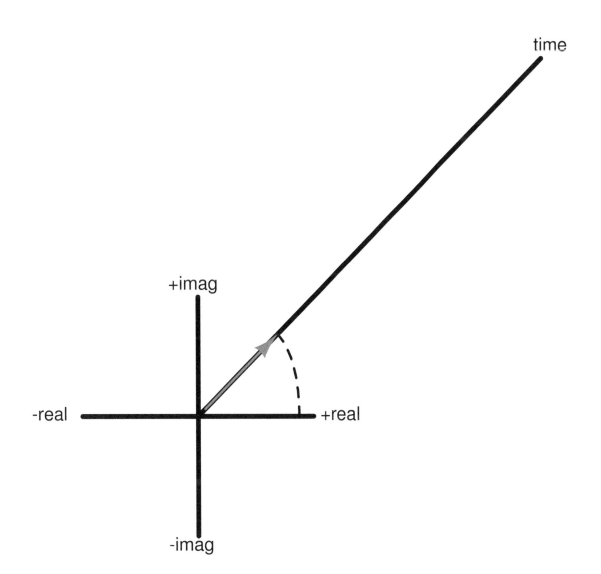

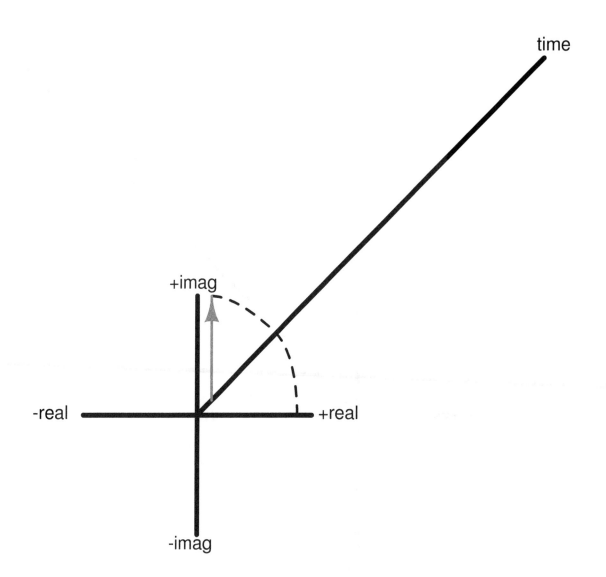

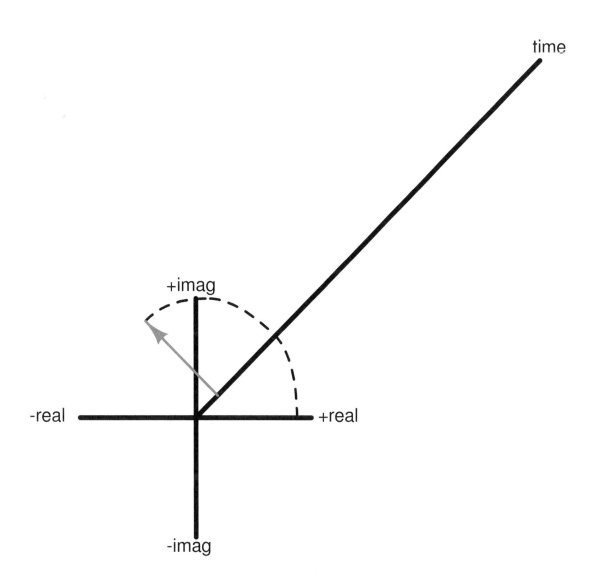

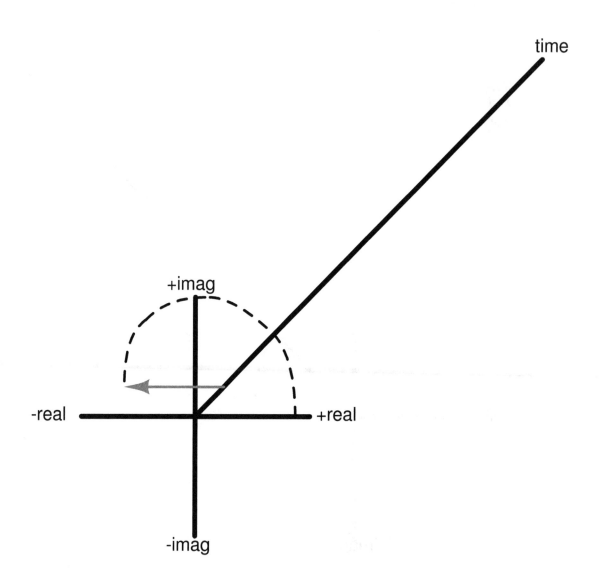

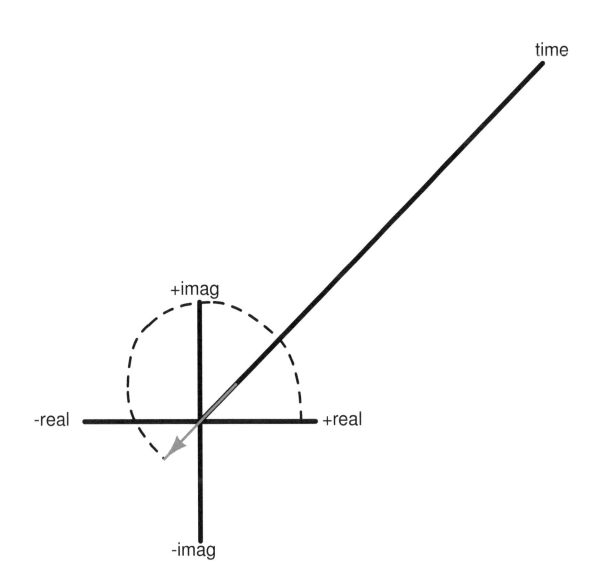

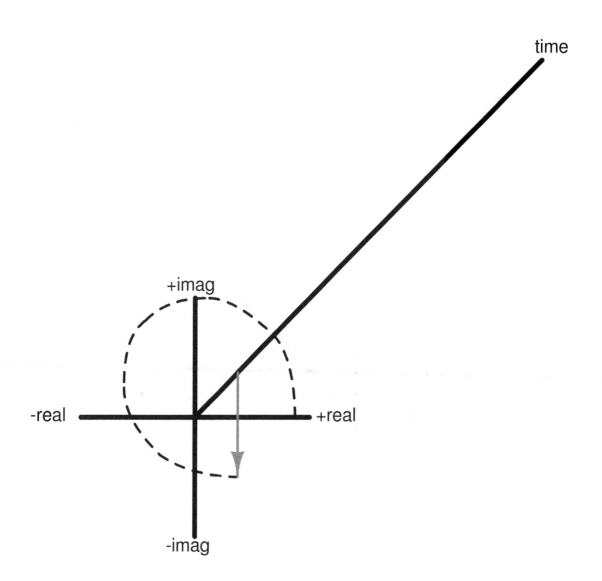

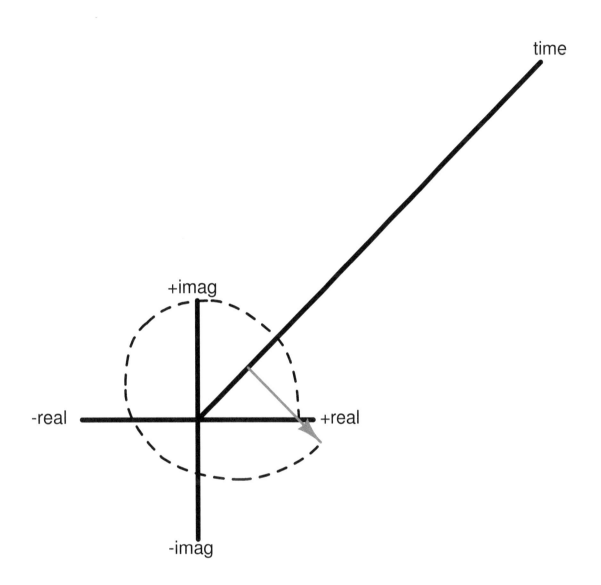

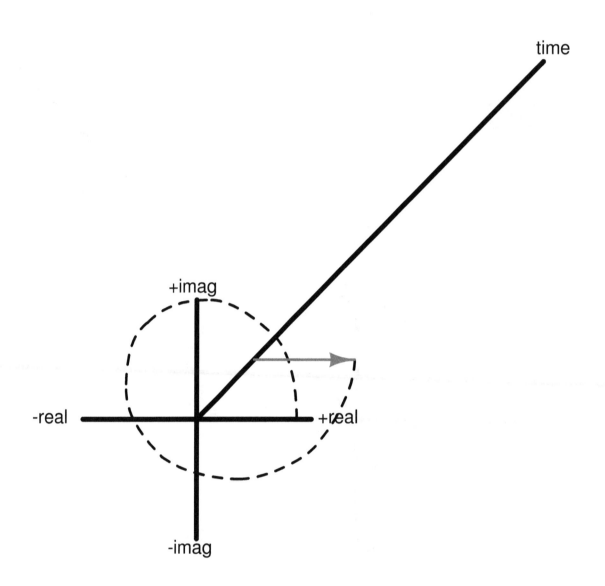

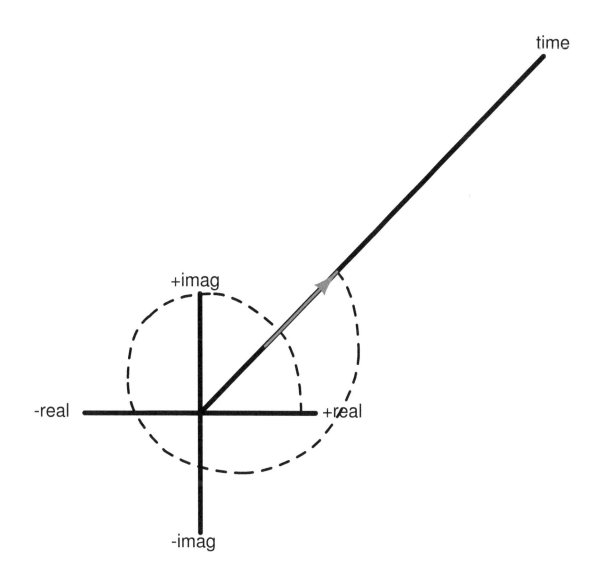

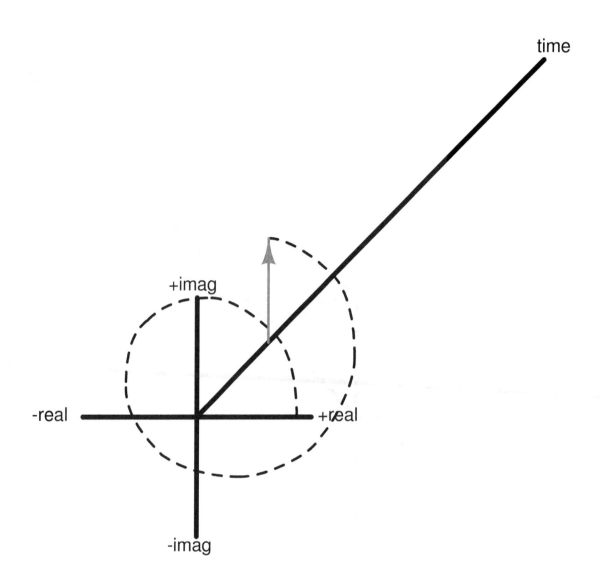

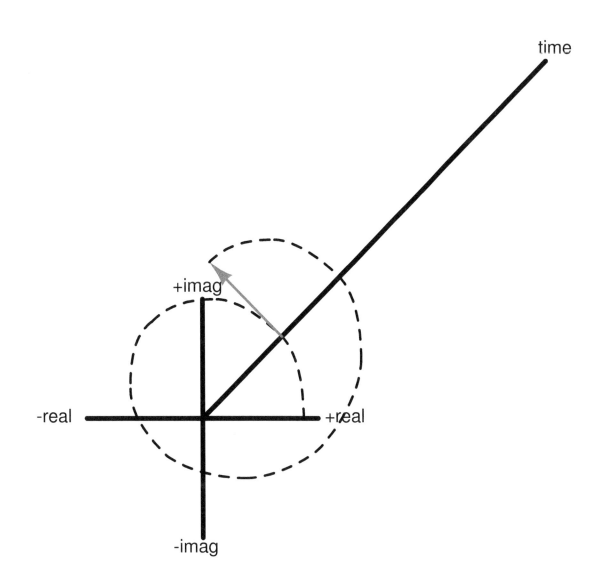

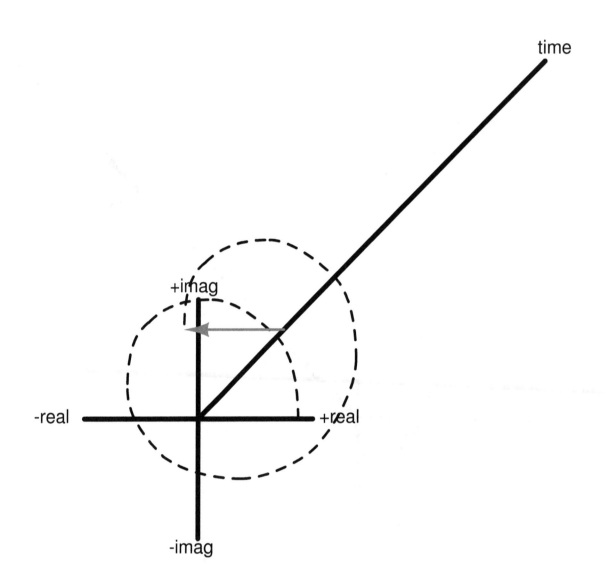

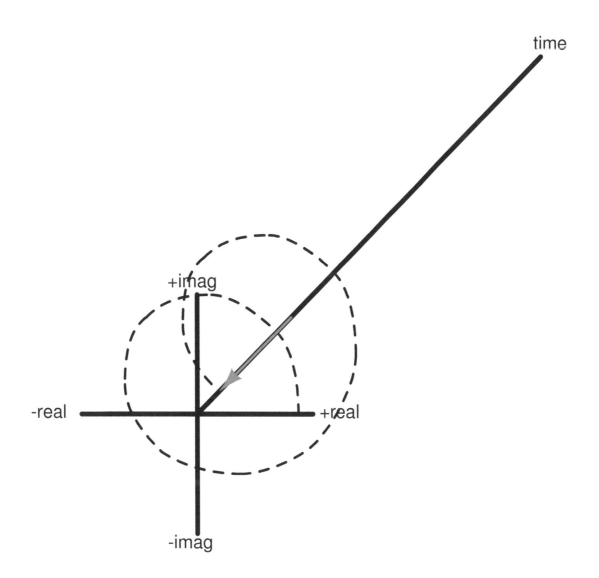

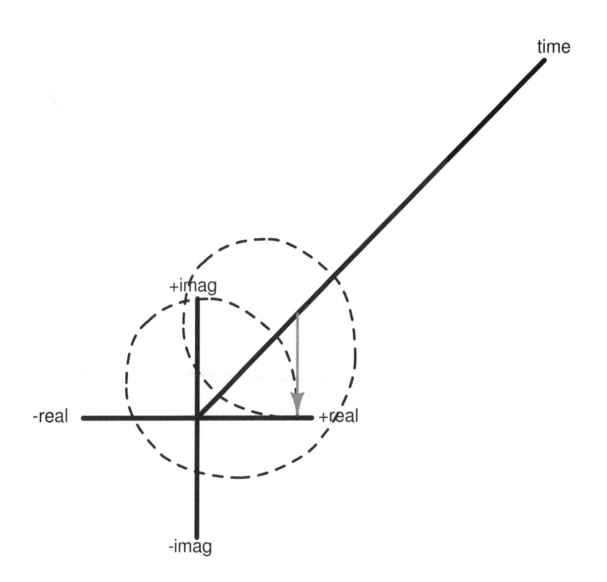

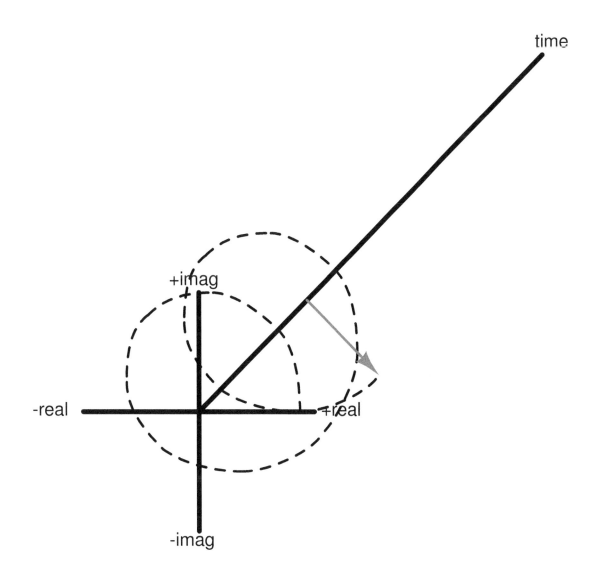

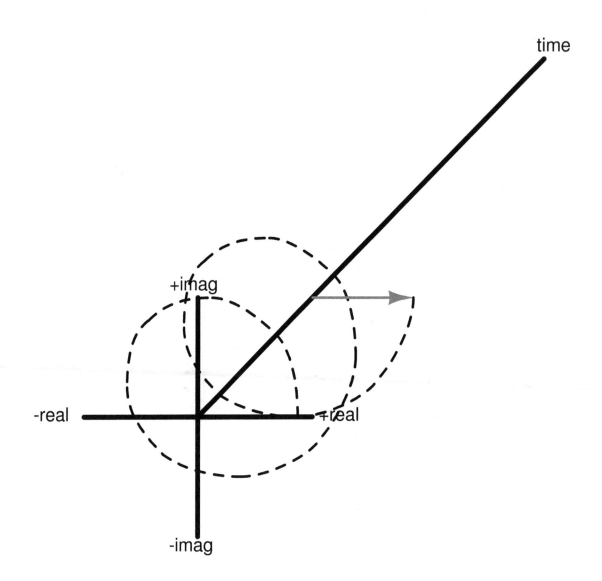

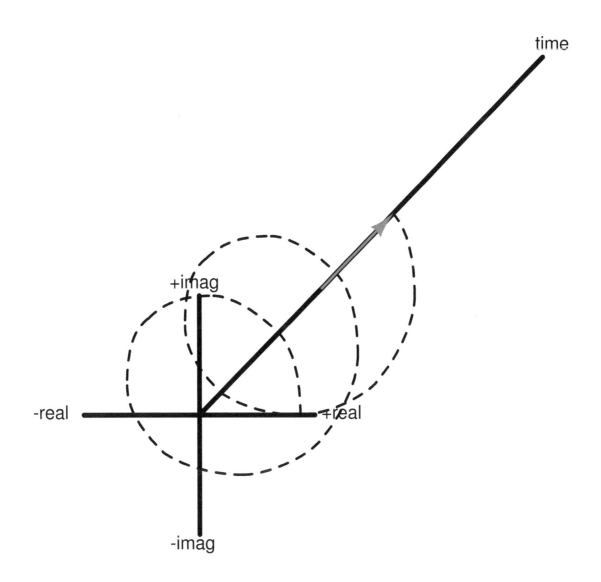

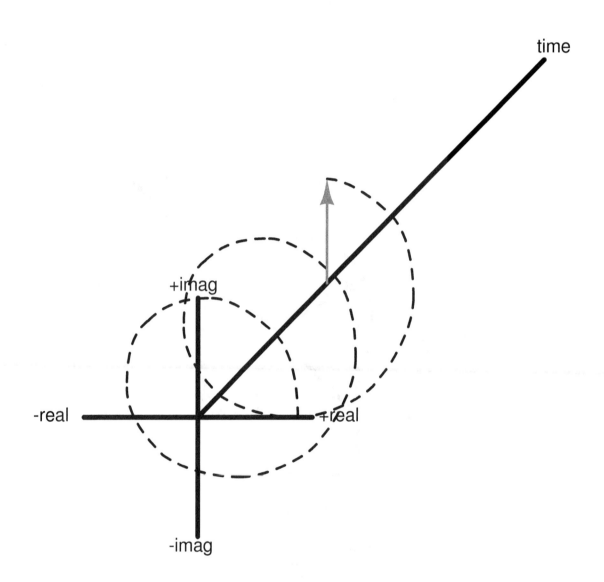

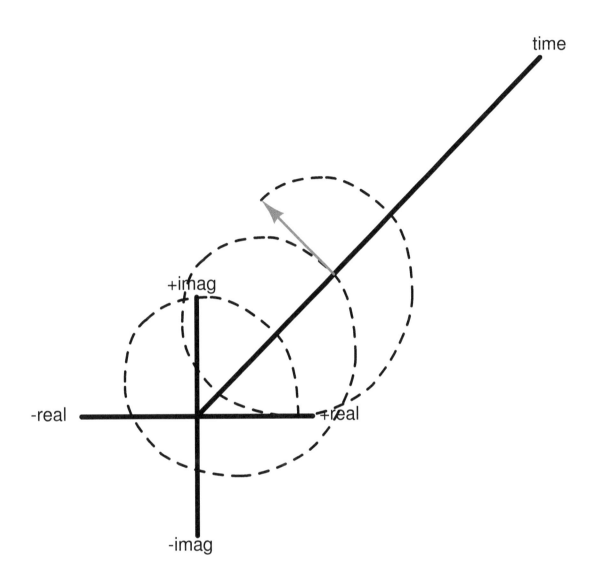

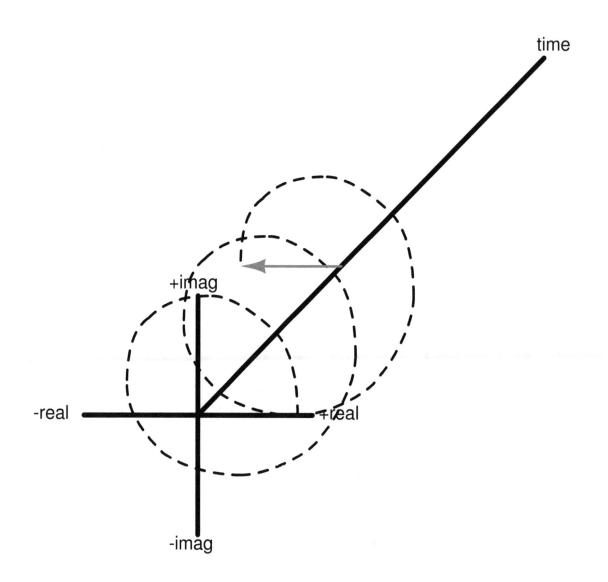

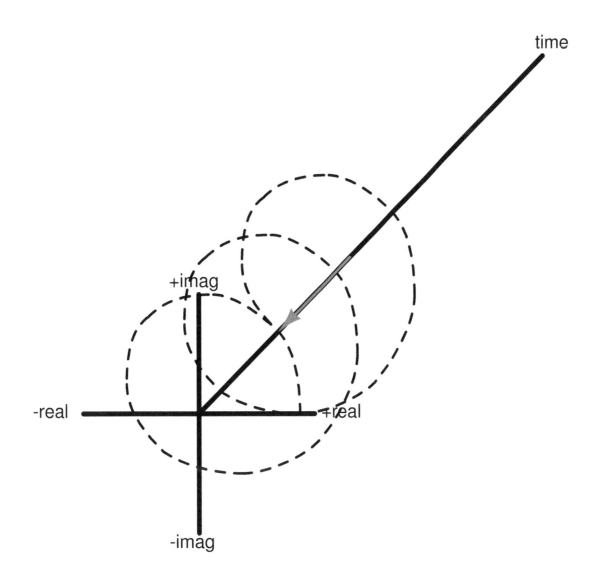

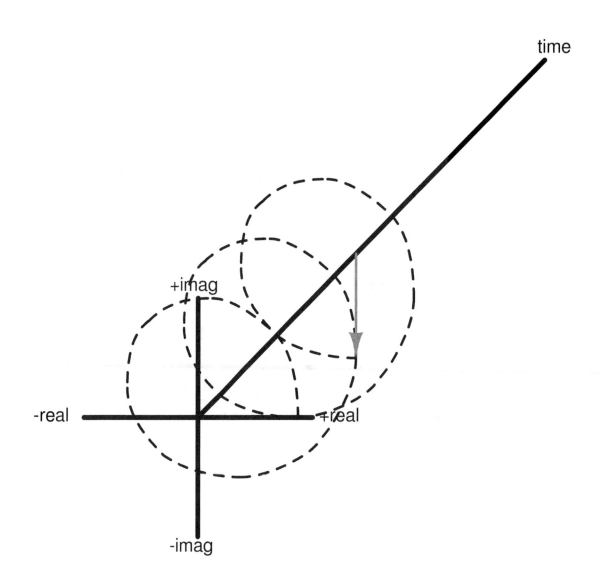

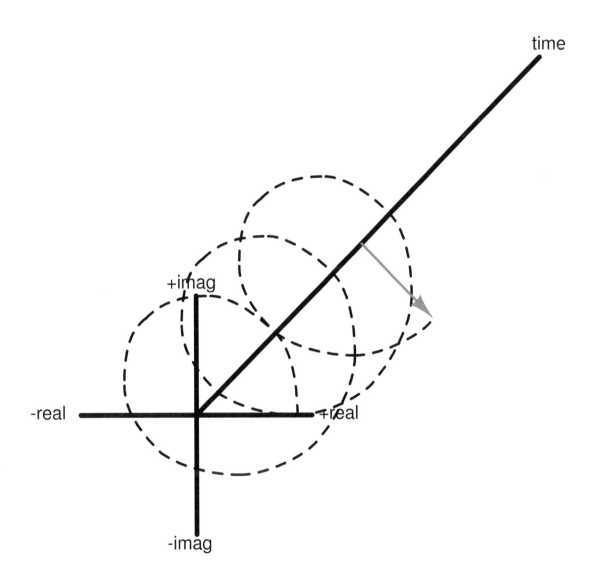

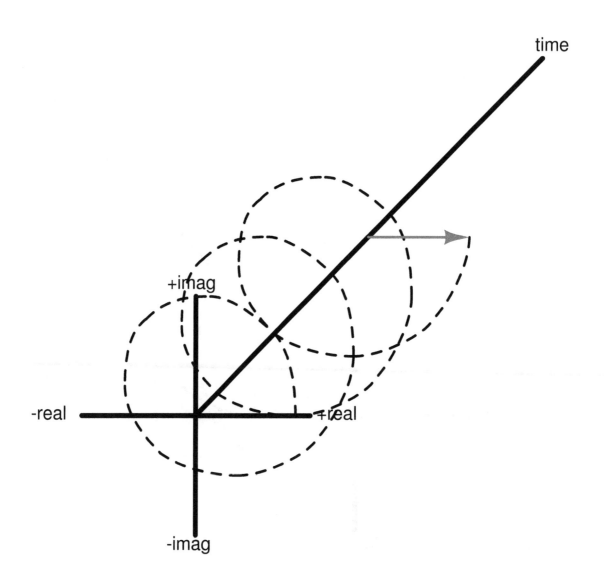

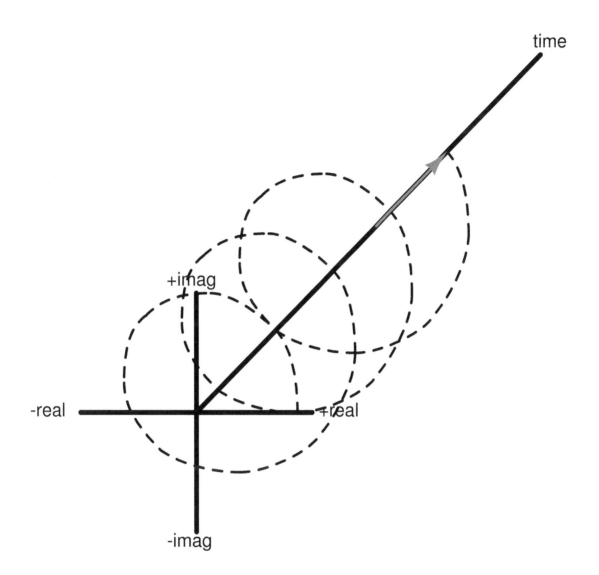

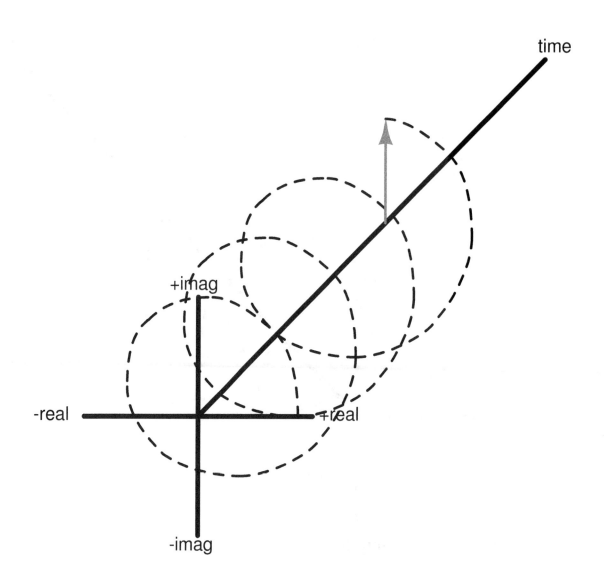

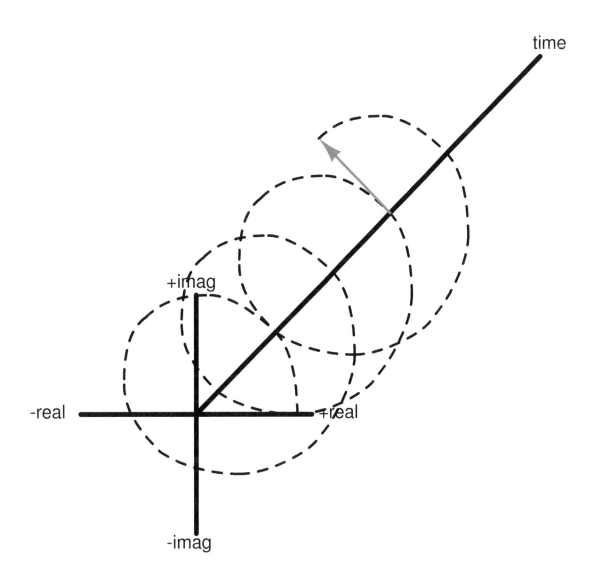

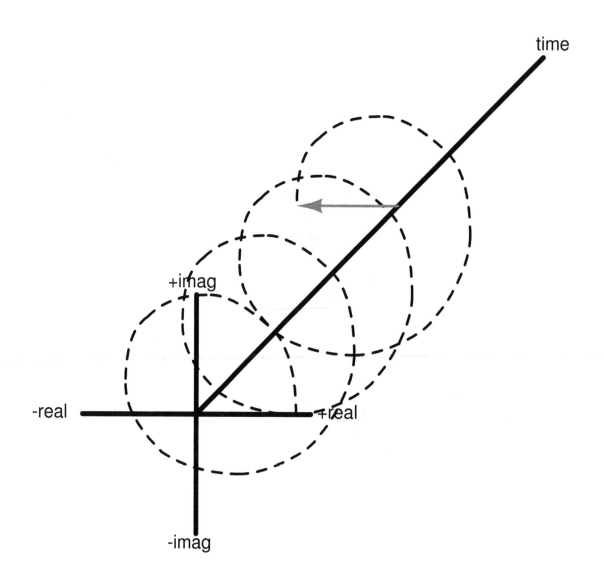

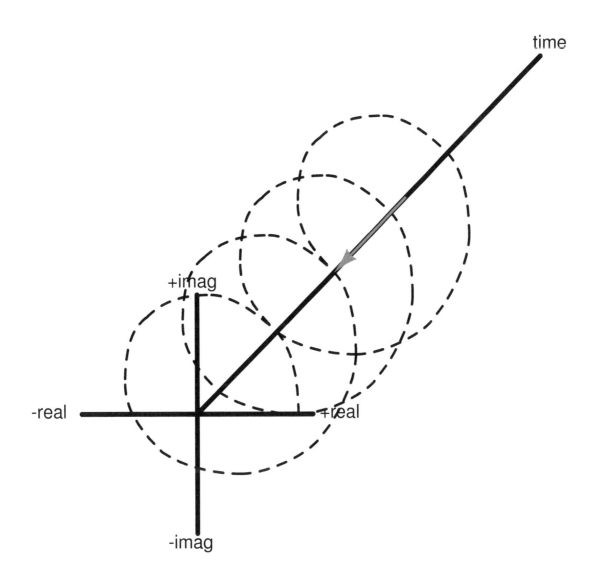

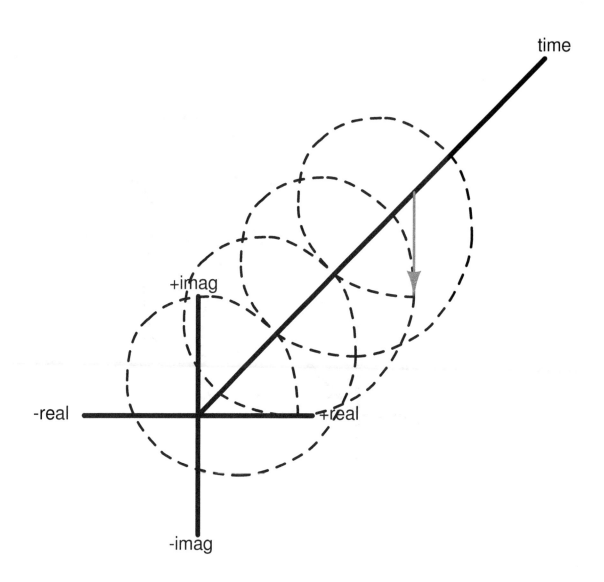

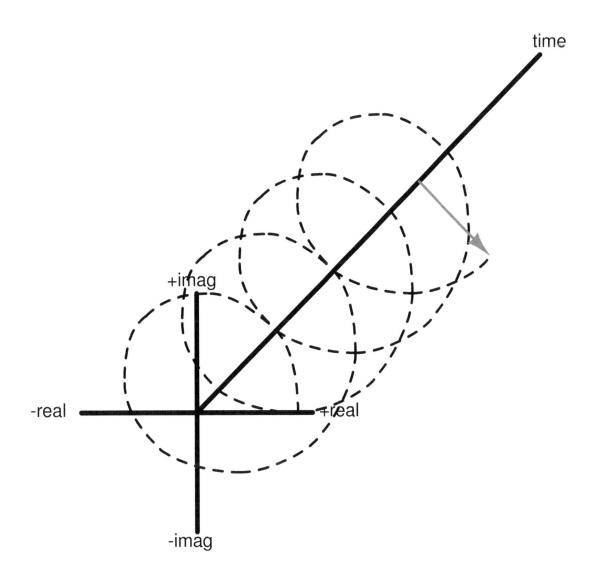

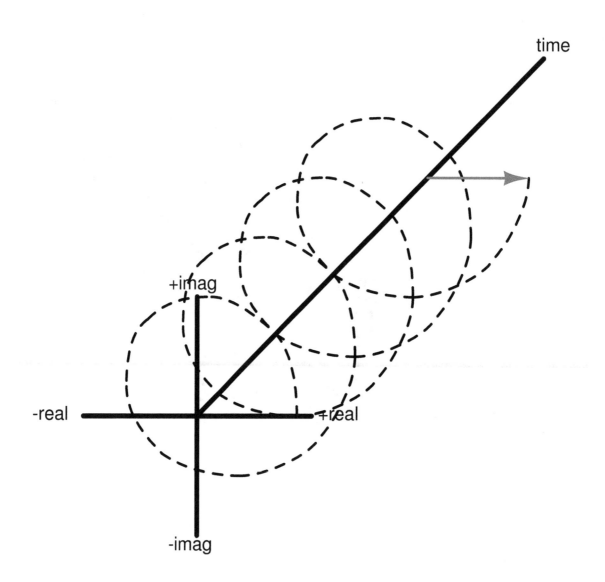

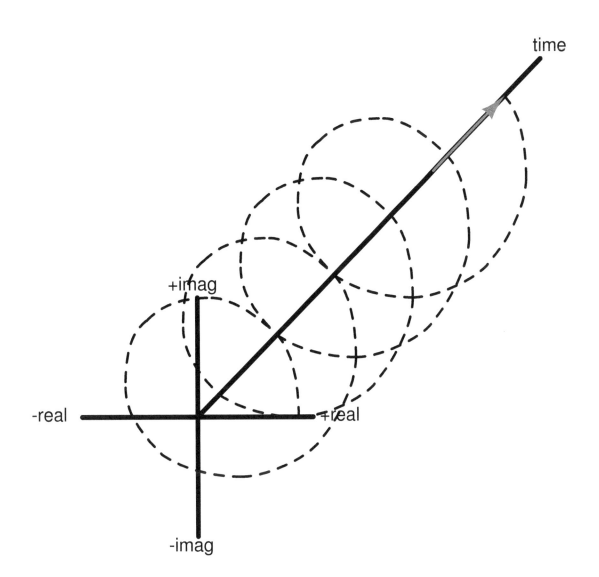

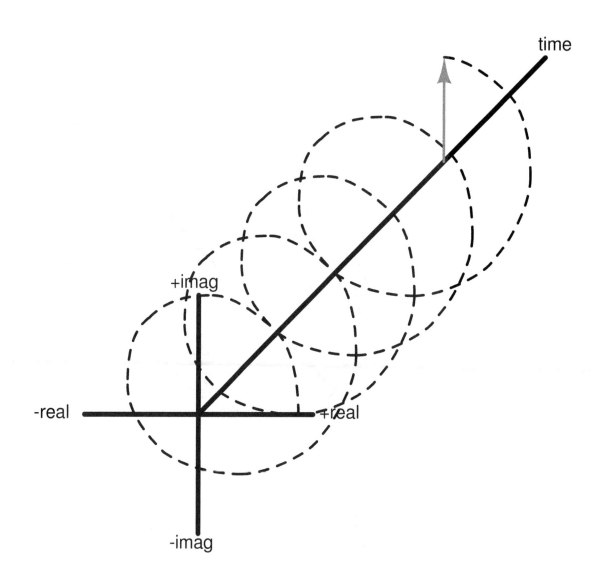

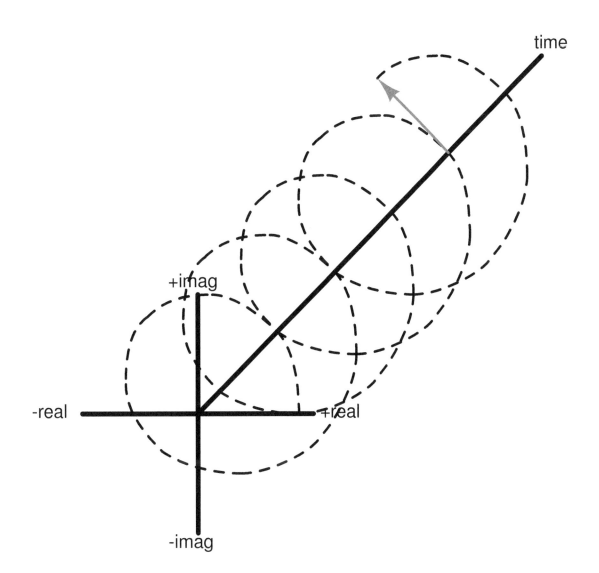

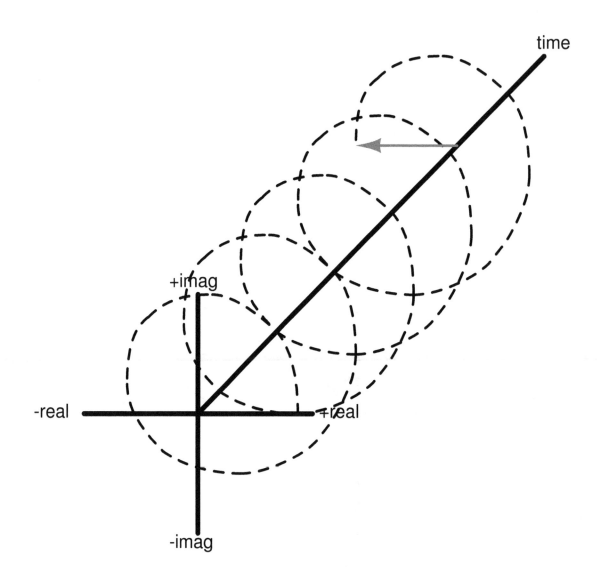

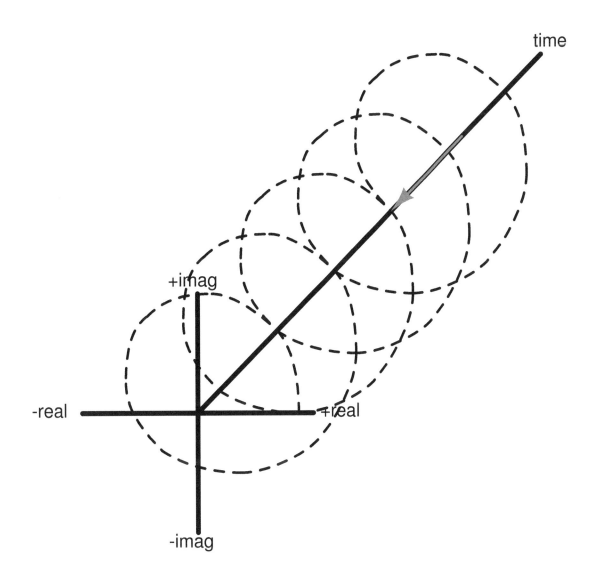

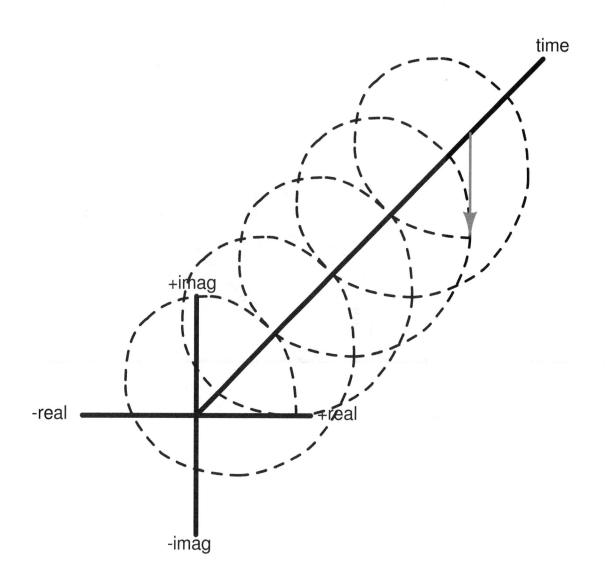

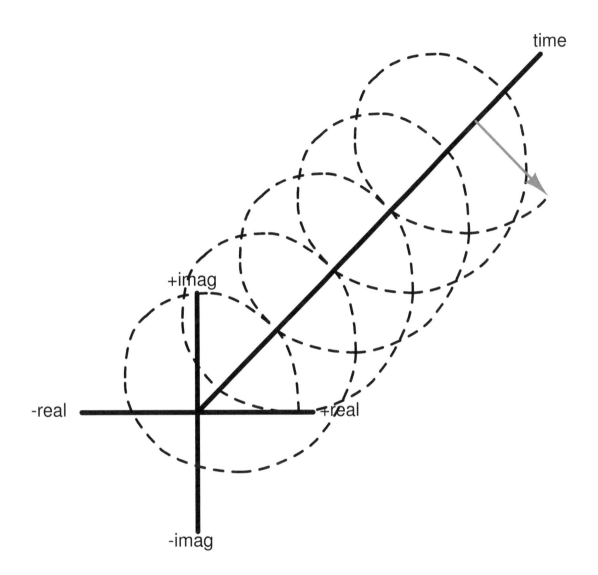

A.4 Differentiation and integration animated

The following animation shows the calculus concepts of differentiation and integration (with respect to time) applied to the filling and draining of a water tank.

The animation shows two graphs relating to the water storage tank: one showing the volume of stored water in the tank (V) and the other showing volumetric flow rate in and out of the tank (Q). We know from calculus that volumetric flow rate is the *time-derivative* of volume:

$$Q = \frac{dV}{dt}$$

We also know that change in volume is the *time-integral* of volumetric flow rate:

$$\Delta V = \int_{t_0}^{t_1} Q \, dt$$

Thus, the example of a water storage tank filling and draining serves to neatly illustrate both concepts in relation to each other.

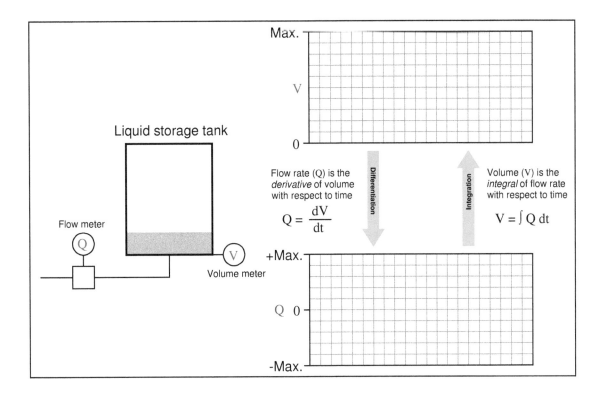

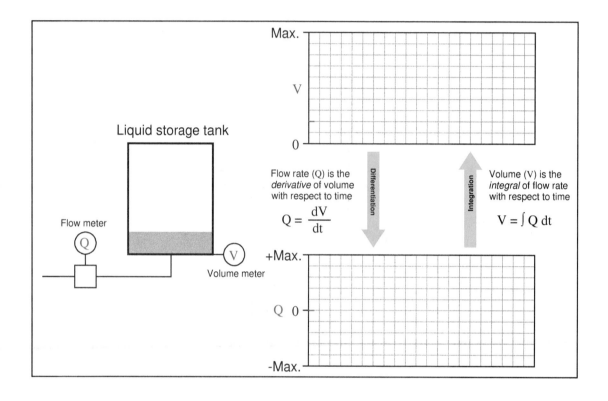

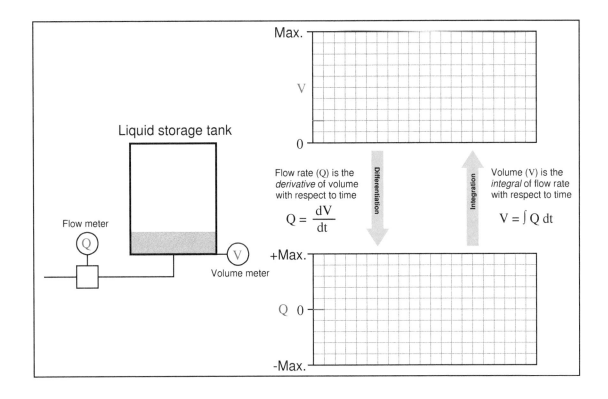

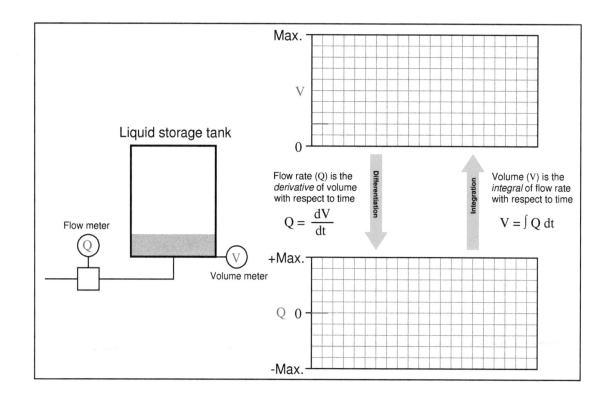

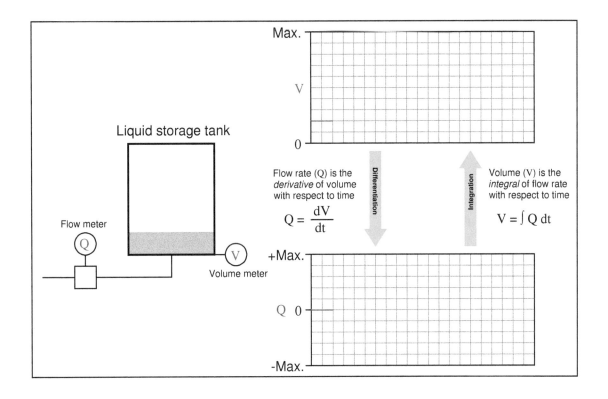

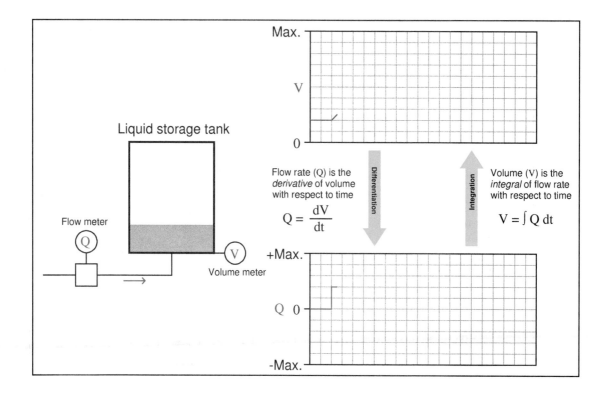

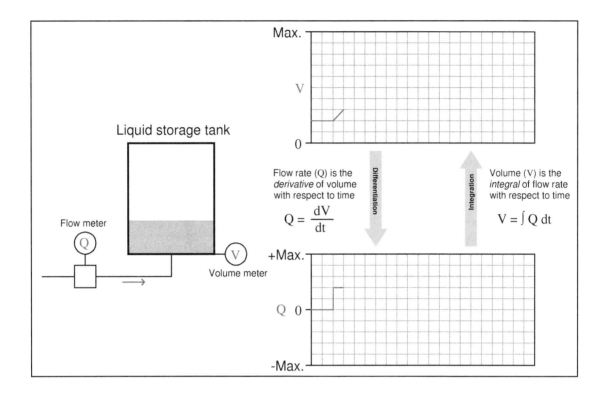

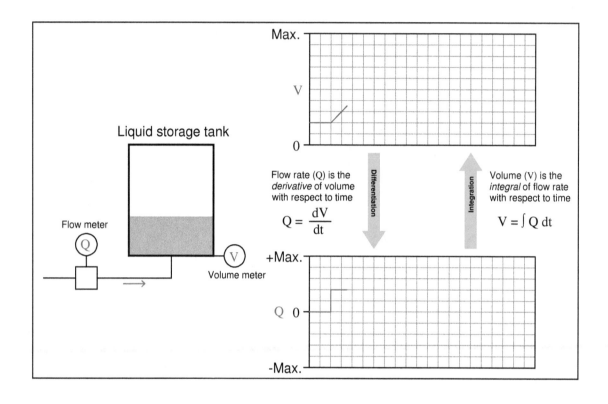

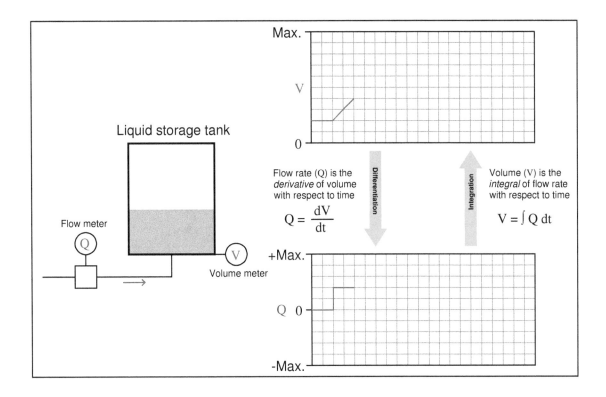

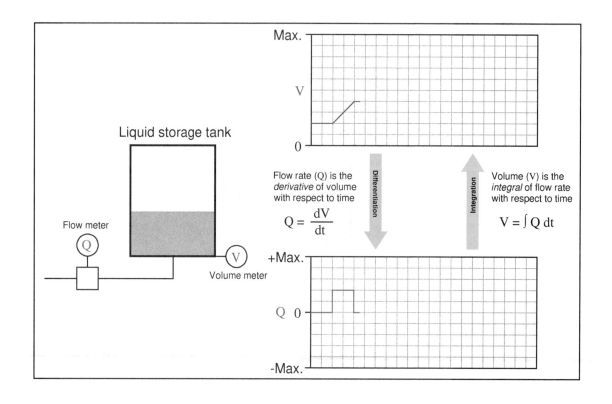

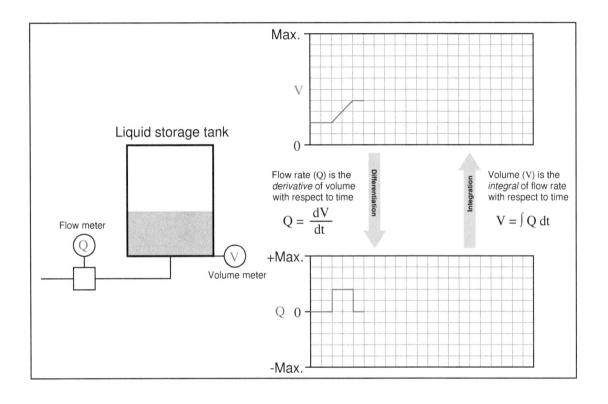

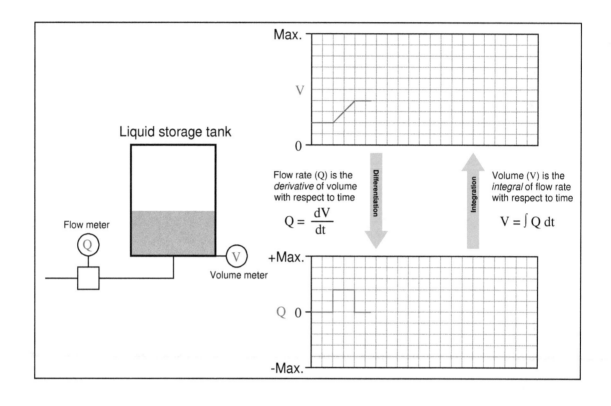

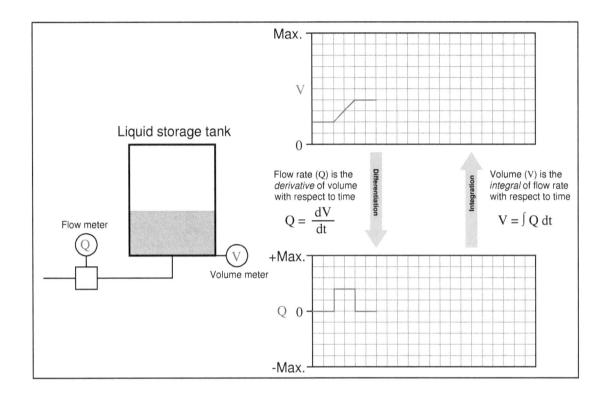

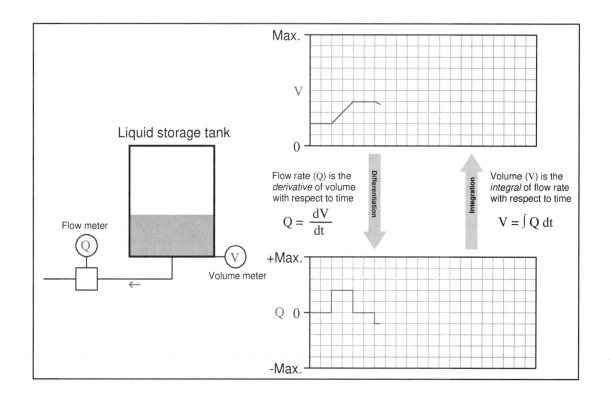

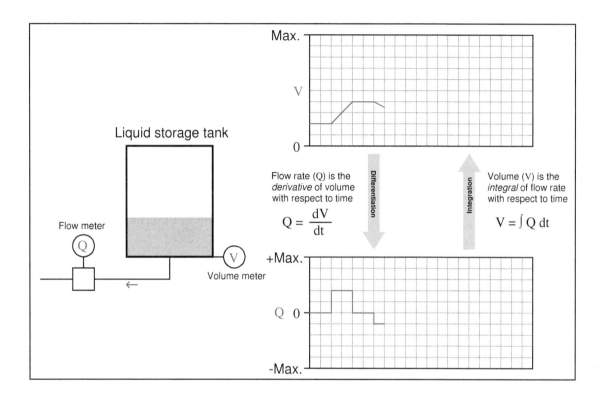

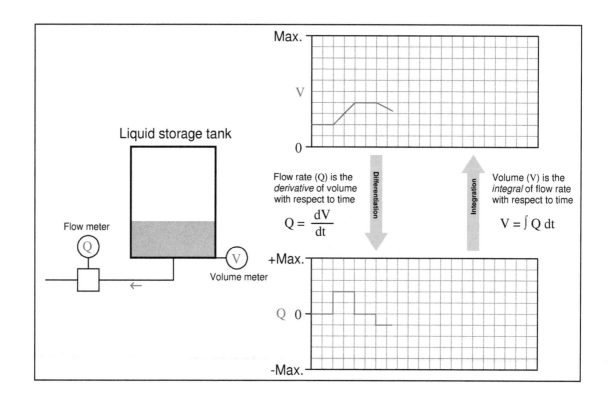

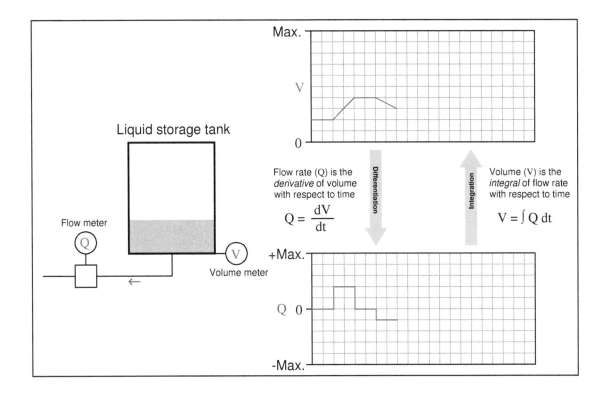

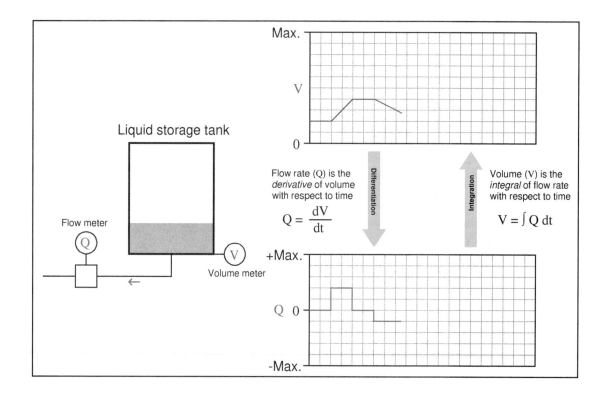

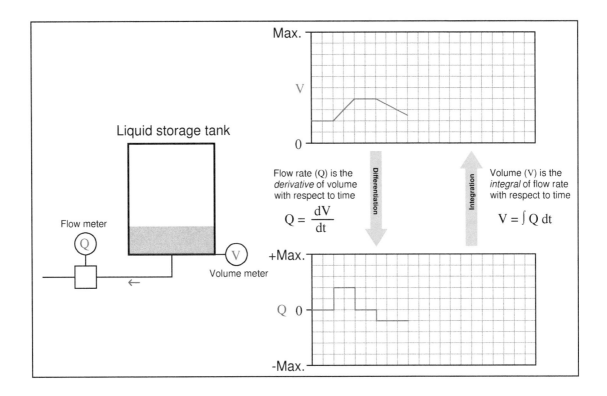

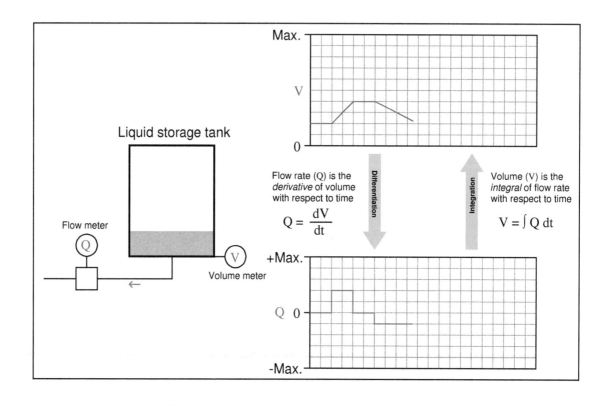

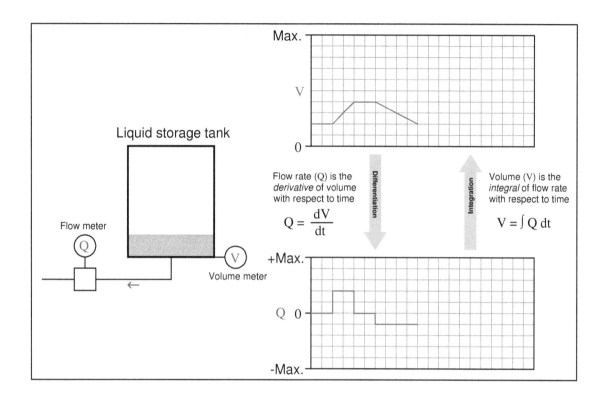

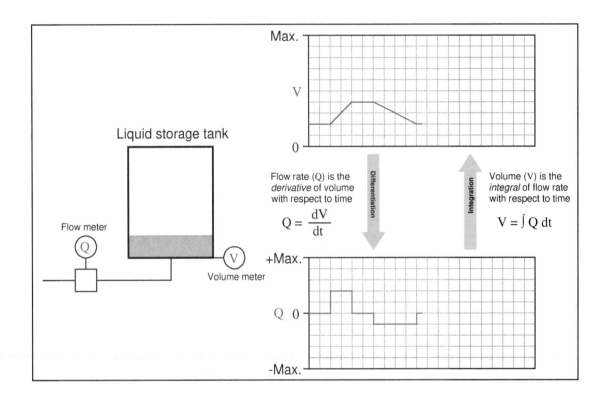

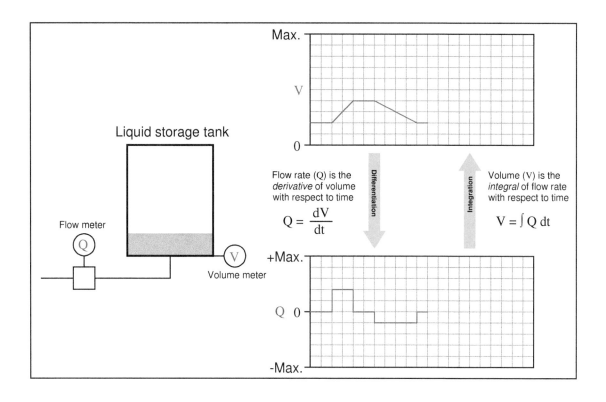

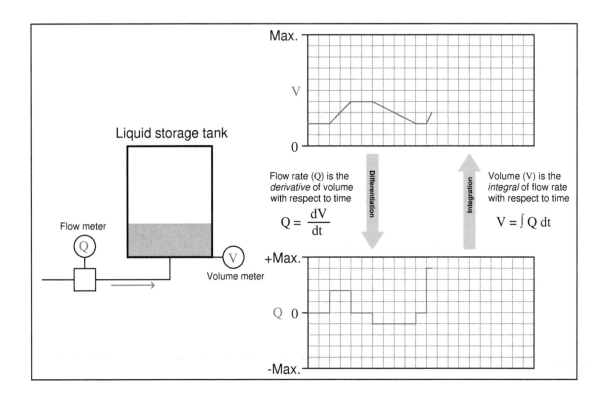

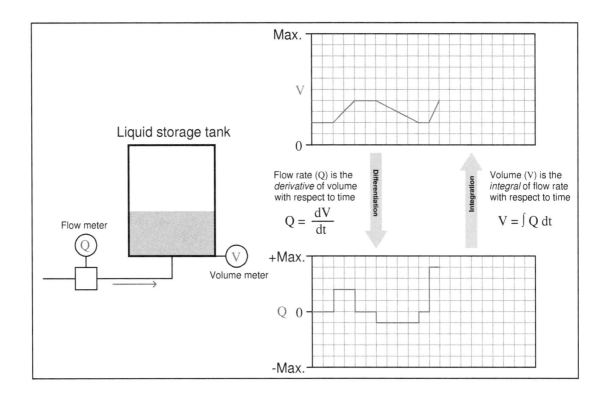

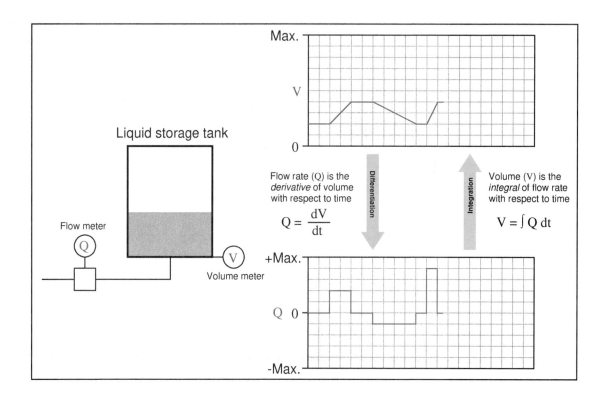

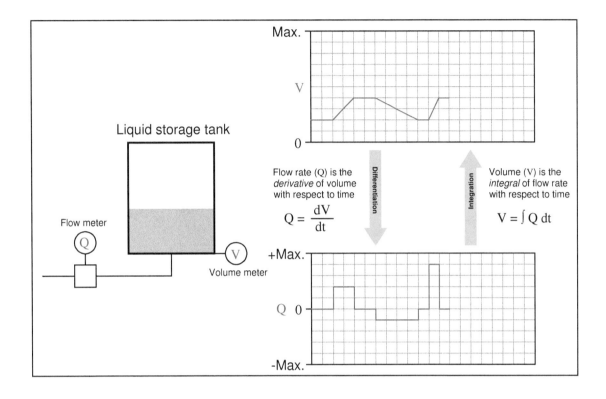

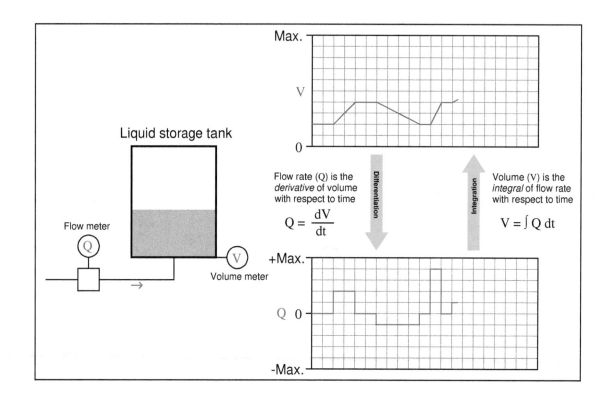

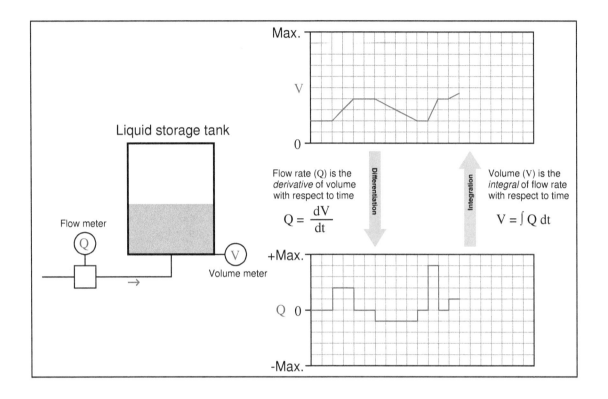

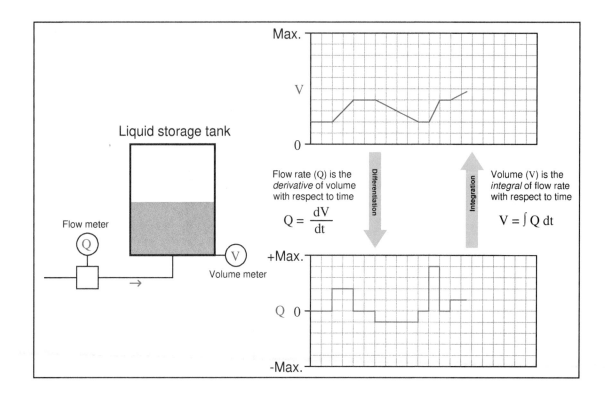

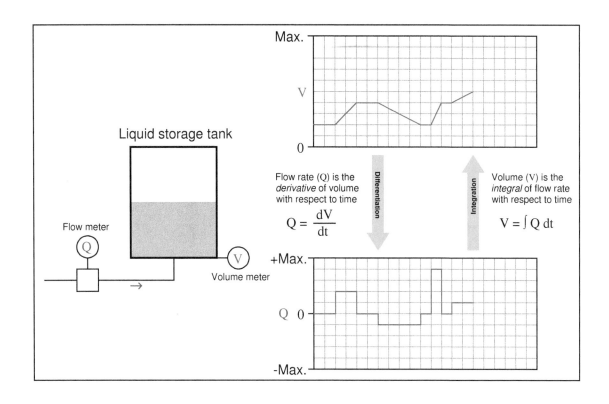

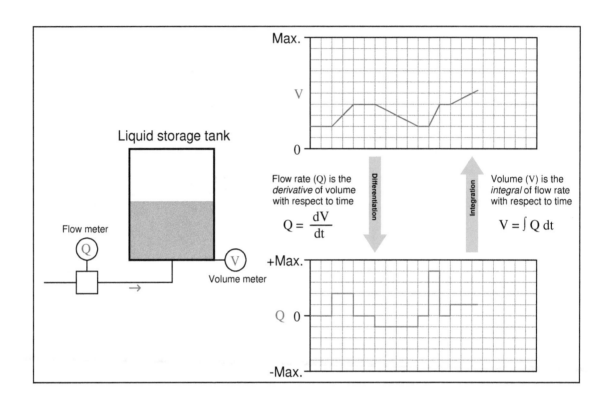

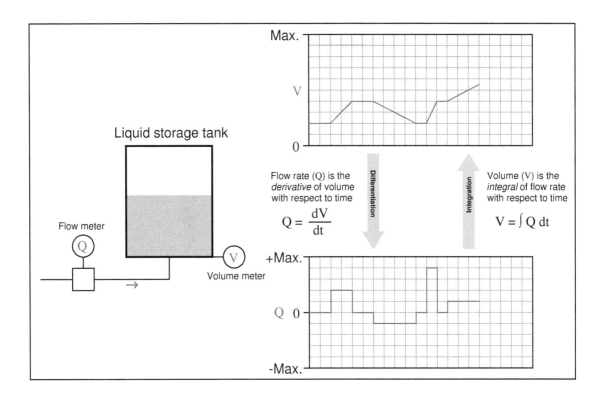

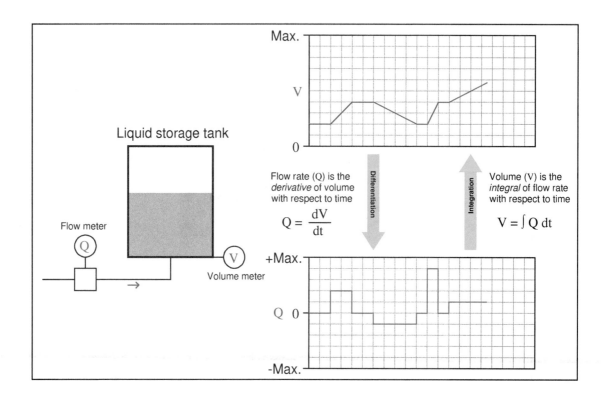

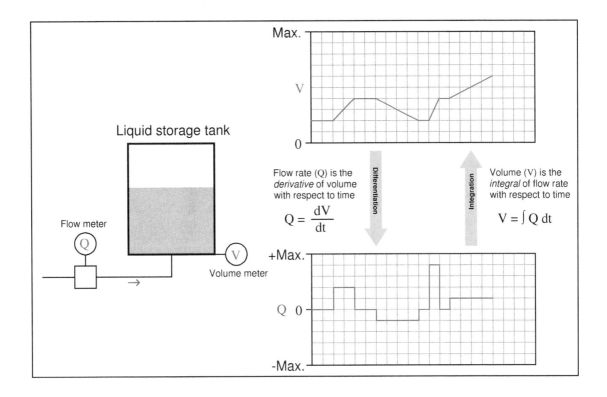

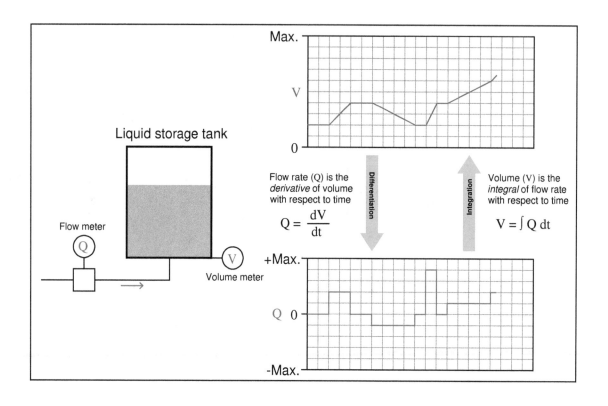

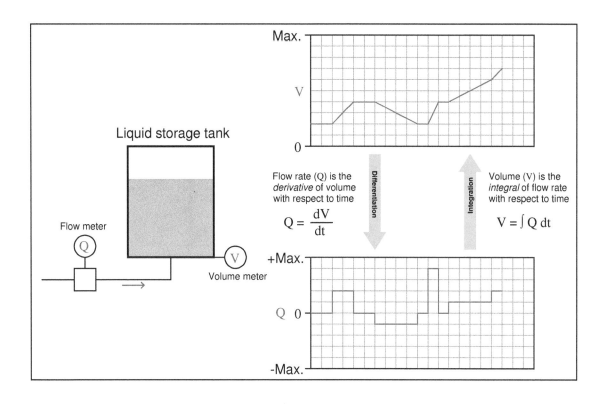

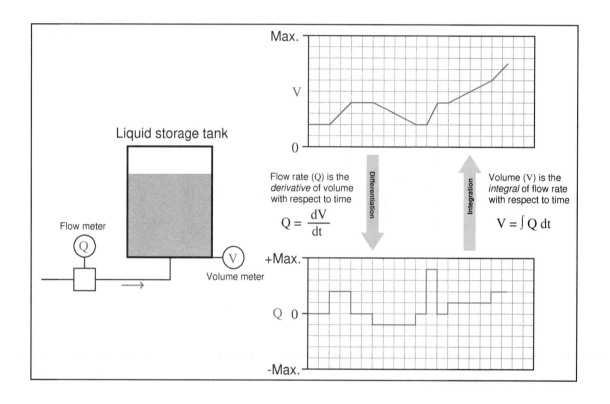

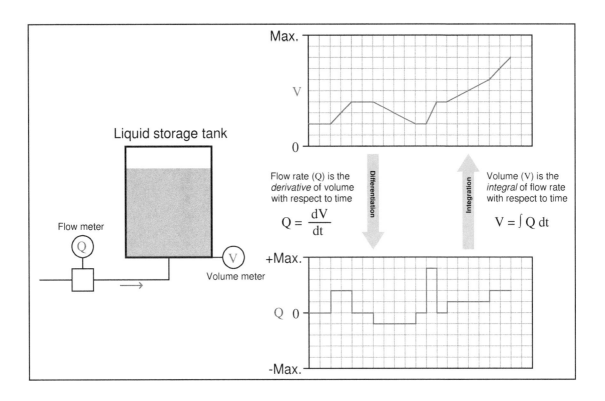

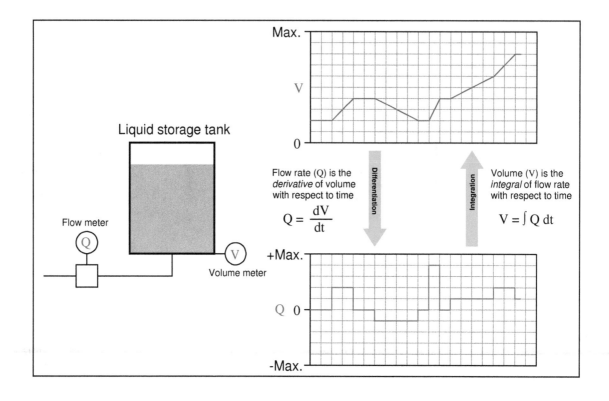

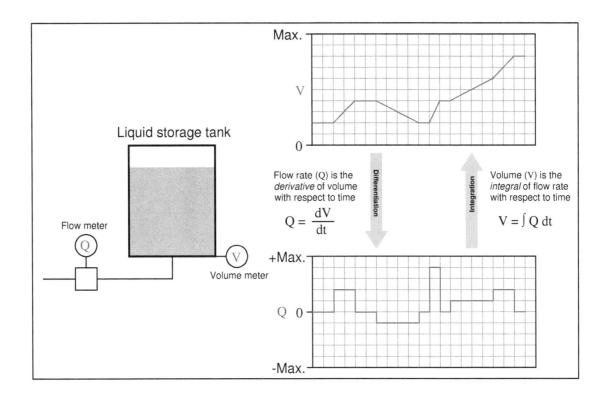

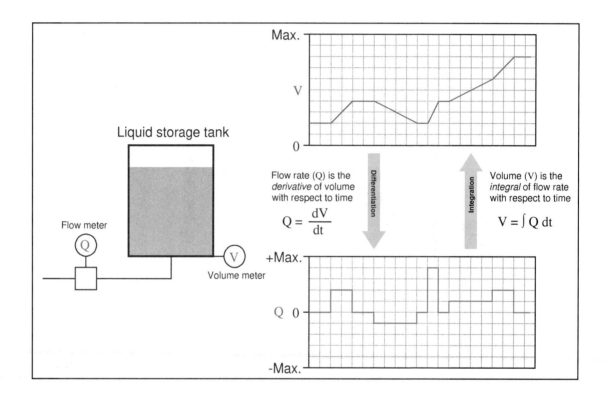

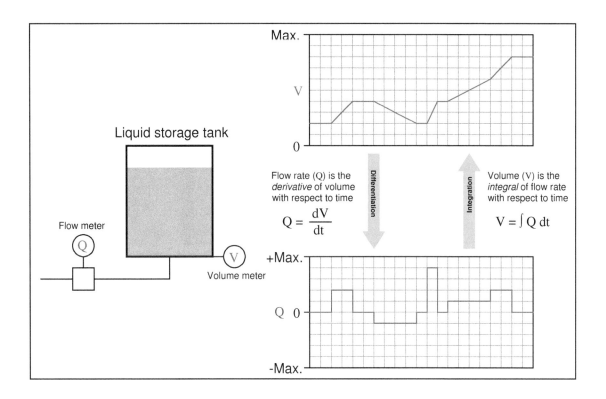

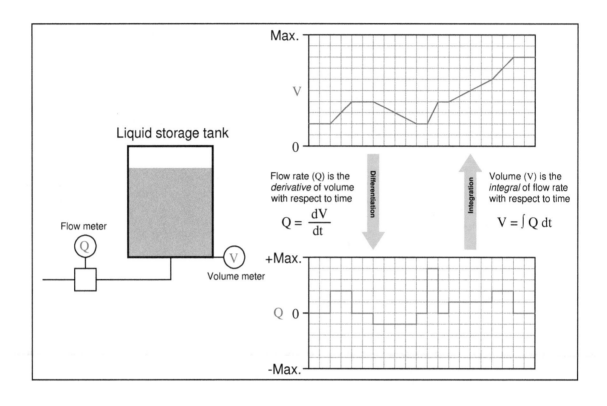

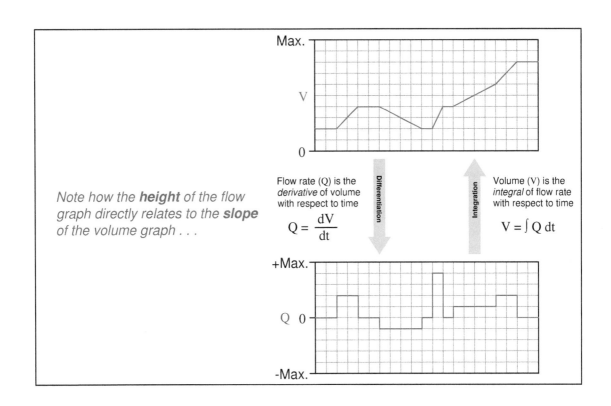

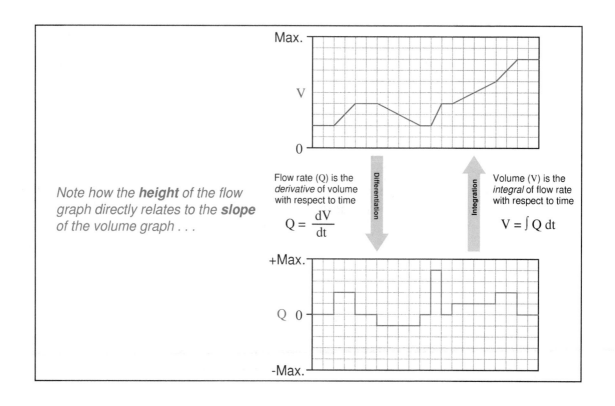

*Note how the **height** of the flow graph directly relates to the **slope** of the volume graph . . .*

Flow rate (Q) is the *derivative* of volume with respect to time

$$Q = \frac{dV}{dt}$$

Volume (V) is the *integral* of flow rate with respect to time

$$V = \int Q \, dt$$

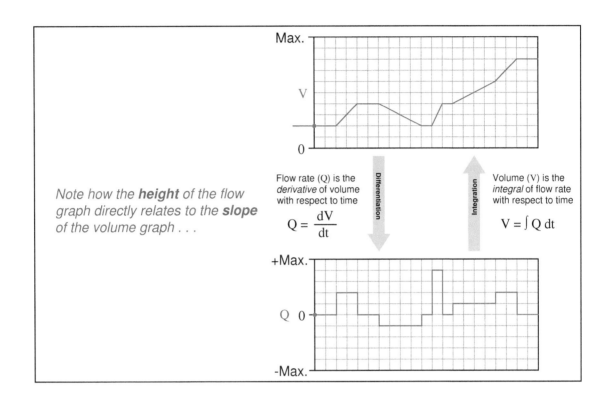

*Note how the **height** of the flow graph directly relates to the **slope** of the volume graph . . .*

Flow rate (Q) is the *derivative* of volume with respect to time

$$Q = \frac{dV}{dt}$$

Volume (V) is the *integral* of flow rate with respect to time

$$V = \int Q \, dt$$

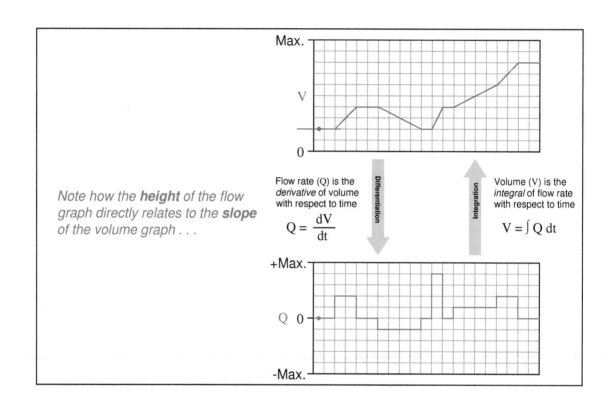

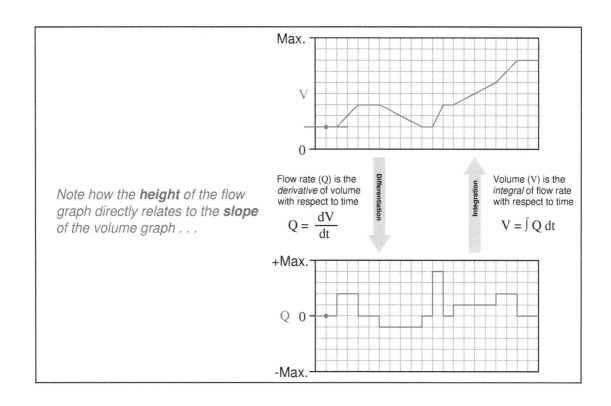

*Note how the **height** of the flow graph directly relates to the **slope** of the volume graph . . .*

Flow rate (Q) is the *derivative* of volume with respect to time

$$Q = \frac{dV}{dt}$$

Volume (V) is the *integral* of flow rate with respect to time

$$V = \int Q \, dt$$

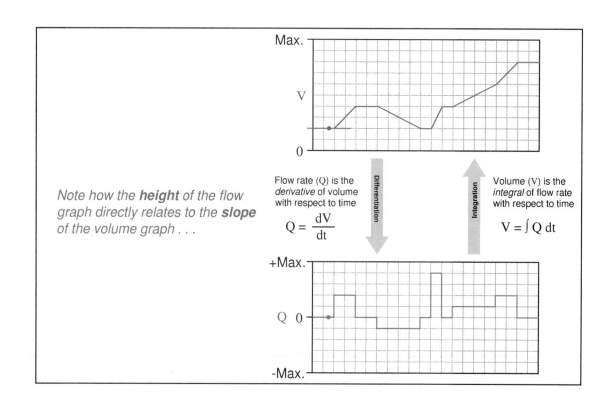

*Note how the **height** of the flow graph directly relates to the **slope** of the volume graph . . .*

Flow rate (Q) is the *derivative* of volume with respect to time

$$Q = \frac{dV}{dt}$$

Volume (V) is the *integral* of flow rate with respect to time

$$V = \int Q \, dt$$

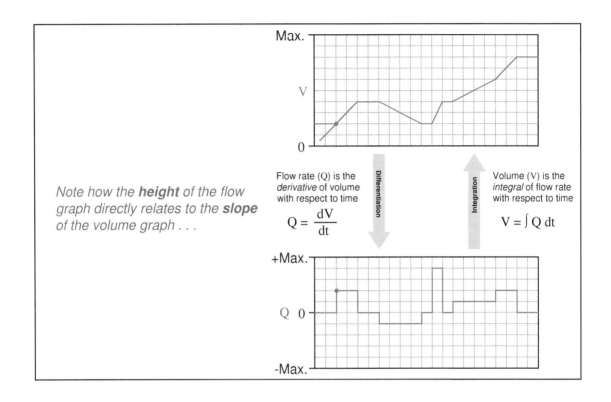

Note how the **height** of the flow graph directly relates to the **slope** of the volume graph . . .

Flow rate (Q) is the *derivative* of volume with respect to time

$$Q = \frac{dV}{dt}$$

Volume (V) is the *integral* of flow rate with respect to time

$$V = \int Q \, dt$$

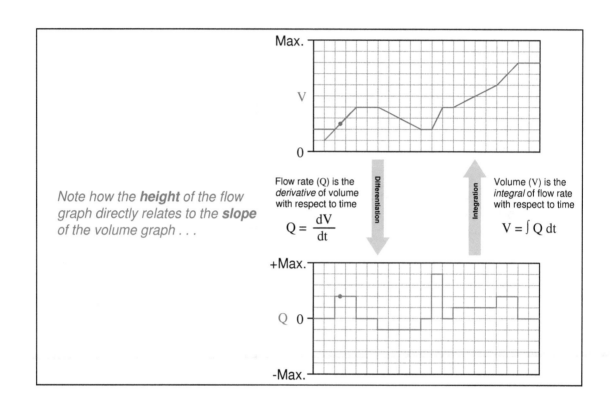

*Note how the **height** of the flow graph directly relates to the **slope** of the volume graph . . .*

Flow rate (Q) is the *derivative* of volume with respect to time

$$Q = \frac{dV}{dt}$$

Volume (V) is the *integral* of flow rate with respect to time

$$V = \int Q \, dt$$

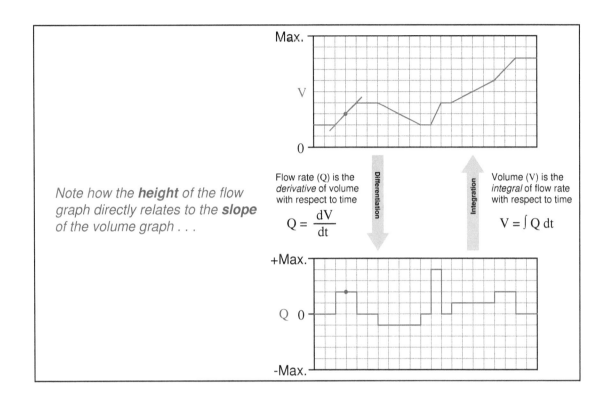

*Note how the **height** of the flow graph directly relates to the **slope** of the volume graph . . .*

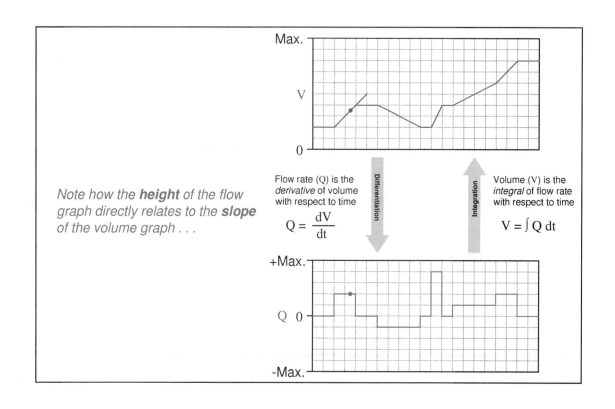

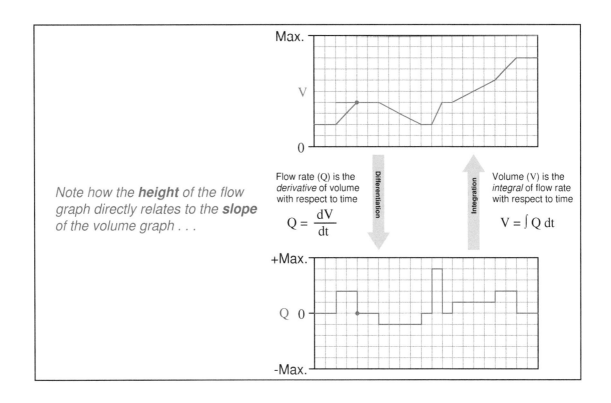

*Note how the **height** of the flow graph directly relates to the **slope** of the volume graph . . .*

Flow rate (Q) is the *derivative* of volume with respect to time

$$Q = \frac{dV}{dt}$$

Volume (V) is the *integral* of flow rate with respect to time

$$V = \int Q \, dt$$

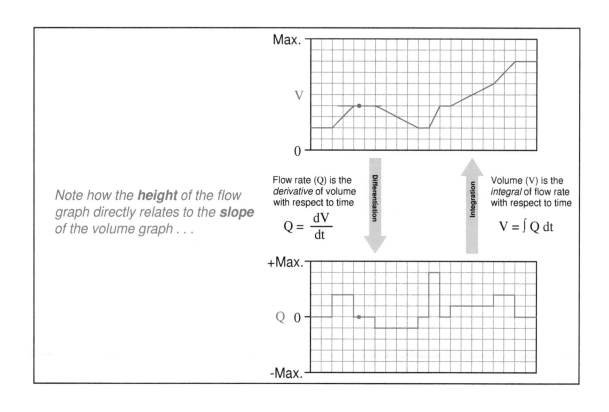

Note how the **height** of the flow graph directly relates to the **slope** of the volume graph . . .

Flow rate (Q) is the *derivative* of volume with respect to time

$$Q = \frac{dV}{dt}$$

Volume (V) is the *integral* of flow rate with respect to time

$$V = \int Q \, dt$$

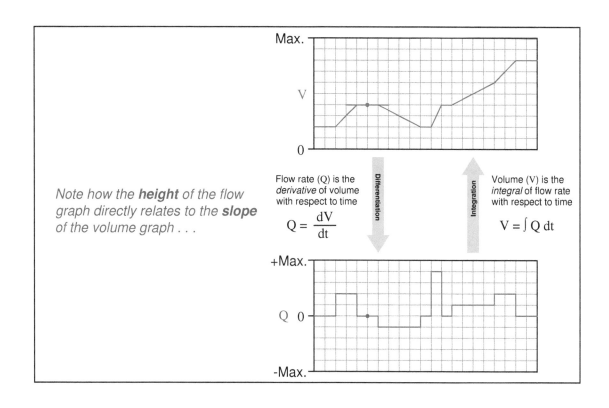

*Note how the **height** of the flow graph directly relates to the **slope** of the volume graph . . .*

Flow rate (Q) is the *derivative* of volume with respect to time

$$Q = \frac{dV}{dt}$$

Volume (V) is the *integral* of flow rate with respect to time

$$V = \int Q \, dt$$

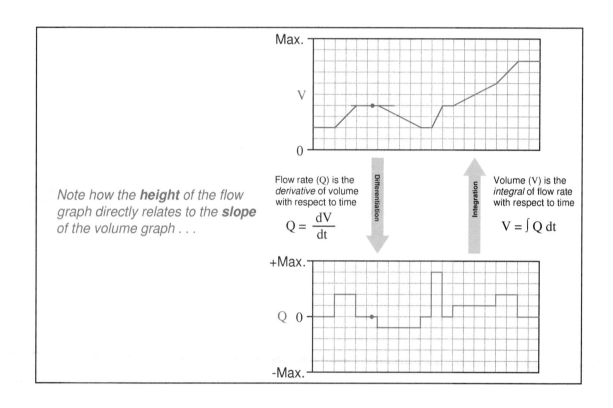

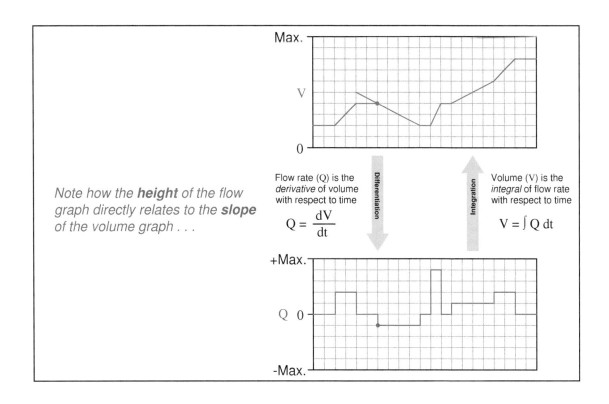

*Note how the **height** of the flow graph directly relates to the **slope** of the volume graph . . .*

Flow rate (Q) is the *derivative* of volume with respect to time

$$Q = \frac{dV}{dt}$$

Volume (V) is the *integral* of flow rate with respect to time

$$V = \int Q \, dt$$

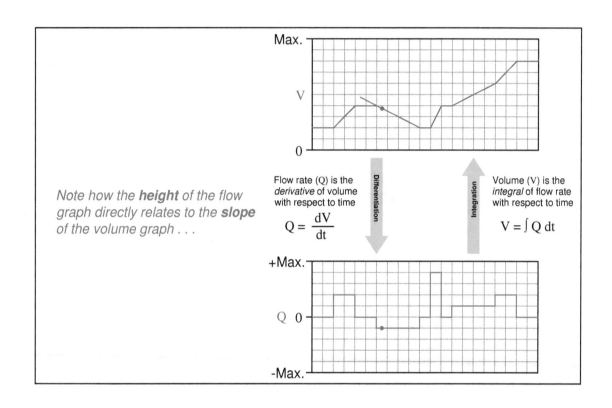

*Note how the **height** of the flow graph directly relates to the **slope** of the volume graph . . .*

Flow rate (Q) is the *derivative* of volume with respect to time

$$Q = \frac{dV}{dt}$$

Volume (V) is the *integral* of flow rate with respect to time

$$V = \int Q \, dt$$

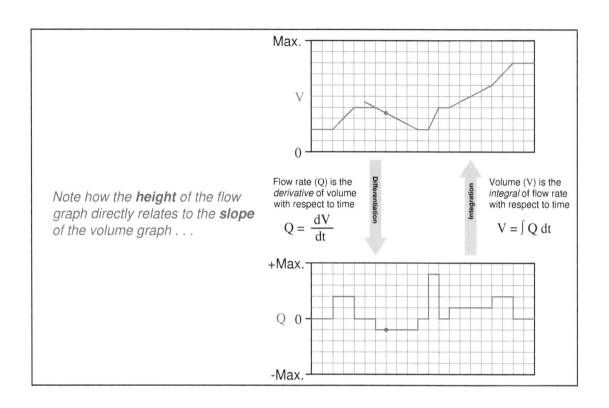

*Note how the **height** of the flow graph directly relates to the **slope** of the volume graph . . .*

Flow rate (Q) is the *derivative* of volume with respect to time

$$Q = \frac{dV}{dt}$$

Volume (V) is the *integral* of flow rate with respect to time

$$V = \int Q \, dt$$

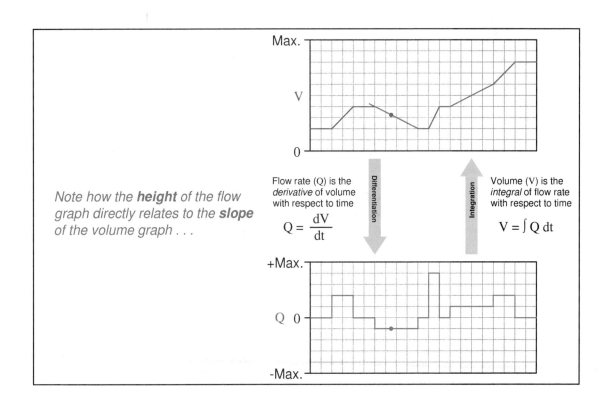

Note how the **height** of the flow graph directly relates to the **slope** of the volume graph . . .

Flow rate (Q) is the *derivative* of volume with respect to time

$$Q = \frac{dV}{dt}$$

Volume (V) is the *integral* of flow rate with respect to time

$$V = \int Q \, dt$$

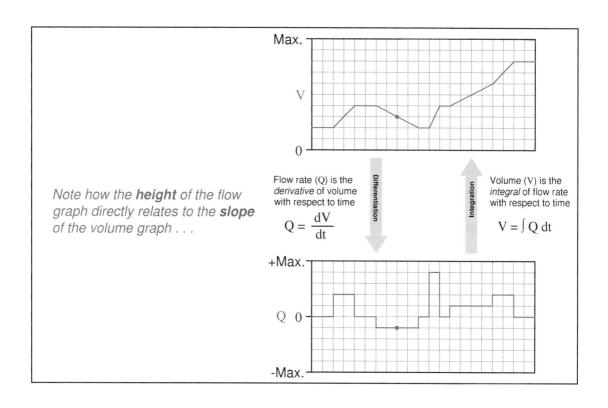

*Note how the **height** of the flow graph directly relates to the **slope** of the volume graph . . .*

Flow rate (Q) is the *derivative* of volume with respect to time

$$Q = \frac{dV}{dt}$$

Volume (V) is the *integral* of flow rate with respect to time

$$V = \int Q \, dt$$

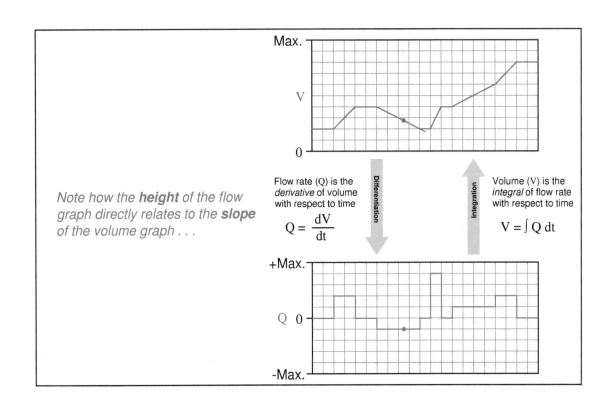

Note how the **height** of the flow graph directly relates to the **slope** of the volume graph . . .

Flow rate (Q) is the *derivative* of volume with respect to time

$$Q = \frac{dV}{dt}$$

Volume (V) is the *integral* of flow rate with respect to time

$$V = \int Q \, dt$$

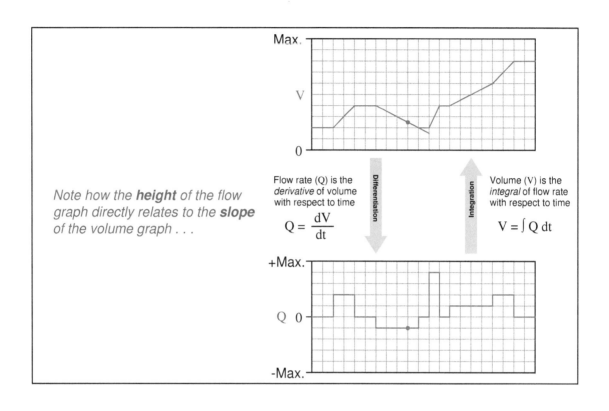

*Note how the **height** of the flow graph directly relates to the **slope** of the volume graph . . .*

Flow rate (Q) is the *derivative* of volume with respect to time

$$Q = \frac{dV}{dt}$$

Volume (V) is the *integral* of flow rate with respect to time

$$V = \int Q \, dt$$

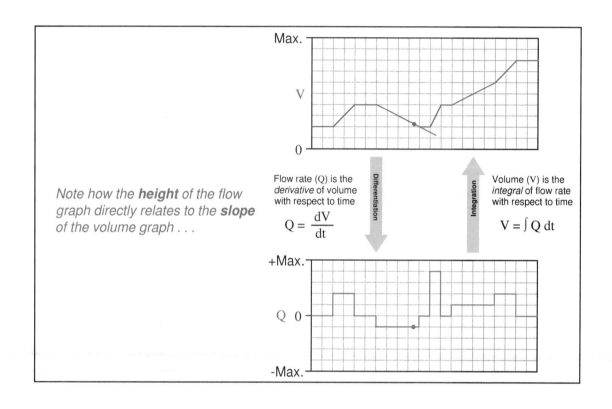

Note how the **height** of the flow graph directly relates to the **slope** of the volume graph . . .

Flow rate (Q) is the *derivative* of volume with respect to time

$$Q = \frac{dV}{dt}$$

Volume (V) is the *integral* of flow rate with respect to time

$$V = \int Q \, dt$$

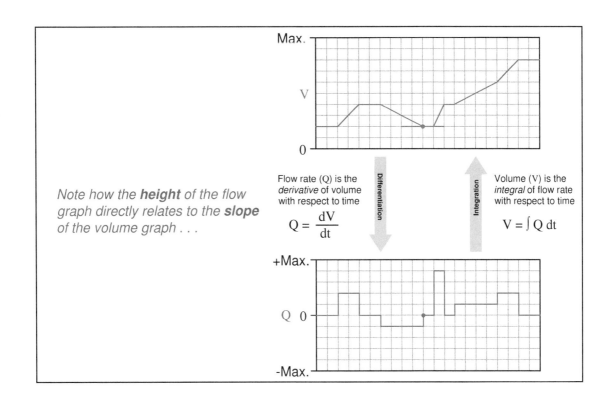

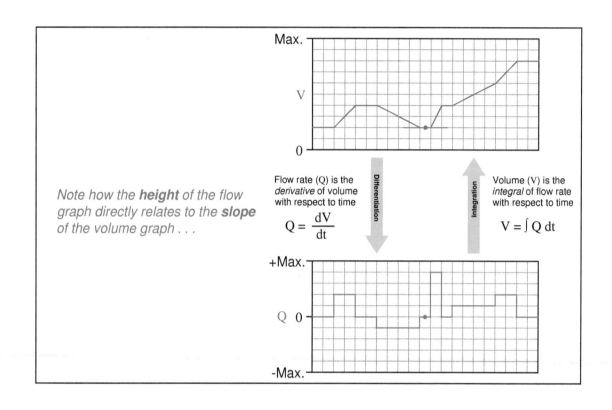

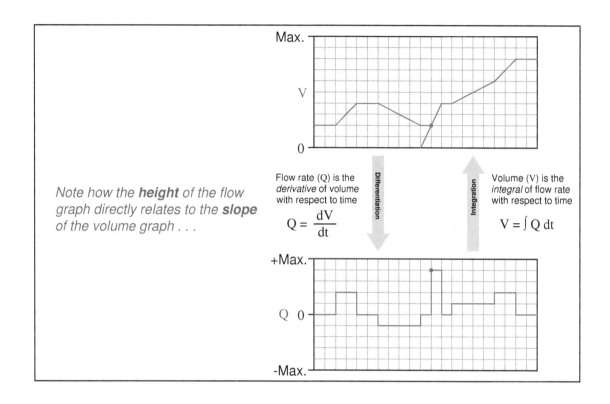

Note how the **height** of the flow graph directly relates to the **slope** of the volume graph . . .

Flow rate (Q) is the *derivative* of volume with respect to time

$$Q = \frac{dV}{dt}$$

Volume (V) is the *integral* of flow rate with respect to time

$$V = \int Q\ dt$$

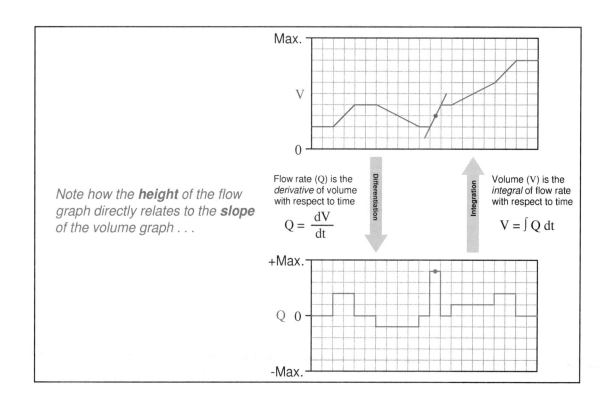

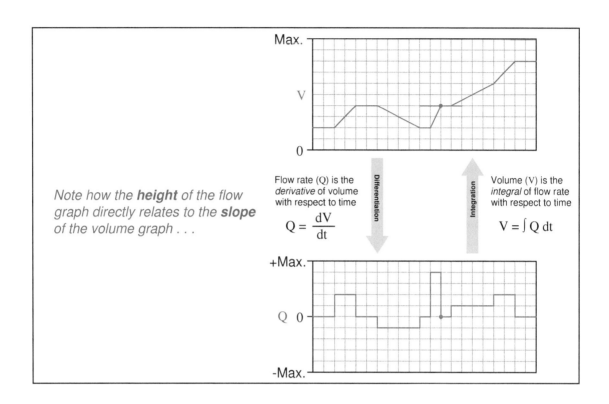

*Note how the **height** of the flow graph directly relates to the **slope** of the volume graph . . .*

Flow rate (Q) is the *derivative* of volume with respect to time

$$Q = \frac{dV}{dt}$$

Volume (V) is the *integral* of flow rate with respect to time

$$V = \int Q \, dt$$

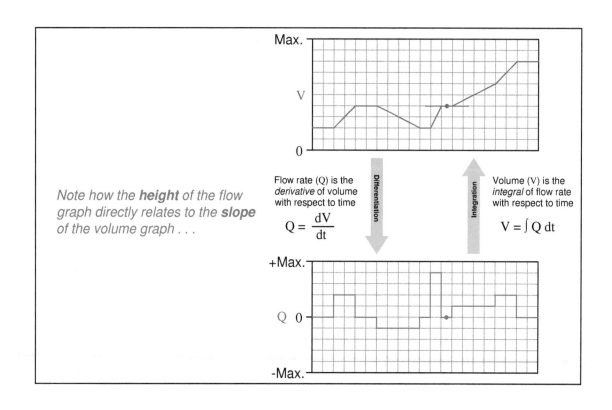

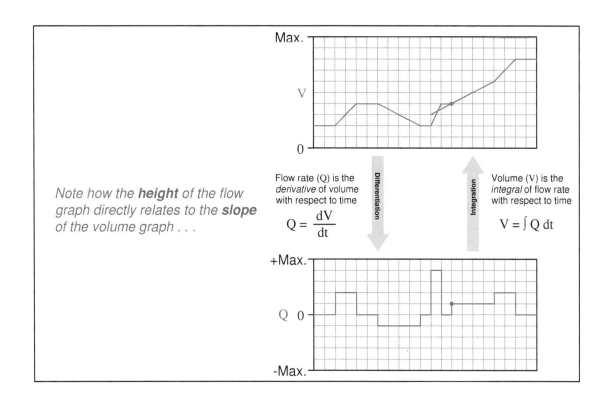

*Note how the **height** of the flow graph directly relates to the **slope** of the volume graph . . .*

Flow rate (Q) is the *derivative* of volume with respect to time

$$Q = \frac{dV}{dt}$$

Volume (V) is the *integral* of flow rate with respect to time

$$V = \int Q \, dt$$

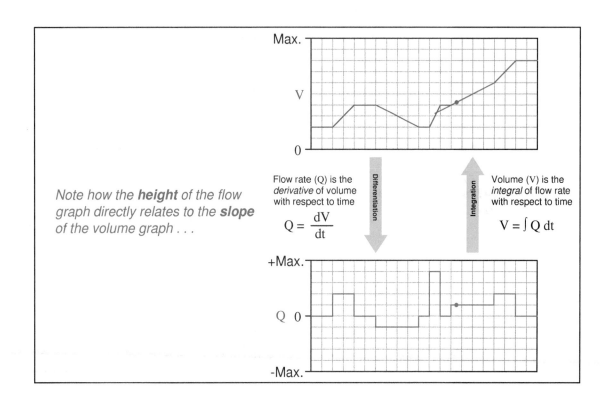

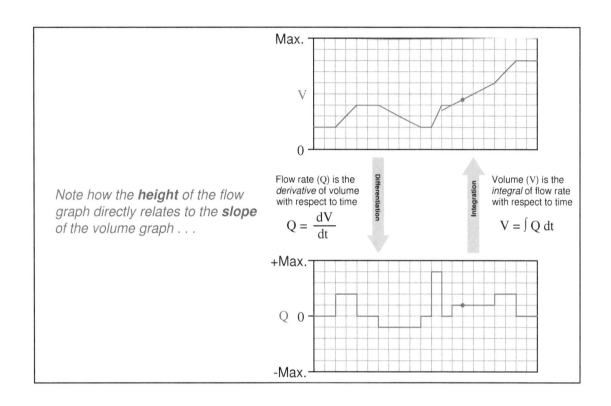

Note how the **height** of the flow graph directly relates to the **slope** of the volume graph . . .

Flow rate (Q) is the *derivative* of volume with respect to time

$$Q = \frac{dV}{dt}$$

Volume (V) is the *integral* of flow rate with respect to time

$$V = \int Q \, dt$$

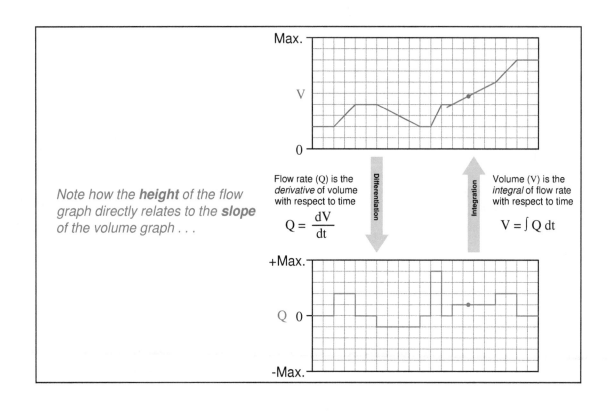

*Note how the **height** of the flow graph directly relates to the **slope** of the volume graph . . .*

Flow rate (Q) is the *derivative* of volume with respect to time

$$Q = \frac{dV}{dt}$$

Volume (V) is the *integral* of flow rate with respect to time

$$V = \int Q \, dt$$

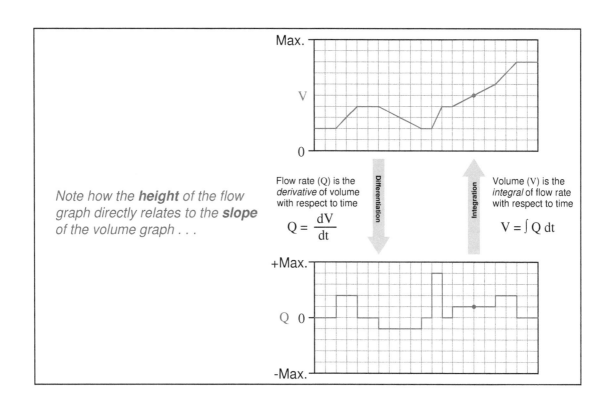

*Note how the **height** of the flow graph directly relates to the **slope** of the volume graph . . .*

Flow rate (Q) is the *derivative* of volume with respect to time

$$Q = \frac{dV}{dt}$$

Volume (V) is the *integral* of flow rate with respect to time

$$V = \int Q \, dt$$

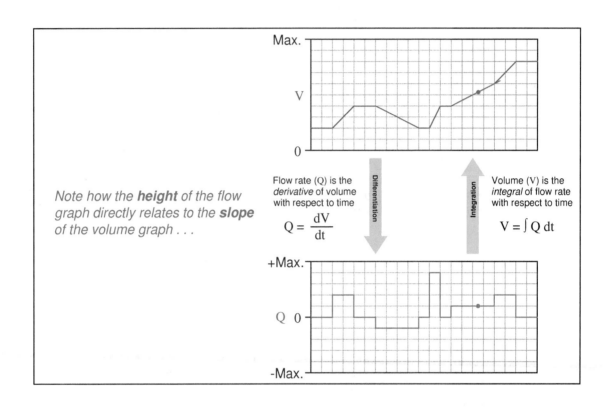

*Note how the **height** of the flow graph directly relates to the **slope** of the volume graph . . .*

Flow rate (Q) is the *derivative* of volume with respect to time

$$Q = \dfrac{dV}{dt}$$

Volume (V) is the *integral* of flow rate with respect to time

$$V = \int Q \, dt$$

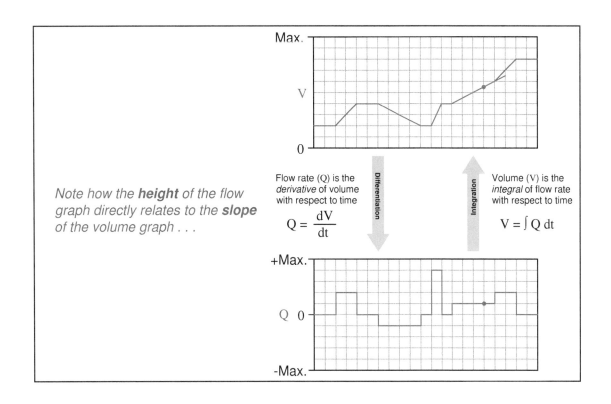

*Note how the **height** of the flow graph directly relates to the **slope** of the volume graph . . .*

Flow rate (Q) is the *derivative* of volume with respect to time

$$Q = \frac{dV}{dt}$$

Volume (V) is the *integral* of flow rate with respect to time

$$V = \int Q \, dt$$

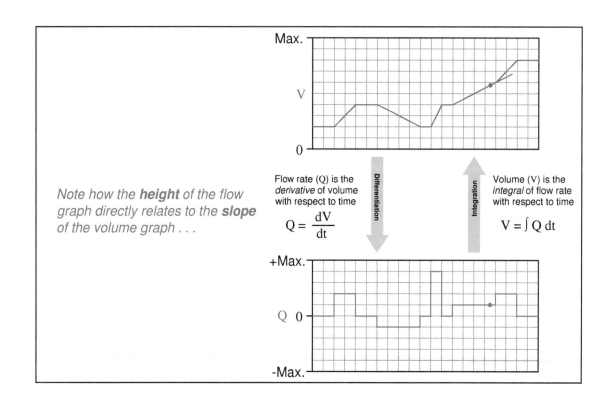

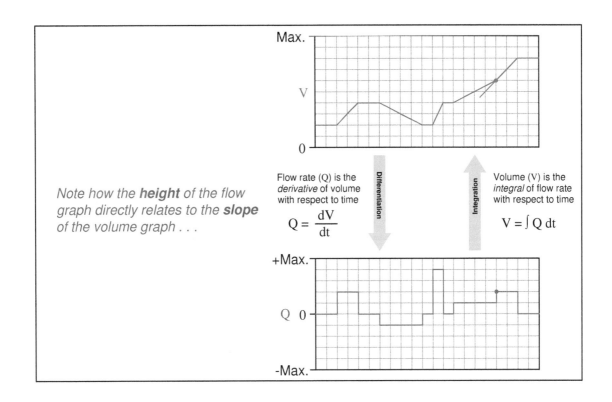

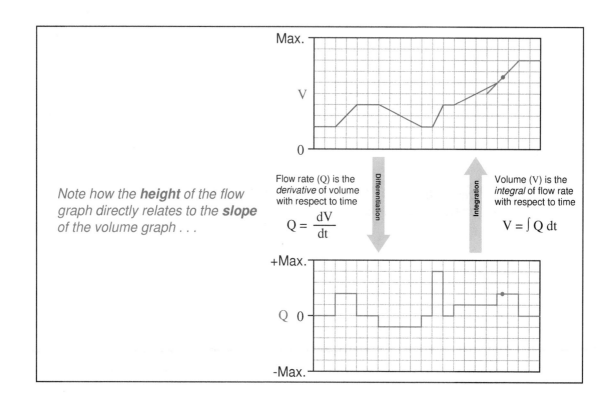

*Note how the **height** of the flow graph directly relates to the **slope** of the volume graph . . .*

Flow rate (Q) is the *derivative* of volume with respect to time

$$Q = \frac{dV}{dt}$$

Volume (V) is the *integral* of flow rate with respect to time

$$V = \int Q \, dt$$

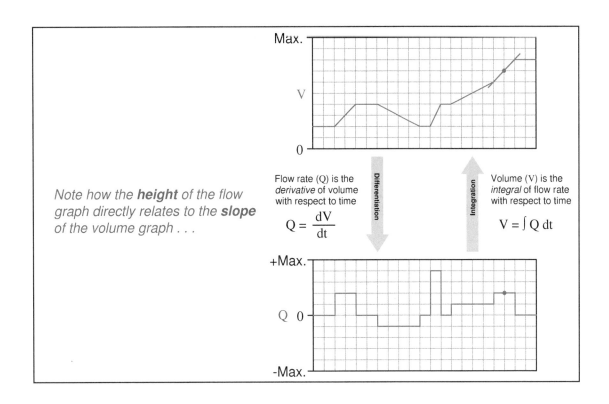

*Note how the **height** of the flow graph directly relates to the **slope** of the volume graph . . .*

Flow rate (Q) is the *derivative* of volume with respect to time

$$Q = \frac{dV}{dt}$$

Volume (V) is the *integral* of flow rate with respect to time

$$V = \int Q \, dt$$

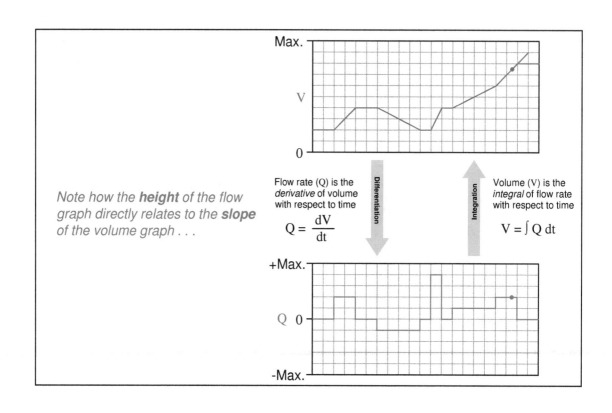

*Note how the **height** of the flow graph directly relates to the **slope** of the volume graph . . .*

Flow rate (Q) is the *derivative* of volume with respect to time

$$Q = \frac{dV}{dt}$$

Volume (V) is the *integral* of flow rate with respect to time

$$V = \int Q \, dt$$

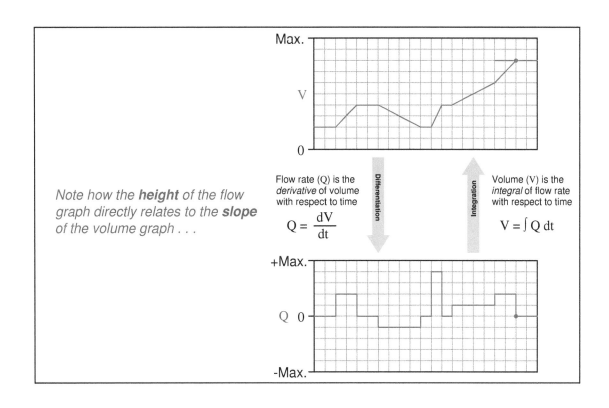

Note how the **height** of the flow graph directly relates to the **slope** of the volume graph . . .

Flow rate (Q) is the *derivative* of volume with respect to time

$$Q = \frac{dV}{dt}$$

Volume (V) is the *integral* of flow rate with respect to time

$$V = \int Q \, dt$$

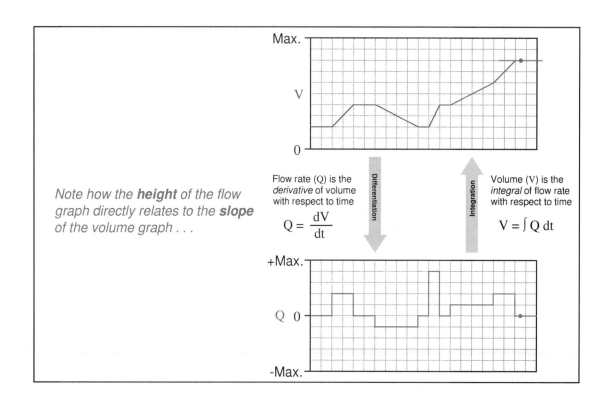

*Note how the **height** of the flow graph directly relates to the **slope** of the volume graph . . .*

Flow rate (Q) is the *derivative* of volume with respect to time

$$Q = \frac{dV}{dt}$$

Volume (V) is the *integral* of flow rate with respect to time

$$V = \int Q \, dt$$

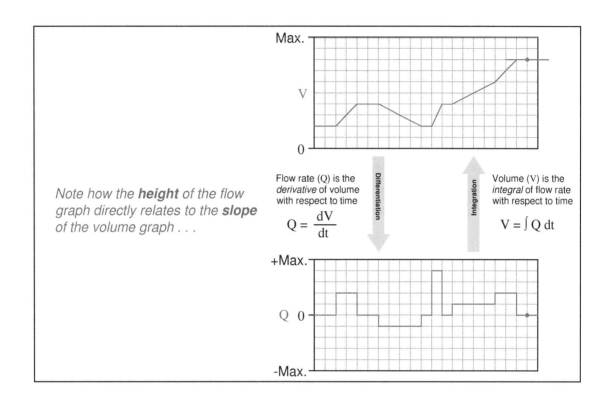

*Note how the **height** of the flow graph directly relates to the **slope** of the volume graph . . .*

Flow rate (Q) is the *derivative* of volume with respect to time

$$Q = \frac{dV}{dt}$$

Volume (V) is the *integral* of flow rate with respect to time

$$V = \int Q \, dt$$

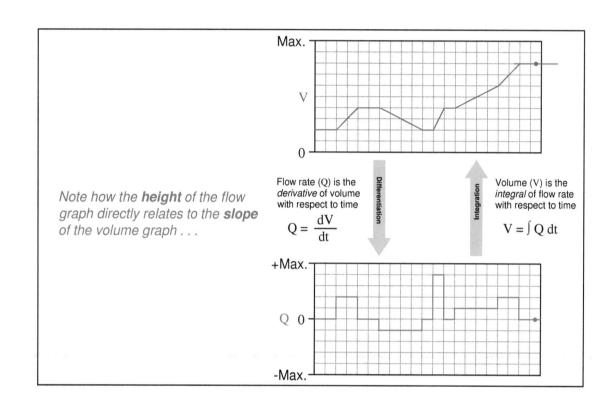

*Note how the **height** of the flow graph directly relates to the **slope** of the volume graph . . .*

Flow rate (Q) is the *derivative* of volume with respect to time

$$Q = \frac{dV}{dt}$$

Volume (V) is the *integral* of flow rate with respect to time

$$V = \int Q \, dt$$

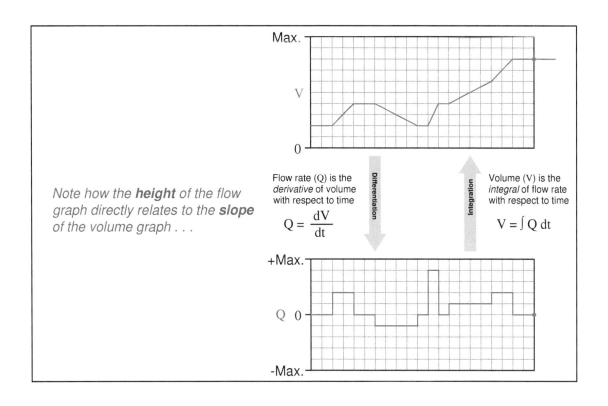

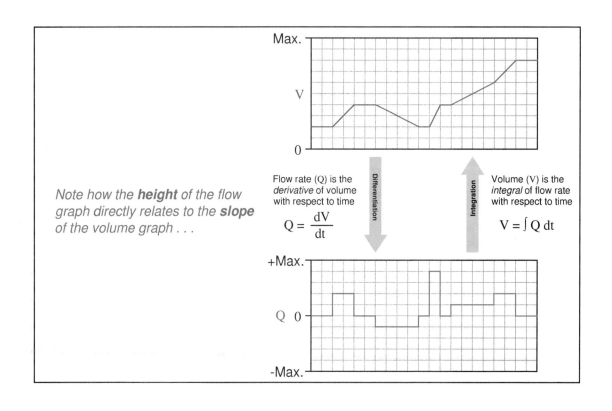

*Note how the **height** of the flow graph directly relates to the **slope** of the volume graph . . .*

Flow rate (Q) is the *derivative* of volume with respect to time

$$Q = \frac{dV}{dt}$$

Volume (V) is the *integral* of flow rate with respect to time

$$V = \int Q\, dt$$

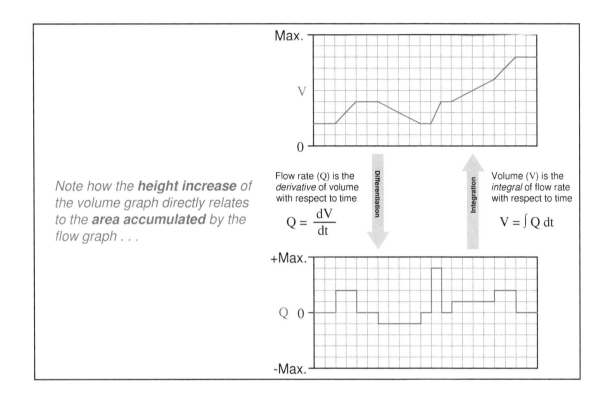

*Note how the **height increase** of the volume graph directly relates to the **area accumulated** by the flow graph . . .*

Flow rate (Q) is the *derivative* of volume with respect to time

$$Q = \frac{dV}{dt}$$

Volume (V) is the *integral* of flow rate with respect to time

$$V = \int Q \, dt$$

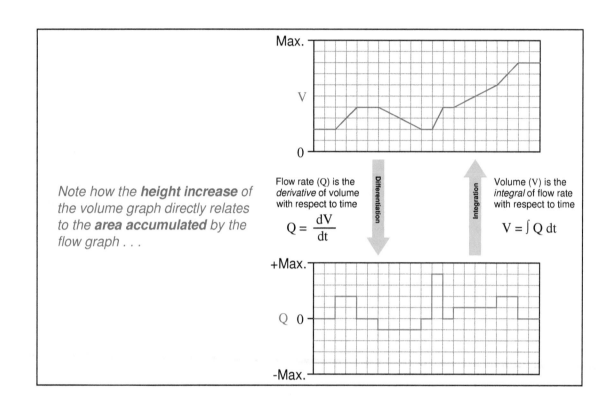

*Note how the **height increase** of the volume graph directly relates to the **area accumulated** by the flow graph . . .*

Flow rate (Q) is the *derivative* of volume with respect to time

$$Q = \frac{dV}{dt}$$

Volume (V) is the *integral* of flow rate with respect to time

$$V = \int Q \, dt$$

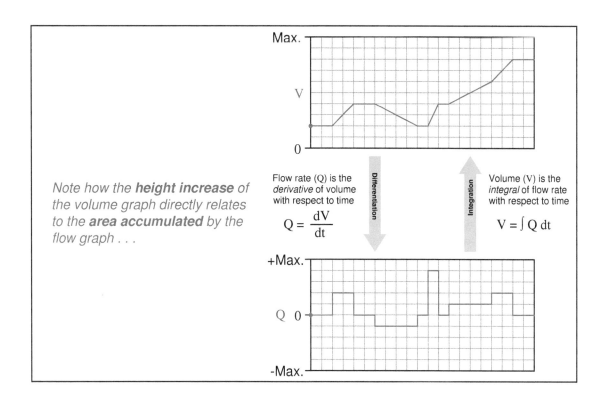

*Note how the **height increase** of the volume graph directly relates to the **area accumulated** by the flow graph . . .*

Flow rate (Q) is the *derivative* of volume with respect to time

$$Q = \frac{dV}{dt}$$

Volume (V) is the *integral* of flow rate with respect to time

$$V = \int Q \, dt$$

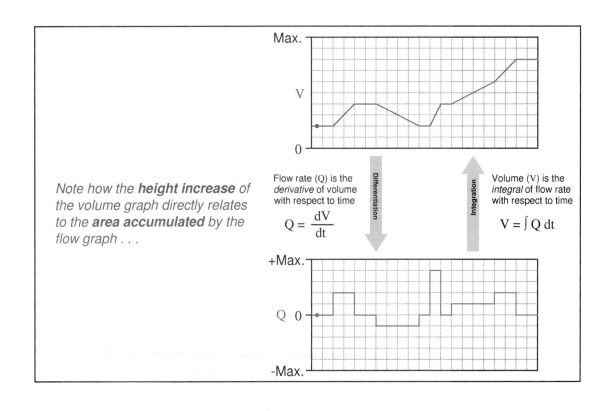

*Note how the **height increase** of the volume graph directly relates to the **area accumulated** by the flow graph . . .*

Flow rate (Q) is the *derivative* of volume with respect to time

$$Q = \frac{dV}{dt}$$

Volume (V) is the *integral* of flow rate with respect to time

$$V = \int Q \, dt$$

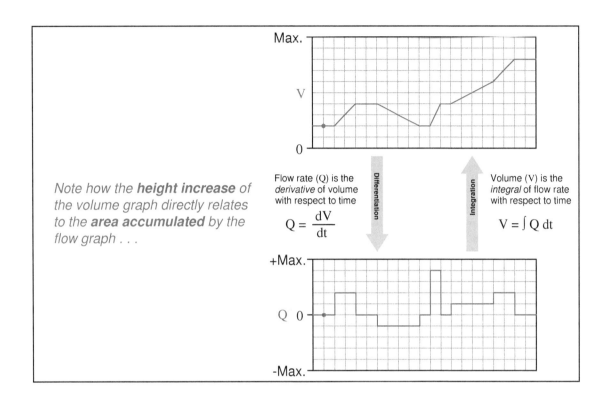

*Note how the **height increase** of the volume graph directly relates to the **area accumulated** by the flow graph . . .*

Flow rate (Q) is the *derivative* of volume with respect to time

$$Q = \frac{dV}{dt}$$

Volume (V) is the *integral* of flow rate with respect to time

$$V = \int Q\, dt$$

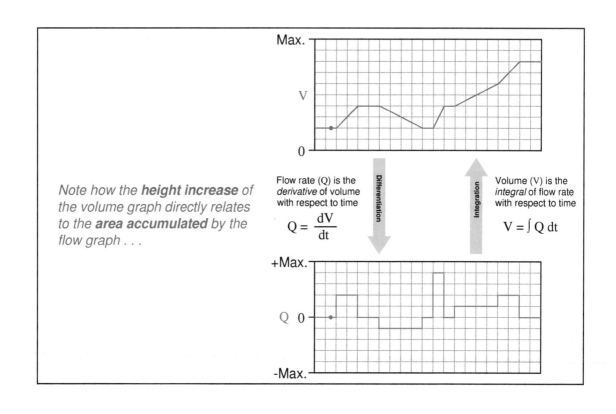

*Note how the **height increase** of the volume graph directly relates to the **area accumulated** by the flow graph . . .*

Flow rate (Q) is the *derivative* of volume with respect to time

$$Q = \dfrac{dV}{dt}$$

Volume (V) is the *integral* of flow rate with respect to time

$$V = \int Q \, dt$$

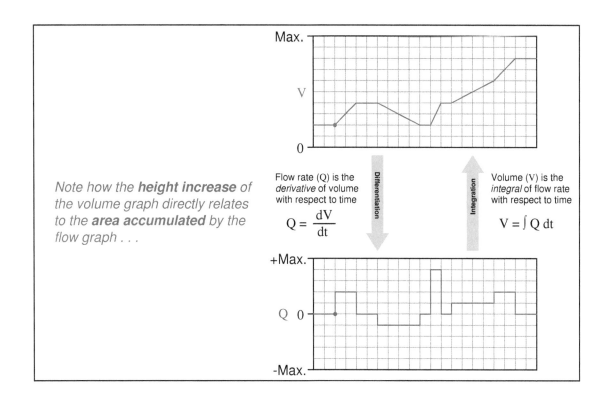

Note how the **height increase** *of the volume graph directly relates to the* **area accumulated** *by the flow graph . . .*

Flow rate (Q) is the *derivative* of volume with respect to time

$$Q = \frac{dV}{dt}$$

Volume (V) is the *integral* of flow rate with respect to time

$$V = \int Q \, dt$$

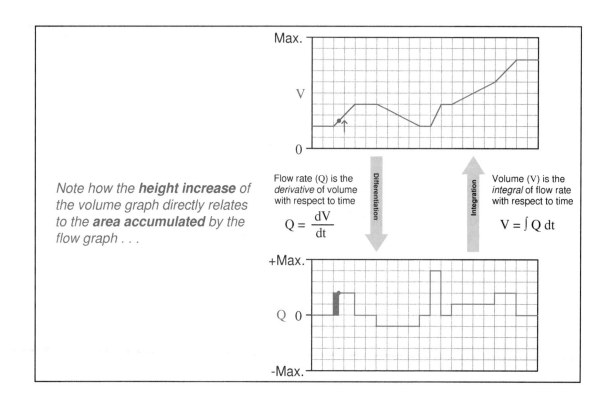

*Note how the **height increase** of the volume graph directly relates to the **area accumulated** by the flow graph . . .*

Flow rate (Q) is the *derivative* of volume with respect to time

$$Q = \frac{dV}{dt}$$

Volume (V) is the *integral* of flow rate with respect to time

$$V = \int Q \, dt$$

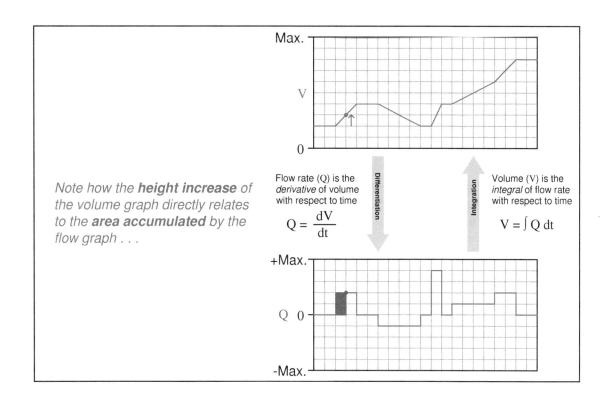

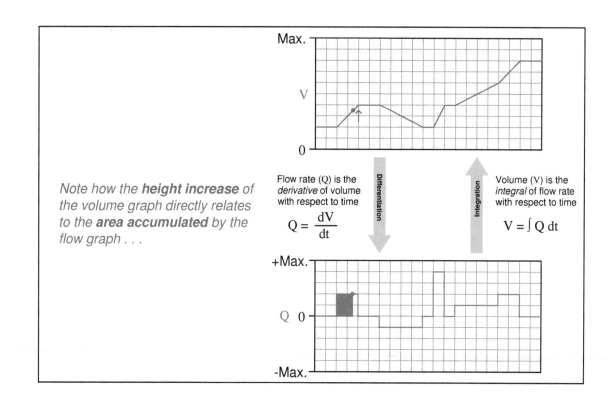

*Note how the **height increase** of the volume graph directly relates to the **area accumulated** by the flow graph . . .*

Flow rate (Q) is the *derivative* of volume with respect to time

$$Q = \frac{dV}{dt}$$

Volume (V) is the *integral* of flow rate with respect to time

$$V = \int Q \, dt$$

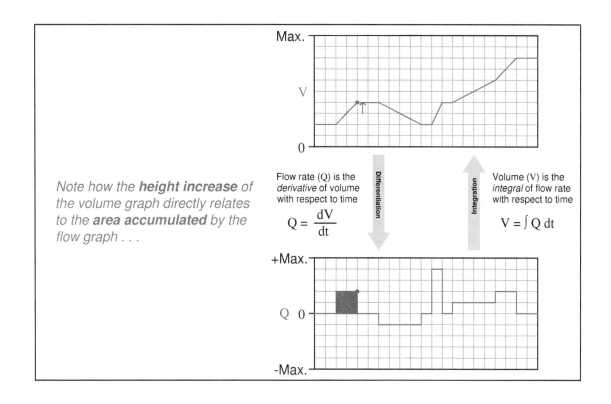

*Note how the **height increase** of the volume graph directly relates to the **area accumulated** by the flow graph . . .*

Flow rate (Q) is the *derivative* of volume with respect to time

$$Q = \frac{dV}{dt}$$

Volume (V) is the *integral* of flow rate with respect to time

$$V = \int Q \, dt$$

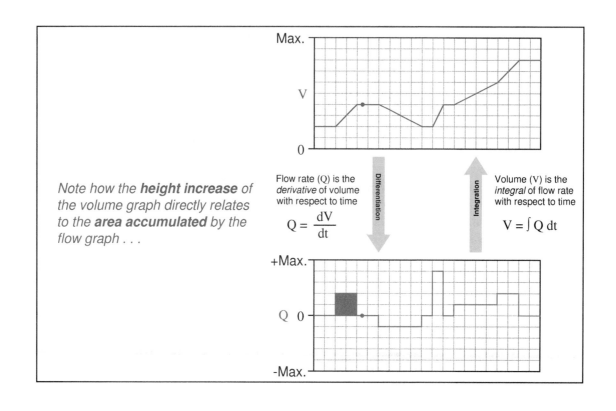

*Note how the **height increase** of the volume graph directly relates to the **area accumulated** by the flow graph . . .*

Flow rate (Q) is the *derivative* of volume with respect to time

$$Q = \frac{dV}{dt}$$

Volume (V) is the *integral* of flow rate with respect to time

$$V = \int Q \, dt$$

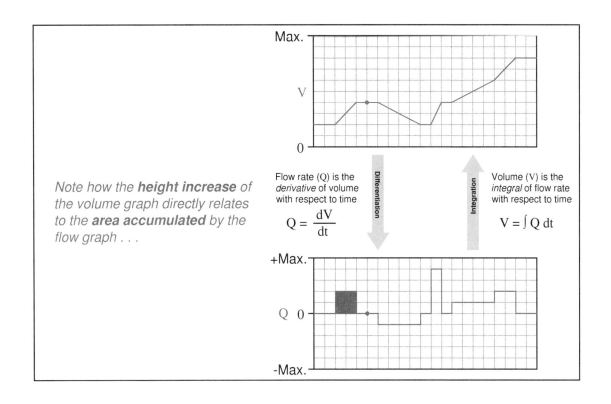

*Note how the **height increase** of the volume graph directly relates to the **area accumulated** by the flow graph . . .*

Flow rate (Q) is the *derivative* of volume with respect to time

$$Q = \frac{dV}{dt}$$

Volume (V) is the *integral* of flow rate with respect to time

$$V = \int Q \, dt$$

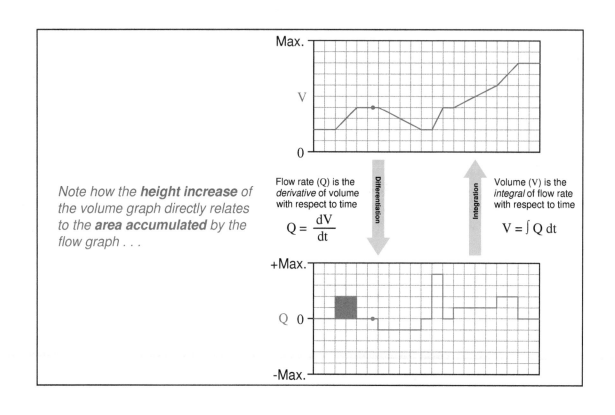

*Note how the **height increase** of the volume graph directly relates to the **area accumulated** by the flow graph . . .*

Flow rate (Q) is the *derivative* of volume with respect to time

$$Q = \frac{dV}{dt}$$

Volume (V) is the *integral* of flow rate with respect to time

$$V = \int Q \, dt$$

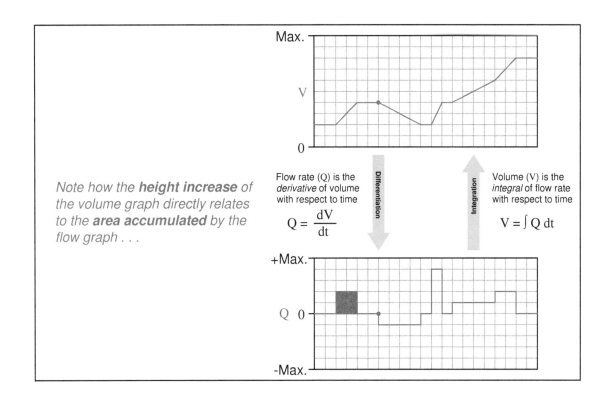

*Note how the **height increase** of the volume graph directly relates to the **area accumulated** by the flow graph . . .*

Flow rate (Q) is the *derivative* of volume with respect to time

$$Q = \frac{dV}{dt}$$

Volume (V) is the *integral* of flow rate with respect to time

$$V = \int Q\, dt$$

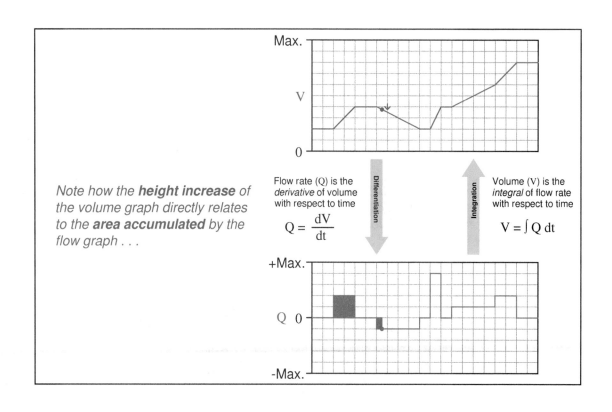

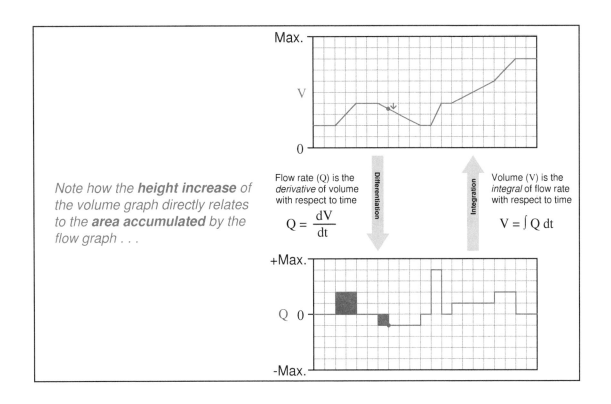

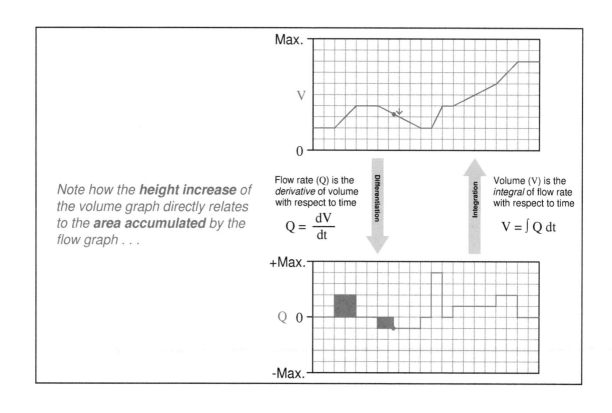

*Note how the **height increase** of the volume graph directly relates to the **area accumulated** by the flow graph . . .*

Flow rate (Q) is the *derivative* of volume with respect to time

$$Q = \frac{dV}{dt}$$

Volume (V) is the *integral* of flow rate with respect to time

$$V = \int Q \, dt$$

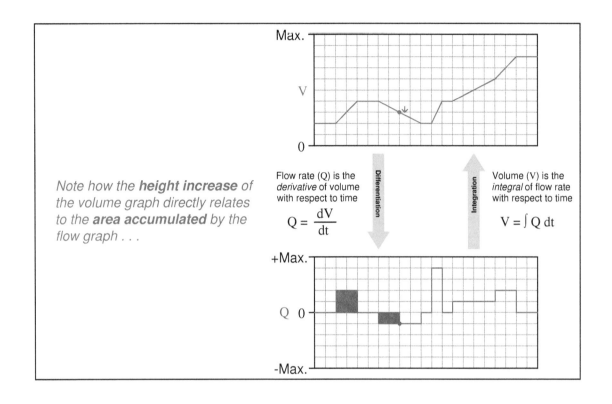

*Note how the **height increase** of the volume graph directly relates to the **area accumulated** by the flow graph . . .*

Flow rate (Q) is the *derivative* of volume with respect to time

$$Q = \frac{dV}{dt}$$

Volume (V) is the *integral* of flow rate with respect to time

$$V = \int Q \, dt$$

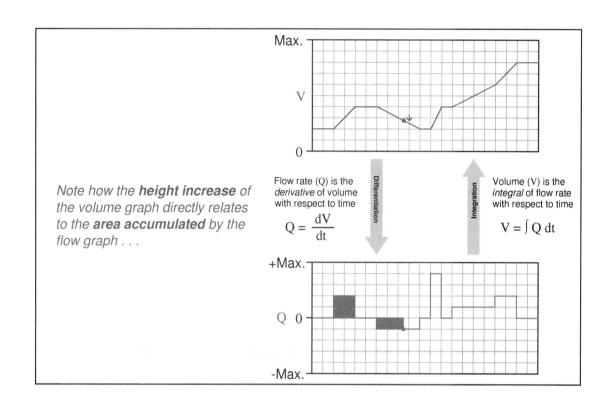

*Note how the **height increase** of the volume graph directly relates to the **area accumulated** by the flow graph . . .*

Flow rate (Q) is the *derivative* of volume with respect to time

$$Q = \frac{dV}{dt}$$

Volume (V) is the *integral* of flow rate with respect to time

$$V = \int Q\, dt$$

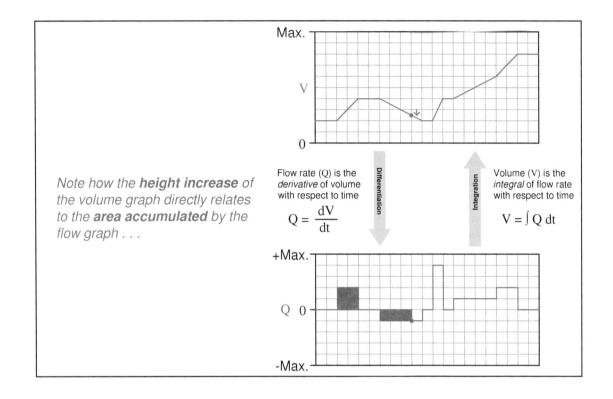

*Note how the **height increase** of the volume graph directly relates to the **area accumulated** by the flow graph . . .*

Flow rate (Q) is the *derivative* of volume with respect to time

$$Q = \frac{dV}{dt}$$

Volume (V) is the *integral* of flow rate with respect to time

$$V = \int Q \, dt$$

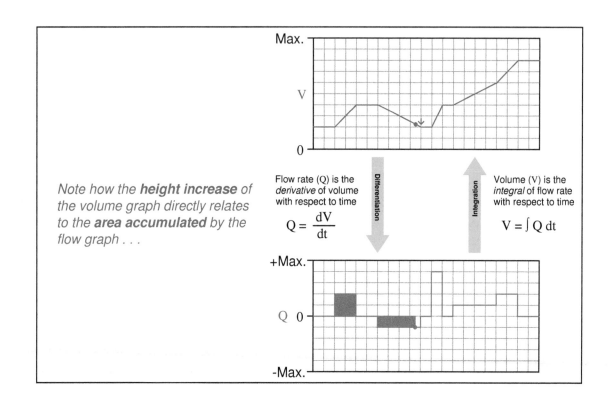

*Note how the **height increase** of the volume graph directly relates to the **area accumulated** by the flow graph . . .*

Flow rate (Q) is the *derivative* of volume with respect to time

$$Q = \frac{dV}{dt}$$

Volume (V) is the *integral* of flow rate with respect to time

$$V = \int Q\, dt$$

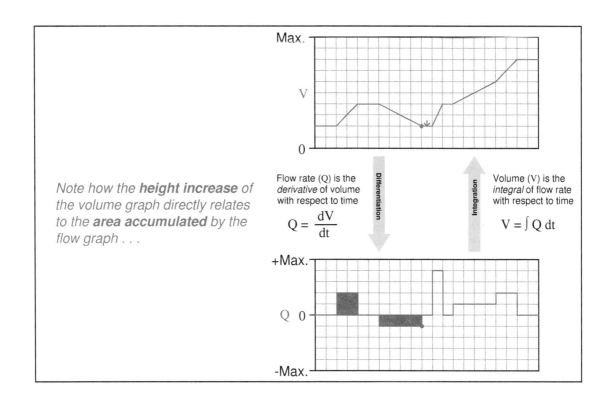

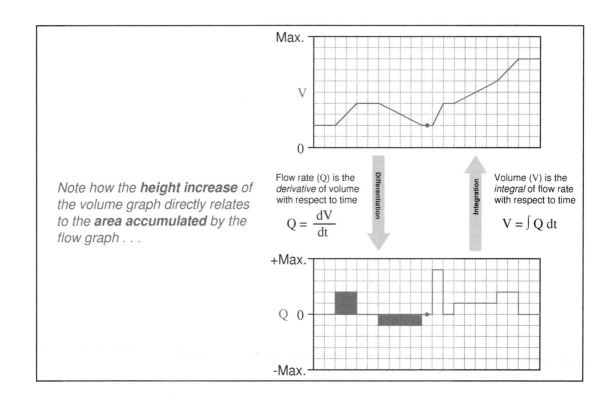

*Note how the **height increase** of the volume graph directly relates to the **area accumulated** by the flow graph . . .*

Flow rate (Q) is the *derivative* of volume with respect to time

$$Q = \frac{dV}{dt}$$

Volume (V) is the *integral* of flow rate with respect to time

$$V = \int Q \, dt$$

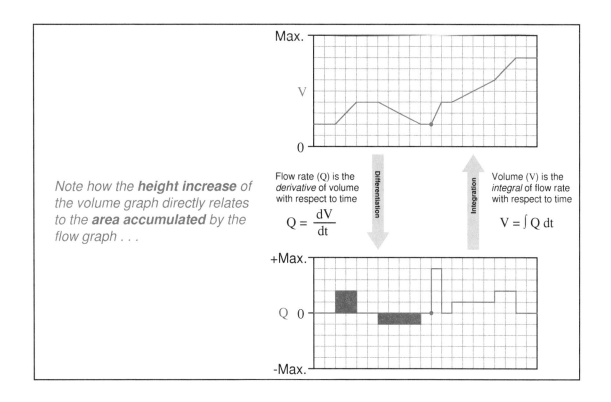

*Note how the **height increase** of the volume graph directly relates to the **area accumulated** by the flow graph . . .*

Flow rate (Q) is the *derivative* of volume with respect to time

$$Q = \frac{dV}{dt}$$

Volume (V) is the *integral* of flow rate with respect to time

$$V = \int Q \, dt$$

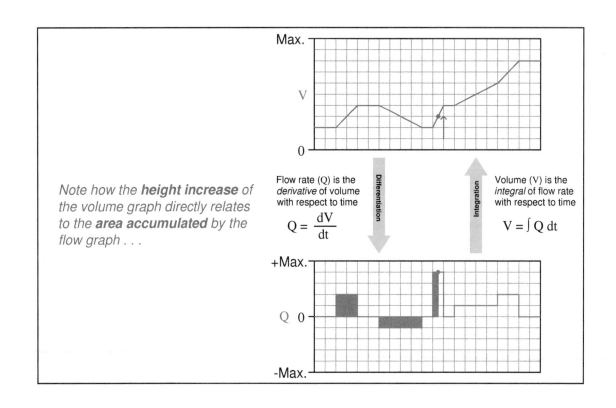

*Note how the **height increase** of the volume graph directly relates to the **area accumulated** by the flow graph . . .*

Flow rate (Q) is the *derivative* of volume with respect to time

$$Q = \frac{dV}{dt}$$

Volume (V) is the *integral* of flow rate with respect to time

$$V = \int Q \, dt$$

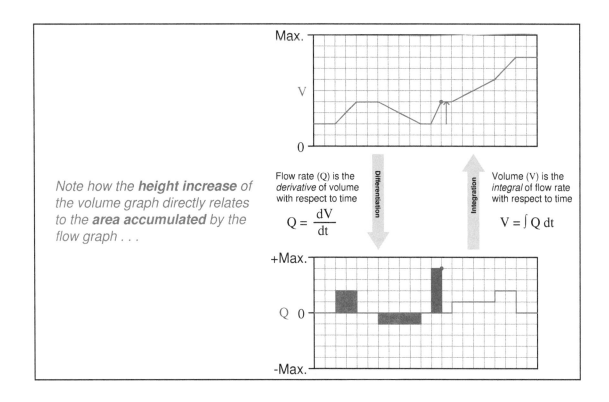

Note how the **height increase** of the volume graph directly relates to the **area accumulated** by the flow graph . . .

Flow rate (Q) is the *derivative* of volume with respect to time

$$Q = \frac{dV}{dt}$$

Volume (V) is the *integral* of flow rate with respect to time

$$V = \int Q \, dt$$

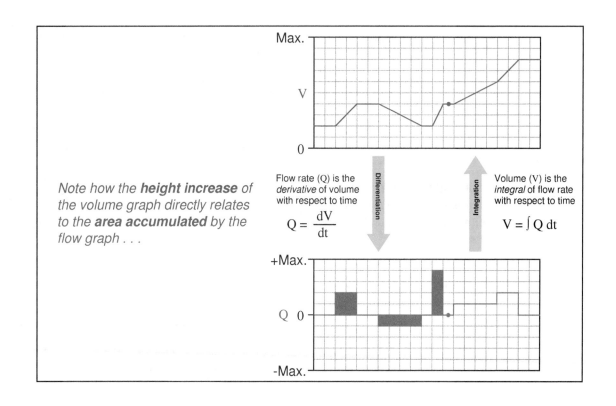

*Note how the **height increase** of the volume graph directly relates to the **area accumulated** by the flow graph . . .*

Flow rate (Q) is the *derivative* of volume with respect to time

$$Q = \frac{dV}{dt}$$

Volume (V) is the *integral* of flow rate with respect to time

$$V = \int Q \, dt$$

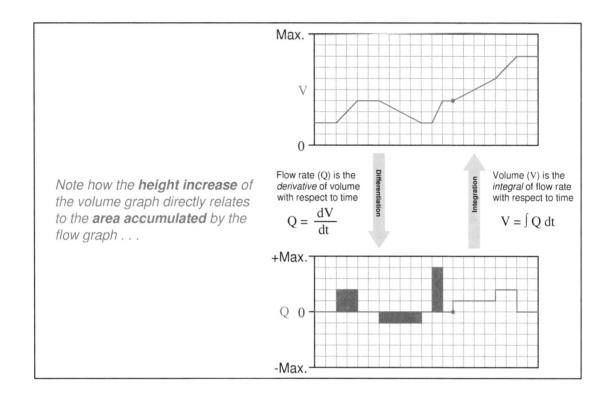

*Note how the **height increase** of the volume graph directly relates to the **area accumulated** by the flow graph . . .*

Flow rate (Q) is the *derivative* of volume with respect to time

$$Q = \frac{dV}{dt}$$

Volume (V) is the *integral* of flow rate with respect to time

$$V = \int Q \, dt$$

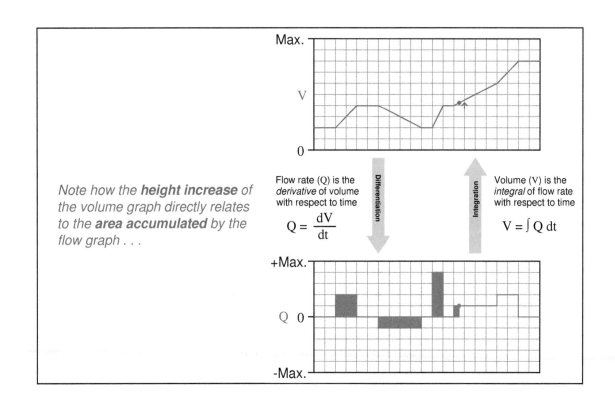

*Note how the **height increase** of the volume graph directly relates to the **area accumulated** by the flow graph . . .*

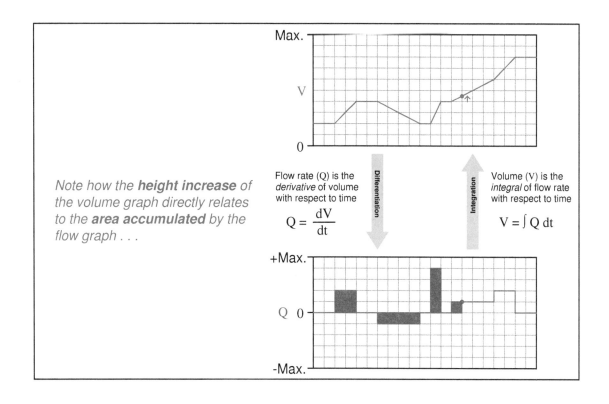

*Note how the **height increase** of the volume graph directly relates to the **area accumulated** by the flow graph . . .*

Flow rate (Q) is the *derivative* of volume with respect to time

$$Q = \frac{dV}{dt}$$

Volume (V) is the *integral* of flow rate with respect to time

$$V = \int Q \, dt$$

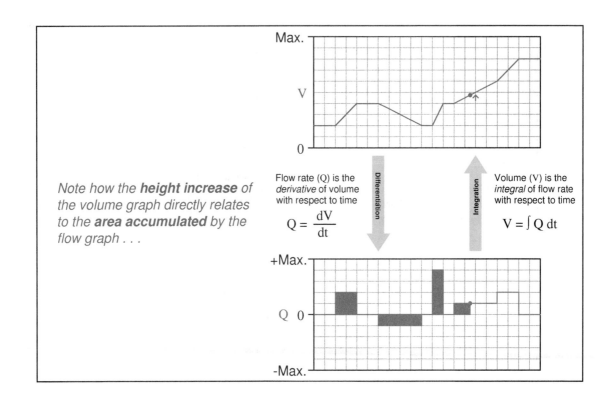

*Note how the **height increase** of the volume graph directly relates to the **area accumulated** by the flow graph . . .*

Flow rate (Q) is the *derivative* of volume with respect to time

$$Q = \frac{dV}{dt}$$

Volume (V) is the *integral* of flow rate with respect to time

$$V = \int Q \, dt$$

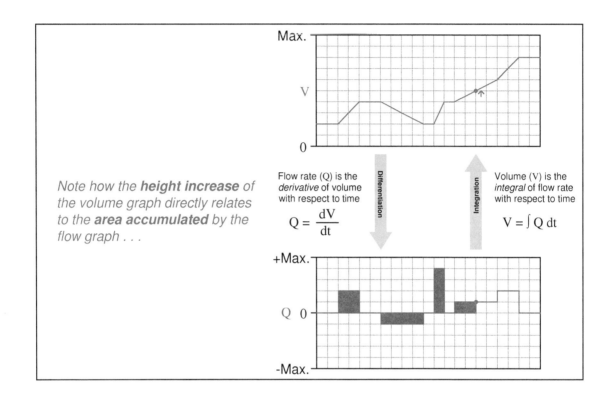

*Note how the **height increase** of the volume graph directly relates to the **area accumulated** by the flow graph . . .*

Flow rate (Q) is the *derivative* of volume with respect to time

$$Q = \frac{dV}{dt}$$

Volume (V) is the *integral* of flow rate with respect to time

$$V = \int Q \, dt$$

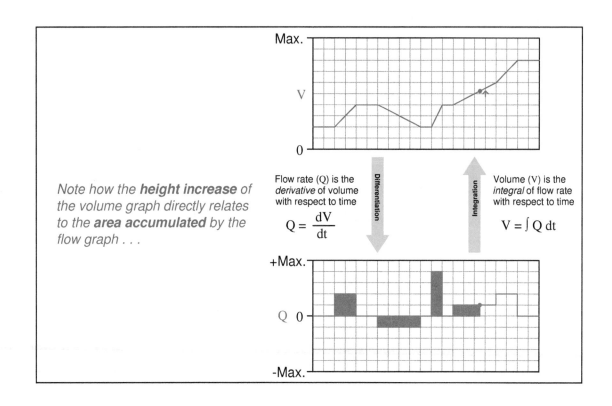

*Note how the **height increase** of the volume graph directly relates to the **area accumulated** by the flow graph . . .*

Flow rate (Q) is the *derivative* of volume with respect to time

$$Q = \frac{dV}{dt}$$

Volume (V) is the *integral* of flow rate with respect to time

$$V = \int Q \, dt$$

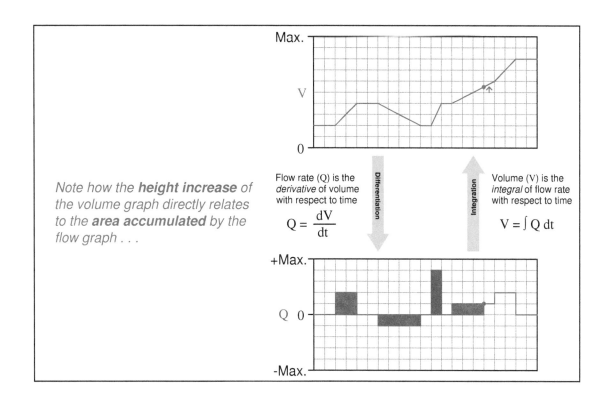

*Note how the **height increase** of the volume graph directly relates to the **area accumulated** by the flow graph . . .*

Flow rate (Q) is the *derivative* of volume with respect to time

$$Q = \frac{dV}{dt}$$

Volume (V) is the *integral* of flow rate with respect to time

$$V = \int Q \, dt$$

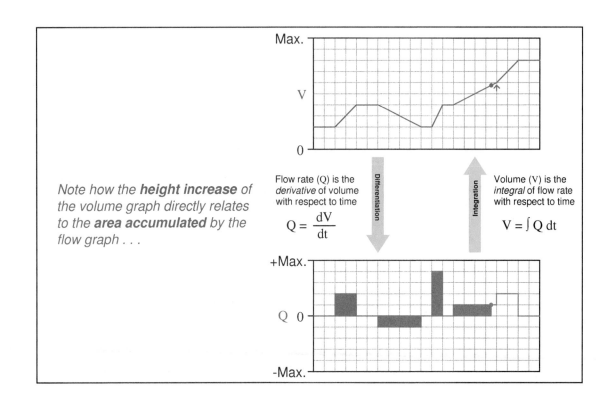

*Note how the **height increase** of the volume graph directly relates to the **area accumulated** by the flow graph . . .*

Flow rate (Q) is the *derivative* of volume with respect to time

$$Q = \frac{dV}{dt}$$

Volume (V) is the *integral* of flow rate with respect to time

$$V = \int Q \, dt$$

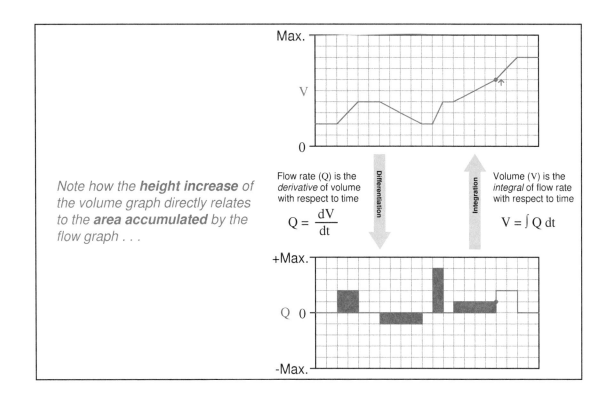

*Note how the **height increase** of the volume graph directly relates to the **area accumulated** by the flow graph . . .*

Flow rate (Q) is the *derivative* of volume with respect to time

$$Q = \frac{dV}{dt}$$

Volume (V) is the *integral* of flow rate with respect to time

$$V = \int Q \, dt$$

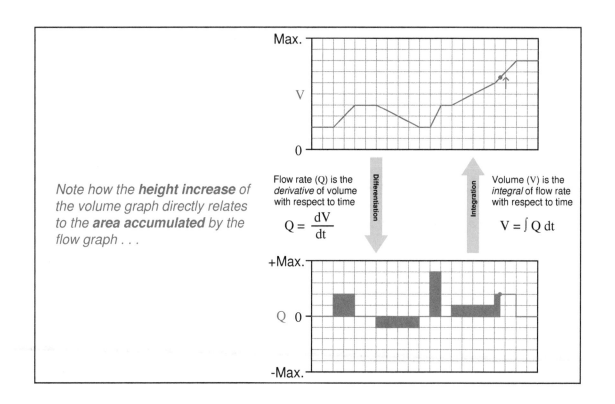

*Note how the **height increase** of the volume graph directly relates to the **area accumulated** by the flow graph . . .*

Flow rate (Q) is the *derivative* of volume with respect to time

$$Q = \frac{dV}{dt}$$

Volume (V) is the *integral* of flow rate with respect to time

$$V = \int Q \, dt$$

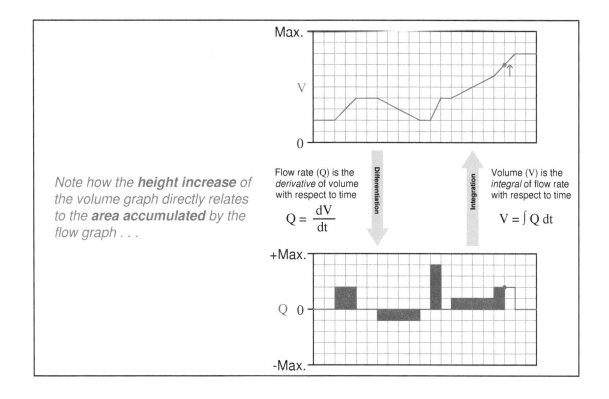

*Note how the **height increase** of the volume graph directly relates to the **area accumulated** by the flow graph . . .*

Flow rate (Q) is the *derivative* of volume with respect to time

$$Q = \frac{dV}{dt}$$

Volume (V) is the *integral* of flow rate with respect to time

$$V = \int Q \, dt$$

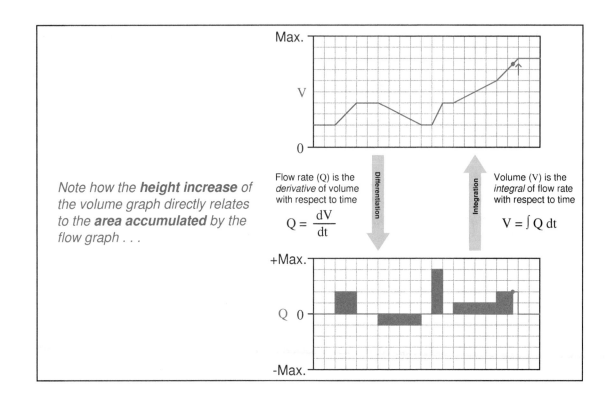

*Note how the **height increase** of the volume graph directly relates to the **area accumulated** by the flow graph . . .*

Flow rate (Q) is the *derivative* of volume with respect to time

$$Q = \frac{dV}{dt}$$

Volume (V) is the *integral* of flow rate with respect to time

$$V = \int Q\, dt$$

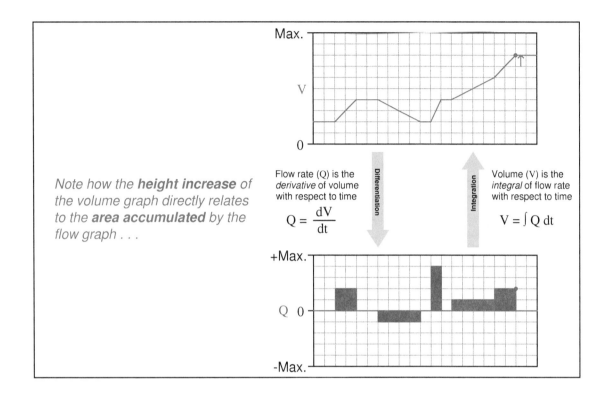

*Note how the **height increase** of the volume graph directly relates to the **area accumulated** by the flow graph . . .*

Flow rate (Q) is the *derivative* of volume with respect to time

$$Q = \frac{dV}{dt}$$

Volume (V) is the *integral* of flow rate with respect to time

$$V = \int Q \, dt$$

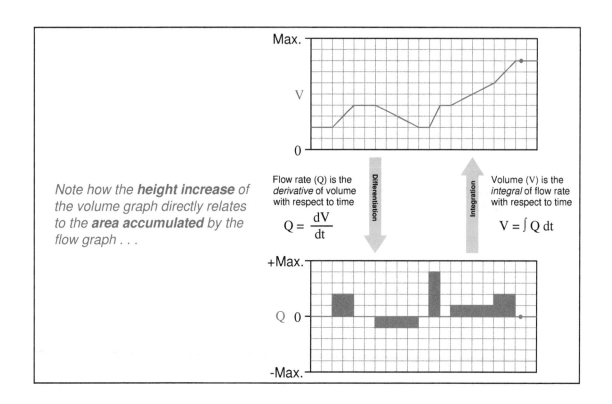

Note how the **height increase** of
the volume graph directly relates
to the **area accumulated** by the
flow graph . . .

Flow rate (Q) is the
derivative of volume
with respect to time

$$Q = \frac{dV}{dt}$$

Volume (V) is the
integral of flow rate
with respect to time

$$V = \int Q\,dt$$

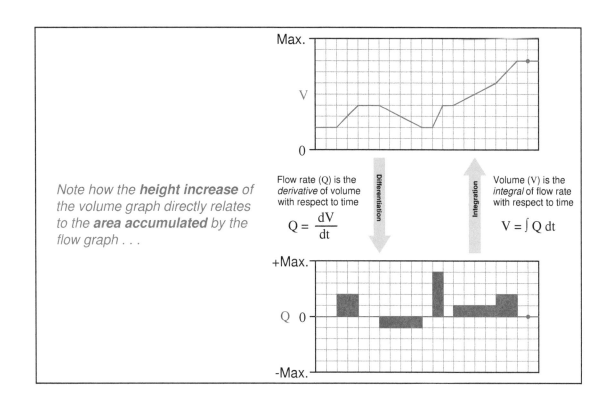

*Note how the **height increase** of the volume graph directly relates to the **area accumulated** by the flow graph . . .*

Flow rate (Q) is the *derivative* of volume with respect to time

$$Q = \frac{dV}{dt}$$

Volume (V) is the *integral* of flow rate with respect to time

$$V = \int Q \, dt$$

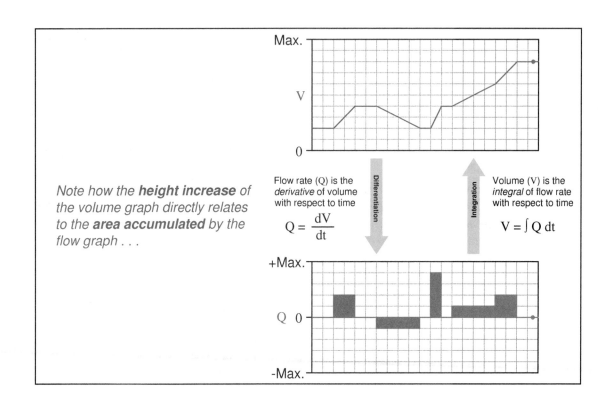

Note how the **height increase** of the volume graph directly relates to the **area accumulated** by the flow graph . . .

Flow rate (Q) is the *derivative* of volume with respect to time

$$Q = \frac{dV}{dt}$$

Volume (V) is the *integral* of flow rate with respect to time

$$V = \int Q \, dt$$

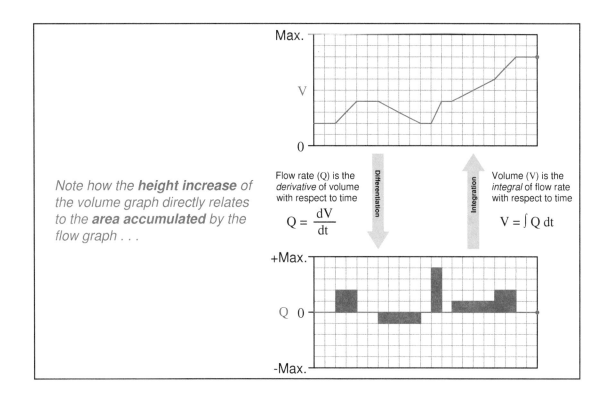

*Note how the **height increase** of the volume graph directly relates to the **area accumulated** by the flow graph . . .*

Flow rate (Q) is the *derivative* of volume with respect to time

$$Q = \frac{dV}{dt}$$

Volume (V) is the *integral* of flow rate with respect to time

$$V = \int Q \, dt$$

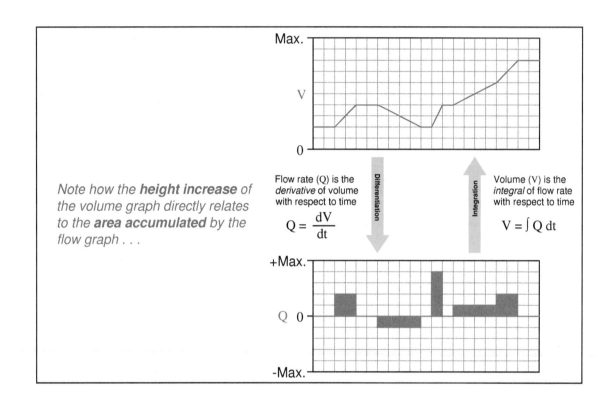

*Note how the **height increase** of the volume graph directly relates to the **area accumulated** by the flow graph . . .*

Flow rate (Q) is the *derivative* of volume with respect to time

$$Q = \frac{dV}{dt}$$

Volume (V) is the *integral* of flow rate with respect to time

$$V = \int Q \, dt$$

A.5 Guided-wave radar level measurement

The following animation shows how a radio-energy pulse travels down and then up the waveguide of a guided-wave radar level instrument, relating the peaks on an *echo curve* to the real-world interfaces inside the process vessel.

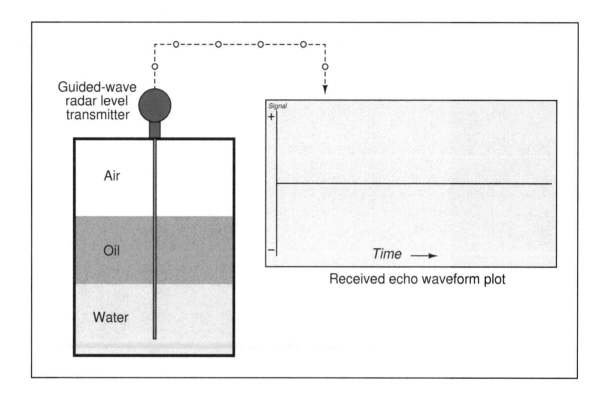

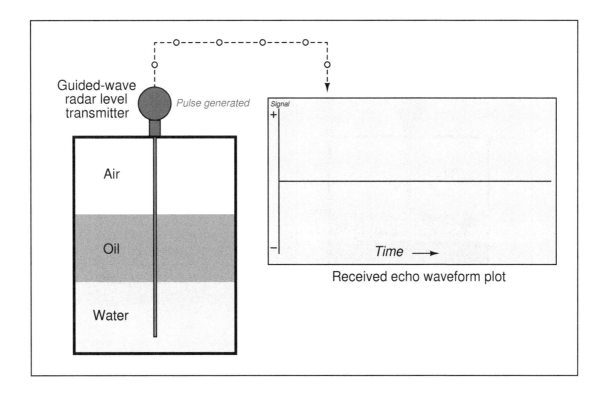

Received echo waveform plot

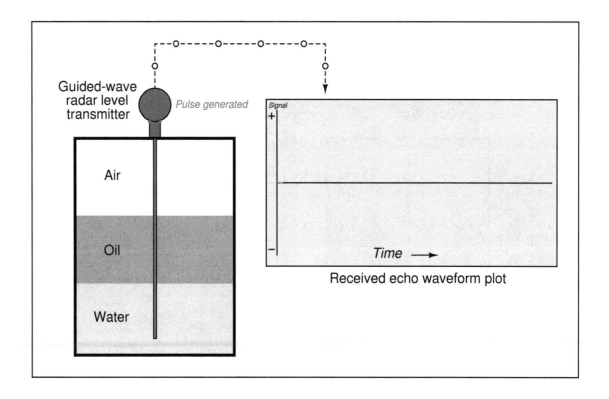

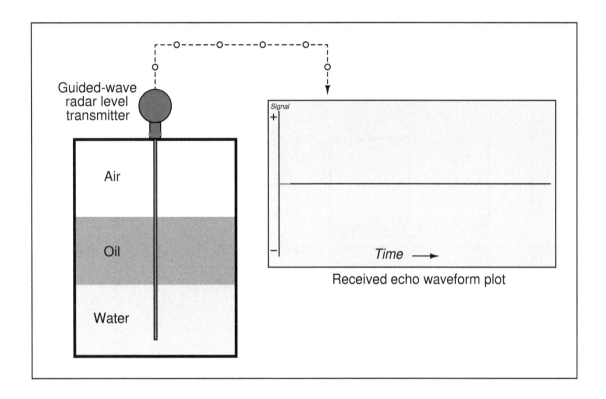

Received echo waveform plot

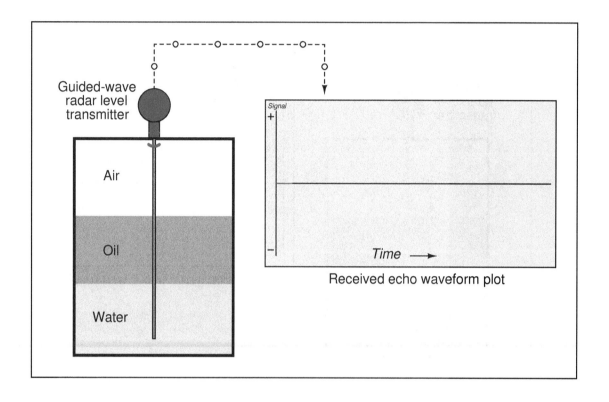

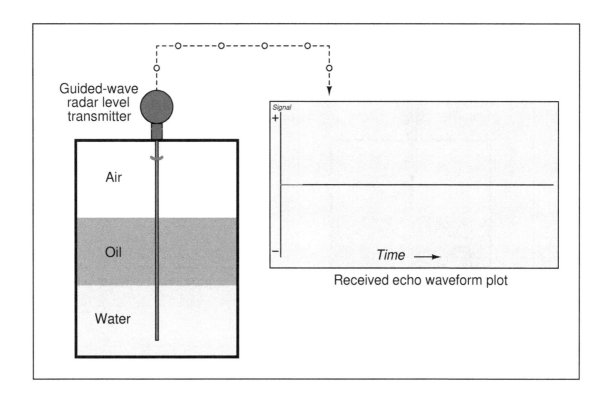

Received echo waveform plot

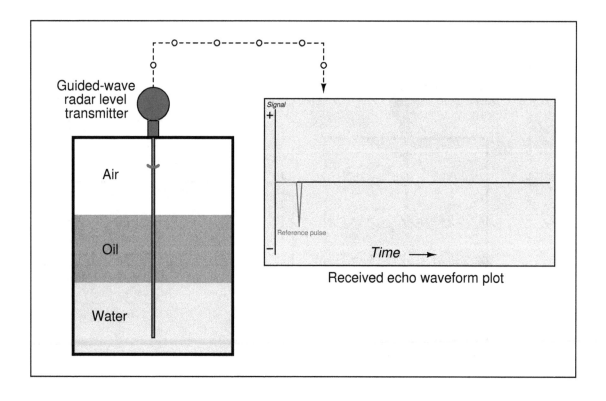

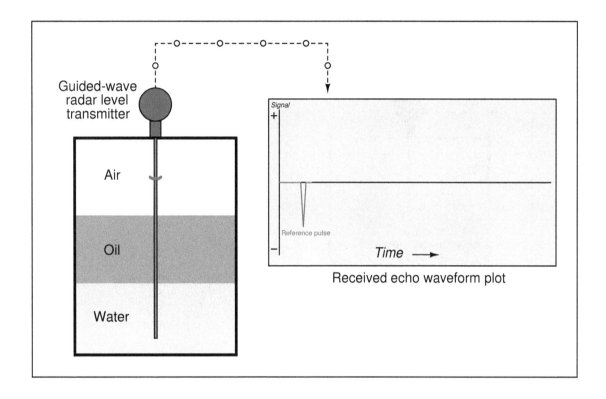

Received echo waveform plot

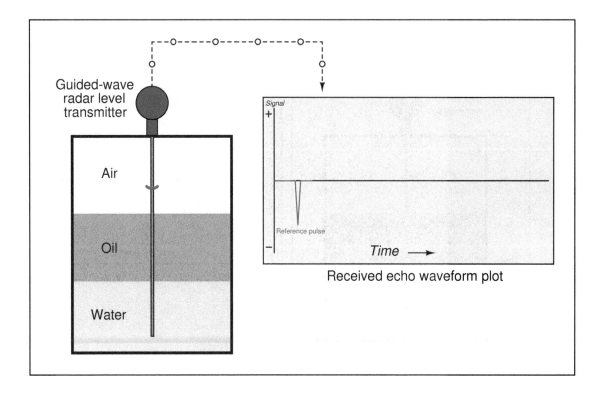

Guided-wave radar level transmitter

Air

Oil

Water

Signal
+

Reference pulse

−

Time →

Received echo waveform plot

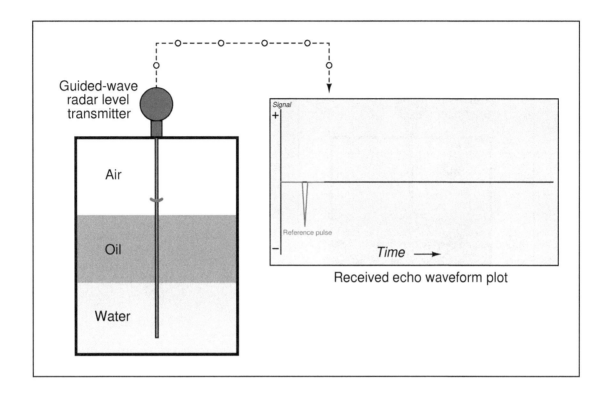

Received echo waveform plot

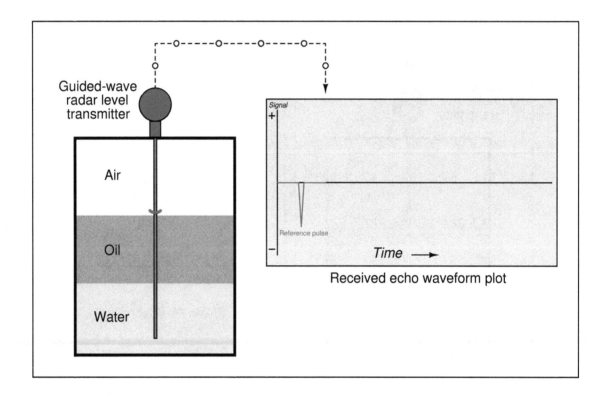

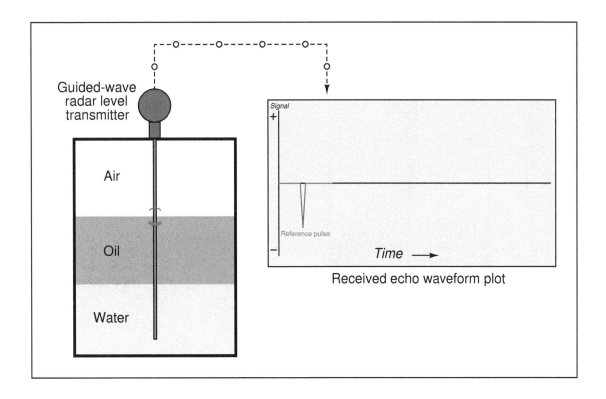

Received echo waveform plot

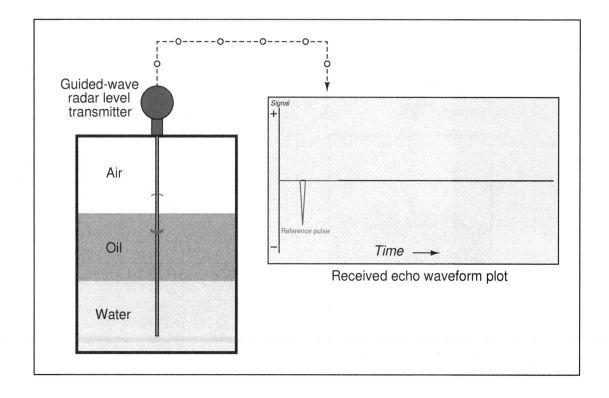

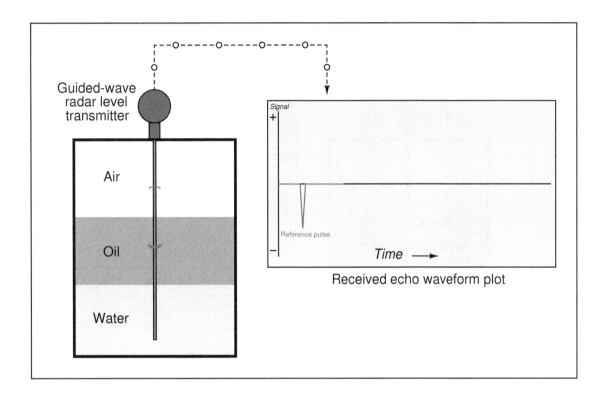

Received echo waveform plot

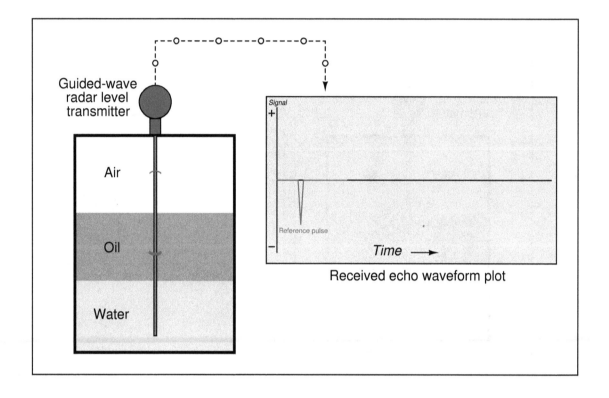

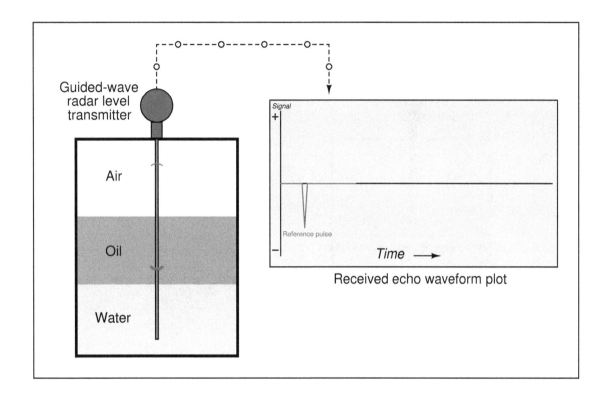

Received echo waveform plot

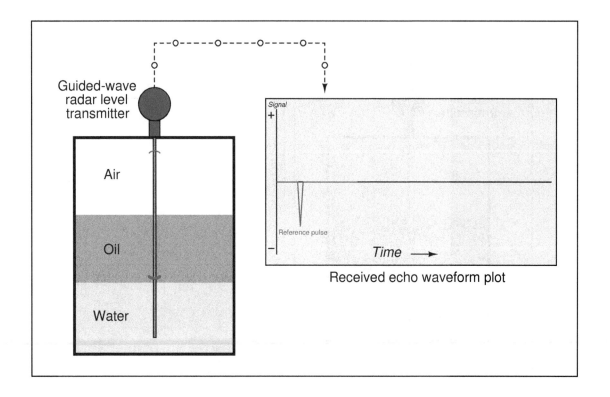

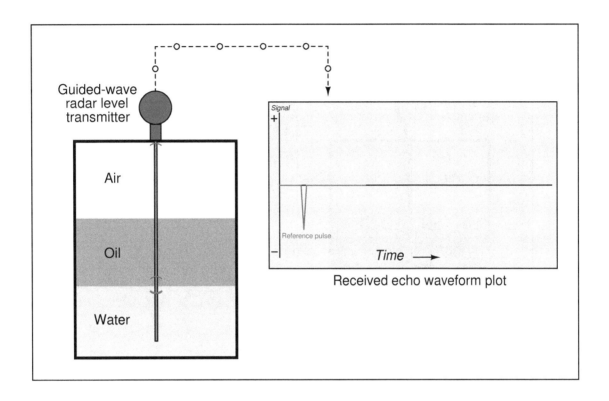

Received echo waveform plot

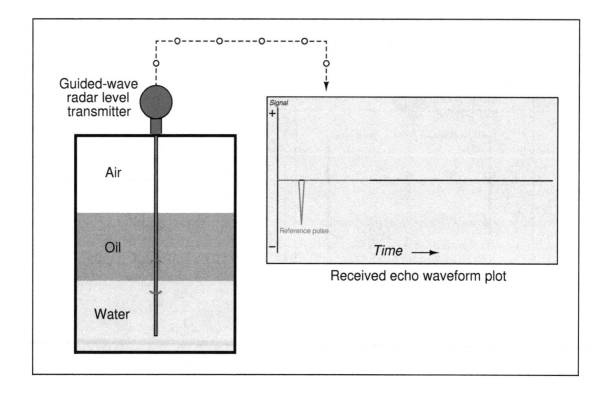

Received echo waveform plot

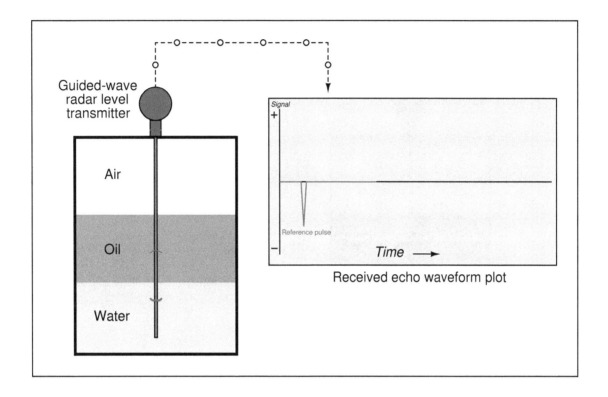

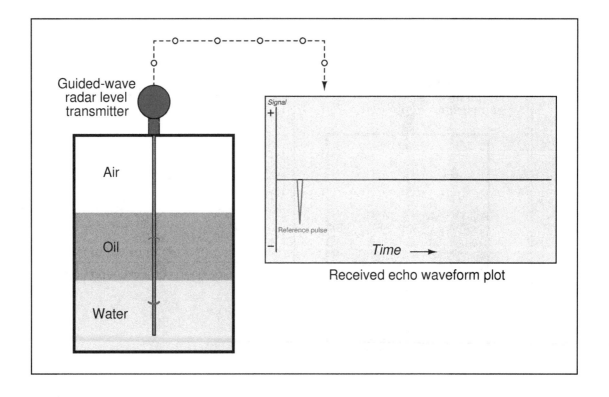

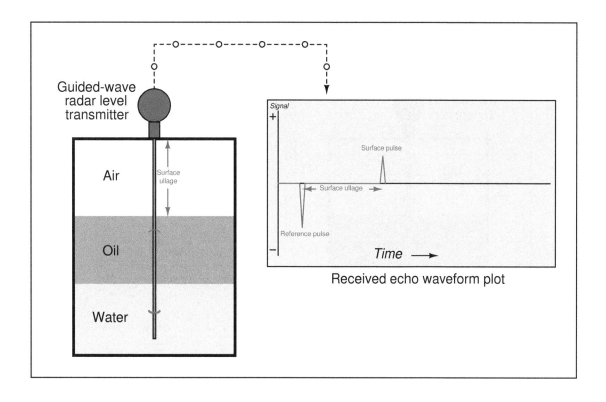

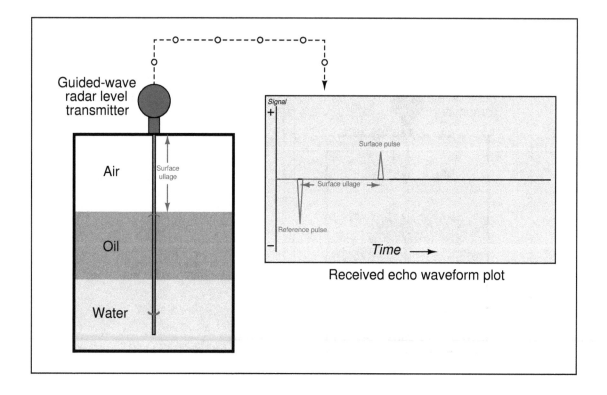

Received echo waveform plot

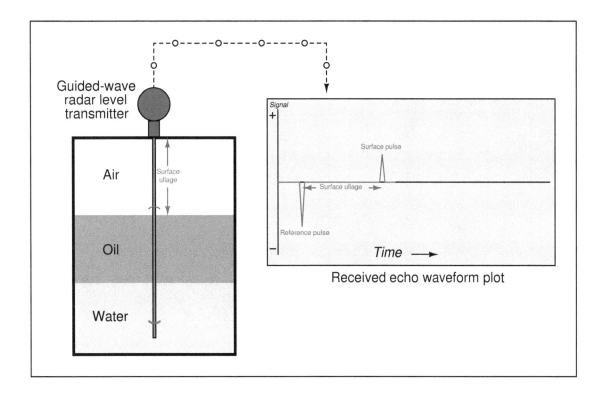

Received echo waveform plot

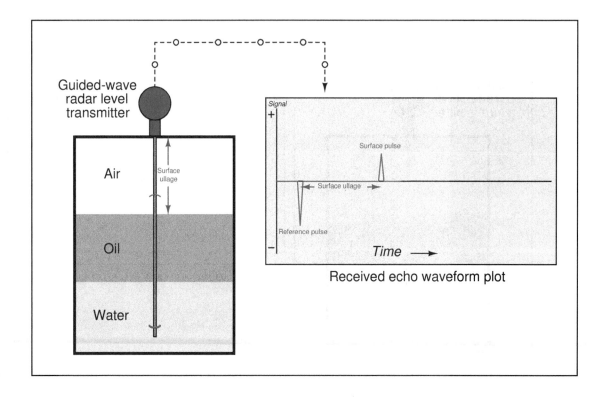

Received echo waveform plot

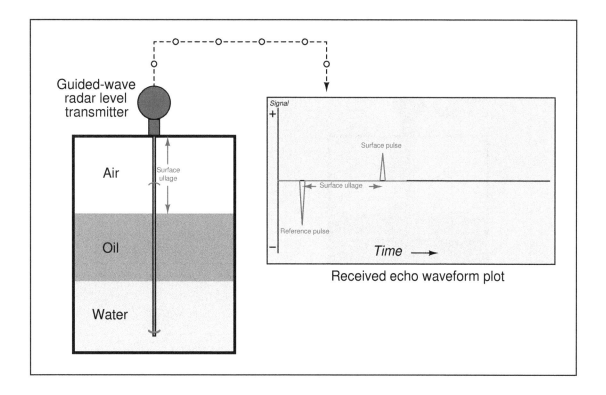

Received echo waveform plot

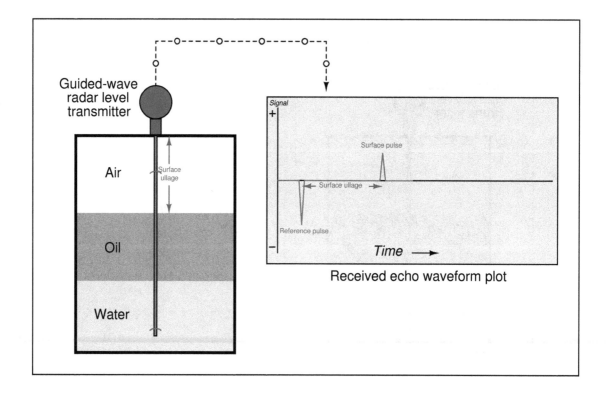

Received echo waveform plot

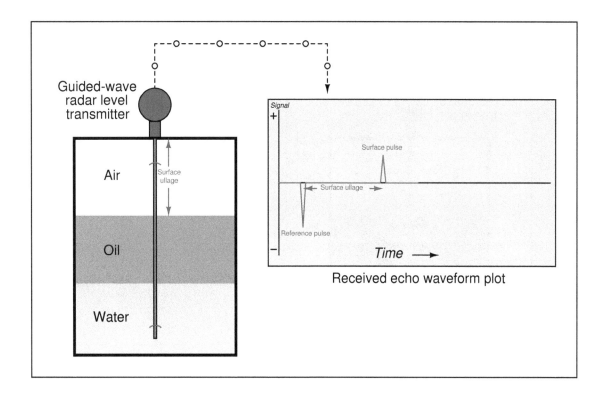

Received echo waveform plot

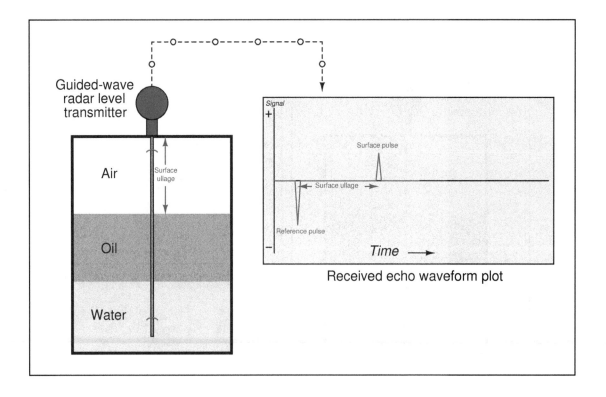

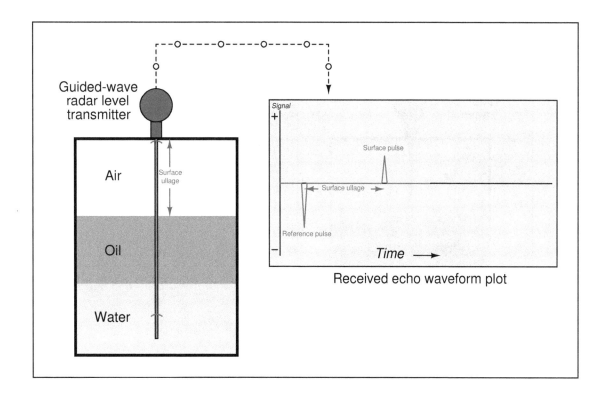

Received echo waveform plot

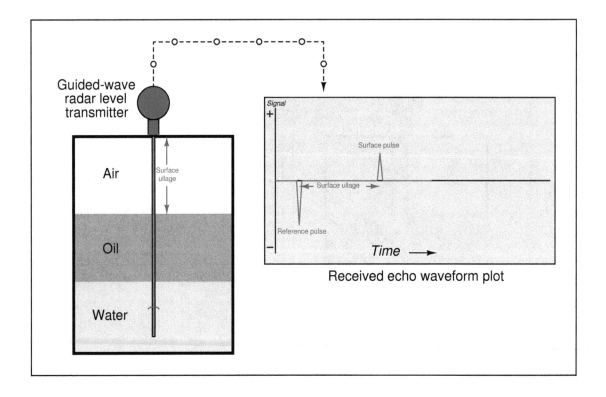

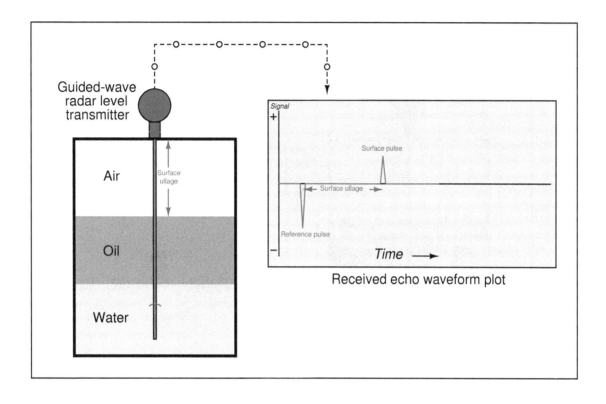

Received echo waveform plot

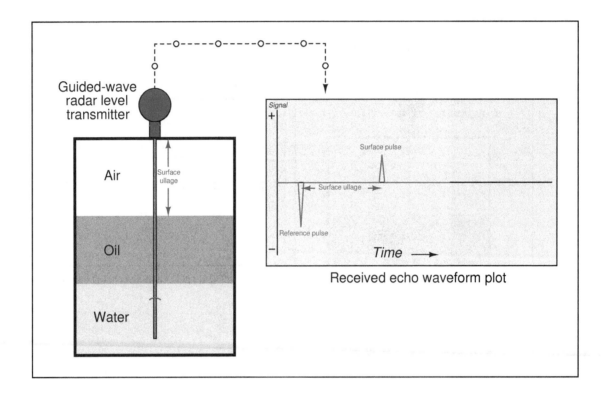

Received echo waveform plot

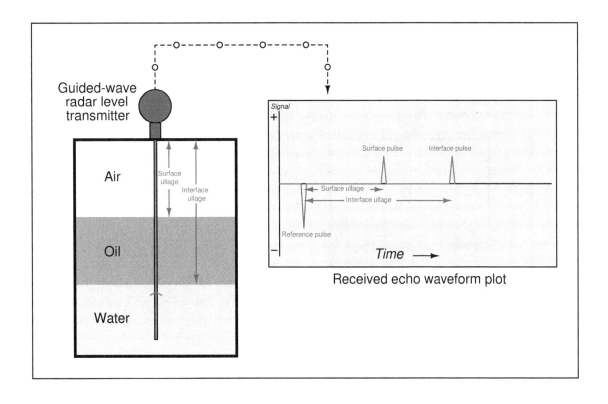

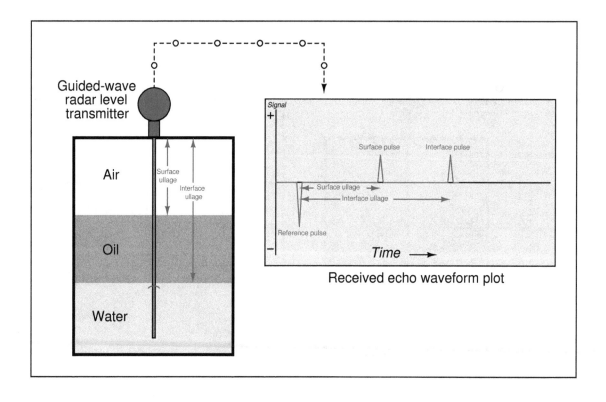

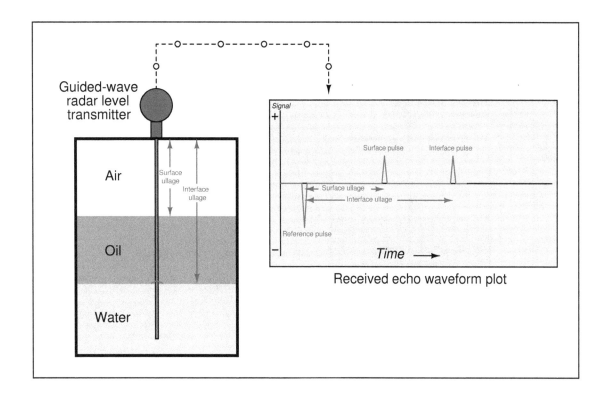

Received echo waveform plot

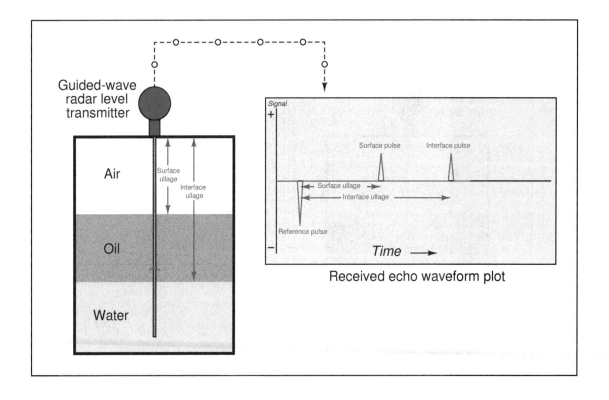

Received echo waveform plot

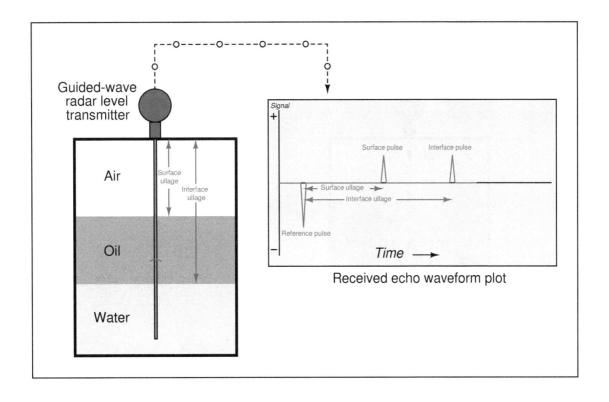

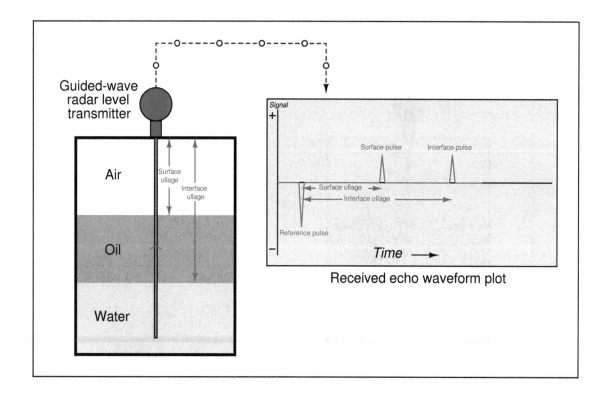

Received echo waveform plot

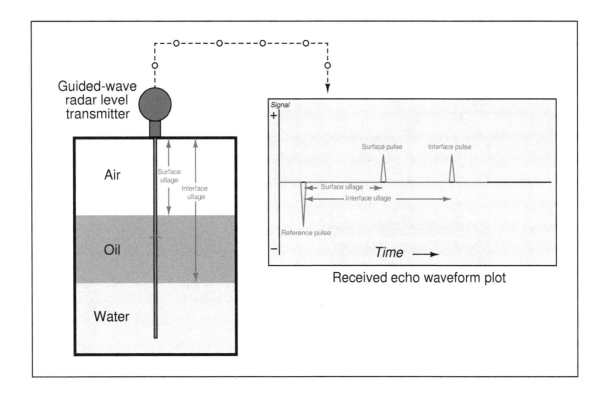

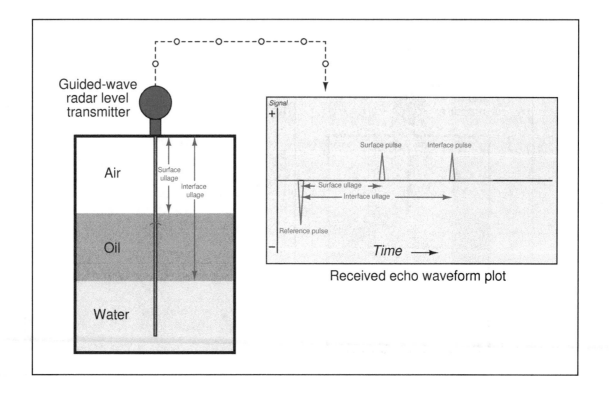

Received echo waveform plot

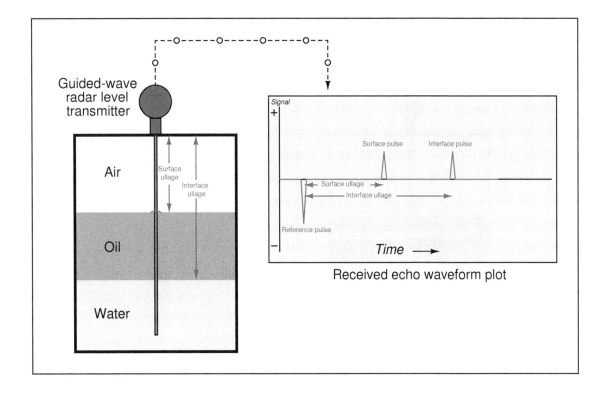

Received echo waveform plot

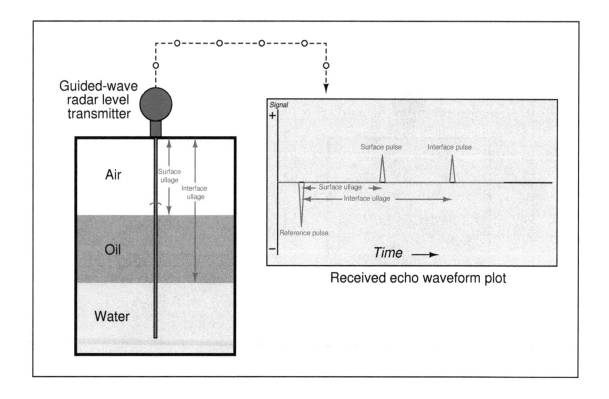

Received echo waveform plot

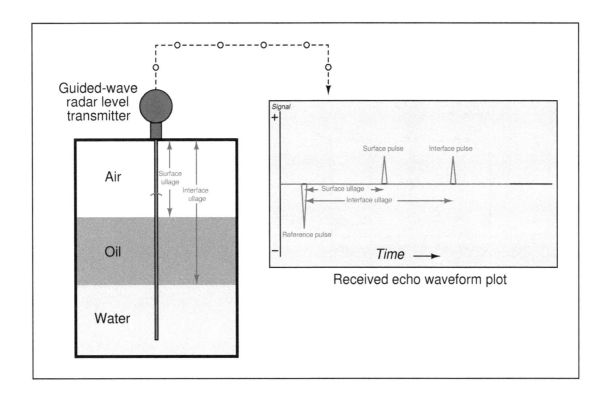

Received echo waveform plot

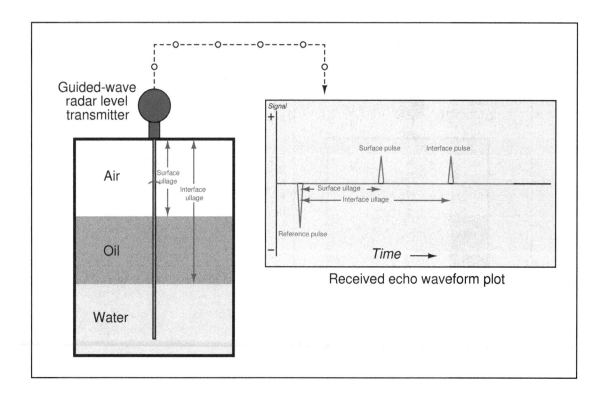

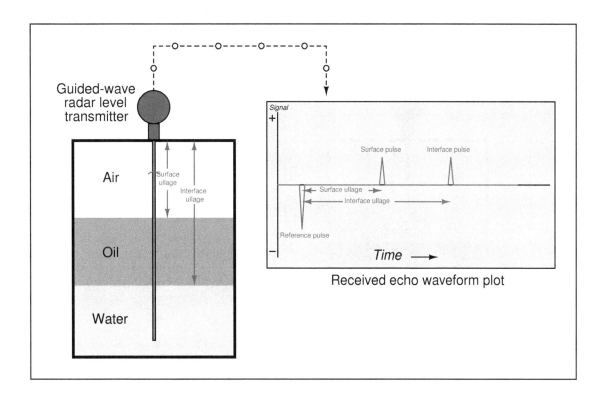

Received echo waveform plot

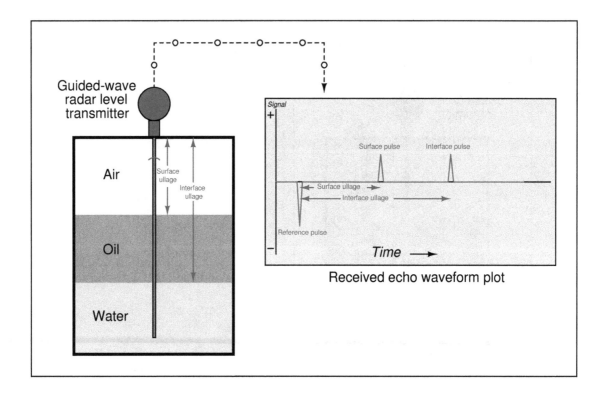

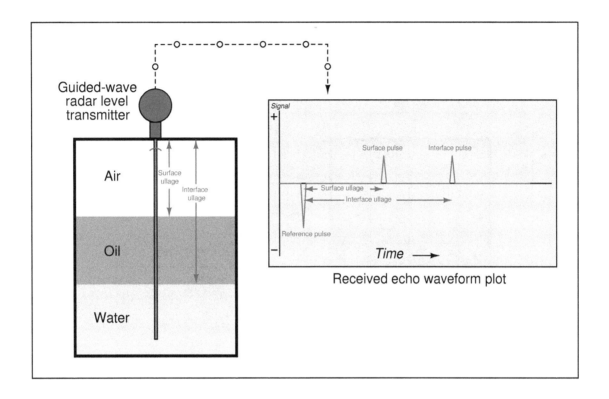

Received echo waveform plot

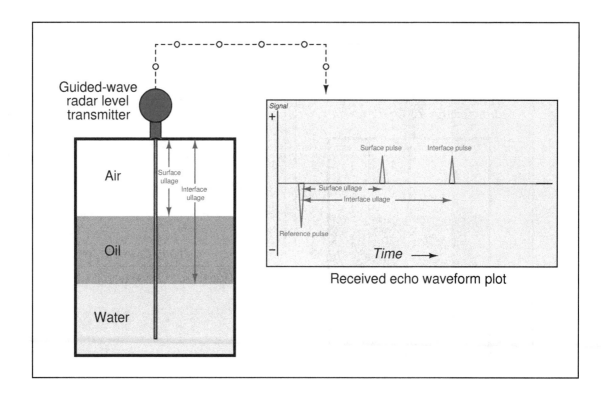

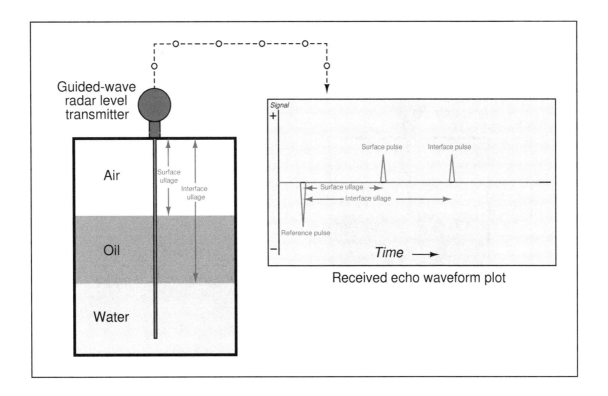

Received echo waveform plot

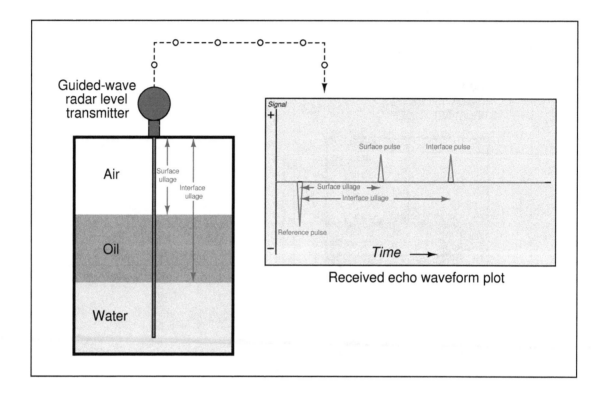

Received echo waveform plot

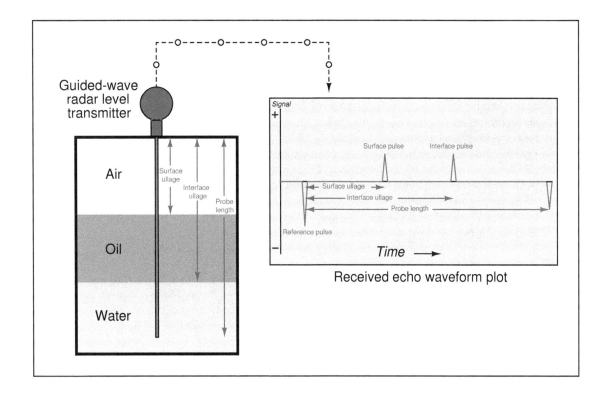

Received echo waveform plot

A.6   Basic chromatograph operation

This animation shows the basic operation of a gas chromatograph, showing the separation of different molecular species in a gas mixture. Each gas type is represented by a different colored dot moving along the tubing.

Carrier gas is represented by orange dots moving constantly through the sample valve and column. Process sample is represented by a cluster of three dots: red (light), green (medium), and blue (heavy) molecules mixed together. These molecules move together at the same rate until they reach the column. There, the light molecules (red) travel fastest, the medium molecules (green) travel slower, and the heavy molecules (blue) travel slowest. Thus, the differing velocities within the chromatograph column performs the task of separation necessary to identify and measure each chemical component in the mixture. All the while, you can see the chromatogram developing, a peak appearing each time one of the components reaches the detector.

Each chemical component (light, medium, heavy) is thus identified by its place in *time* when its peak appears on the chromatogram, while the concentration (quantity) of each component is discernible by the *area* integrated underneath each peak.

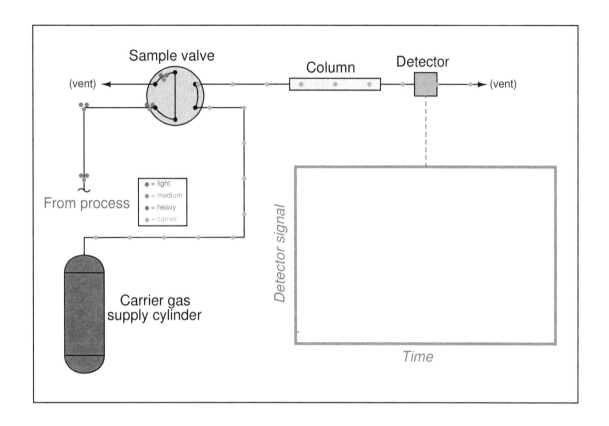

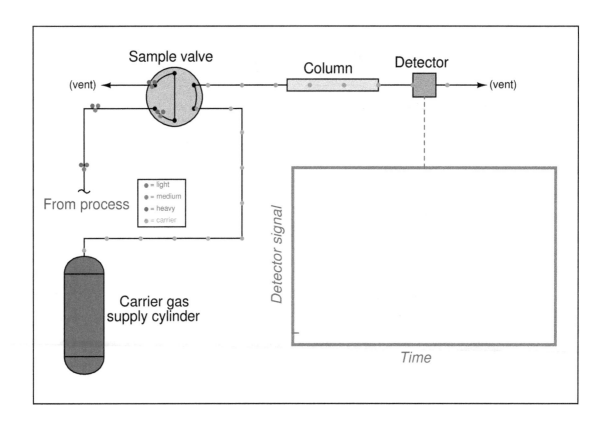

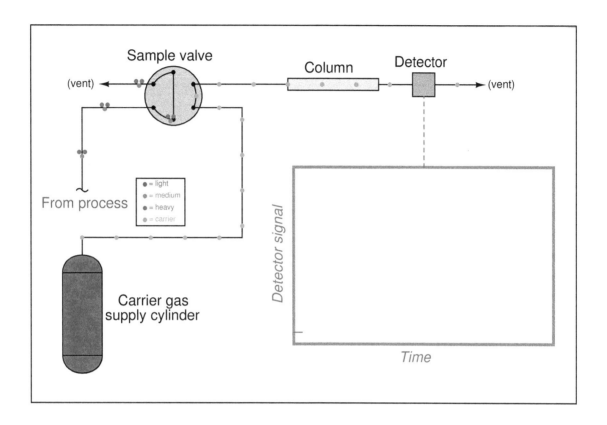

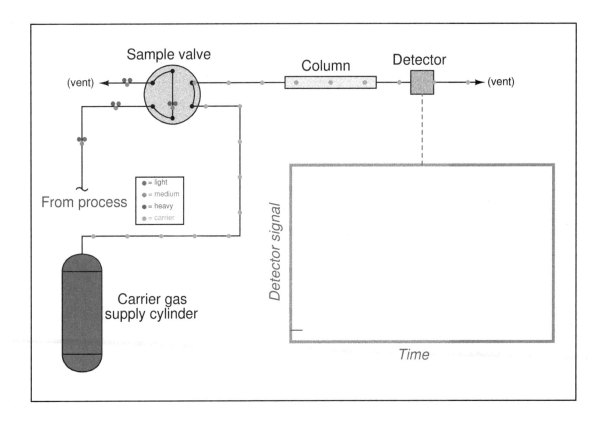

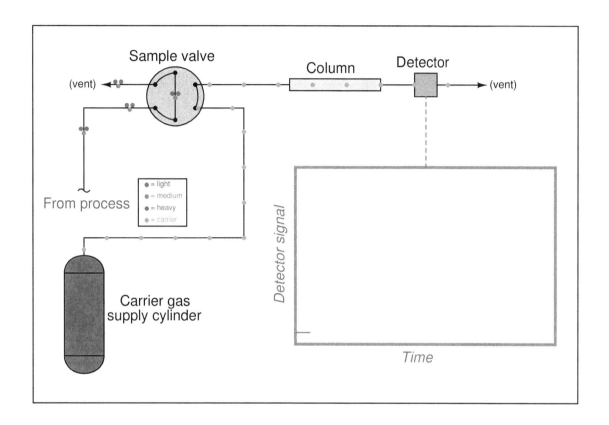

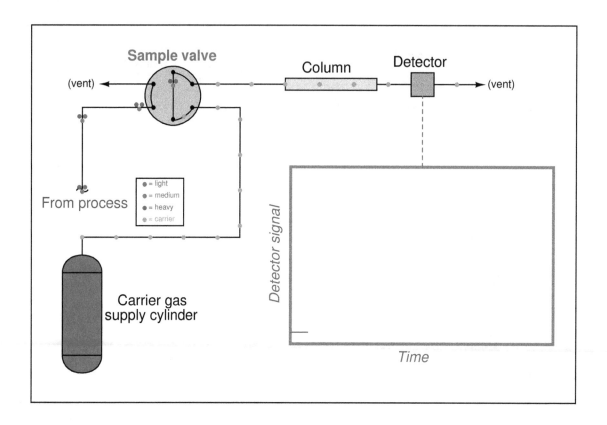

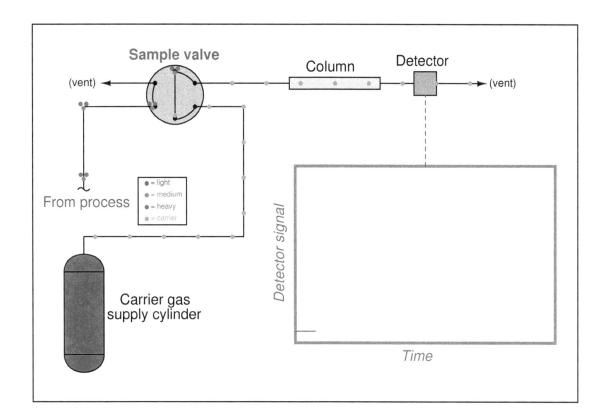

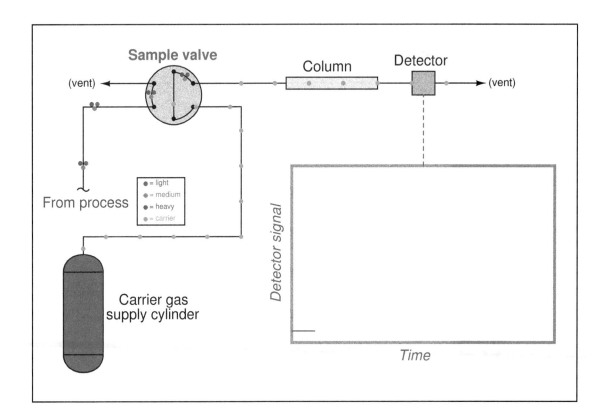

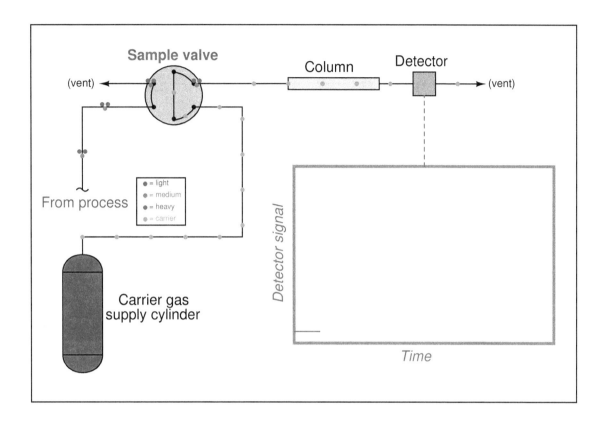

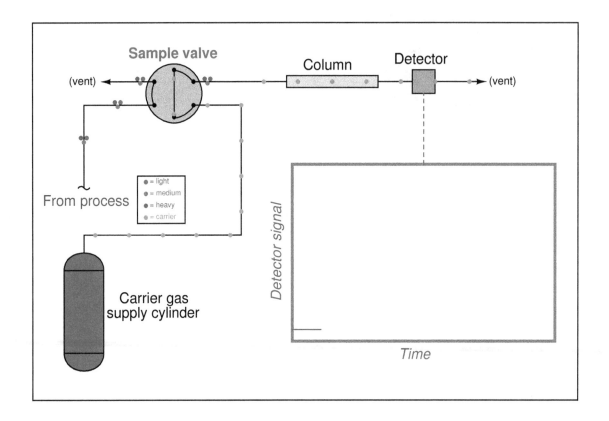

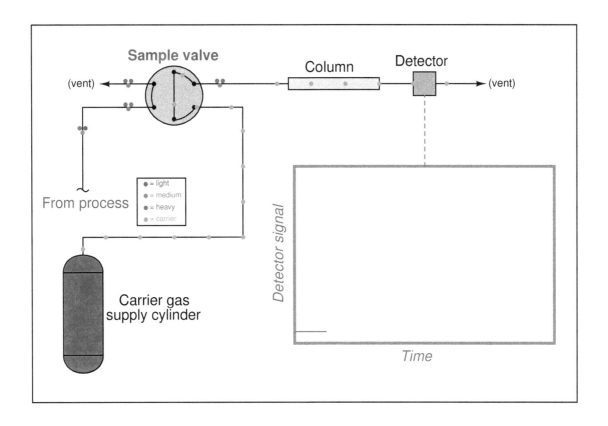

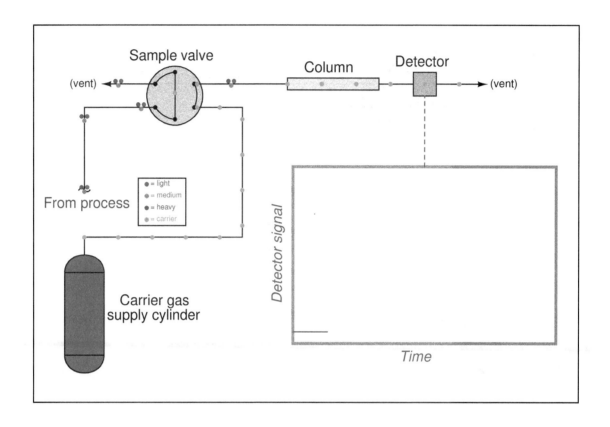

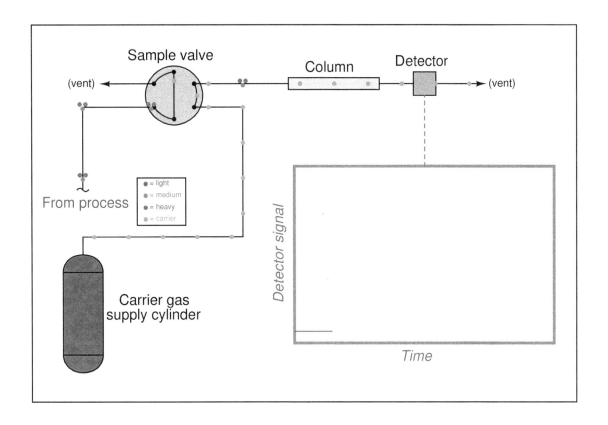

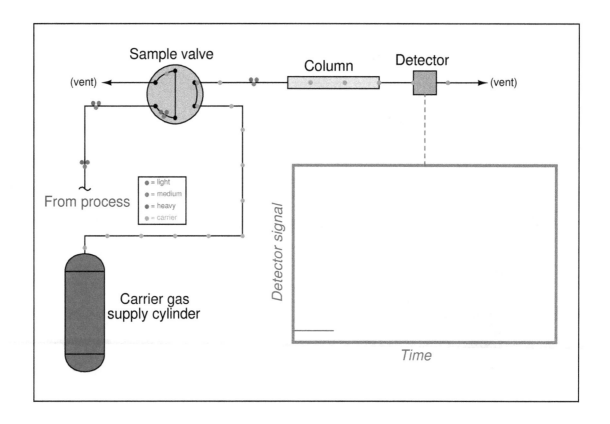

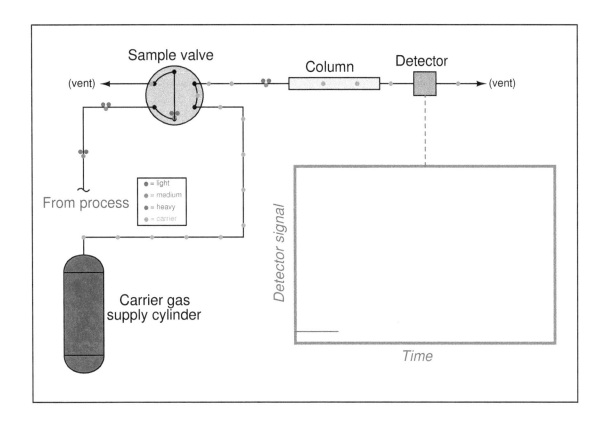

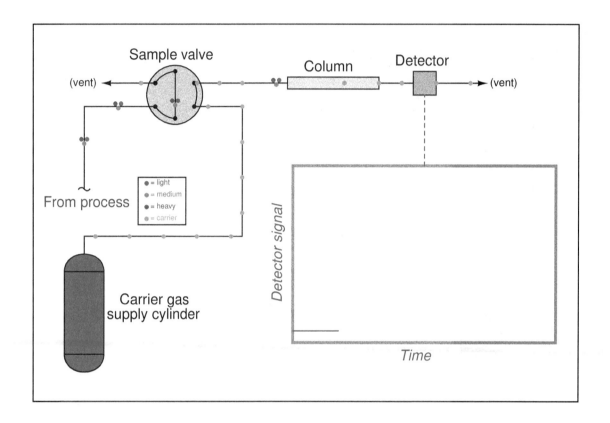

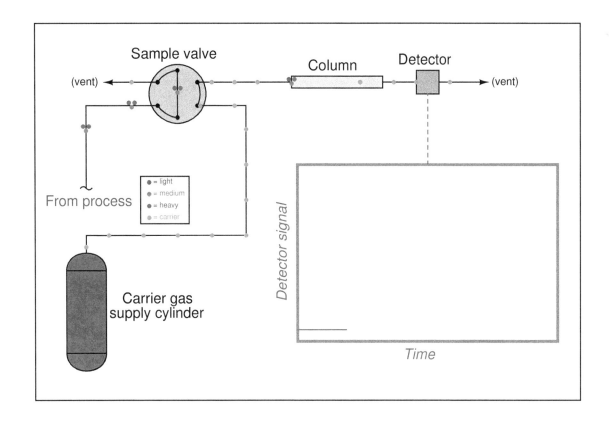

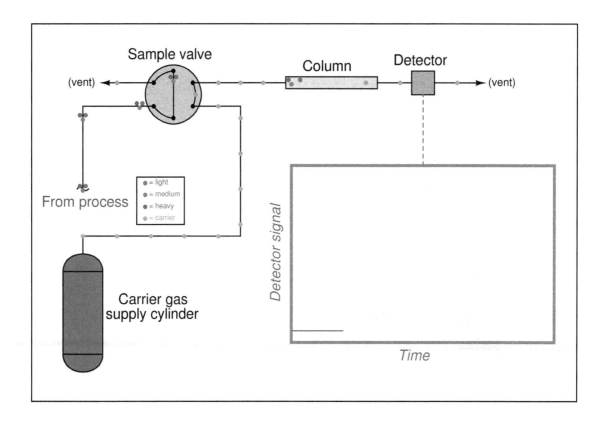

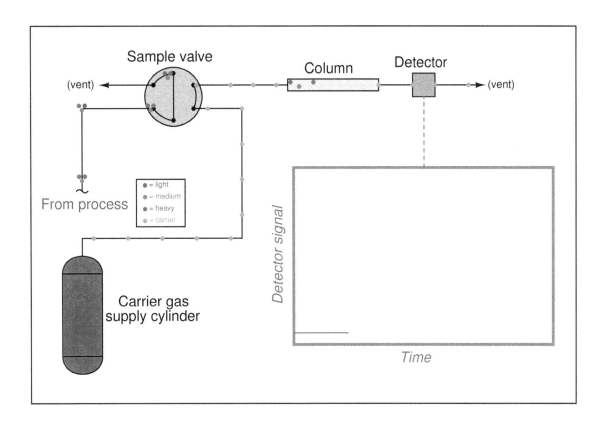

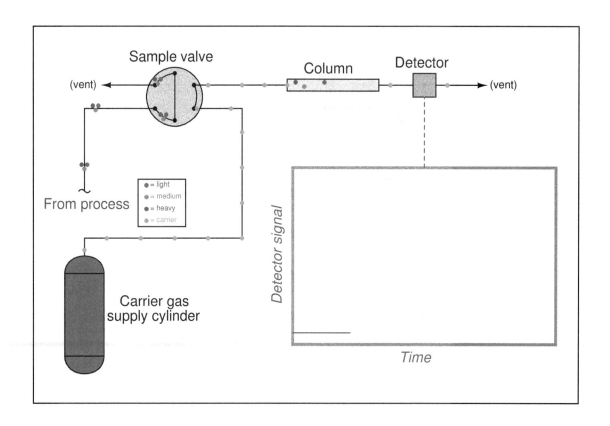

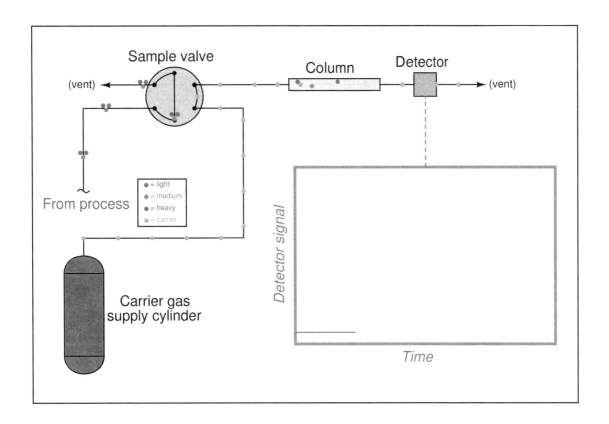

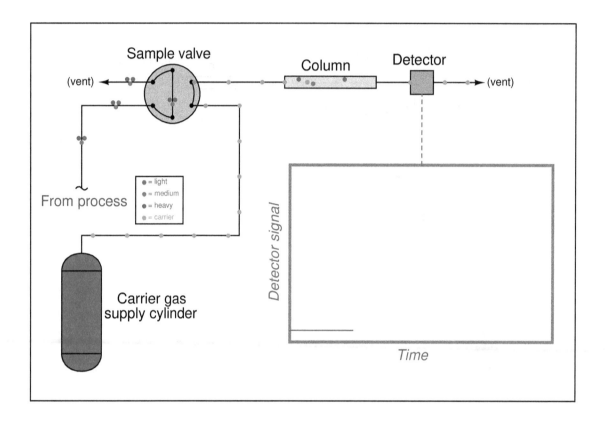

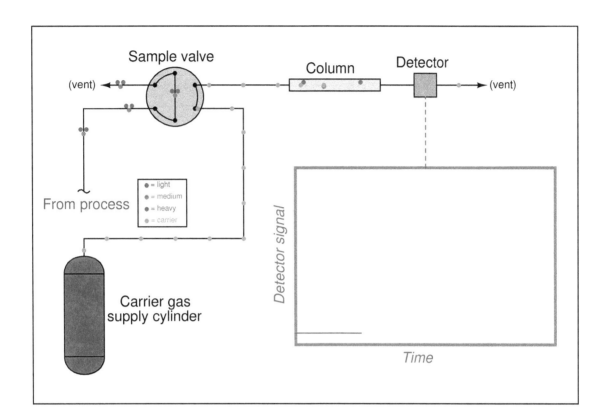

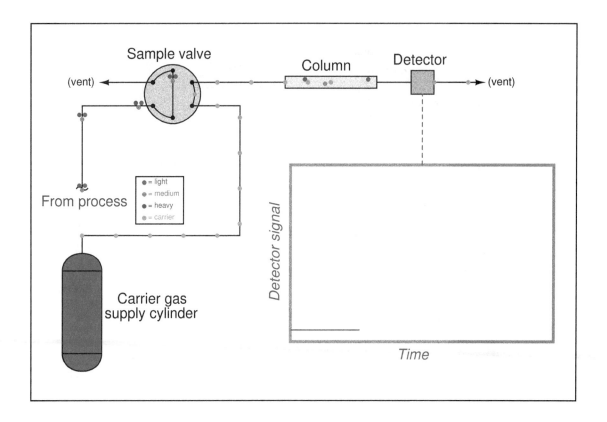

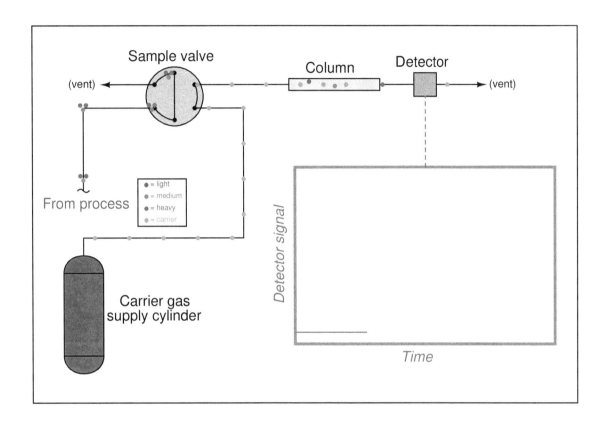

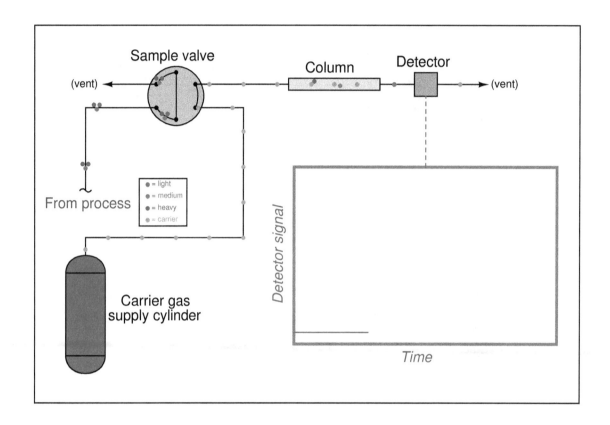

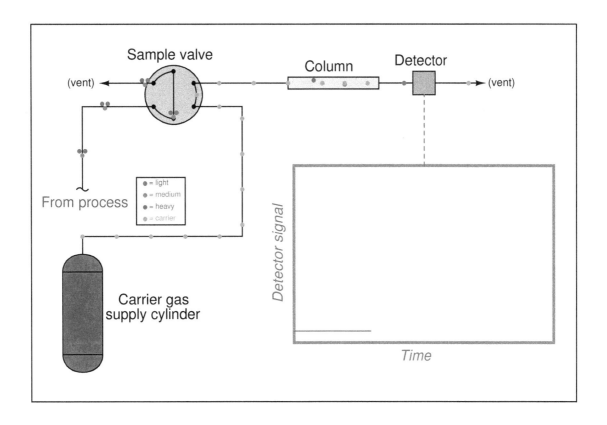

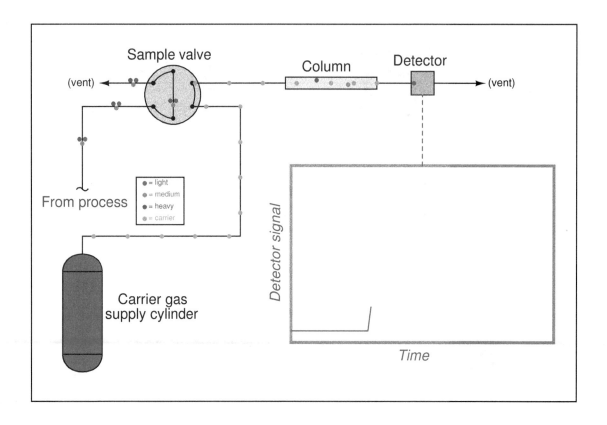

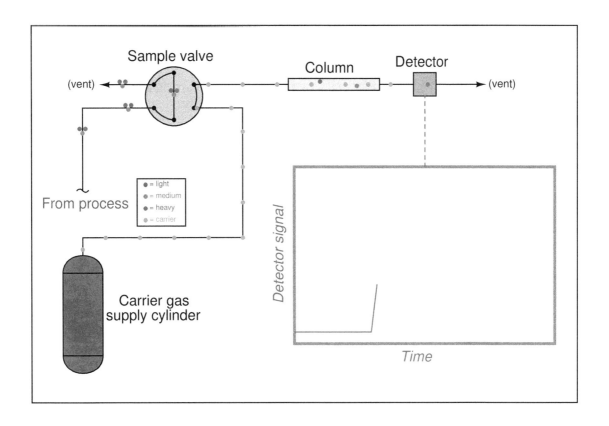

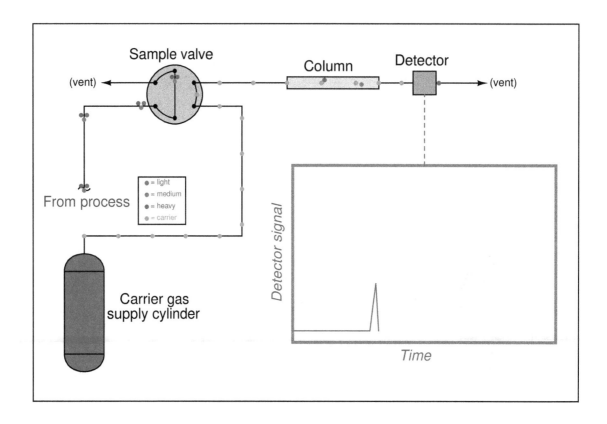

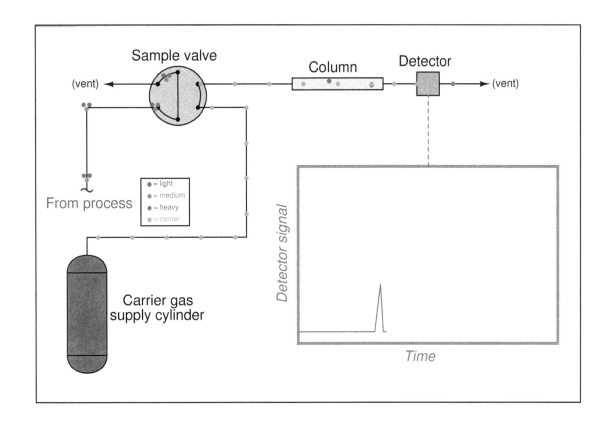

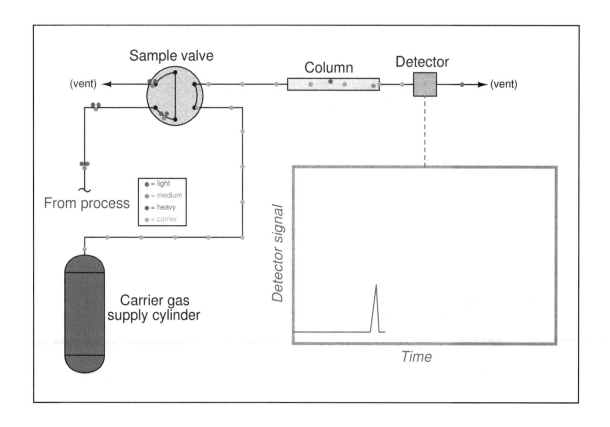

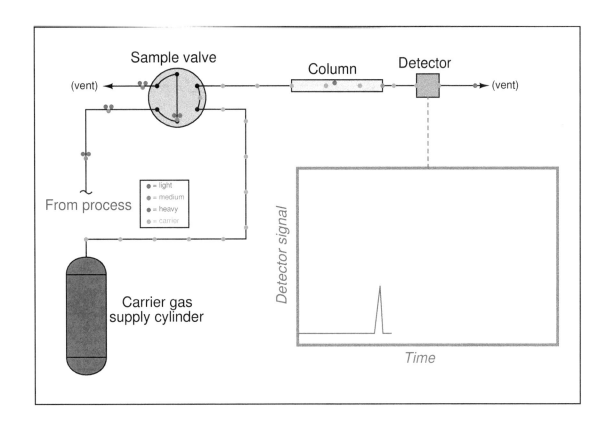

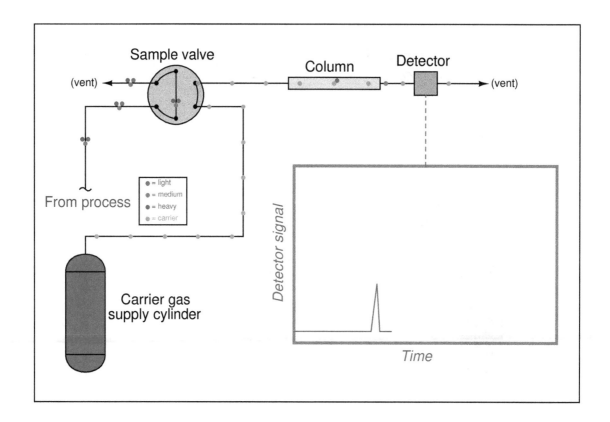

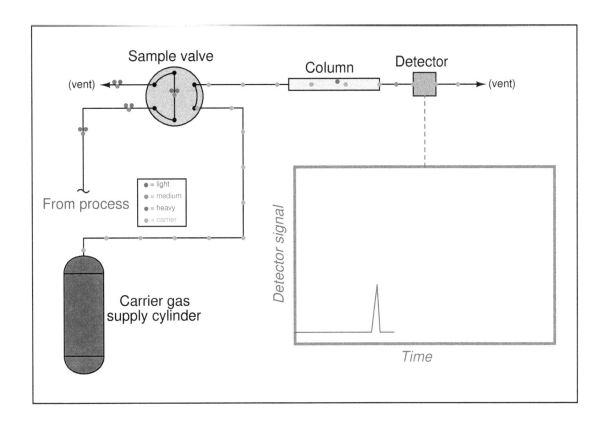

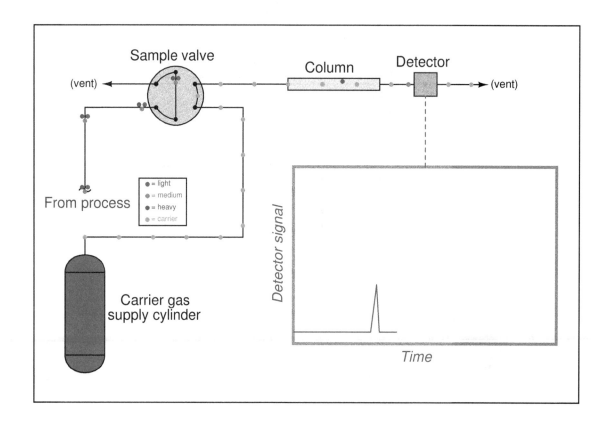

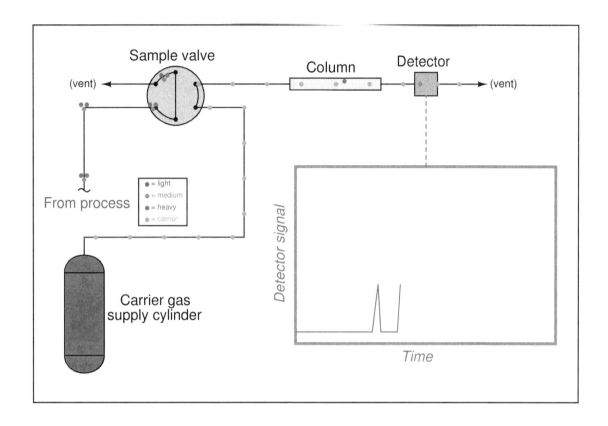

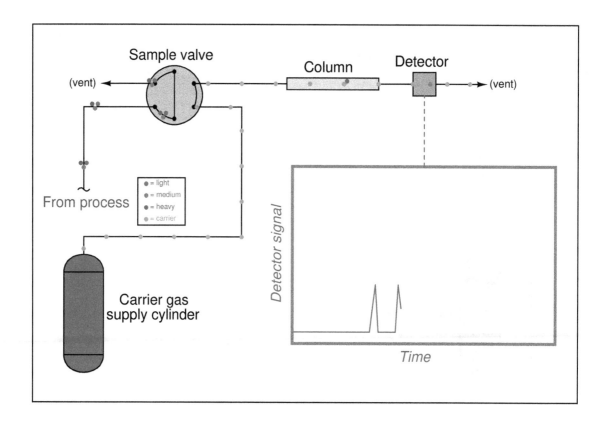

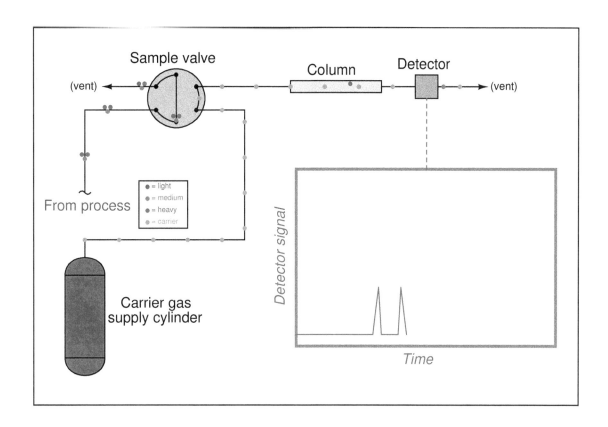

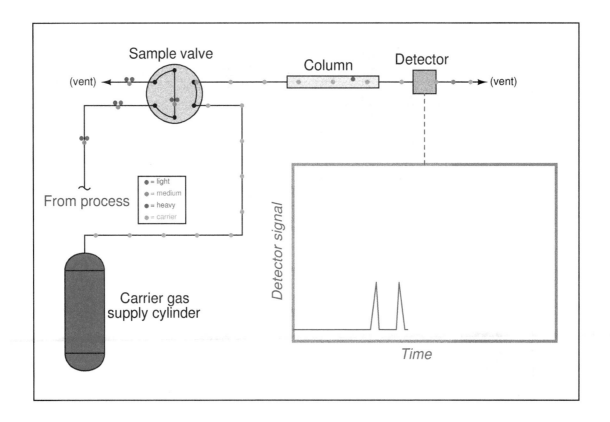

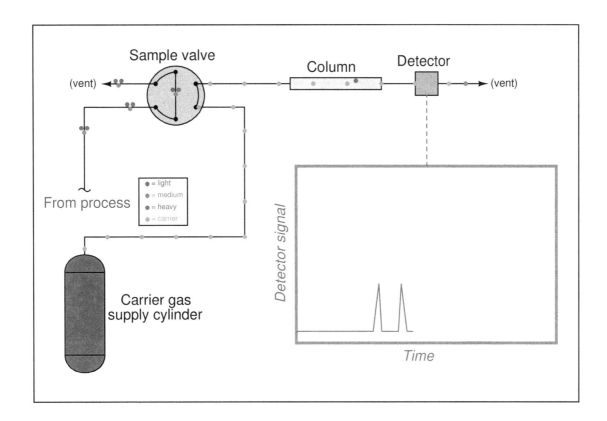

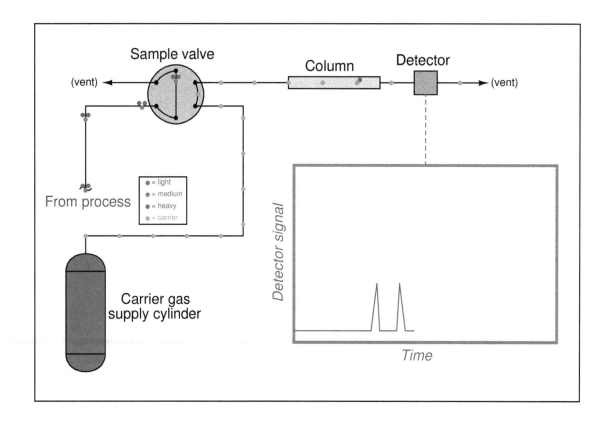

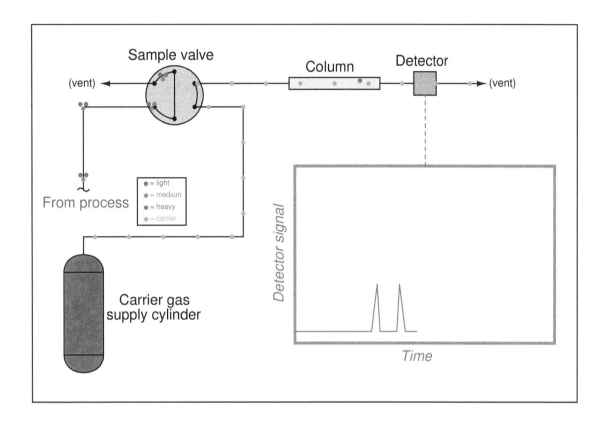

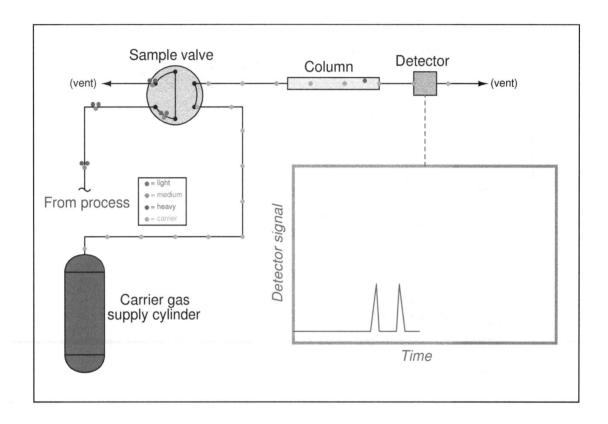

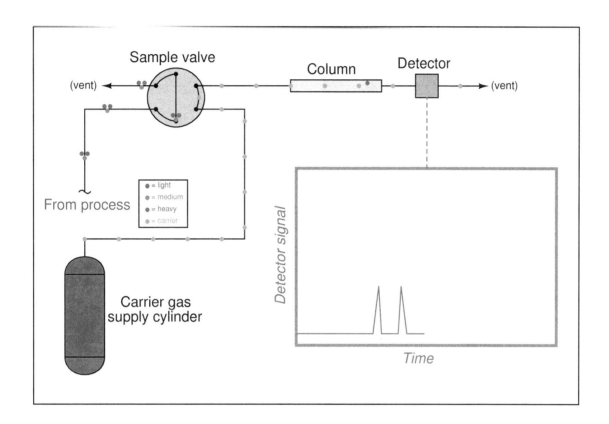

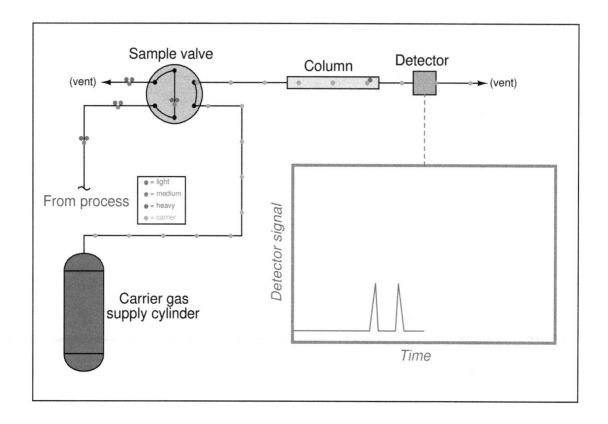

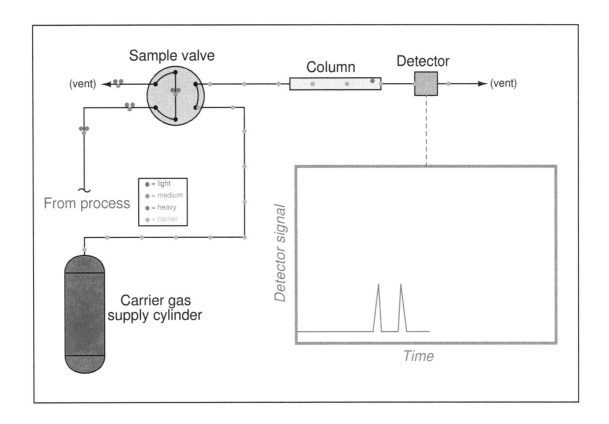

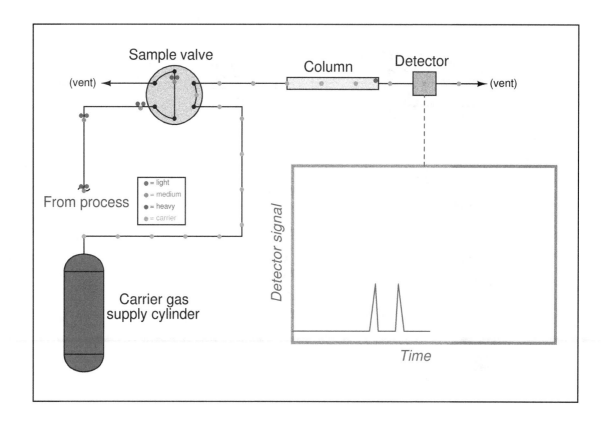

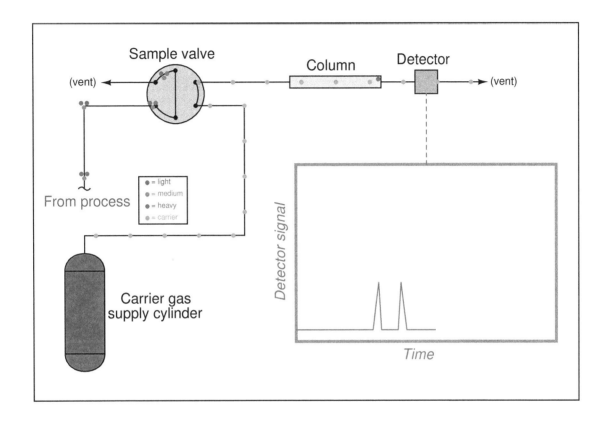

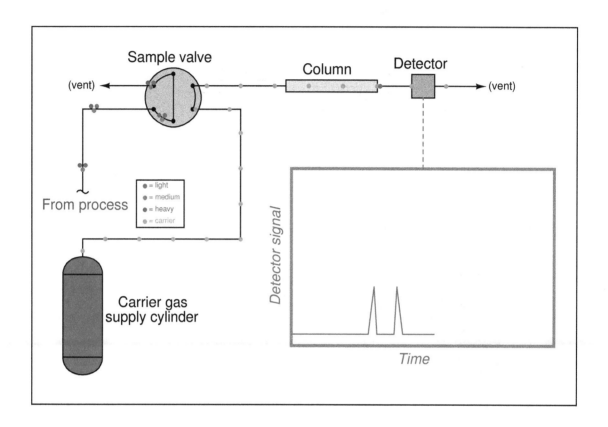

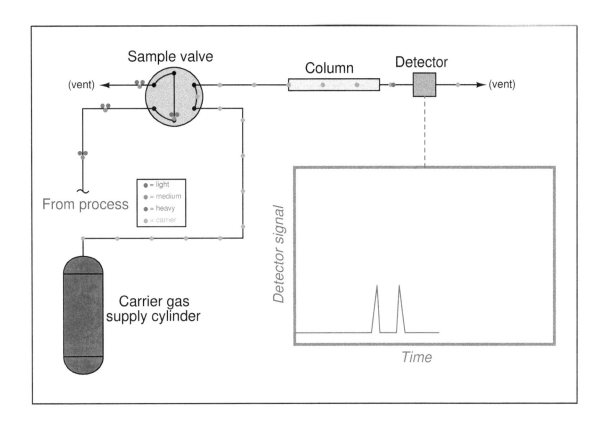

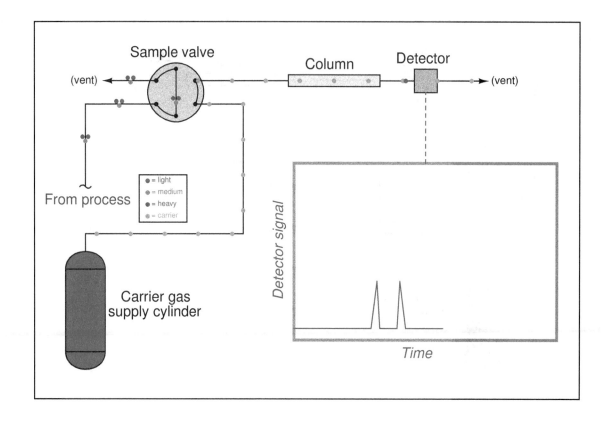

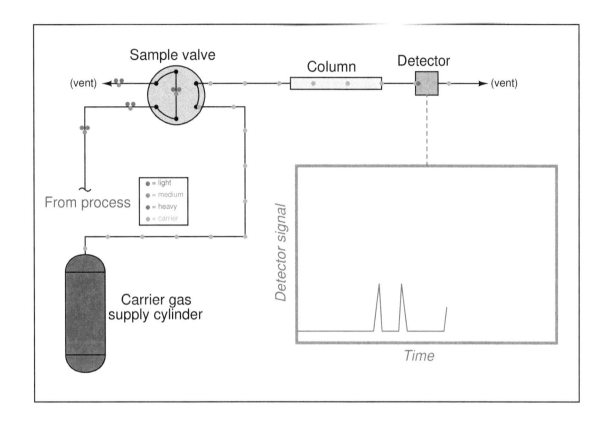

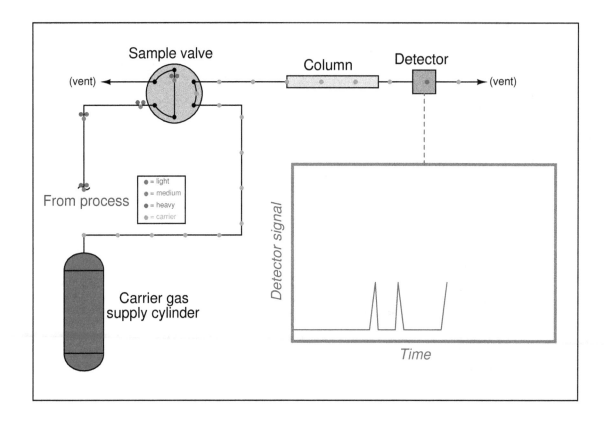

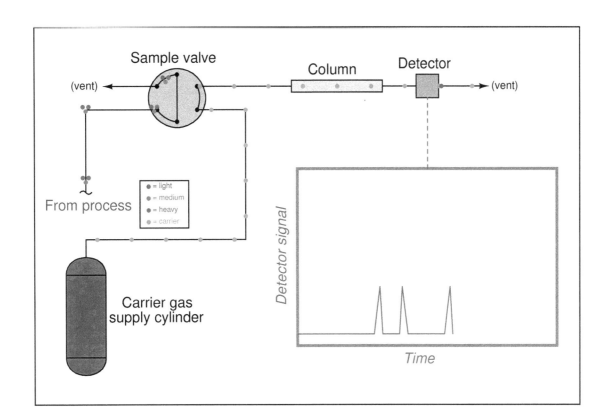

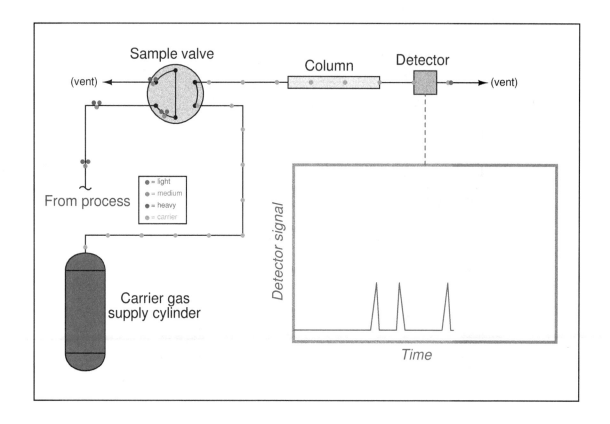

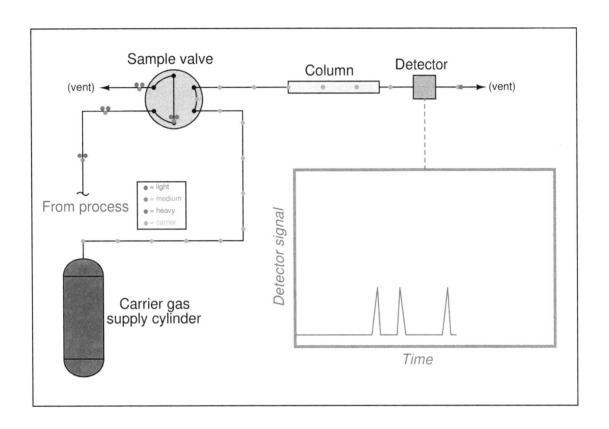

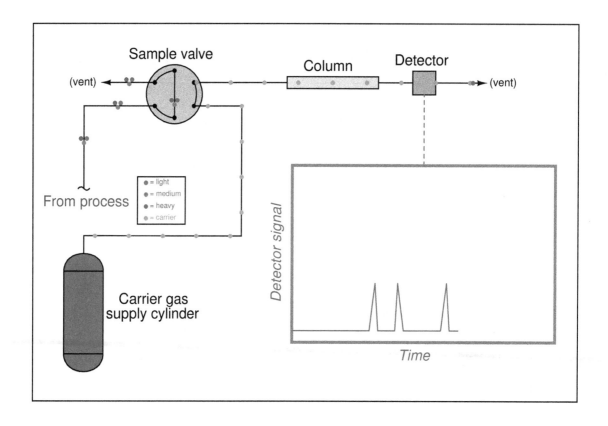

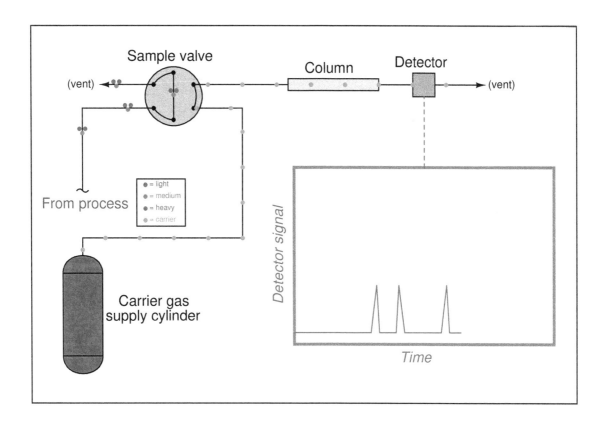

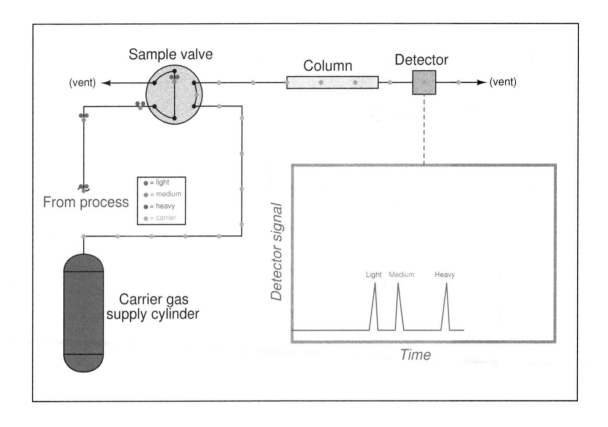

Appendix B

Doctor Strangeflow, or how I learned to relax and love Reynolds numbers

Of all the non-analytical (non-chemistry) process measurements students encounter in their Instrumentation training, flow measurement is one of the most mysterious. Where else would we have to *take the square root* of a transmitter signal just to measure a process variable in the simplest case? Since flow measurement is so vital to many industries, it cannot go untouched in an Instrumentation curriculum. Students must learn how to measure flow, and how to do it accurately. The fact that it is a fundamentally complex thing, however, often leads to oversimplification in the classroom. Such was definitely the case in my own education, and it lead to a number of misunderstandings that were corrected after a lapse of 15 years, in a sudden "Aha!" moment that I now wish to share with you.

The orifice plate is to flow measurement what a thermocouple is to temperature measurement: an inexpensive yet effective primary sensing element. The concept is disarmingly simple. Place a restriction in a pipe, then measure the resulting pressure drop (ΔP) across that restriction to infer flow rate.

You may have already seen a diagram such as the following, illustrating how an orifice plate works:

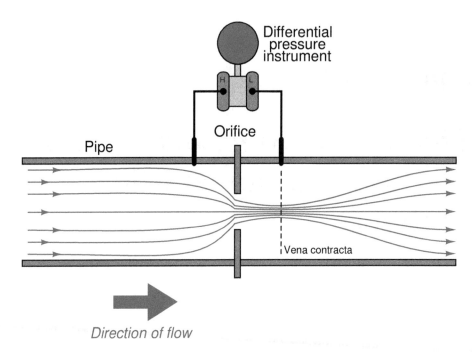

Now, the really weird thing about measuring flow this way is that the resulting ΔP signal does not linearly correspond to flow rate. Double the flow rate, and the ΔP quadruples. Triple the flow rate and the ΔP increases by a factor of nine. To express this relationship mathematically:

$$Q^2 \propto \Delta P$$

In other words, differential pressure across an orifice plate (ΔP) is proportional to the *square* of the flow rate (Q^2). To be more precise, we may include a coefficient (k) with a precise value that turns the proportionality into an equality:

$$Q^2 = k(\Delta P)$$

Expressed in graphical form, the function looks like one-half of a parabola:

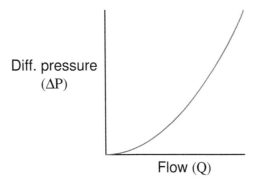

To obtain a linear flow measurement signal from the differential pressure instrument's output signal, we must "square root" that signal, either with a computer inside the transmitter, with a computer inside the receiving instrument, or a separate computing instrument (a "square root extractor"). We may see mathematically how this yields a value for flow rate (Q), following from our original equation:

$$Q^2 = k(\Delta P)$$

$$\sqrt{Q^2} = \sqrt{k(\Delta P)}$$

$$Q = \sqrt{k(\Delta P)}$$

. . . substituting a new coefficient value k[1] . . .

$$Q = k\sqrt{\Delta P}$$

Students are taught that the differential pressure develops as a consequence of energy conservation in the flowing liquid stream. As the liquid enters a constriction, its velocity must increase to account for the same volumetric rate through a reduced area. This results in kinetic energy increasing, which must be accompanied by a corresponding decrease in potential energy (i.e. pressure) to conserve total fluid energy.

---

[1]Since we get to choose whatever k value we need to make this an equality, we don't have to keep k inside the radicand, and so you will usually see the equation written as it is shown in the last step with k outside the radicand.

Pressure measurements taken in a venturi pipe confirm this:

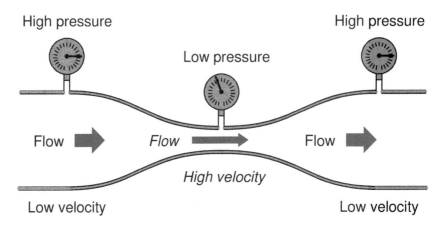

In all honesty, this did not make sense to me when I heard this. My "common sense" told me the fluid pressure would *increase* as it became crammed into the constriction, not decrease. Even more, "common sense" told me that whatever pressure was lost through the constriction would never be regained, contrary to the pressure indication of the gauge furthest downstream. Accepting this principle was an act of faith on my part, putting preconceived notions aside for something new. A leap of faith, however, is not the same as a leap in understanding. I believed what I was told, but I really didn't understand *why* it was true.

The problem intensified when my teacher showed a more detailed flow equation. This new equation contained a term for fluid density (ρ):

$$Q = k\sqrt{\frac{\Delta P}{\rho}}$$

What this equation showed us is that orifice plate flow measurement depended on density. If the fluid density changed, our instrument calibration would have to change in order to maintain good accuracy of measurement. Something disturbed me about this equation, though, so I raised my hand. The subsequent exchange between my teacher and I went something like this:

Me: What about viscosity?

Teacher: What?

Me: Doesn't fluid viscosity have an effect on flow measurement, just like density?

Teacher: You don't see a variable for viscosity in the equation, do you?

Me: Well, no, but it's *got* to have some effect on flow measurement!

Teacher: How come?

Me: Imagine clean water flowing through a venturi, or through the hole of an orifice plate. At a certain flow rate, a certain amount of ΔP will develop across the orifice. Now imagine that same orifice flowing an equal rate of liquid honey: approximately the same density as water, but much thicker. Wouldn't the increased "thickness," or viscosity, of the honey result in more friction through the orifice, and thus more of a pressure drop than what the water would create?

Teacher: I'm sure viscosity has some effect, but it must be minimal since it isn't in the equation.

Me: Then why is honey so hard to suck through a straw?

Teacher: Come again?

Me: A straw is a narrow pipe, similar to the throat of a venturi or the hole of an orifice, right? The difference in pressure between the suction in my mouth and the atmosphere is the ΔP across that orifice. The result is flow through the straw. If viscosity is of such little effect, then why is liquid honey so much harder to suck through a straw than water? The pressure is the same, the density is about the same, then why isn't the flow rate the same according to the equation you just gave us?

Teacher: In industry, we usually don't measure fluids as thick as honey, and so it's safe to ignore viscosity in the flow equation . . .

My teacher's smokescreen – that thick fluid flow streams were rare in industry – did nothing to alleviate my confusion. Despite my ignorance of the industrial world, I could very easily imagine liquids that were more viscous than water, honey or no honey. Somewhere, somehow, someone had to be measuring the flow rate of such liquids, and there the effects of viscosity on orifice ΔP must be apparent. Surely my teacher knew this. But then why did the flow equation not have a variable for viscosity in it? How could this parameter be unimportant? Like most students, though, I could see that arguing would get me nowhere and it was better for my grade to just go along with what the teacher said than to press for answers he couldn't give. In other words, I swept my doubts under the carpet of "learning" and made a leap of faith.

After that, we studied different types of orifice plates, different types of pressure tap locations, and other inferential primary sensing elements (Pitot tubes, target meters, pipe elbows, etc.). They all worked on Bernoulli's principle of decreased pressure through a restriction, and they all required square root extraction of the pressure signal to obtain a linearized flow measurement. In fact, this became the sole criterion for determining whether or not we needed square root extraction on the signal: did the flow measurement originate from a differential pressure instrument? If so, then we needed to "square root" the signal. If not, we didn't. A neat and clean distinction, separating ΔP-based flow measurements from all the others (magnetic, vortex shedding, Coriolis effect, thermal, etc.). Nice, clean, simple, neat, and only 95% correct, as I was to discover later.

Fast-forward fifteen years. I was now a teacher in a technical college, teaching Instrumentation to students just like myself a decade and a half ago. It was my first time preparing to teach flow measurement, and so I brushed up on my knowledge by consulting one of the best technical references I could get my hands on: Béla Lipták's *Process Measurement and Analysis*, third edition. Part of the *Instrument Engineers' Handbook* series, this wonderful work was to be our primary text as we explored the world of process measurement during the 2002-2003 academic year.

It was in reading this book that I had an epiphany. Section 2.8 of the text discussed a type of flowmeter I had never seen or heard of before: the *laminar* flowmeter. As I read this section of the book, my jaw hit the floor. Here was a differential-pressure-based flowmeter that was linear! That is, there was no square root extraction required at all to convert the ΔP measurement into a flow measurement. Furthermore, its operation was based on some weird equation called the *Hagen-Poiseuille* Law rather than Bernoulli's Law.

Early in the section's discussion of this flowmeter, a couple of paragraphs explained the meaning of something called *Reynolds number* of a flow stream, and how this was critically important to laminar flowmeters. Now, I had heard of Reynolds number before when I worked in industry, but I never knew what it meant. All I knew is that it had something to do with the selection of flowmeter types: one must know the Reynolds number of a fluid before one could properly select which type of flow-measuring instrument to use in a particular application. Since this determination typically fell within the domain of instrument engineers and not instrument technicians (as I was), I gave myself permission to remain ignorant about it and blissfully went on my way. Little did I know that Reynolds number held the key to understanding my "honey-through-a-straw" question of years ago, as well as comprehending (not just believing) how orifice plates actually worked.

According to Lipták, laminar flowmeters were effective only for low Reynolds numbers, typically below 1200. Cross-referencing the orifice plate section of the same book told me that Reynolds numbers for typical orifice-plate flow streams were much greater (10000 or higher). Furthermore, the orifice plate section contained an insightful passage on page 152 which I will now quote here. Italicized words indicate my own emphasis, locating the exact points of my "Aha!" moments:

> The basic equations of flow assume that the velocity of flow is uniform across a given cross-section. In practice, flow velocity at any cross section approaches zero in the boundary layer adjacent to the pipe wall, and varies across the diameter. *This flow velocity profile has a significant effect on the relationship between flow velocity and pressure difference developed in a head meter.* In 1883, Sir Osborne Reynolds, an English scientist, presented a paper before the Royal Society, proposing a single, dimensionless ratio now known as Reynolds number, as a criterion to describe this phenomenon. This number, *Re*, is expressed as
>
> $$Re = \frac{VD\rho}{\mu}$$
>
> where V is velocity, D is diameter, ρ is density, and μ is absolute viscosity. Reynolds number expresses the ratio of inertial forces to viscous forces. At a very low Reynolds number, viscous forces predominate, and the inertial forces have little effect. *Pressure difference approaches direct proportionality to average flow velocity and to viscosity.* At high Reynolds numbers, inertial forces predominate and viscous drag effects become negligible.

What the second paragraph is saying is that for slow-moving, viscous fluids (such as honey in a straw), the forces of friction (fluid "dragging" against the pipe walls) are far greater than the forces of inertia (fluid momentum). This means that the pressure difference required to move such a fluid through a pipe primarily works to overcome the friction of that fluid against the walls of the pipe. For most industrial flows, where the flow velocities are fast and the fluids have little viscosity (like clean water), flow through an orifice plate is assumed to be frictionless. Thus, the pressure dropped across a constriction is *not* the result of friction between the fluid and the pipe, but rather it is a consequence of having to *accelerate* the fluid from a low velocity to a high velocity through the narrow orifice.

My mistake, years ago, was in assuming that water flowing through an orifice generated substantial friction, and that this is what created the ΔP across an orifice plate. This is what my "common sense" told me. In my mind, I imagined the water having to rub past the walls of the pipe, past the face of the orifice plate, and through the constriction of the orifice at a very high speed, in order to make it through to the other side. I memorized what my teacher told us about energy exchange and how pressure had to drop as velocity increased, but I never really internalized it because I still held to my faulty assumption of friction being the dominant mechanism of pressure drop in an orifice plate. In other words, while I could parrot the doctrine of kinetic and potential energy exchange, I was still *thinking* in terms of friction, which is a totally different phenomenon. The difference between these two phenomena is the difference between energy *exchanged* and energy *dissipated*. To use an electrical analogy, it is the difference between *reactance* (X) and *resistance* (R). Incidentally, many electronics students experience the same confusion when they study reactance, mistakenly thinking it is the same thing as resistance where in reality it is quite different in terms of energy, but that is a subject for another essay!

In a frictionless flow stream, fluid pressure decreases as fluid velocity increases in order to conserve energy. Another way to think of this is that a pressure differential must develop in order to provide the "push" needed to *accelerate* the fluid from a low speed to a high speed. Conversely, as the fluid slows back down after having passed through the constriction, a reverse pressure differential must develop in order to provide the "push" needed for that *deceleration*:

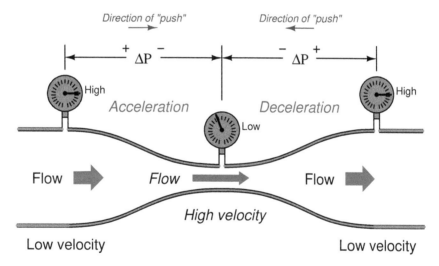

A moving mass does not simply slow down on its own! There must be some opposing force to decelerate a mass from a high speed to a low speed. This is where the pressure recovery downstream of the orifice plate comes from. If the pressure differential across an orifice plate originated primarily from friction, as I mistakenly assumed when I first learned about orifice plates, then there would be no reason for the pressure to *ever* recover downstream of the constriction. The presence of friction means energy *lost*, not energy *exchanged*. Although both inertia and friction are capable of creating pressure drops, the lasting effects of these two different phenomena are definitely not the same.

There is a quadratic ("square") relationship between velocity and differential pressure precisely because there is a quadratic relationship between velocity and kinetic energy as all first-quarter physics students learn ($E_k = \frac{1}{2}mv^2$). This is why ΔP increases with the square of flow rate (Q^2)

and why we must "square-root" the ΔP signal to obtain a flow measurement. This is also why fluid density is so important in the orifice-plate flow equation. The denser a fluid is, the more work will be required to accelerate it through a constriction, resulting in greater ΔP, all other conditions being equal:

$$Q = k \sqrt{\frac{\Delta P}{\rho}} \qquad \text{(Our old friend, the "orifice plate" equation)}$$

This equation is only accurate, however, when fluid friction is negligible: when the viscosity of the fluid is so low and/or its speed is so high that the effects of potential and kinetic energy exchange completely overshadow[2] the effects of friction against the pipe walls and against the orifice plate. This is indeed the case for most industrial flow applications, and so this is what students first study as they learn how flow is measured. Unfortunately, this is often the *only* equation two-year Instrumentation students study with regard to flow measurement.

In situations where Reynolds number is low, fluid friction becomes the dominant factor and the standard "orifice plate" equation no longer applies. Here, the ΔP generated by a viscous fluid moving through a pipe really does depend primarily on how "thick" the fluid is. And, just like electrons moving through a resistor in an electric circuit, the pressure drop across the area of friction is directly proportional to the rate of flow ($\Delta P \propto Q$ for fluids, $V \propto I$ for electrons). This is why laminar flowmeters – which work only when Reynolds number is low – yield a nice *linear* relationship between ΔP and flow rate and therefore do not require square root extraction of the ΔP signal. These flowmeters do, however, require temperature compensation (and even temperature *control* in some cases) because flow measurement accuracy depends on fluid viscosity, and fluid viscosity varies according to temperature. The Hagen-Poiseuille equation describing flow rate and differential pressure for laminar flow (low Re) is shown here for comparison:

$$Q = k \left(\frac{\Delta P D^4}{\mu L} \right)$$

Where,

Q = Flow rate (gallons per minute)
k = Unit conversion factor = 7.86×10^5
ΔP = Pressure drop (inches of water column)
D = Pipe diameter (inches)
μ = Liquid viscosity (centipoise) – this is a temperature-dependent variable!
L = Length of pipe section (inches)

Note that if the pipe dimensions and fluid viscosity are held constant, the relationship between flow and differential pressure is a direct proportion:

$$Q \propto \Delta P$$

---

[2]In engineering, this goes by the romantic name of *swamping*. We say that the overshadowing effect "swamps" out all others because of its vastly superior magnitude, and so it is safe (not to mention simpler!) to ignore the smaller effect(s). The most elegant cases of "swamping" are when an engineer intentionally designs a system so the desired effect is many times greater than the undesired effect(s), thereby forcing the system to behave more like the ideal. This application of swamping is prevalent in electrical engineering, where resistors are often added to circuits for the purpose of overshadowing the effects of stray (undesirable) resistance in wiring and components.

In reality, there is no such thing as a frictionless flow (excepting superfluidic cases such as helium II which are well outside the bounds of normal experience), just as there is no such thing as a massless flow (no inertia). In normal applications there will always be both effects at work. By not considering fluid friction for high Reynolds numbers and not considering fluid density for low Reynolds numbers, engineers draw simplified models of reality which allow us to more easily measure fluid flow. As in so many other areas of study, we exchange accuracy for simplicity, precision for convenience. Problems arise when we forget that we've made this Faustian exchange and wander into areas where our simplistic models are no longer reasonable.

Perhaps the most practical upshot of all this for students of Instrumentation is to realize exactly why and how orifice plates work. Bernoulli's equation does *not* include any considerations of friction. To the contrary, we must assume the fluid to be completely frictionless in order for the concept to make sense. This explains several things:

- There is pressure recovery downstream of an orifice: most of the pressure lost at the vena contracta is regained further on downstream as the fluid decelerates to its original (slow) speed. Permanent pressure drop will occur only where there is energy *lost* through the constriction, such as in cases where fluid friction is substantial. Where the fluid is frictionless there is no mechanism in an orifice to dissipate energy, and so with no energy lost there must be full pressure recovery as the fluid returns to its original speed.

- Pressure tap location makes a difference: to ensure that the downstream tap is actually sensing the pressure at a point where the fluid is moving significantly faster than upstream (the "vena contracta"), and not just anywhere downstream of the orifice. If the pressure drop were due to friction alone, it would be permanent and the downstream tap location would not be as critical.

- Standard orifice plates have knife-edges on their upstream sides: to minimize contact area (friction points) with the high-speed flow.

- Care must be taken to ensure Reynolds number is high enough to permit the use of an orifice plate: if not, the linear $Q/\Delta P$ relationship for viscous flow will assert itself along with the quadratic potential/kinetic energy relationship, causing the overall $Q/\Delta P$ relationship to be polynomial rather than purely quadratic, and thereby corrupting the measurement accuracy.

- Sufficient upstream pipe length is needed to condition flow for orifice plate measurement, not to make it "laminar" as is popularly (and wrongly) believed, but to allow natural turbulence to "flatten" the flow profile for uniform velocity. *Laminar flow* is something that only happens when viscous forces overshadow inertial forces (e.g. flow at low Reynolds numbers), and is totally different from the *fully developed turbulent flow* that orifice plates need for accurate measurement.

In a more general sense, the lesson we should learn here is that blind faith is no substitute for understanding, and that a sense of confusion or disagreement during the learning process is a sign of one or more misconceptions in need of correction. If you find yourself disagreeing with what you are being taught, either you are making a mistake and/or your teacher is. Pursuing your questions to their logical end is the key to discovery, while making a leap of faith (simply believing what you are told) is an act of avoidance: escaping the discomfort of confusion and uncertainty at the expense of a deeper learning experience. This is an exchange no student should ever feel they must make.

References

Lipták, Béla G. et al., *Instrument Engineers' Handbook – Process Measurement and Analysis Volume I*, Third Edition, CRC Press, New York, NY.

Appendix C

Disassembly of a sliding-stem control valve

The following collection of photographs chronicles the complete disassembly of a Fisher E-body globe valve with pneumatic diaphragm actuator. This control valve design is quite mature, but nevertheless enjoys wide application in modern industrial settings.

An important safety note when disassembling pneumatic control valves is to first relieve all tension from the actuator spring so that its stored energy cannot harm you or anyone else. These springs may be quite large, exerting *thousands of pounds* of force during normal operation.

Spring tension may be relieved by moving the spring adjuster until it turns easily by hand without further aid of tools, or in the procedure shown in the following photographs by loosening the spanner nut attaching the actuator yoke to the valve bonnet.

This is the complete control valve, without a positioner attached. What you see here is the actuator (painted green) and the valve body (painted grey), mounted on a steel plate for student learning in a laboratory setting. The left-hand photograph shows the complete control valve assembly, while the right-hand photograph shows a student loosening the spanner nut holding the valve actuator yoke to the valve body:

The next step is to un-couple the actuator stem from the valve stem. On Fisher sliding-stem valves, this connection is made by a split block with threads matching those on each stem. Removing two bolts from the block allows it to be taken apart (left-hand photograph). Nuts threaded on to the valve stem, jammed up against the coupling block, must also be loosened before the stems may be uncoupled (right-hand photograph):

It is very important that no spring tension exists on the stem prior to disassembly of the stem coupler, or else the two stems will slip past each other with great force once the coupler is removed. Spring tension must be released, either by loosening the spring adjuster or by loosening the spanner nut holding the actuator yoke to the bonnet.

A close-up photograph of this stem connector block, with the front half removed for inspection, shows how it engages both threaded stems (valve and actuator) in a single nut-like assemblage. The solid valve stem (below) slides into the hollow actuator stem (above), while the split connector "nut" engages the threads of both, holding the two stems together so they move up and down as one piece:

Once the actuator and valve body stems have been uncoupled, the actuator may be removed from the valve body entirely:

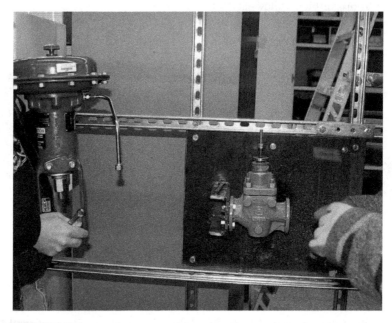

The bonnet is held to the rest of the valve body (in this case) by four large studs. Removing the nuts on these studs allows the bonnet to be lifted off the body, exposing the valve trim for view:

Seats in Fisher E-body globe valves rest in the bottom of the body, held in place by the cage surrounding the valve plug. Once the bonnet is removed from the body, the seat may be removed without need of any specialized tools (left-hand photograph). A view inside the body shows the place where the seat normally rests (right-hand photograph):

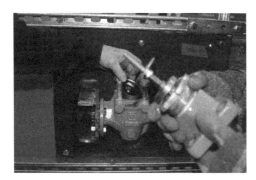 

With the bonnet removed, the plug and cage may be easily removed for inspection:

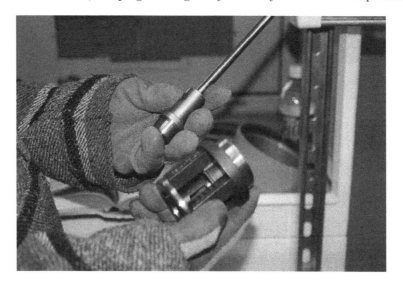

The packing follower (between the student's fingers) has been removed from the valve bonnet, and you can see the upper Teflon packing rings within the bonnet. The student is also holding the packing flange in the same hand as the packing follower (left-hand photograph). In the right-hand photograph, we see the student using a screwdriver to gently push the Teflon packing rings out of the bonnet, from the bottom side. Care should be taken not to gouge or otherwise damage these rings during removal:

The left-hand photograph shows all the packing components stacked on top of each other on the concrete floor, next to the bonnet. From top to bottom you see the following components: a felt wiper, the packing follower, five (5) Teflon packing rings, a coil spring, and the packing box ring. The right-hand photograph shows the same packing components stacked on the valve stem:

 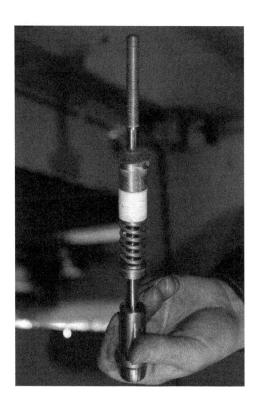

Turning to the actuator, we begin disassembly by loosening the diaphragm hold-down bolts (left-hand photograph) and removing the upper half of the diaphragm casing (right-hand photograph). A single bolt secures the upper diaphragm plate to the top of the actuator stem:

In the left-hand photograph you see the student removing the spring seat, having previously loosened the spring adjustor nut. With the spring seat removed, the spring may be removed from the actuator assembly. In the right-hand photograph the spring adjustor and spring seat have been removed from the actuator stem. The student is now pointing at the valve spring, partially removed:

Sliding the actuator diaphragm, plate, and stem out of the actuator assembly from the top of the actuator makes it easy to remove the large actuator spring (left-hand photograph). The right-hand photograph shows all the moving actuator components re-assembled in their proper order outside of the yoke:

The left-hand photograph shows the lower half of the actuator casing, with the student removing six (6) hold-down bolts joining this casing half to the actuator yoke. The right-hand photograph shows the actuator casing half completely removed from the yoke, revealing a gasket and the bronze stem bushing (which serves to both guide the actuator stem and seal air pressure, since this is a reverse-acting actuator):

A circular spring clip holds the stem bushing in the yoke casting. The left-hand photograph shows the student using pliers to squeeze this spring clip and remove it from its groove cut into the metal of the yoke. In the right-hand photograph, we see the student using the wooden handle of a hammer to gently tap the bushing out of the yoke. The bushing has rubber O-ring seals between it and the yoke casting, so a small amount of force will be necessary to dislodge it. Using the hammer's wooden handle to drive the bushing instead of a metal tool protects the relatively soft bronze bushing from impact damage. Note how the student's right hand is waiting to catch the bronze bushing when it emerges from the hole, to protect it from falling against the hard concrete floor:

The final photograph shows the bushing removed from its hole:

Appendix D

How to use this book – some advice for teachers

If you would like to maximize your students' learning in a field of study that emphasizes critical thinking as much as Instrumentation, I have one simple piece of advice: *engage your students, don't just present information to them*. Do not make the mistake so many teachers do, of thinking it is their role in the classroom to provide information in pre-digested form to their students, and that it is each student's responsibility to passively absorb this information.

High achievement happens only in an atmosphere of high expectations. If you design coursework allowing students to expend minimal effort, your students will achieve minimal learning. Alternatively, if you require students to think deeply about their subject of study, challenge them with interesting and relevant assignments, and hold them accountable to rigorous standards of demonstrated competence, your students can and will move mountains.

In this appendix I present to you some concepts and models for achieving high standards of learning in the field of Instrumentation. The ideas documented here have all been proven to work in my own instruction, and I continue to use them on a daily basis. However, this is not a rigid blueprint for success – I invite and encourage others to experiment with variations on the same themes. More than anything else, I hope to encourage educators with examples of unconventional thinking and unconventional curricula, to show what may be accomplished if you allow yourself to be creative and results-driven in your instructional design.

D.1   Teaching technical theory

Learning is not merely a process of information transfer. It is first and foremost a transformation of one's thinking. When we learn something substantial, it alters the way we perceive and interact with the world around us. Learning any subject also involves a substantial accumulation of facts in one's memory, but memorization alone is not really learning (at least it isn't learning at the college level). Transmitting facts into a student's memory is easy, and does not require a live human presenter. A well-written book or a well-edited video can do a far better and far more consistent job of conveying facts and concepts than any live presentation[1]. The greatest value of a human teacher is to help students develop higher-order thinking skills. This includes (but is not limited to) problem-solving, logical reasoning, diagnostic techniques, and metacognition (critiquing one's own thinking).

Rather than devote most of your classroom time to lecture-style presentations – where the flow of information goes primarily from you to your students – place the responsibility for fact-gathering on your students. Have them read books such as this[2] and arrive at the classroom *prepared* to discuss what they have already studied.

When students are with you in the classroom and in the lab, probe their understanding with questions – lots of questions. Give them realistic problems to solve. Challenge them with projects requiring creative thought. Get your students to reveal how they think, both to you and to their peers. This will transform your classroom atmosphere from a monologue into a dialogue, where you engage with the minds of your students as partners in the learning process instead of lecturing to them as subordinates.

A format I have used with great success is to assign homework exploring new topics, so students much research those topics in advance of our coverage of it in class. The pedagogical term for this is an *inverted classroom*, where communication of facts occurs outside of class, and higher-order activities such as problem-solving occur inside the classroom. This stands in contrast to traditional learning structures, where the instructor spends most of the class time transmitting facts and working example problems, while subsequent homework questions (ostensibly completed on the students' own time) stimulate the development of problem-solving skills.

Homework in my "inverted" classroom comes in the form of question sets designed to lead the student on a path to acquiring the necessary facts and exposing them to certain problem-solving techniques. Some of these questions point students directly to specific texts to read, while others allow students to choose their own research material. Many of the questions are simply conceptual or quantitative problems to solve, without reference to source materials. When my students arrive for class, they first take a quiz on the material they should have studied for that day. After the quiz, students working alone or in small groups solving the assigned problems. My role as instructor is to assist students in their problem-solving efforts, observing the students' attempts, offering advice, and helping to identify and correct misconceptions. Students are much more engaged, less distracted,

---

[1]To be sure, there are some gifted lecturers in the world. However, rather than rely on a human being's live performance, it is better to capture the brilliance of an excellent presentation in static form where it may be peer-reviewed and edited to perfection, then placed into the hands of an unlimited number of students in perpetuity. In other words, if you think you're great at explaining things, do us all a favor and translate that brilliance into a format capable of reaching more people!

[2]It would be arrogant of me to suggest my book is the best source of information for your students. Have them research information on instrumentation from other textbooks, from manufacturers' literature, from whitepapers, from reference manuals, from encyclopedia sets, or whatever source(s) you deem most appropriate. If you possess knowledge that your students need to know that isn't readily found in any book, *publish it for everyone's benefit!*

and comfortable raising questions while working together in these groups than they ever are as one large group.

Students are considered finished with the class session (and free to leave) when they are able to successfully demonstrate to me their grasp of the day's material. This happens in the form of a "summary quiz" which may be done one-on-one (my preference) or as a whole class. If done individually, the summary quizzes must be varied enough so I am adequately challenging each student even though they have overheard classmates answering similar summary quiz questions.

An inverted classroom structure shifts the burden of transmitting facts and concepts from a live teacher to static sources such as textbooks[3]. This shift in responsibility frees valuable class time for more important tasks, namely the refinement of higher-order thinking skills. It makes no sense to have a subject-matter expert (the instructor) spend most or all of the students' valuable time at school doing what a book or a video could do just as well. Any instructor who can be replaced with a book or a video *should* be replaced by a book or a video! It is far better to apply that same subject-matter expertise to challenges no book or video can meet: actively identifying student misconceptions, dispensing targeted advice for overcoming difficulties, and stimulating students' minds with follow-up questions designed to illuminate concepts in further detail.

Several important advantages are realized by managing a classroom in this way. First, the instructor gets to have a very close-up and personal view of each students' comprehension, struggles, and misconceptions. If you are accustomed to teaching in a lecture or other "stand-up-in-front" classroom format, you will be utterly amazed to see what your students do and do not comprehend when you watch them dialogue and problem-solve in small groups. Much of what goes on inside your students' minds is hidden from you when they are seated in neat rows watching you in the front of the classroom. When students are free to work together in more intimate settings, you get to see how they think, what they understand, and most importantly what they mis-understand.

Another important advantage of an inverted classroom is how it maximizes your contact with those students who need it most. Faster students are able to finish their work quickly and leave (or stay in the class to tutor their peers), while the slower students stay to the very end with you until their work is done at their pace. *When conversing with small groups (3 to 4 students) in the classroom, I spend an average of only 5 to 6 minutes per student to query them on their understanding of the day's material and to challenge them with at least one problem to solve.* The time saved is incredibly useful, as it allows you to re-structure the topic coverage to include more review, cover additional material, or devote more time to hands-on labwork and projects.

Perhaps the most important benefit of an inverted classroom is that students learn how to independently research, which is no small feat. In a complex field where technology advances on a daily basis, your students will need to be able to learn new facts on their own (without your assistance!) after they graduate. Employers have consistently advised me that this is the single most important skill any person can learn in school: how to independently acquire new knowledge and new abilities. Such a skill not only prepares them for excellence in their chosen career, but it also brings great benefit to every other area of life where the acquisition of new information is essential to decision-making (e.g. participation in the democratic process, legal proceedings, medical decision-making, investing, parenting, etc.). In this way, the inverted classroom is not just a "better mousetrap" but in fact is really catching a "better mouse."

---

[3]And multimedia resources, too! With all the advances in multimedia presentations, there is no reason why an instructor cannot build a library of videos, computer simulations, and other engaging resources to present facts and concepts to students outside of class time.

D.1.1 The problem with lecture

I speak negatively of lecture as a teaching tool because I have suffered its ineffectiveness from a teacher's perspective for several years. The problem is not that students cannot learn from an instructor's eloquent presentation; it's that following the lead of an expert's presentation obscures students' perception of their own learning. Stated in simple terms, *lecture forces every student into the role of spectator when they should be participants.* Students observing a lecture cannot tell with certainty whether they are actually learning from an expert presentation, or whether they are merely being stimulated. This is not an obvious concept to grasp, so allow me to elaborate in more detail.

When I began my professional teaching career, I did what every other teacher I knew did: I lectured daily to my students. My goal was to transfer knowledge into my students' minds, and so I chose the most direct method I knew for that necessary transference of information.

My first year of teaching, like most teachers' first year, was a trial by fire. Many days I was lecturing to my students on some subject I had just reviewed (for the first time in a long time) no more than a few days before. My lesson plans were chaotic to non-existent. By my second and third years, however, I had developed lesson plans and was thereby able to orchestrate my lectures much more efficiently. These lesson plans were complete enough to support live demonstrations of concepts during almost every lecture, listing all the materials, components, and equipment I would need to set up in preparation. If an extensive amount of set-up was required for some demonstration, instructions would be found in lesson plan(s) multiple days in advance in order to give me adequate time. The result was a very smooth and polished presentation in nearly every one of my lectures. I was quite proud of the work I had done.

However, I noticed a strange and wholly unintended consequence of all this preparation: with each passing year, my students' long-term recall of concepts presented in lecture seemed to grow worse. Even my best students, who demonstrated an obvious commitment to their education by their regular study habits, outstanding attendance, and quality work, would shock me by asking me to re-explain basic concepts we had covered in extensive detail months before. They never complained about the lectures being bad – quite to the contrary, their assessment of my lectures was always "excellent" in my performance surveys.

An increasingly common lament of students as they tried to do the homework was *"I understand things perfectly when you lecture, but for some reason I just can't seem to figure it out on my own."* This baffled me, since I had made my presentations as clear as I could, and students seemed engaged and attentive throughout. It was clear to me as I later worked with these students that often they were missing crucial concepts and/or harbored severe misconceptions, and that there was *no way* things should have made sense to them during lecture given these misconceptions.

Another detail that caught my attention was the fine condition of their textbooks. In fact, their textbooks were looking better and better with each passing year. At the conclusion of my first teaching year, my students textbooks looked as though they had been dragged behind a moving vehicle: pages wrinkled, binding worn, and marks scribbled throughout the pages. As my lectures became more polished, the textbooks appeared less and less used. My reading assignments were no less thorough than before, so why should the books be used less?

One day I overheard a student's comment that made sense of it all. I was working in my office, and just outside my door were two students conversing who didn't think I could hear them talking. One of them said to the other, *"Isn't this class the best? The lectures are so good, you don't even have to read the book!"* At the sound of this, my heart sunk. I began to realize what the problem was, what was needed to fix it, and how I had unwittingly created a poor learning environment despite the best of intentions.

The fundamental problem is this: students observing an expert presentation are fooled into thinking the concepts are easier to grasp and the processes easier to execute than they actually are. The mastery and polish of the lecturer actually hinders student learning by veiling the difficulty of the tasks. Matters are no different watching a professional athlete or musician at work: a master makes any task look effortless. It isn't until you (the spectator) actually try to do the same thing (as a participant) that you realize just how challenging it is, just how much you have to learn, and how much effort you must invest before you achieve a comparable level of proficiency.

When students told me "for some reason" they just couldn't seem to solve the same problems I did during lecture "even though they understood it perfectly" as I lectured, they were being honest. This was not some excuse made up to cover a lack of effort. From their perspective, they truly believed they grasped the concepts while watching me work through them in front of class, and were genuinely mystified why it was so hard for them to perform the same problem-solving tasks on their own.

The simple fact of the matter was that my students did *not* actually grasp the concepts as they watched me lecture. If they had, the solution of similar problems after lecture should have presented little trouble for them. Lecture had generated a *false sense of understanding* in their minds. This made my lectures worse than useless, for not only did they fail to convey the necessary knowledge and skill, but they actually *created an illusion of proficiency* in the minds of my students powerful enough to convince them they did not need to explore the concepts further (by reading their textbooks). This served to hinder learning rather than foster learning.

What I needed to do was shatter this illusion if my students were to learn from me more effectively, and especially if they were ever to become independent learners. Thus began my own personal quest of educational reform.

D.1.2   A more accurate model of learning

A humbling fact every teacher eventually learns is that the depth of a student's learning is primarily a function of the student's effort and not their own. Even the most dedicated and talented instructor cannot make a student learn if that student does not invest the necessary time and effort. Conversely, even an unmotivated or incompetent instructor cannot prevent a self-dedicated student from learning on their own.

However, a great many students enter college with the belief that learning is a passive activity: *"It's the instructor's job to give us information – all we're supposed to do is attend and observe."* Unfortunately, this flawed model of learning seems embedded in modern American culture, anchored in students' minds from years of compulsory lecture-based education. The role of teacher as expert presenter is so relentlessly reinforced that we have difficulty recognizing its flaws, much less conceiving better alternatives. Teachers choosing to depart from this model invite suspicion and even anger from students accustomed to the status quo of lecture.

One way to help see past one's own biases on a subject is to consider the same (or similar) subject in a different context. Here, the absurdity of passive learning becomes obvious if we simply switch contexts from academic instruction to athletic instruction. It would be laughable if a coach or fitness trainer were the one performing all the weight-lifting, sprints, stretching, and practice movements while the student never did anything but observe. It would be only slightly less humorous if the trainer spent the whole of every session modeling these activities, leaving the student to practice those activities on their own time as "homework." Instead, effective physical training sessions *always* place the student in an active role as soon and as often as possible, so that the instructor's valuable expertise may applied toward identifying errors and recommending corrections. Instructor-led demonstration is minimized in order to maximize time spent with the student *practicing* their craft.

Academic learning really isn't much different. If we want students to learn new skills and acquire new mental abilities, students must *practice* those skills and mental processes, and it is during this practice time that an instructor's expertise becomes most valuable. Accurate self-perception is another reason to immediately engage students in practicing the skills and processes they seek to learn. When students directly experience just how challenging any concept or task is to master, they immediately recognize how much they have to learn. This self-awareness is a vital first step to learning, proving the need for committed action on their part. Only after this recognition is the student psychologically prepared for the hard work of learning.

This is why I favor the "inverted classroom" approach. Students must engage with the new subject(s) *prior* to every classroom session. From the very moment they arrive, they recognize the challenges of the subject matter, and where they need help understanding it. With the presentation of facts occurring before class time, the bulk of that time may be spent actually *applying* the concepts instead of *encountering* them. The relatively mundane task of fact-gathering is relegated to students' time outside of class, while the more challenging and meaningful tasks of problem-solving and analysis happen where the instructor can actually observe and coach.

In order for an inverted classroom structure to work, though, each student must have access to the necessary information in static form (e.g. textbooks, videos, etc.), and be held accountable for doing the necessary preparations. Access and accountability are absolutely essential to ensuring an inverted classroom will work. Without the former, students will become frustrated; without the latter, some students won't engage.

D.1.3 The ultimate goal of education

As I have transitioned from a traditional lecturer to a "reform-minded" educator, my general philosophy of education has shifted. My teaching techniques and classroom organization changes preceded this philosophical shift, to be sure, but more and more I am realizing just how important it is to have an educational philosophy, and how such a philosophy helps to guide future reforms.

When I began teaching, my belief was that teaching was a matter of knowledge and skills transference: it was my job as an educator first and foremost to transfer information into my students' minds. Now, it is my belief that my primary task is to help my students become *autonomous*: able to analyze complex data, turn their thoughts into practical action, and continue learning long after they have left my classroom. If all I accomplish is helping my students memorize facts, procedures, and formulae about instrumentation, I have utterly failed them. My real job is to challenge them to become autonomous, critical thinkers and doers, so they will be able to fully take responsibility for their own lives and their own careers.

This shift in philosophy happened as a result of contact with many employers of my students, who told me the most important thing any student could learn in school was *how to learn*. In life, learning is not an option but a necessity, especially in highly complex fields such as instrumentation and control. Any instrument technician or engineer who stagnates in their learning is destined for obsolescence. Conversely, those with the ability and drive to continually learn new things will find opportunities opening for them all throughout their careers.

A former classmate of mine I studied instrumentation with told me of his path to success in this field. While never the smartest person in class (he struggled mightily with math concepts), he was always very determined and goal-oriented. His first job placed him in the field of automotive research and development, where he was responsible for "instrumenting" heavy trucks with sensors to perform both destructive and non-destructive tests on them. The instrumentation he used at this job was often quite different from the industrial instruments he learned in school, and so he found himself having to constantly refer to textbooks and equipment manuals to learn how the technology worked. This was true even when he was "on the road" doing field-service work. He told me of many evenings spent in some tavern or pub, a beer in one hand and an equipment manual in the other, learning how the equipment was supposed to work so he could fix the customer's problem the next day.

This hard work and self-directed learning paid off handsomely for my friend, who went on to set up an entire testing department for a major motorcycle manufacturer, and then started his own vehicle testing company (specializing in power-sport vehicles) after that. None of this would have been possible for my friend had he relied exclusively on others to teach him what he needed to know, taking the passive approach to learning so many students do. The lesson is very general and very important: *continual learning is a necessary key to success.*

One of the corollaries to my philosophy of education is that individual learning styles are ultimately not to be accommodated. This may come as a shock to many educators, who have learned about the various styles of learning (auditory, kinesthetic, visual, etc.) and how the acquisition of new information may be improved by teaching students according to their favored modes. Please understand that I am not denying the fact different people prefer learning in different ways. What I am saying is that we fail to educate our students (i.e. empower them with *new* abilities) if all we ever do is teach to their preferences, if we never challenge them to do what is novel or uncomfortable.

The well-educated person can learn by listening, learn by watching others, *and* learn by direct hands-on experimentation. A truly educated person may still retain a preference for one of these modes over the others, but that preference does not *constrain* that person to learning in only one way. This is our goal as educators: nothing short of expanding each student's modes of learning.

If a student experiences difficulty learning in a particular way, the instructor needs to engage with that student in whatever mode makes the most sense for them *with the goal of strengthening the areas where that student is weak.* For example, a student who is weak in reading (visual/verbal) but learns easily in a hands-on (kinesthetic) environment should be shown how to relate what they perceive kinesthetically to the words they read in a book. Spending time with such students examining an instrument to learn how it functions, then reading the service manual or datasheet for that instrument to look for places where it validates the same principles, is one example of how an instructor might help a student build connections between their strong and weak modes of learning. Investigating subjects through multiple approaches such as this also shows students the value of each learning mode: a student might find they easily grasp "how" an instrument works by directly observing and experimenting with it, but that they more readily grasp "why" it was built that way by reading the manufacturer's "theory of operation" literature.

Keeping the goal of life-long learning in mind, we must ask ourselves the question of how our students will need to learn new things once they are no longer under our tutelage. The obvious answer to this question is that they will need to be able to learn in any mode available to them, if they are to flourish. Life is indifferent to our needs: reality does not adapt itself to favor our strengths or to avoid challenging our weaknesses. Education must therefore focus on the well-rounded development of learning ability.

By far the greatest amount of resistance I encounter from students in terms of learning styles is learning by reading. It is rare to find a student who reads well, for example, but struggles at learning in a hands-on environment (kinesthetic) or struggles to understand spoken information (auditory). The reason for this, I believe, is that reading is a wholly unnatural skill. Entire cultures exist without a written language, but there is not a culture in the world that lacks a spoken one. Interpreting the written word, to the level of proficiency required for technical learning, is a skill born of much practice.

Unfortunately, the popular application of learning styles in modern education provides students with a ready-made, officially-sanctioned excuse for not only their inability ("That's just not how my brain works"), but also for continued tolerate of that inability ("I shouldn't have to learn in a way I'm not good at"). The challenge for the instructor is helping students develop their ability to learn in non-favored modes despite this psychological resistance.

My general advice for educators is to never compromise the "big picture" philosophy of empowering your students' thinking. Some key points I always try to keep in mind are:

- **Lead by example:** Regularly showcase for your students your own excitement for the subject and your own continual learning adventures. Likewise, you need to model the same learning modes you ask them to develop: let them see you learn new things, demonstrating how multiple modes are necessary to be an effective self-educator.

- **Teach by asking questions:** Socrates had the right idea – if you want to make people examine their assumptions and discover misconceptions, ask lots of challenging questions. This is how I have conditioned myself to respond to student questions: I generally answer with a question of my own seeking the heart of the student's confusion. Posing "thought experiments" for students to conduct is another form of questioning that not only clarifies concepts, but also builds good critical-thinking habits. Anyone familiar with Socrates' fate knows, however, that people tend to react defensively when their assumptions are challenged by persistent questions. A helpful hint for avoiding this kind of reaction is to give the student adequate time and personal space to contemplate your questions. If the student ever feels uncomfortable either with your observation of their efforts or the rapidity of your questioning, they will "lock up" and refuse to engage. Sometimes the best way to manage this behavior is to pose a question to the student, then tell them you will get back to them after a few minutes, rather than to watch them struggle answering your question(s).

- **Be willing to provide the help they need:** If students struggle at certain tasks or with thinking in certain ways, devote extra time with them to practice these skills. Let them know in very practical ways how you are willing to work just as hard as you are asking them to work. Note that this does not mean giving in to demands for lecture. That would be giving students what they *want*, rather than what they *need*. Instead, it means focusing directly on whatever weaknesses are hindering their growth as learners, and aggressively working to strengthen those weaknesses. If it means reserving time to read with students who say they can't understand the text, then that is your job as their teacher.

- **Nothing builds confidence and dissolves apprehension like success:** Remind your students of the challenges they have already overcome, and the progress they have already made.

- **Be patient:** That same student who complains now about having to read, to think independently, and tackle challenging problems will come back to thank you years later. Just as you expected them to think long-term while they were in your class, so you need to think long-term with regard to their appreciation for your standards and efforts. Transformative education is a marathon, not a sprint!

D.2   Teaching technical practices (labwork)

Labwork is an essential part of any science-based curriculum. Here, much improvement may be made over the "standard" educational model to improve student learning. In my students' Instrumentation courses, I forbid the use of pre-built "trainer" systems and lab exercises characterized by step-by-step instructions. Instead, I have my students construct real working instrumentation systems. The heart of this approach is a "multiple-loop" system spanning as large a geographic area as practically possible, with instruments of all kinds connecting to a centralized control room area. None of the instruments need perform any practical purpose, since the goal of the multiple-loop system is for students to learn about the instruments themselves.

A model for a multiple-loop system might look something like this:

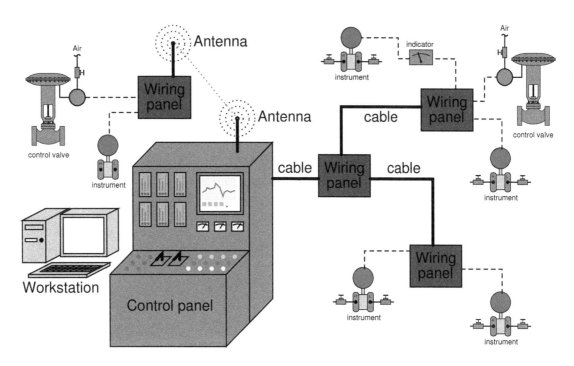

Instruments may or may not be grouped together to form complete control systems, since process control is not necessarily the purpose of this system. The primary purpose of a multiple-loop instrument system is to provide an infrastructure for students to investigate instrumentation apart from the dynamics of a functioning process. The separation of controls from process may seem counter-productive at first, but it actually provides a rich and flexible learning experience. Students are able to measure instrument signals and correlate them with actual physical measurements, take instruments in and out of service, check instrument calibration, see the effects of calibration on measurement accuracy and resolution, practice lock-out and tag-out procedures, diagnose instrument problems introduced by the instructor, practice installing and removing instruments, remove old wire and pull new wire into place, practice sketching and editing loop diagrams, and many other practical tasks without having to balance the needs of a working process. The system may be altered at any time as needed, since there are no process operating constraints to restrict maintenance operations.

In another area of the lab room is a pneumatic control panel and a cabinet housing the distributed control system (DCS) I/O rack:

The rest of the lab room is dedicated as a "field area" where field instruments are mounted and wires (or tubes) run to connect those instruments to remote indication and/or control devices:

Note the use of metal strut hardware to form a frame which instruments may be mounted to, and the use of flexible liquid-tight conduit to connect field instruments to rigid conduit pieces so loop wiring is never exposed.

A less expensive alternative[5] to metal strut is standard *industrial pallet racking*, examples shown here with 2 inch pipe attached for instrument mounting, and enclosures attached for instrument cable routing and termination:

The multiple-loop system is designed to be assembled, disassembled, and reassembled repeatedly as each student team works on a new instrument. As such, it is in a constant state of flux. It is not really a *system* so much as it is an *infrastructure* for students to build working loops and control systems within.

---

[5]When I built my first fully-fledged educational loop system in 2006 at Bellingham Technical College in Washington state (I built a crude prototype in 2003), I opted for Cooper B-Line metal strut because it seemed the natural choice for the application. It wasn't until 2009 when I needed to expand and upgrade the loop system to accommodate more students that I happened to come up with the idea of using pallet racking as the framework material. Used pallet racking is plentiful, and very inexpensive compared to building a comparable structure out of metal strut. As these photographs show, I still used Cooper B-Line strut for some portions, but the bulk of the framework is simply pallet racking adapted for this unconventional application.

In addition to the multiple-loop system, my students' lab contains working processes (also student-built!) which we improve upon every year. One such process is a water flow/level/temperature control system, shown here:

Another is a turbocompressor system, built around a diesel engine turbocharger (propelled by the discharge of a 2 horsepower air blower) and equipped with a pressurized oil lubrication system and temperature/vibration monitor:

Yet another permanent process is this electrical power monitoring unit, where protective (overcurrent) relay operation may be demonstrated:

Measurements of voltage and current in this particular system may be integrated into the rest of the multi-loop system by using voltage and current transducers with 4-20 mA output signals. Digital protective relays may be connected to the multi-loop system using serial data communication (RS-232, RS-485) signals.

The process piping and equipment on these permanent systems are altered only when necessary, but the control systems on these processes may undergo major revisions each year when a new group of students takes the coursework relevant to those systems. Having a set of functioning process systems present in the lab at all times also gives students examples of working instrument systems to study as they plan construction of their temporary loops in the multiple-loop system.

D.3   Teaching diagnostic principles and practices

Diagnostic ability is arguably the most difficult skill to develop within a student, and also the most valuable skill a working technician can possess[6]. In this section I will outline several principles and practices teachers may implement in their curricula to teach the science and art of troubleshooting to their students.

First, we need to define what "troubleshooting" is and what it is not. It is *not* the ability to follow printed troubleshooting instructions found in equipment user's manuals[7]. It is *not* the ability to follow one rigid sequence of steps ostensibly applicable to any equipment or system problem[8]. Troubleshooting is first and foremost the practical application of *scientific thinking* to repair of malfunctioning systems. The principles of hypothesis formation, experimental testing, data collection, and re-formulation of hypotheses is the foundation of any detailed cause-and-effect analysis, whether it be applied by scientists performing primary research, by doctors diagnosing their patients' illnesses, or by technicians isolating problems in complex electro-mechanical-chemical system. In order for anyone to attain mastery in troubleshooting skill, they need to possess the following traits:

- A rock-solid understanding of relevant, fundamental principles (e.g. how electric circuits work, how feedback control loops work)

- Close attention to detail

- An open mind, willing to pursue actions led by data and not by preconceived notions

The first of these points is addressed by any suitably rigorous curriculum. The other points are habits of thought, best honed by months of practice. Developing diagnostic skill requires much time and practice, and so the educator must plan for this in curriculum design. It is not enough to sprinkle a few troubleshooting activities throughout a curriculum, or (worse yet!) to devote an isolated course to the topic. Troubleshooting should be a topic tested on every exam, present in every lab activity, and (ideally) touched upon in every day of the student's technical education.

Scientific, diagnostic thinking is characterized by a repeating cycle of *inductive* and *deductive* reasoning. Inductive reasoning is the ability to reach a general conclusion by observing specific details. Deductive reasoning is the ability to predict details from general principles. For example, a student engages in deductive reasoning when they conclude an "open" fault in a series DC circuit will cause current in that circuit to stop. That same student would be thinking inductively if they measured zero current in a DC series circuit and thus concluded there was an "open" fault somewhere in it. Of these two cognitive modes, inductive is by far the more difficult because multiple solutions exist for any one set of data. In our zero-current series circuit example, inductive reasoning might lead the troubleshooter to conclude an open fault existed in the circuit. However, an unpowered source could also be at fault, or for that matter a malfunctioning ammeter falsely registering zero

---

[6]One of the reasons diagnostic skill is so highly prized in industry is because so few people are actually good at it. This is a classic case of supply and demand establishing the value of a commodity. Demand for technicians who know how to troubleshoot will always be high, because technology will always break. Supply, however, is short because the skill is difficult to teach. This combination elevates the value of diagnostic skill to a very high level.

[7]Yes, I have actually heard people make this claim!

[8]The infamous "divide and conquer" strategy of troubleshooting where the technician works to divide the system into halves, isolating which half the problem is in, is but *one particular procedure: merely one tool in the diagnostician's toolbox*, and does not constitute the whole of diagnostic method.

current when in fact there is current. Inductive conclusions are *risky* because the leap from specific details to general conclusions always harbor the potential for error. Deductive conclusions are *safe* because they are as secure as the general principles they are built on (e.g. *if* an "open" exists in a series DC circuit, there will be *no* current in the circuit, guaranteed). This is why inductive conclusions are always validated by further deductive tests, not vice-versa. For example, if the student induced that an unpowered voltage source might cause the DC series circuit to exhibit zero current, they might elect to test that hypothesis by measuring voltage directly across the power supply terminals. If voltage is present, then the hypothesis of a dead power source is incorrect. If no voltage is present, the hypothesis is provisionally true[9].

Scientific method is a cyclical application of inductive and deductive reasoning. First, an hypothesis is made from an observation of data (inductive). Next, this hypothesis is checked for validity – an experimental test to see whether or not a prediction founded on that hypothesis is correct (deductive). If the data gathered from the experimental test disproves the hypothesis, the scientist revises the hypothesis to fit the new data (inductive) and the cycle repeats.

Since diagnostic thinking requires both deductive and inductive reasoning, and deductive is the easier of the two modes to engage in, it makes sense for teachers to focus on building deductive skill first. This is relatively easy to do, simply by adding on to the theory and practical exercises students already engage in during their studies.

Both deductive and inductive diagnostic exercises lend themselves very well to Socratic discussions in the classroom, where the instructor poses questions to the students and the students in turn suggest answers to those questions. The next two subsections demonstrate specific examples showing how deductive and inductive reasoning may be exercised and assessed, both in a classroom environment and in a laboratory environment.

D.3.1 Deductive diagnostic exercises

Deductive reasoning is where a person applies general principles to a specific situation, resulting in conclusions that are logically necessary. In the context of instrumentation and control systems, this means having students predict the consequence(s) of specified faults in systems. The purpose of building this skill is so that students will be able to quickly and accurately test "fault hypotheses" in their minds as they analyze a faulted system. If they suppose, for example, that a cable has a break in it, they must be able to deduce what effects a broken cable will have on the system in order to formulate a good test for proving or disproving that hypothesis.

---

[9]Other things could be at fault. An "open" test lead on the multimeter for example could account for both the zero-current measurement and the zero-voltage measurement. This scientific concept eludes many people: it is far easier to *disprove* an hypothesis than it is to *prove* one. To quote Albert Einstein, "No amount of experimentation can ever prove me right; a single experiment can prove me wrong."

Example: predicting consequence of a single fault

For example, consider a simple three-resistor series DC circuit, the kind of lab exercise one would naturally expect to see within the first month of education in an Instrumentation program. A typical lab exercise would call for students to construct a three-resistor series DC circuit on a solderless breadboard, predict voltage and current values in the circuit, and validate those predictions using a multimeter. A sample exercise is shown here:

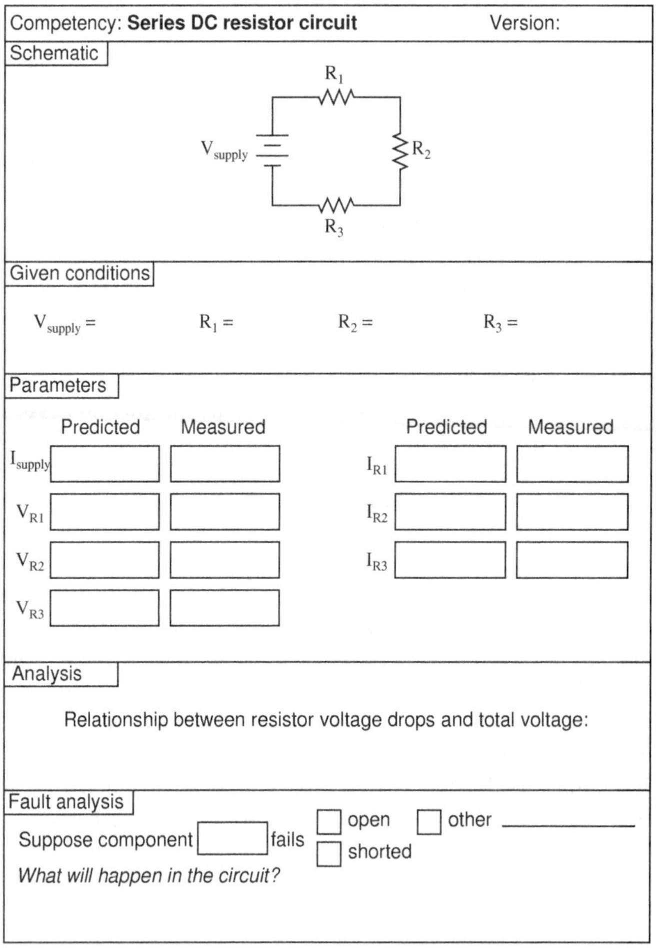
Competency: **Series DC resistor circuit** Version:

Schematic

Given conditions

V_{supply} = R_1 = R_2 = R_3 =

Parameters

	Predicted	Measured		Predicted	Measured
I_{supply}			I_{R1}		
V_{R1}			I_{R2}		
V_{R2}			I_{R3}		
V_{R3}					

Analysis

Relationship between resistor voltage drops and total voltage:

Fault analysis

Suppose component [ ] fails ☐ open ☐ other ________ ☐ shorted

What will happen in the circuit?

Note the **Fault Analysis** section at the end of this page. Here, after the instructor has verified the correctness of the student's mathematical predictions and multimeter measurements, he or she would then challenge the student to predict the effects of a random component fault (either quantitatively or qualitatively), perhaps one of the resistors failing open or shorted. The student makes their predictions, then the instructor simulates that fault in the circuit (either by pulling the resistor out of the solderless breadboard to simulate an "open" or placing a jumper wire in parallel with the resistor to simulate a "short"). The student then uses his or her multimeter to verify the predictions. If the predicted results do not agree with the real measurements, the instructor works with the student to identify why their prediction(s) were faulty and hopefully correct any misconceptions leading to the incorrect result(s). Finally, a different component fault is chosen by the instructor, predictions made by the student, and verification made using a multimeter. The actual amount of time added to the instructor's validation of student lab completion is relatively minor, but the benefits of exercising deductive diagnostic processes are great.

Example: predicting consequences of multiple faults

An example of a more advanced deductive diagnostic exercise appropriate to later phases of a student's Instrumentation education appears here. A loop diagram shows a pressure recording system for an iso-butane distillation column:

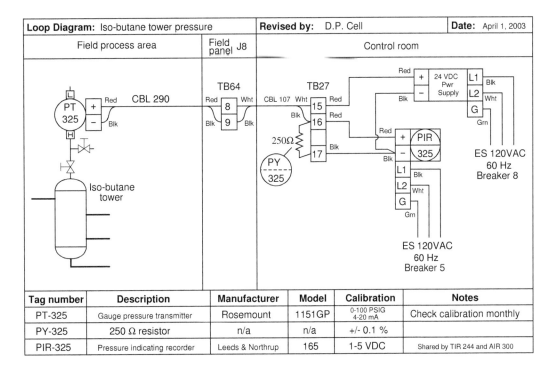

Tag number	Description	Manufacturer	Model	Calibration	Notes
PT-325	Gauge pressure transmitter	Rosemount	1151GP	0-100 PSIG 4-20 mA	Check calibration monthly
PY-325	250 Ω resistor	n/a	n/a	+/- 0.1 %	
PIR-325	Pressure indicating recorder	Leeds & Northrup	165	1-5 VDC	Shared by TIR 244 and AIR 300

A set of questions accompanying this diagram challenge each student to predict effects in the instrument system resulting from known faults, such as:

- PT-325 block valve left shut and bleed valve left open (*predict voltage between TB27-16 and TB27-17*)

- Loose wire connection at TB64-9 (*predict pressure indication at PIR-325*)

- Circuit breaker #5 shut off (*predict loop current at applied pressure of 50 PSI*)

Given each hypothetical fault, there is only one correct conclusion for any given question. This makes deductive exercises unambiguous to assess.

Example: identifying possible faults

A more challenging type of deductive troubleshooting problem easily given in homework or on exams appears here. It asks students to examine a list of potential faults, marking each one of them as either "possible" or "impossible" based on whether or not each fault is independently capable of accounting for all symptoms in the system:

Suppose a voltmeter registers 6 volts between test points **C** and **B** in this series-parallel circuit:

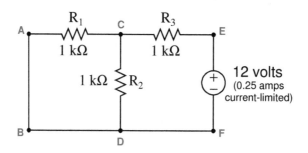

Fault	Possible	Impossible
R_1 failed open		
R_2 failed open		
R_3 failed open		
R_1 failed shorted		
R_2 failed shorted		
R_3 failed shorted		
Voltage source dead		

This is still a *deductive* thinking exercise because each of the faults is given to the student, and it is a matter of deduction to determine whether or not each one of these proposed faults is capable of accounting for the symptoms. Students need only apply the general rules of electric circuits to tell whether or not each of these faults would cause the reported circuit behavior.

True to form for any deductive problem, there can only be one correct answer for each proposed fault. This makes the exercise easy and unambiguous to grade, while honing vitally important diagnostic skills.

One of the benefits of this kind of fault analysis problem is that it requires students to consider *all* consequences of a proposed fault. In order for one of the faults to be considered "possible," it must account for all symptoms, not just one symptom. An example of this sort of problem is seen here:

Suppose the voltmeter in this circuit registers a strong *negative* voltage. A test using a digital multimeter (DMM) shows the voltage between test points **D** and **B** to be 6 volts:

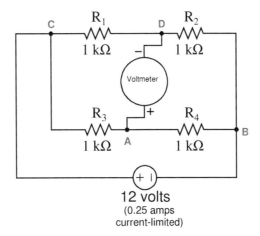

Fault	Possible	Impossible
R_1 failed open		
R_2 failed open		
R_3 failed open		
R_4 failed open		
R_1 failed shorted		
R_2 failed shorted		
R_3 failed shorted		
R_4 failed shorted		
Voltage source dead		

Several different faults are capable of causing the meter to read strongly negative (R_1 short, R_2 open, R_3 open, R_4 short), but only two are capable of this while not affecting the normal voltage (6 volts) between test points D and B: R_3 open or R_4 short. This simple habit of checking to see that the proposed fault accounts for *all* apparent conditions and not just some of them is essential for effective troubleshooting.

This same question format may be easily applied to most any system, not just electrical circuits. Consider this example, determining possible versus impossible faults on an exhaust scrubber system:

After years of successful operation, the level control loop in this exhaust scrubbing system begins to exhibit problems. The liquid level inside the scrubbing tower mysteriously drops far below setpoint, as indicated by the level gauge (LG) on the side of the scrubber. The operators have tried to rectify this problem by increasing the setpoint adjustment on the level controller (LC), to no avail. The level transmitter (LT) is calibrated 3 PSI at 0% (low) level and 15 PSI at 100% (high) level:

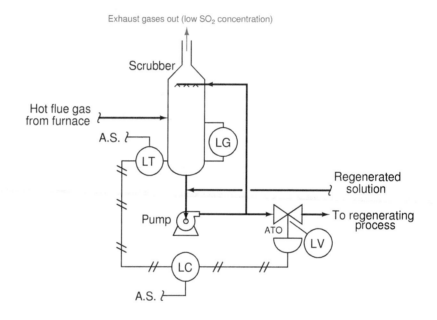

Fault	Possible	Impossible
Air supply to LT shut off		
Air supply to LC shut off		
Pump shut off		
Broken air line between LT and LC		
Broken air line between LC and LV		
Plugged nozzle inside LC		
Plugged orifice inside LC		
Leak in bottom of scrubber		

This exercise is particularly good because it requires the student to determine the action of the level controller (LC) before some of the proposed faults may be analyzed. In this case, the level controller must be *direct-acting*, so that an increasing liquid level inside the scrubber will cause an increasing air signal to the air-to-open (ATO) valve. letting more liquid out of the scrubber to stabilize the level. Without knowing that the level controller is direct-acting, it would be impossible to conclude the effect of a failed air supply to the level transmitter (LT), the first fault proposed in the table.

Example: assessing value of multiple diagnostic tests

A variation on this theme of determining the possibility of proposed faults is to assess the usefulness of proposed diagnostic tests. In other words, the student is presented with a scenario where something is amiss with a system, but instead of selecting a set of proposed faults as being either possible or impossible, the student must determine whether or not a set of proposed *tests* would be diagnostically relevant. An example of this in a simple series-parallel resistor circuit is shown here:

> Suppose a voltmeter registers 0 volts between test points **E** and **F** in this circuit. Determine the diagnostic value of each of the following tests. Assume only one fault in the system, including any single component or any single wire/cable/tube connecting components together. If a proposed test could provide new information to help you identify the location and/or nature of the one fault, mark "yes." Otherwise, if a proposed test would not reveal anything relevant to identifying the fault (already discernible from the measurements and symptoms given so far), mark "no."

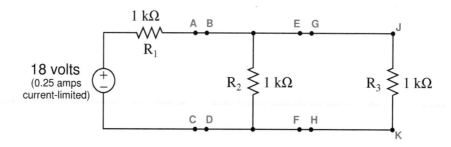

Diagnostic test	Yes	No
Measure V_{AC} with power applied		
Measure V_{JK} with power applied		
Measure V_{CK} with power applied		
Measure I_{R1} with power applied		
Measure I_{R2} with power applied		
Measure I_{R3} with power applied		
Measure R_{AC} with source disconnected from R_1		
Measure R_{DF} with source disconnected from R_1		
Measure R_{EG} with source disconnected from R_1		
Measure R_{HK} with source disconnected from R_1		

This form of diagnostic problem tends to be much more difficult to solve than simply determining the possibility of proposed faults. To solve this form of problem, the student must first determine all possible component faults, and then assess whether or not each proposed test would provide *new information* useful in identifying which of these possible faults is the actual fault.

In this example problem, there are really only a few possible faults: a dead source, an open resistor R_1, a shorted resistor R_2, a shorted resistor R_3, or a broken wire (open connection) somewhere in the loop E-B-A-C-D-F.

The first proposed test – measuring voltage between points A and B – would be useful because it would provide different results given a dead source, open R_1, shorted R_2, or shorted R_3 versus an open between A E or between C-F. Any of the former faults would result in 0 volts between A and B, while any of the latter faults would result in full source voltage between A and B.

The next proposed test – measuring voltage between points J and K – would be useless because we already know what the result will be: 0 volts. This result of this proposed test will be the same no matter which of the possible faults causing 0 voltage between points E and F exists, which means it will shed no new light on the nature or location of the fault.

Despite being very challenging, this type of deductive diagnostic exercise is nevertheless easy to administer and unambiguous to grade, making it very suitable for written tests.

D.3.2 Inductive diagnostic exercises

Inductive reasoning is where a person derives general principles from a specific situation. In the context of instrumentation and control systems, this means having students propose faults to account for specific symptoms and data measured in systems. This is actual troubleshooting, as opposed to deductive diagnosis which is an enabling skill for effective troubleshooting.

While real hands-on exercises are best for developing inductive diagnostic skill, much learning and assessment may be performed in written form as well.

Example: proposing faults in loop diagram

This exam question is a sample of an inductive diagnosis exercise presented in written form:

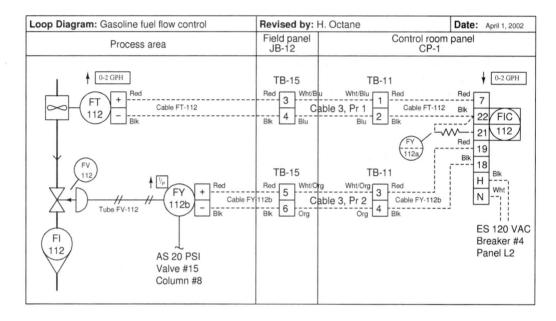

This system used to work just fine, but now it has a problem: the controller registers zero flow, and its output signal (to the valve) is saturated at 100% (wide open) as though it were trying to "ask" the valve for more flow. Your first diagnostic step is to check to see if there actually is gasoline flow through the flowmeter and valve by looking at the rotameter. The rotameter registers a flow rate in excess of 2 gallons per hour.

Identify possible faults in this system that could account for the controller's condition (no flow registered, saturated 100% output), depending on what you find when you look at the rotameter:

- Possible fault:
- Possible fault:

Here, the student must identify two probably faults to account for all exhibited symptoms. More than two different kinds of faults are possible[10], but the student need only identify two faults independently capable of causing the controller to register zero flow when it should be registering more than 2 GPH.

---

[10] Jammed turbine wheel in flowmeter, failed pickup coil in flowmeter, open wire in cable FT-112 or pair 1 of cable 3 (assuming the flow controller's display was not configured to register below 0% in an open-loop condition), etc.

Example: "virtual troubleshooting"

An excellent supplement to any hands-on troubleshooting activities is to have students perform "virtual troubleshooting" with you, the instructor. This type of activity cannot be practiced alone, but requires the participation of someone who knows the answer. It may be done with individual students or with a group.

A "virtual troubleshooting" exercise begins with a schematic diagram of the system such as this, containing clearly labeled test points (and/or terminal blocks) for specifying the locations of diagnostic tests:

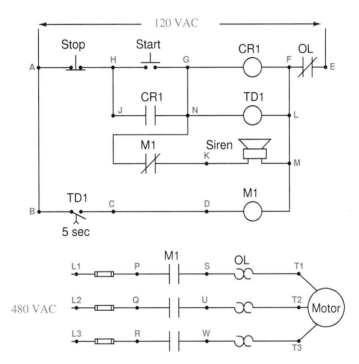

Each student has their own copy of the diagram, as does the instructor. The instructor has furthermore identified a realistic fault within this system, and has full knowledge of that fault's effects. In other words, the instructor is able to immediately tell a student how much voltage will be read between any two test points, what the effect of jumpering a pair of test points will be, what will happen when a pushbutton is pressed, etc.

The activity begins with a brief synopsis of the system's malfunction, narrated by the instructor. Students then propose diagnostic tests to the instructor, with the instructor responding back to each student the results of their tests. As students gather data on the problem, they should be able to narrow their search to find the fault, choosing appropriate tests to identify the precise nature and location of the fault. The instructor may then assess each student's diagnostic performance based on the number of tests and their sequence.

When performed in a classroom with a large group of students, this is actually a lot of fun!

Example: realistic faults in solderless breadboards

Solderless breadboards are universally used in the teaching of basic electronics, because they allow students to quickly and efficiently build different circuits using replaceable components. As wonderful as breadboards are for fast construction of electronic circuits, however, it is virtually impossible to create a realistic component fault without the fault being evident to the student simply by visual inspection. In order for a breadboard to provide a realistic *diagnostic* scenario, you must find a way to hide the circuit while still allowing access to certain test points in the circuit.

A simple way to accomplish this is to build a "troubleshooting harness" consisting of a multi-terminal block connected to a multi-conductor cable. Students are given instructions to connect various wires of this cable to critical points in the circuit, then cover up the breadboard with a five-sided box so that the circuit can no longer be seen. Test voltages are measured between terminals on the block, not by touching test leads to component leads on the breadboard (since the breadboard is now inaccessible).

The following illustration shows what this looks like when applied to a single-transistor amplifier circuit:

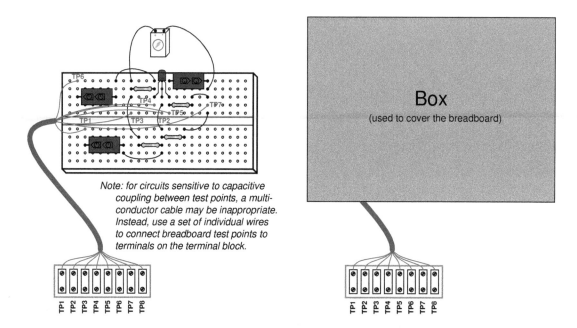

Note: for circuits sensitive to capacitive coupling between test points, a multi-conductor cable may be inappropriate. Instead, use a set of individual wires to connect breadboard test points to terminals on the terminal block.

If students cannot visually detect a fault, they must rely on voltage measurements taken from terminals on the block. This is quite challenging, as not even the shapes of the components may be seen with the box in place. The only guide students have for relating terminal block test points to points in the circuit is the schematic diagram, which is good practice because it forces students to interpret and follow the schematic diagram.

Example: realistic faults in a multi-loop instrument system

Whole instrumentation systems may also serve to build and assess individual diagnostic competence. In my lab courses, students work in teams to build functioning measurement and control loops using the infrastructure of a multiple-loop system (see Appendix section D.2 beginning on page 3172 for a detailed description). Teamwork helps expedite the task of constructing each loop, such that even an inexperienced team is able to assemble a working loop (transmitter connected to an indicator or controller, with wires pulled through conduits and neatly landed on terminal blocks) in just a few hours.

Each student creates their own loop diagram showing all instruments, wires, and connection points, following ISA standards. These loop diagrams are verified by doing a "walk-through" of the loop with all student team members present. The "walk-through" allows the instructor to inspect work quality and ensure any necessary corrections are made to the diagrams. After each team's loop has been inspected and all student loop diagrams edited, the diagrams are placed in a document folder accessible to all students in the lab area.

Once the loop is wired, calibrated, inspected, and documented, it is ready to be faulted. When a student is ready to begin their diagnostic exercise, they gather their team members and approach the instructor. The instructor selects a loop diagram from the document folder *not* drawn by that student, ideally of a loop constructed by another team. The student and teammates leave the lab room, giving the instructor time to fault the loop. Possible faults include:

- Loosen wire connections

- Short wire connections (loose strands of copper strategically placed to short adjacent terminals together)

- Cut cables in hard-to-see locations

- Connect wires to the wrong terminals

- Connect wire pairs backward

- Mis-configure instrument calibration ranges

- Insert square root extraction where it is not appropriate

- Mis-configure controller action or display

- Insert unrealistically large damping constants in either the transmitter, indicator, or final element

- Plug pneumatic signal lines with foam earplugs

- Turn off hand valves

- Trip circuit breakers

After the fault has been inserted, the instructor calls the student team back into the lab area (ideally using a hand-held radio, simulating the work environment of a large industrial facility where technicians carry two-way radios) to describe the symptoms. This part of the exercise works best when the instructor acts the part of a bewildered operator, describing what the system is not doing correctly, without giving any practical advice on the location of the problem or how to fix it[11]. An important detail for the instructor to include is the "history" of the fault: is this a new loop which has never worked correctly, or was it a working system that failed? Faults such as mis-connected wires are realistic of improper installation (new loop), while faults such as loose connections are perfectly appropriate for previously working systems. Whether the instructor freely offers this "history" or waits for the student to ask, it is important to include in the diagnostic scenario because it is an extremely useful piece of information to know while troubleshooting actual systems in industry. Virtually anything may be wrong (including multiple faults) in a brand-new installation, whereas previously working systems tend to fail in fewer ways.

After this introduction, the one student begins his or her diagnosis, with the other team members acting as scribes to document the student's steps. The diagnosing student may ask a teammate for manual assistance (e.g. operating a controller while the student observes a control valve's motion), but no one is allowed to help the one student diagnose the problem. The instructor observes the student's procedure while the student explains the rationale motivating each action, with only a short time given (typically 5 minutes) to determine the general location of the fault causing the problem (e.g. located in the transmitter, control valve, wiring, controller/DCS, tubing, etc.). If after that time period the student is unable to correctly isolate the general location, the exercise is aborted and the instructor reviews the student's actions (as documented by the teammates) to help the student understand where they went wrong in their diagnosis. Otherwise, the student is given more time[12] to pinpoint the nature of the fault.

Depending on the sequencing of your students' coursework, some diagnostic exercises may include components unfamiliar to the student. For example, a relatively new student familiar only with the overall function of a control loop but intimately familiar with the workings of measurement devices may be asked to troubleshoot a loop where the fault is in the control valve positioner rather than in the transmitter. I still consider this to be a fair assessment of the student's diagnostic ability, so long as the expectations are commensurate with the student's knowledge. I would not expect a student to precisely locate the nature of a positioner fault if they had never studied the function or configuration of a valve positioner, but I would expect them to be able to broadly identify the location of the fault (e.g. "it's somewhere in the valve") so long as they knew how a control signal is supposed to command a control valve to move. That student should be able to determine by manually adjusting the controller output and measuring the output signal with the appropriate loop-testing tools that the valve was not responding as it should despite the controller properly performing its function. The ability to diagnose problems in instrument systems where some components of the system are mysterious "black boxes" is a very important skill, because your students *will* have to do exactly that when they step into industry and work with specific pieces of equipment they never had time to learn about in school[13].

---

[11]I must confess to having a lot of fun here. Sometimes I even try to describe the problem incorrectly. For instance, if the problem is a huge damping constant, I might tell the student that the instrument simply does not respond, because that it what it looks like if you do not take the time to watch it respond *very slowly*.

[12]The instructor may opt to step away from the group at this time and allow the student to proceed unsupervised for some time before returning to observe.

[13]I distinctly remember a time during my first assignment as an industrial instrument technician that I had to troubleshoot a problem in a loop where the transmitter was an oxygen analyzer. I had no idea how this particular analyzer functioned, but I realized from the loop documentation that it measured oxygen concentration and output a signal corresponding to the percentage concentration (0 to 21 percent) of O_2. By subjecting the analyzer to known concentrations of oxygen (ambient air for 21%, inert gas for 0%) I was able to determine the analyzer was responding

I find it nearly impossible to fairly assign a letter or percentage grade to any particular troubleshooting effort, because no two scenarios are quite the same. Mastery assessment (either pass or fail, with multiple opportunities to re-try) seems a better fit. Mastery assessment with no-penalty retries also enjoys the distinct advantage of directing needed attention toward and providing more practice for weaker students: the more a student struggles with troubleshooting, the more they must exercise those skills.

Successfully passing a troubleshooting exercise requires not only that the fault be correctly identified and located in a timely manner, but that all steps leading to the diagnosis are logically justified. Random "trial and error" tests by the student will result in a failed attempt, even if the student was eventually able to locate the fault. A diagnosis with no active tests such as multimeter or test gauge measurements, or actions designed to stimulate system components, will also fail to pass. For example, a student who successfully locates a bad wiring connection by randomly tugging at every wire connection should *not* pass the troubleshooting exercise because such actions do not demonstrate diagnostic thinking[14].

To summarize key points of diagnostic exercises using a multiple-loop system:

- Students work in teams to build each loop

- Loop inspection and documentation finalized by a "walk-through" with the instructor

- Instructor placement of faults (it is important no student knows what is wrong with the loop!)

- Each student individually diagnoses a loop, with team members acting merely as scribes

- Students must use loop diagrams drawn by someone else, ideally diagnosing a loop built by a different team

- Brief time limit for each student to narrow the scope of the problem to a general location in the system

- Passing a diagnostic exercise requires:

    $\rightarrow$ Accurate identification of the problem

    $\rightarrow$ Each diagnostic step logically justified by previous results

    $\rightarrow$ Tests (measurements, component response checks) performed before reaching conclusions

- Mastery (pass/fail) assessment of each attempt, with multiple opportunities for re-tries if necessary

---

quite well, and that the problem was somewhere else in the system. If the analyzer had failed my simple calibration test, I would have known there was something wrong with it, which would have led me to either get help from other technicians working at that facility or simply replace the analyzer with a new unit and try to learn about and repair the old unit in the shop. In other words, my ignorance of the transmitter's specific workings did not prevent me from diagnosing the loop in general.

[14]Anyone can (eventually) find a fault if they check every detail of the system. Randomly probing wire connections or aimlessly searching through a digital instrument's configuration is not troubleshooting. I have seen technicians waste incredible amounts of time on the job randomly searching for faults, when they could have proceeded much more efficiently by taking a few multimeter measurements and/or stimulating the system in ways revealing what and where the problem is. One of your tasks as a technical educator is to discourage this bad habit by refusing to tolerate random behavior during a troubleshooting exercise!

D.4   Practical topic coverage

Nearly every technical course teaches and tests students on definitions, basic concepts, and at least some form of quantitative analysis. If you really intend to prepare your students for the challenges of a career like instrumentation, however, you must cover far more than this. The following is a list of topics that should be represented in your curriculum every bit as prevalently as definitions, basic concepts, and math:

- *Qualitative* analysis of instrument systems (e.g. "Predict how the control system will respond if the flow rate *increases*")

- *Qualitative* analysis of processes (e.g. "Predict what will happen to the pressure in reactor vessel R-5 if valve LV-21 *closes*")

- Spatial relations (e.g. mapping wires in a schematic diagram to connection points in a pictorial diagram)

- Evaluating the validity of someone else's diagnosis of a problem (e.g. "The last instrument technician to examine this system concluded the problem was a shorted cable. Based on the data presented here, do you agree or disagree with that conclusion?")

- Identification of safety hazards and possible means of mitigation

- Documentation, both creating it and interpreting it

- Basic project management principles (e.g. scheduling of time, budgeting material and fiscal resources, limiting project scope, following through on "loose ends")

- Mental math (e.g. approximate calculation without the use of computing equipment)

- Evaluation of real-life case studies (e.g. students read and answer questions on industry accident reports such as those published by the US Chemical Safety and Hazard Investigation Board)

These topics can and should be an explicit – not implicit – part of theory and lab (practical) instruction alike. I do not recommend teaching these topics in separate courses, but rather embedding them within each and every course taught in an Instrumentation program. By "explicit" I mean that these topics should be scheduled for discussion within lesson plans, included within student homework questions, appear as actual questions on exams, and individually demonstrated during labwork.

D.5 Principles, not procedures

One of the marks of a successful problem-solver is a habit of applying general principles to every new problem encountered. One of the marks of an ineffective problem-solver is a fixation on procedural steps. Sadly, most of the students I have encountered as a technical college teacher fall into this latter category, as well as a number of working instrument technicians.

Teachers share a large portion of the blame for this sad state of affairs. In an effort to get our students to a place where they are able to solve problems on their own, there is the temptation to provide them with step-by-step procedures for each type of problem they encounter. This is a fundamentally flawed approach to teaching, because a set of rigid procedures only works on a very specific set of problems. To be sure, your students might learn how to solve problems falling within this narrow field by following your algorithmic procedures, but they will be helpless when faced with problems not precisely fitting that same mold. In other words, they might be able to pass your exams but they will flounder when faced with real-world challenges, and you are utterly wasting their time if you are not preparing them for real-world challenges.

I am as guilty of this as any other teacher. When I first began teaching (the subject of electronics), I was dismayed at how difficult it was for students to grasp certain fundamental concepts, such as the analysis of series-parallel resistor circuits. Knowing that I had a very limited amount of time to get my students ready to pass the upcoming exam on series-parallel circuits, I decided to make things simpler for my students by repeatedly demonstrating a set of simple steps by which one could analyze and solve any series-parallel resistor circuit. Fellow instructors did the same thing, and gladly shared their procedures with me, including tips such as the use of different pen colors (black for drawing wires and components, red for writing current values and directional arrows, and blue for writing voltage values and braces) to help organize all the work. The procedure could be long-winded depending on how many nested levels of series-parallel resistors were in the circuit, but precisely followed it would never fail to yield the correct answers. Students greatly appreciated me giving them a set of step-by-step instructions they could follow.

The fallacy of this approach became increasingly evident to me as students would request repeated demonstrations on more and more example problems. I remember one particular classroom session, after having applied this procedure to at least a half-dozen example problems, that one of the students asked me to do one more example. *"Are you kidding?"* was the unspoken thought rushing through my mind, *"Just how many times must I demonstrate the same steps over and over before you can do it on your own?"* It suddenly occurred to me that my students were not learning how to solve problems – instead, they were merely memorizing a sequence of steps including keystrokes on their calculators. Despite all my effort, the only thing I was preparing them to do successfully was pass the upcoming exam, and that was only because the exam contained exactly the same types of problems I was beating to death on the whiteboard in front of class.

What I should have been doing instead was presenting to my students *only* the general principles of resistor circuits, which may be neatly summarized as such:

- Ohm's Law ($V = IR$, where V, I, and R must all refer to the same resistor or same subset of resistors)

- Resistances in series add to make a larger total resistance ($R_{series} = R_1 + R_2 + \cdots R_n$)

- Resistances in parallel diminish to make a smaller total resistance ($R_{parallel} = \frac{1}{\frac{1}{R_1} + \frac{1}{R_2} + \cdots \frac{1}{R_n}}$)

- Current is the same through all series-connected components

- Voltage is the same across all parallel-connected components

It was not as though I had failed to present these principles often enough, nor that I had failed to demonstrate where these principles applied in the procedure. My fault was in giving students a comprehensive procedure in the first place, which had the unintended consequence of drawing their attention away from the fundamental principles. *The simple reason why a step-by-step procedure makes any problem easier to solve is because it eliminates the need for the student to apply general principles to that problem, which is the very thing I my students actually needed to learn.* To put it bluntly, a comprehensive procedure "does the thinking" for the student, because the application of general principles is already pre-determined and encoded into the steps of the procedure itself. What we get by robotically following the procedure is only an illusion of problem-solving competence. The real test of whether or not students have mastered the principles (rather than the procedure) is to check their performance on solving similar problems of different form, where the rote procedure is not applicable.

In order to teach students to approach problem-solving from a conceptual rather than procedural perspective, you must insist students show you how they make the links between general principles and the specifics of given problems. A useful tool for doing this is to have students maintain a notebook identifying and explaining general principles in their own words. You may choose to allow students the use of their own notepage or notecard on exams, as an incentive to tersely summarize all the major principles they will need to solve problems on exams.

An inverted classroom structure is well-suited for the encouragement of principle-based problem solving, in that it affords you the opportunity to see how students approach problems and to continually emphasize principles over procedures.

D.6   Assessing student learning

As a general rule, *high achievement only takes place in an atmosphere of high expectations.* Sometimes these expectations come from within: a self-motivated individual pushes himself or herself to achieve the extraordinary. Most often, the expectations are externally imposed: someone else demands extraordinary performance. One of your responsibilities as a teacher is to hold the standard of student performance high (but reasonable!), and this is done through valid, rigorous assessment.

When the time comes to assess your students' learning, prioritize performance assessment over written or verbal response. In other words, require that your students *demonstrate* their competence rather than merely explain it. Performance assessment takes more time than written exams, but the results are well worth it. Not only will you achieve a more valid measurement of your students' learning, but they will experience greater motivation to learn because they know they must put their learning into action.

Make liberal use of *mastery* assessments in essential knowledge and skill domains, where students must repeat a demonstration of competence as many times as necessary to achieve perfect performance. Not only does this absolutely guarantee students will learn what they should, but the prospect of receiving multiple opportunities to demonstrate knowledge or skill has the beneficial effect of relieving psychological stress for the student. Mastery assessment lends itself very well to the measurement of diagnostic ability.

An idea I picked up through a discussion on an online forum with someone from England regarding engineering education is the idea of breaking written exams into two parts: a *mastery* exam and a *proportional* exam. Students must pass the mastery exam(s) with 100% accuracy in order to receive a passing grade for each course, while the proportional exam is graded like any regular exam (with a score between 0% and 100%) and contributes to their letter grade. Students are given multiple opportunities to pass each mastery exam, with different versions of the mastery exam given at each re-take. Mastery exams cover all the basic concepts, with very straight-forward questions (no tricks or ambiguous wording). The proportional exam, by contrast, is a single-effort test filled with challenging problems requiring high-level thinking. By dividing exams into two parts, it is possible to guarantee the entire class has mastered basic concepts while challenging even the most capable students.

Another unconventional assessment strategy is to create multi-stage exams, where the grade or score received for the exam depends on the highest level passed. I have applied this to the subject of PLC programming: a large number of programming projects are provided as examples, each one fitting into one of four categories of increasing difficulty. The first level is the minimum required to pass the course, while the fourth level is so challenging that only a few students will be able to pass it in the time given. For each of these levels, the student is given the design parameters (e.g. "program a motor start-stop system with a timed lockout preventing a re-start until at least 15 seconds has elapsed"); a micro-PLC; a laptop computer with the PLC programming software; the necessary switches, relays, motors, and other necessary hardware; and 1 hour of time to build and program a working system. There are too many example projects provided for any student to memorize solutions to them all, especially when no notes are allowed during the assessment (only manufacturer's documentation for the PLC and other hardware). This means the student must demonstrate both mastery of the basic PLC programming and wiring elements, as well as creative design skills to arrive at their own solution to the programming problem. There is no limit to the number of attempts a student may take to pass a given level, and no penalty for failed efforts. Best of all, this assessment method demands little of the instructor, as the working project "grades" itself.

My philosophy on assessment is that good assessment is actually more important than good instruction. If the assessments are valid and rigorous, student learning (and instructor teaching!) will rise to meet the challenge. However, even the best instruction will fail to produce consistently high levels of student achievement if students know their learning will never be rigorously assessed. In a phrase, *assessment drives learning.*

For those who might worry about an emphasis on assessment encouraging teachers to "teach to the test," I offer this advice: there is nothing wrong with teaching to the test so long as the test is valid! Educators usually avoid teaching to the test out of a fear students might pass the test(s) without actually learning what they are supposed to gain from taking the course. If this is even possible, it reveals a fundamental problem with the test: it does not actually measure what you want students to know. A valid test is one that cannot be "foiled" by teaching in any particular way. Valid tests challenge students to think, and cannot be passed through memorization. Valid tests avoid asking for simple responses, demanding students articulate reasoning in their answers. *Valid tests are passable only by competence.*

Another important element of assessment is long-term review. You should design the courses in such a way that important knowledge and skill areas are assessed on an ongoing basis up through graduation. Frequent review of foundational concepts is a best practice for attaining mastery in any domain.

D.7 Summary

To summarize some of the key points and concepts for teaching:

- Do not waste class time transmitting facts to students – let the students research facts outside of class

- Use class time to develop high-level thinking skills (e.g. problem-solving, diagnostic techniques, metacognition)

- Use Socratic dialogue to challenge each and every student on the subject matter

- Focus on general principles, not specific procedures

- Make labwork as realistic as possible

- Build diagnostic skill by first exercising deductive reasoning, as a prelude to inductive reasoning

- Incorporate frequent troubleshooting exercises in the lab, with students diagnosing realistic faults in instrument systems

- Include a broad range of practical topics and aspects in all coursework rather than fall into the convention of focusing on memorizing definitions, stating concepts, and performing quantitative calculations

- Assess student learning validly and rigorously

- Review important knowledge and skill areas continually until graduation – build this review into the program courses themselves (homework, quizzes, exams) rather than relying on ad hoc review

One final piece of advice for educators at every level: *it is better to teach a few things well than to teach many things poorly!* If external constraints force you to "cover" too much material in too little time, focus on making each learning exercise as integrative as possible, so students will experience different topics in ways that reinforce and give context to each other.

Appendix E

Contributors

This is an open-source book, which means everyone has a legal write to modify it to their liking. As the author, I freely accept input from readers that will make this book better. This appendix exists to give credit to those readers who have made substantial contributions to this book.

Sadly, this list does not show the names of *every* person who has helped me identify and correct minor typographical and grammatical errors. The list of names and errors would be quite substantial, I must admit. Those persons who are listed for their identification of typographical errors have earned a place on the list through sheer volume of errors found. I am indebted to my students, and to readers around the world for their careful reading of the text and their considerate feedback.

If you find this textbook exceeds your expectations, know that it is primarily because of this detailed reader feedback. When a book is edited by its author, based on daily feedback from the people using it, errors as well as unclear explanations get identified and corrected at a much more rapid pace than what is typical during a traditional (published) textbook editing cycle. As an author, you might think you have explained something perfectly well, only to find your readers mystified. It is only by discovering your readers are baffled by your "good" explanation that you will know to edit that explanation, or to take a completely different approach in explaining the concept. Then, you need to get more feedback after the edit(s), to see whether or not you've made an improvement.

This kind of tight feedback loop between authors and readers simply doesn't happen in a traditional textbook publishing model. In an open-source, self-published model, however, this kind of feedback is not only possible, but is also typical. For this reason, I strongly encourage teachers everywhere to consider writing their own open-source textbooks, using their students as editors to prove whether or not their writing is effective. The world is unfortunately filled with poor-quality texts. Let's change this sad state of affairs, one open-source textbook at a time!

E.1   Error corrections

Abdelaziz, Ahmed (June 2014)

- Identified numerous typographical errors throughout the book, mostly *repeated repeated* words.

Ames, Ben (January 2015)

- Identified typographical error in the *Problem-Solving and Diagnostic Strategies* chapter.

Balbaa, Amr (May 2015)

- Identified mathematical error in the *Continuous Fluid Flow Measurement* chapter.

Barton, Bruce (January 2015, March 2015, April 2015)

- Identified typographical errors in the *Discrete Control Elements* chapter regarding solenoid valve construction, in the *Process Dynamics and PID Controller Tuning* chapter, and also in the *Digital Data Acquisition and Networks* chapter.

Brierly, Ryan (April 2016)

- Identified typographical error in the *Physics* chapter regarding transition points between laminar and turbulent flow regimes.

Brown, Kevin (February 2011)

- Identified typographical error in the *Control Valves* chapter.

Bultsma, Brent (September 2011-June 2012)

- Identified multiple typographical errors throughout the book.

Edwards, William (February 2013)

- Identified error in the *Calculus* chapter, where Δt was confused with Δm.

Esher, Cynthia A. (December 2009)

- Identified error of referring to "SAMA" diagrams. I changed these references to "functional" diagrams instead (*Instrumentation Documents* chapter).

Floyd, Wade (November 2015)

- Identified an error the distance between Earth and the sun *Chemistry* chapter: the distance is 93 *million* miles, not 93 *billion* miles as I originally had it.

Gassman, Jessica (February 2013)

- Suggested clarification in the *Closed-Loop Control* chapter regarding the auto/manual transfer of pneumatic controllers.

Glundberg, Blake (February 2011, April 2011)

- Identified typographical errors (*Control Valves* chapter, *Digital Data Acquisition and Networks* chapter).

Ireland, Chandler (February 2015)

- Identified typographical error (*Closed-Loop Control* chapter).

Joe, Savio (February 2013)

- Identified typographical error (*Instrument Connections* chapter).

Ketteridge, Tim and Dieckman, Lloyd (April 2013)

- Identified conceptual errors in the IP and TCP sections of the *Digital Data Acquisition and Network* chapter. The task of portioning large data blocks into segments and packets is the job of TCP (or Application-layer protocols), not IP. IP does, however, support fragmentation of large packets into smaller packets and reassembly into large packets again if needed for communication along network types that cannot support large packet sizes. I had these concepts conflated.

Ketteridge, Tim (December 2013)

- Identified an error in measurement units for the Ideal Gas Law ($PV = NkT$). When using this form of the Ideal Gas Law, P needs to be in units of Pascals (not atmospheres), and V needs to be in units of cubic meters (not liters).

Marques, Tiago (July 2012)

- Identified lack of origin lines in linear graphs where 4-20 mA signals are related to practical measurement ranges (*Analog Electronic Instrumentation* chapter).

Mhyre, Phil (January 2012)

- Identified spelling error on the word "desiccant" (*Discrete Control Elements* chapter).

Nobiling, David (July 2016)

- Identified typographical error in the (*Digital data acquisition and networks* chapter), where the terms "PLC" and "VFD" were interchanged.

Pulido, Alberto (May 2012)

- Identified conceptual error regarding Fieldbus terminator resistor packages (*FOUNDATION Fieldbus Instrumentation* chapter).

Sangani, Champa (September 2009) and Brainard, Ben (January 2012)

- Identified calculation error in milliamp-to-pH scaling problem (*Analog Electronic Instrumentation* chapter).

Saxton, Trevor (August 2015)

- Identified mathematical errors in the "Phasor Arithmetic" section of the (*AC Electricity* chapter).

Schultz, Steven (February 2011)

- Identified error in a pilot-loaded pressure regulator design depicted in the *Control Valves* chapter.

Sloan, Wyatt (October 2012)

- Identified typographical error in the *Continuous Temperature Measurement* chapter. Later (January 2013) identified multiple places where the word "identify" was used instead of "identity."

Sobczak, Jacek (October 2015)

- Identified missing files in the .tar archive, suggested changes to the LaTeX source code producing better page numbering in the book's table of contents, and identified broken links on the website.

Thompson, Brice (June 2009)

- Identified errors in high/low select and high/low limit function illustrations (*Basic Process Control Strategies* chapter).

TooToonchy, Hossein (June 2014)

- Identified typographical errors in the (*FOUNDATION Fieldbus Instrumentation* chapter).

Tsiporenko, Michael (June 2009)

- Identified typographical error in the *Introduction to Industrial Instrumentation* chapter.

Villajulca, Jose Carlos (August 2011)

- Identified error in PID tuning response for a "generic" simulated process in the *Process Dynamics and PID Controller Tuning* chapter.

Wingerter, John (January 2017)

- Identified an oversight in explaining safety considerations during control valve disassembly, in the *Disassembly of a Sliding-Stem Control Valve* appendix.

Young, Jeanne (April-May 2012)

- Identified errors in proximity switch wire color codes and suggested correction in the *Programmable Logic Controllers* chapter.

- Identified pattern inconsistent with industry convention in motor control circuit schematic diagrams within the *Discrete Control Elements* chapter.

E.2   New content

Coy, John J. et. al. (December 2012)

- Sampled graphic illustrations from his NASA Reference Publication "Gearing" (1152), showing different types of gear trains for inclusion in the *Simple Machines* section of the *Physics* chapter. This NASA document, being a US government publication unclassified and approved for "unlimited" distribution, lies within the public domain and so may be freely sampled into other works such as this one.

Dennis, Japheth (2011-2012 academic year)

- Suggested additional examples of PID controller responses to graph, to help illustrate the unique features of each action.

Faydor, L. Litvin et. al. (December 2012)

- Sampled 5-planet planetary gear set illustration from his NASA publication "New Design and Improvement of Planetary Gear Trains", for inclusion in the *Simple Machines* section of the *Physics* chapter. This NASA document, being a US government publication unclassified and approved for "unlimited" distribution, lies within the public domain and so may be freely sampled into other works such as this one. Also sampled helicopter transmission gear set illustration from his NASA publication "Handbook on Face Gear Drives With a Spur Involute Pinion".

Goertz, Kevin (2006-2007 academic year)

- Took photographs of various flowmeters, control valves, and an insertion pH probe assembly.

Marshall, Travis (2012-2013 academic year)

- Contributed analogy of a "shadow" to the explanation of absorption spectroscopy, after helping to install a laser stack gas analyzer during a summer internship and helping to explain that instrument's operating principle to himself.

Poelma, John (2010-2011 academic year)

- Took photographs of various pressure vessels, instruments, control valves, and other process hardware at NASA's Stennis Space Center in Mississippi.

Toner, Aaron (August 2012)

- Took photographs of his self-constructed MicroLogix 1100 / C-More Micro PLC trainer for inclusion in this book.

Appendix F

Creative Commons Attribution License

F.1   A simple explanation of your rights

This is an "open-source" textbook, which means the entirety of it is freely available for public perusal, reproduction, distribution, and even modification. All digital "source" files comprising this textbook reside at the following website:

http://ibiblio.org/kuphaldt/socratic/sinst/book/

The Creative Commons Attribution license grants you (the recipient), as well as anyone who might receive my work from you, the right to freely use it. This license also grants you (and others) the right to modify my work, so long as you properly credit my original authorship and declare these same rights (to my original work) for your own readers. My work is copyrighted under United States law, but this license grants everyone else in the world certain freedoms not customarily available under full copyright. This means no one needs to ask my permission, or pay any royalties to me, in order to read, copy, distribute, publish, or otherwise use this book.

If you choose to modify my work, you will have created what legal professionals refer to as a *derivative work*. The Creative Commons license broadly groups derivative works under the term *adapted material*. In simple terms, the fundamental restriction placed on you when you do this is you must properly credit me for the portions of your adaptation that are my original work. Otherwise, you may treat your adaptation the same way you would treat a completely original work of your own. This means you are legally permitted to enjoy full copyright protection for your adaptation, up to and including exclusive rights of reproduction and distribution. In other words, this license does *not* bind your derivative work under the same terms and conditions I used to release my original work, although it does require you notify your readers of this free source text and the License under which it is offered.

The practical upshot of this is you may modify my work and re-publish it as you would any other book, with the full legal right to demand royalties, restrict distributions, etc. This does not compromise the freedom of my original work, because that is still available to everyone under the terms and conditions of the Attribution license[1]. It does, however, protect the investment(s) you make in creating the adaptation by allowing you to release the adaptation under whatever terms you see fit (so long as those terms comply with current intellectual property laws, of course).

In summary, the following "legalese" is actually a very good thing for you, the reader of my book. It grants you permission to do so much more with this text than what you would be legally allowed to do with any other (traditionally copyrighted) book. It also opens the door to open collaborative development, so it might grow into something far better than what I alone could create.

---

[1]You *cannot* pass my original work to anyone else under different terms or conditions than the Attribution license. That is called *sublicensing*, and the Attribution license forbids it. In fact, any re-distribution of my original work must come with a notice to the Attribution license, so anyone receiving the book through you knows their rights.

F.2 Legal code

Creative Commons Attribution 4.0 International Public License

By exercising the Licensed Rights (defined below), You accept and agree to be bound by the terms and conditions of this Creative Commons Attribution 4.0 International Public License ("Public License"). To the extent this Public License may be interpreted as a contract, You are granted the Licensed Rights in consideration of Your acceptance of these terms and conditions, and the Licensor grants You such rights in consideration of benefits the Licensor receives from making the Licensed Material available under these terms and conditions.

Section 1 – Definitions.

a. **Adapted Material** means material subject to Copyright and Similar Rights that is derived from or based upon the Licensed Material and in which the Licensed Material is translated, altered, arranged, transformed, or otherwise modified in a manner requiring permission under the Copyright and Similar Rights held by the Licensor. For purposes of this Public License, where the Licensed Material is a musical work, performance, or sound recording, Adapted Material is always produced where the Licensed Material is synched in timed relation with a moving image.

b. **Adapter's License** means the license You apply to Your Copyright and Similar Rights in Your contributions to Adapted Material in accordance with the terms and conditions of this Public License.

c. **Copyright and Similar Rights** means copyright and/or similar rights closely related to copyright including, without limitation, performance, broadcast, sound recording, and Sui Generis Database Rights, without regard to how the rights are labeled or categorized. For purposes of this Public License, the rights specified in Section 2(b)(1)-(2) are not Copyright and Similar Rights.

d. **Effective Technological Measures** means those measures that, in the absence of proper authority, may not be circumvented under laws fulfilling obligations under Article 11 of the WIPO Copyright Treaty adopted on December 20, 1996, and/or similar international agreements.

e. **Exceptions and Limitations** means fair use, fair dealing, and/or any other exception or limitation to Copyright and Similar Rights that applies to Your use of the Licensed Material.

f. **Licensed Material** means the artistic or literary work, database, or other material to which the Licensor applied this Public License.

g. **Licensed Rights** means the rights granted to You subject to the terms and conditions of this Public License, which are limited to all Copyright and Similar Rights that apply to Your use of the Licensed Material and that the Licensor has authority to license.

h. **Licensor** means the individual(s) or entity(ies) granting rights under this Public License.

i. **Share** means to provide material to the public by any means or process that requires permission under the Licensed Rights, such as reproduction, public display, public performance, distribution, dissemination, communication, or importation, and to make material available to the public including in ways that members of the public may access the material from a place and at a time individually chosen by them.

j. **Sui Generis Database Rights** means rights other than copyright resulting from Directive 96/9/EC of the European Parliament and of the Council of 11 March 1996 on the legal protection of databases, as amended and/or succeeded, as well as other essentially equivalent rights anywhere in the world.

k. **You** means the individual or entity exercising the Licensed Rights under this Public License. **Your** has a corresponding meaning.

Section 2 – Scope.

a. License grant.

1. Subject to the terms and conditions of this Public License, the Licensor hereby grants You a worldwide, royalty-free, non-sublicensable, non-exclusive, irrevocable license to exercise the Licensed Rights in the Licensed Material to:

A. reproduce and Share the Licensed Material, in whole or in part; and

B. produce, reproduce, and Share Adapted Material.

2. Exceptions and Limitations. For the avoidance of doubt, where Exceptions and Limitations apply to Your use, this Public License does not apply, and You do not need to comply with its terms and conditions.

3. Term. The term of this Public License is specified in Section 6(a).

4. Media and formats; technical modifications allowed. The Licensor authorizes You to exercise the Licensed Rights in all media and formats whether now known or hereafter created, and to make technical modifications necessary to do so. The Licensor waives and/or agrees not to assert any right or authority to forbid You from making technical modifications necessary to exercise the Licensed Rights, including technical modifications necessary to circumvent Effective Technological Measures. For purposes of this Public License, simply making modifications authorized by this Section 2(a)(4) never produces Adapted Material.

5. Downstream recipients.

A. Offer from the Licensor – Licensed Material. Every recipient of the Licensed Material automatically receives an offer from the Licensor to exercise the Licensed Rights under the terms and conditions of this Public License.

B. No downstream restrictions. You may not offer or impose any additional or different terms or conditions on, or apply any Effective Technological Measures to, the Licensed Material if doing so restricts exercise of the Licensed Rights by any recipient of the Licensed Material.

6. No endorsement. Nothing in this Public License constitutes or may be construed as permission

to assert or imply that You are, or that Your use of the Licensed Material is, connected with, or sponsored, endorsed, or granted official status by, the Licensor or others designated to receive attribution as provided in Section 3(a)(1)(A)(i).

b. Other rights.

1. Moral rights, such as the right of integrity, are not licensed under this Public License, nor are publicity, privacy, and/or other similar personality rights; however, to the extent possible, the Licensor waives and/or agrees not to assert any such rights held by the Licensor to the limited extent necessary to allow You to exercise the Licensed Rights, but not otherwise.

2. Patent and trademark rights are not licensed under this Public License.

3. To the extent possible, the Licensor waives any right to collect royalties from You for the exercise of the Licensed Rights, whether directly or through a collecting society under any voluntary or waivable statutory or compulsory licensing scheme. In all other cases the Licensor expressly reserves any right to collect such royalties.

Section 3 – License Conditions.

Your exercise of the Licensed Rights is expressly made subject to the following conditions.

a. Attribution.

1. If You Share the Licensed Material (including in modified form), You must:

A. retain the following if it is supplied by the Licensor with the Licensed Material:

i. identification of the creator(s) of the Licensed Material and any others designated to receive attribution, in any reasonable manner requested by the Licensor (including by pseudonym if designated);

ii. a copyright notice;

iii. a notice that refers to this Public License;

iv. a notice that refers to the disclaimer of warranties;

v. a URI or hyperlink to the Licensed Material to the extent reasonably practicable;

B. indicate if You modified the Licensed Material and retain an indication of any previous modifications; and

C. indicate the Licensed Material is licensed under this Public License, and include the text of, or the URI or hyperlink to, this Public License.

2. You may satisfy the conditions in Section 3(a)(1) in any reasonable manner based on the medium, means, and context in which You Share the Licensed Material. For example, it may be

reasonable to satisfy the conditions by providing a URI or hyperlink to a resource that includes the required information.

3. If requested by the Licensor, You must remove any of the information required by Section 3(a)(1)(A) to the extent reasonably practicable.

4. If You Share Adapted Material You produce, the Adapter's License You apply must not prevent recipients of the Adapted Material from complying with this Public License.

Section 4 – Sui Generis Database Rights.

Where the Licensed Rights include Sui Generis Database Rights that apply to Your use of the Licensed Material:

a. for the avoidance of doubt, Section 2(a)(1) grants You the right to extract, reuse, reproduce, and Share all or a substantial portion of the contents of the database;

b. if You include all or a substantial portion of the database contents in a database in which You have Sui Generis Database Rights, then the database in which You have Sui Generis Database Rights (but not its individual contents) is Adapted Material; and

c. You must comply with the conditions in Section 3(a) if You Share all or a substantial portion of the contents of the database.

For the avoidance of doubt, this Section 4 supplements and does not replace Your obligations under this Public License where the Licensed Rights include other Copyright and Similar Rights.

Section 5 – Disclaimer of Warranties and Limitation of Liability.

a. Unless otherwise separately undertaken by the Licensor, to the extent possible, the Licensor offers the Licensed Material as-is and as-available, and makes no representations or warranties of any kind concerning the Licensed Material, whether express, implied, statutory, or other. This includes, without limitation, warranties of title, merchantability, fitness for a particular purpose, non-infringement, absence of latent or other defects, accuracy, or the presence or absence of errors, whether or not known or discoverable. Where disclaimers of warranties are not allowed in full or in part, this disclaimer may not apply to You.

b. To the extent possible, in no event will the Licensor be liable to You on any legal theory (including, without limitation, negligence) or otherwise for any direct, special, indirect, incidental, consequential, punitive, exemplary, or other losses, costs, expenses, or damages arising out of this Public License or use of the Licensed Material, even if the Licensor has been advised of the possibility of such losses, costs, expenses, or damages. Where a limitation of liability is not allowed in full or in part, this limitation may not apply to You.

c. The disclaimer of warranties and limitation of liability provided above shall be interpreted in a manner that, to the extent possible, most closely approximates an absolute disclaimer and waiver of all liability.

Section 6 – Term and Termination.

a. This Public License applies for the term of the Copyright and Similar Rights licensed here. However, if You fail to comply with this Public License, then Your rights under this Public License terminate automatically.

b. Where Your right to use the Licensed Material has terminated under Section 6(a), it reinstates:

1. automatically as of the date the violation is cured, provided it is cured within 30 days of Your discovery of the violation; or

2. upon express reinstatement by the Licensor.

For the avoidance of doubt, this Section 6(b) does not affect any right the Licensor may have to seek remedies for Your violations of this Public License.

c. For the avoidance of doubt, the Licensor may also offer the Licensed Material under separate terms or conditions or stop distributing the Licensed Material at any time; however, doing so will not terminate this Public License.

d. Sections 1, 5, 6, 7, and 8 survive termination of this Public License.

Section 7 – Other Terms and Conditions.

a. The Licensor shall not be bound by any additional or different terms or conditions communicated by You unless expressly agreed.

b. Any arrangements, understandings, or agreements regarding the Licensed Material not stated herein are separate from and independent of the terms and conditions of this Public License.

Section 8 – Interpretation.

a. For the avoidance of doubt, this Public License does not, and shall not be interpreted to, reduce, limit, restrict, or impose conditions on any use of the Licensed Material that could lawfully be made without permission under this Public License.

b. To the extent possible, if any provision of this Public License is deemed unenforceable, it shall be automatically reformed to the minimum extent necessary to make it enforceable. If the provision cannot be reformed, it shall be severed from this Public License without affecting the enforceability of the remaining terms and conditions.

c. No term or condition of this Public License will be waived and no failure to comply consented to unless expressly agreed to by the Licensor.

d. Nothing in this Public License constitutes or may be interpreted as a limitation upon, or waiver of, any privileges and immunities that apply to the Licensor or You, including from the legal processes of any jurisdiction or authority.

Creative Commons is not a party to its public licenses. Notwithstanding, Creative Commons may elect to apply one of its public licenses to material it publishes and in those instances will be considered the "Licensor." Except for the limited purpose of indicating that material is shared under a Creative Commons public license or as otherwise permitted by the Creative Commons policies published at creativecommons.org/policies, Creative Commons does not authorize the use of the trademark "Creative Commons" or any other trademark or logo of Creative Commons without its prior written consent including, without limitation, in connection with any unauthorized modifications to any of its public licenses or any other arrangements, understandings, or agreements concerning use of licensed material. For the avoidance of doubt, this paragraph does not form part of the public licenses.

Creative Commons may be contacted at `creativecommons.org`.

Index

C_d factor, 2175
C_v factor, 2167
K_v factor, 2167
Q (quality factor of RLC filter circuit), 434
ΔT, meaning of, 128
α particle radiation, 1489
β particle radiation, 1489
γ ray radiation, 1489
λ (failure rate), 2614
μ metal, 619
$\times 10$ oscilloscope probe, 2543
s variable, 402, 404, 435
"Desktop Process" PID tuning tool, 2479
10 to 50 mA, 889
2-wire RTD circuit, 1512
2-wire transmitter, 311, 887
21 relay, distance protection, 2012
3 to 15 PSI, 912
3-element boiler feedwater control, 2517
3-valve manifold, 1355
3-way globe valve, 2076, 2149
3-wire RTD circuit, 1514, 1515
4 to 20 mA, 857, 911
4-wire RTD circuit, 1513
4-wire resistance measurement circuit, 1513, 1748
4-wire transmitter, 885
4-wire transmitter, active output, 889
4-wire transmitter, passive output, 889
4-wire transmitter, sinking output, 889
4-wire transmitter, sourcing output, 889
5-point calibration, 1274
5-valve manifold, 1358
50 relay, instantaneous overcurrent protection, 1983
51 relay, time overcurrent protection, 1983
67 relay, directional current protection, 2010
802.11, 1040

802.3, 1027, 1033
802.4, 1037
802.5, 1037
86 relay, auxiliary/lockout, 2036
87 relay, differential protection, 1992

ABB 800xA distributed control system (DCS), 2364
Absolute addressing, 797
Absolute pressure, 186, 1308
Absolute viscosity, 199
Absolute zero, 123
Absorption spectroscopy, 256
AC, 355
AC excitation, magnetic flowmeter, 1678
AC generator synchronization, 1901
AC induction motor, 718, 2799
Accelerometer, 2716
Acceptance testing, 640
Acid, 286
Acid, strong, 286
Acid, weak, 286
Acre, defined, 57
Action, controller, 506, 2266, 2571
Activation energy, 275
Active intrinsic safety barrier, 2594
Active reading, 2730
Actuator, 2096
Actuator, valve, 2067
Acyclic communication, Fieldbus, 1151, 1155
Adaptive gain controller, 2416
ADC, 780, 973, 986
Additive transformer windings, 1946
Address, 1099
Address Resolution Protocol, 1069, 1078
Administrator privilege, computer, 2714
Admittance, 360

Aerosol, 229

Aeroswage SX-1 tubing tool, 586

AGA Report #11, 1717

AGA Report #3, 1632, 1639

AGA Report #7, 1658

AGA Report #9, 1685

Air dryer, 694

Air-to-close valve, 2116

Air-to-open valve, 2116

Alarm, annunciator, 521

Alarm, process, 2374

Algorithm, 2260

Algorithm, control, 536

Aliasing, 991

Alkaline, 286

Alkaline, strong, 287

Alkaline, weak, 287

Allen-Bradley ControlLogix 5000 PLC, 767

Allen-Bradley Data Highway (DH) network, 1053

Allen-Bradley GuardLogix DCSRT safety instruction, 2667

Allen-Bradley GuardLogix PLC, 2667

Allen-Bradley MicroLogix 1000 PLC, 851

Allen-Bradley PLC-5, 764

Allen-Bradley SLC 500 PLC, 765, 874

Alpha particle radiation, 1489

Altek model 334A loop calibrator, 902

Alternating current, 355

Alternating motor control, 829

American alpha value, RTD, 1511

American Gas Association, 1632, 1639, 1658, 1685, 1717

Ametek model 1500 induction relay, 676

Ammeter, clamp-on, 341

Amp-turn, 340

Ampère, André, 299

Ampere, 299

Amplifier, instrumentation, 1001

amu, 262

Analog input, single-ended, 1008

Analog-to-digital converter, 780, 973, 986

Analyzer, 1297

Anderson, Norman A., 3

Andrew, William G, 3

Angle of repose, 1464, 2060

Anion, 228, 1743

Annubar, 1618, 1729

Annunciator, 521

Anode, 1743

ANPT pipe threads, 568

ANSI codes for power systems, 1979

ANSI pressure classes for flanges, 563

ANSI/IEEE standard 754-1985 for floating-point numbers, 978, 1121

Antenna, half-wave, 488, 1204

Antenna, quarter-wave, 489, 1204

Antenna, whip, 489, 1204

Antenna, Yagi, 490, 1204

Anti-aliasing filter, 993

Anti-resonance resistor, 429

Anti-virus, computer, 2714

API, degrees, 180

Application blacklisting, computer, 2714

Application whitelisting, computer, 2714

Arbitration, channel, 1035

Arc blast, 725

Arc chute, circuit breaker, 1916

Arc flash, 631, 725

Archimedes' Principle, 191, 1445

Area, defined, 57

Armature, 341

ARP, 1069, 1078, 1079

As-found calibration, 1256

As-left calibration, 1256

Asbestos valve stem packing, 2092

ASCII, 984

ASCII Modbus frames, 1106

ASCO model NH90 "Hydramotor" linear actuator/positioner, 931

ASCO solenoid valve, 704

Ashcroft temperature switch, 679

Assignment, computer programming, 2377

Asynchronous data transfer, 1021, 1027

Atmospheres, 188

Atom, 228

Atomic clock, 1278, 1281

Atomic mass, 228, 230, 237

Atomic mass units, 262

Atomic number, 228, 230, 236

Atomic weight, 228, 230, 237

Aufbau order, 247

Auto-tuning PID controller, 2444

Automatic mode, 498, 2371

Autotransformer, 470

Autotransformer, boost, 470

Autotransformer, buck, 471

Auxiliary contact, 735, 790

Averaging Pitot tube, 1617

Avogadro's number, 262

B.I.F. Universal Venturi tube, 1621

B/W Controls model 1500 induction relay, 676

Babbage, Charles, 1174

Back-calculation variable, FOUNDATION Fieldbus function block programming, 1160, 1166, 2556, 2559

Background substances, 1802

Backpressure, nozzle, 917

Backup (remote), protective relay, 2033

BACnet, 2352

Baffle, 917

Bailey Infi90 distributed control system, 2364

Bailey Net90 distributed control system, 2364

Balance beam scale, 919

Ball valve, 687, 2080

Ball valve, characterized, 2081

Ball valve, segmented, 2081

Balling, degrees, 181

Balmer series, 255

Bang-bang control, 2263

Bar, 188

Bara, 188

Barg, 188

Bark, degrees, 181

Barkhausen criterion (loop oscillation), 2427, 2431

Barometer, 1308

Barrier circuit, intrinsic safety, 2593

Base, 286

Base unit, 72

Base, strong, 287

Base, weak, 287

basisk Siemens PLC exploit, 2715

Bastard voltage, 454

Bathtub curve, 2620

Baud rate, 1026

Baudot code, 983, 1020, 1022

Baumé, degrees, 180

Bazovsky, Igor, 2599, 2604

Beer-Lambert Law, 1802

Bel, 1206

Bell 202 FSK standard, 1092

Bell 202 standard, 1024

Belleville washer spring, 1924, 2089

Bellingham Technical College, 3177

Bellows, 920, 1310

Bellows packing seal, 2093

Belt, 117

Bench set, control valve, 2125

Bend radius, optical fiber, 638

Bently-Nevada model 1701 FieldMonitor vibration monitor, 1878

Bently-Nevada model 3300 vibration monitor, 1876

Bently-Nevada vibration monitoring equipment, 1872

BER (bit error rate), 1222

Beresford, Dillon, 2715

Bernoulli's equation, 208, 1582

Bernoulli, Daniel, 208

Beta particle radiation, 1489

Beta ratio of flow element, 1601, 1638

Bethlehem flow tube, 1621

Bevel gear, 111

Bi-metal strip, 1503

Biddle Versa-Cal loop calibrator, 907

Binary numeration, 974

Biological oxygen demand, 2260

Bit, 773, 974

Bit error rate, 1222

Bit rate, 1025

Black, Harold, 938

Blackbody, 1562

Blackbody calibrator, 1290

Blacklisting, application, 2714

Bleed port adapter test accessory, 1363

Bleed valve fitting, 1361

Bleeding pneumatic relay, 933

BLEVE, 161

Blind, pipe, 565

Blocking, protective relay, 2006, 2027, 2035, 2037

Blowdown pressure, safety valve, 2644, 2651

Bluetooth, 2722

Bluff body, 1663

BMS, 836, 2673
BNC style connector, 625, 1877
BNU, Fieldbus, 1157
BOD, 2260
Bode plot, 414
Body, valve, 2067
Boiling Liquid Expanding Vapor Explosion (BLEVE), 161
Boiling point of water, 1287
Boiling Water Reactor (BWR), 2684
Bonneville Power Administration, 1897
Boolean variable, 2377
Booster relay, 2158
Boosting autotransformer, 470
Bourdon tube, 571, 1310
Boyle's Law, 198
bps, 1025
Branch tee fitting, 582
Branching, computer programming, 2379
Brazosport College, 1613
Brick, Fieldbus coupler, 1141
Bridge circuit, 327
British Thermal Unit, 125
Brix, degrees, 181
Broadcast address, IP, 1075
Brush, DC motor, 2225
Brute force attack, password, 2713
BSPP pipe threads, 570
BSPT pipe threads, 569
BTU, 125
Bubble tube, 1416
Bubble-tight valve shut-off, 2096
Bucket, motor control, 734
Bucking autotransformer, 471
Buffer solution, 1297, 1757, 1773
Bulk modulus, 1459, 1461, 1479, 1636, 1681, 1682
Bulkhead, tube, 581
Buoyancy, 191, 1445
Buoyant test of density, 192
Burden, instrument transformer, 1960
Burgess, George, 2732
Burn-in period, 2620, 2627
Burner Management System (BMS), 836, 2673
Burnout, thermocouple, 1552
Burst mode, HART, 1100
Bus duct, 607

Bus, electrical power, 1886
Bus, power system, 1904
Busway, 607
Butterfly valve, 687, 2080
BWR, 2684
Byte, 975

Cable pulling, 602
Cable tray, 604
Cache, ARP, 1078
Cage-guided globe valve, 2074
Cageless displacer level instrument, 1444
Calculus, Fundamental Theorem of, 26
Calibration, 1247, 1297
Calibration gas, 264, 1258, 1299, 1833, 1847
Calibration management software, 1260
Calibration, dry versus wet, 1445
Calibration, pH instrument, 1773
Calibrator, loop, 902
Callendar-van Dusen formula, 1510
calorie, 125
Calorie, dietary, 125
Cam, 2138
CAN network, 1039
CANDU nuclear reactor, 2693
Cap, tube, 583
Capacitance, 346
Capacitive level switch, 675
Capacitively-coupled voltage transformer (CCVT), 1935
Capacitor, 346
Capacitor-start AC induction motor, 720
Capacity tank, 1263, 2316
Capillary tube, 1368, 1371, 1641
Captive flow, 2501, 2561
Captive variable, 2501
Carrier Sense Multiple Access (CSMA), 1039
Cascade control strategy, 2371, 2490
Cathode, 1743
Cation, 228, 1743
Caustic, 286
Caustic, strong, 287
Caustic, weak, 287
Cavitation corrosion, 2203, 2217
Cavitation, control valve, 2201

CCVT, capacitively-coupled voltage transformer, 1935
Cell constant, conductivity sensor, 1747
Celsius, 123, 1287
CEMS, 1841
Centigrade, 1287
Centrifugal force, 1704
Centrifuge, 2694
Centripetal force, 1704
cgs, 72
Chain, 120
Chain reaction, nuclear, 2693
Channel arbitration, 1035
Channel, FOUNDATION Fieldbus AI block parameter, 1192
Characteristic impedance, 474
Characteristic, control valve, 2178
Characterized ball valve, 2081
Characterizing process dynamics, 2396, 2430
Charging current, power line, 1999
Charles's Law, 198
Chart recorder, 512, 2444
Chemical seal, 1368
Chemical versus nuclear reaction, 265
Chemiluminescence, 1835
Chemistry, 226
Chernobyl nuclear reactor accident, 2411
Choked flow, 2208
Chopper valve, 2668
Chord, ultrasonic flowmeter, 1684
Chromatogram, 1779
Chromatograph, dual-column, 1793
Chromatography, 1776
CIP, 571, 1367
Cippoletti weir, 1645, 2052
Circuit breaker, 1909
Circuit, electric, 299
Circular chart recorder, 512
Cistern manometer, 1306
Clamp-on ammeter, 341
Clamp-on milliammeter, 894
Clamp-on ultrasonic flowmeter, 1686
Class I filled system, 1505
Class II filled system, 162, 1507
Class III filled system, 1506
Class V filled system, 1505

Class, hazardous area, 2585
Classified location, 2584
Clean-In-Place, 571, 1367
Clutch mechanism, electric valve actuator, 2111
Cold junction compensation, 1532
Cold junction, thermocouple, 1523, 1532
Collision, 1039
Collision domain, Ethernet, 1059
Colloid, 229
Column, chromatograph, 1776
Combination electrode, 1760
Combustion, 273
Common logarithm, 1206
Common-mode rejection, 1341
Common-mode voltage, 618, 998, 1049
Communicator, HART, 1090
Commutating diode, 757, 2480
Commutator, DC motor, 2225
Compel Data (CD) token, Fieldbus, 1155
Compensating leg, 1423
Compensating probe, capacitive level instrument, 1488
Complex number, 369, 375
Complex plane, 373
Composition probe, capacitive level instrument, 1488
Compositional chemical formula, 235, 270
Compound, 228
Compressed air dryer, 2629
Compressibility, 198
Compression fitting, 1345
Compression terminal, 595
Compression terminal "crimping" tool, 597
Computer, mainframe, 984
Concentration cell, 1755
Condensate boot, 1443
Conductance, 318
Conduction, heat, 1502
Conduction, heat transfer, 128
Conductivity cell, 1745
Conductivity sensor, 1745
Conductor, electrical, 298
Confidence, of scientific data, 1173
Confined space, 1844
Conical-entrance orifice plate, 1609
Connector, tube, 581

Conservation of Electric Charge, 205, 316
Conservation of electric Charge, 323
Conservation of Energy, 73, 75, 89, 99, 172, 208, 226, 248, 278, 317, 319, 450, 1216, 1536, 1580, 2197, 2239, 2402, 2518
Conservation of Mass, 73, 205, 226, 265, 322, 1580, 1700, 2402, 2518
Constant of proportionality, 1585
Constraint, hard versus soft, 2570
Contactor, 463, 724, 758, 762, 763, 790, 806, 826, 1913
Continental Code, 1019
Continuous Emissions Monitoring System (CEMS), 1841
Control algorithm, 2260
Control characters, 984
Control valve, 2067
Control valve noise, 2167, 2174, 2176, 2210
Controlled rectifier, 2229
Controller, 497
Controller action, direct vs. reverse, 506, 2266, 2571
Controller gain, 2267
Convection, 1502
Convection, heat transfer, 130
Conventional flow, 660
Converter, 497
Coolants, 138
Cooling tower, evaporative, 168
Cooling tower, forced-draft, 169
Cooling tower, induced-draft, 169
Coordination, protective relay, 1990
Coplanar DP sensor, 1328
Coriolis force, 1705
Coriolis mass flowmeter, 1706
Corner taps (orifice plate), 1611
Coulomb, 293, 299
Count value, ADC, 973, 987
Counter instruction, PLC programming, 818
Counter, Geiger, 1492
Counterpropagation ultrasonic flowmeter, 1682
Coupling device, Fieldbus, 2368
Coupling device, fieldbus, 1141
Covalent substance, 282
cps, 1287
Crest, weir, 1646

Crimp terminal, 595
Crimping tool, 597
Critical damping, 428
Critical flow, 2208
Critical flow nozzle, 2209
Critical speed, rotating machine, 2699
Critical temperature, 153
Crosby pressure safety relief valve, 2652
Cross product, 93, 304, 1705
Cross-torquing, 564
Crossover cable, 1061
CS Fieldbus function block, 2559
CSMA, 1039
CSMA/BA channel arbitration, 1039
CSMA/CA, 1040
CSMA/CD channel arbitration, 1039, 1056
CT, 732, 1888, 1936
Current, 298, 299
Current sinking, 304, 774, 777
Current sourcing, 304, 774, 777
Current transformer (CT), 732, 1888, 1904, 1936
Current transformer (CT) safety, 1951
Curved arrow notation, showing DC voltage, 308
Curved manometer, 2048
Custody transfer, 1433, 1589, 1630, 1658, 1691, 1700, 1717
Cycles per second, 1287
Cyclic communication, Fieldbus, 1151, 1155

DAC, 780, 973, 986
Dall flow tube, 1621
Daltons, 262
Damper, 2083
Damping, 1261
Damping, anti-resonance resistor, 429
Damping, automobile suspension, 429
Damping, critical, 428
Damping, digital algorithm, 1263
Damping, feedback control loop, 429
Danfoss pressure switch, 667
Daniel four-beam ultrasonic flowmeter, 1684
DAQ, 515, 1002
DAQ input, single-ended, 1008
Data acquisition module, 515, 1002
Data Communications Equipment, 1046, 1058
Data diode, 2723

Data Terminal Equipment, 1046, 1058

dB, 481, 1206, 1216

dBm, 1210, 1218

dBW, 1211, 1218

DC, 355

DC excitation, magnetic flowmeter, 1678

DC injection, AC motor braking, 2238, 2239

DCE, 1046, 1058

DCS, 2362

DD file, 1088

DDC, 2349

DDL, 1177

de Broglie, Louis, 252

Dead leg, 571

Dead time, 514, 991, 1792, 2430, 2451, 2528

Dead time function, 2529, 2538

Dead time function used for dynamic compensation, 2529, 2531

Dead-test unit, 1291

Deadband setting, flow switch, 683

Deadband setting, pressure switch, 667

Deadband setting, temperature switch, 679

Deadband, control valve response, 2194

Deadband, integral, 2282, 2438

Deadband, reset, 2282, 2438

Deadweight tester, 1291

Deadweight tester, pneumatic, 1293

Dean effect, 204

Decade box, resistance, 1285

Decibel, 481, 1206, 1216

Decimal numeration, 974

Decrement (counter), 818

Decryption, 2722

Defense-in-depth, 2583, 2711

Degrees API, 180

Degrees Balling, 181

Degrees Bark, 181

Degrees Baumé, 180

Degrees Brix, 181

Degrees Oleo, 181

Degrees Plato, 181

Degrees Soxhlet, 181

Degrees Twaddell, 180

Delta model DSC-1280 DDC controller, 2352

Denial-of-service attack, 2708, 2719

Density, influence on hydrostatic level measurement accuracy, 1411

Dependability, 2596, 2597, 2671

Dependability, versus security, 2656

Dependent current source, 882

Derivative control, 2283

Derivative control action, 2287

Derivative notation, calculus, 1726, 1861

Derivative, calculus, 8, 12

Derived unit, 72

Desiccant, 694

desiccant, 2629

Destination host unreachable, error message, 1074

Desuperheater, 2507

Determinism, network, 1152, 1232

Determinism, network communication, 1040

Deviation alarm, 2374

Device Description file, 1088

Device Description Language, 1177

Device Join Key, WirelessHART, 1242

DeviceNet, 1039

Dew point, 695

DHCP, 1071

Diaphragm, 928, 1310

Diaphragm valve, 687, 2068

Diaphragm, isolating, 572, 1320, 1322, 1325, 1367

Dictionary attack, password, 2713

Dielectric constant, 1467

Dielectric constant, influence on radar level measurement accuracy, 1472

Dielectric heating, RF cable, 1212

Dietary Calorie, 125

Differential, 2283

Differential capacitance pressure sensor, 1322

Differential current protection, 1992

Differential equation, 27, 2419

Differential measurement mode on an oscilloscope, 1199

Differential notation, calculus, 1726

Differential pressure, 186, 1308

Differential pressure switch, 668

Differential setting, flow switch, 683

Differential setting, pressure switch, 667

Differential setting, temperature switch, 680

Differential temperature sensing circuit, 334

Differential voltage protection, 1992

Differential voltage signal, 618

Differential voltmeter, 339

Differential, calculus, 8

Differentiation, applied to capacitive voltage and current, 404, 405

Differentiator circuit, 37

Diffraction grating, 1804

Digital damping, 1263

Digital multimeter, 515, 1285, 2789

Digital-to-analog converter, 780, 973, 986

Dimensional analysis, 61, 71, 90, 129, 179, 192, 202, 206, 209, 312

DIN rail, "top hat", 599

DIN rail, G, 599

Diode, commutating, 757, 2480

Diode, in current loop circuit, 895

Dip tube, 1416

Dipole antenna, half-wave, 488, 1204

Direct addressing, 797

Direct current, 355

Direct digital control (DDC), 2349

Direct valve actuator, 2116

Direct valve body, 2116

Direct-acting controller, 506, 2266, 2571

Direct-acting pneumatic relay, 929

Direct-acting transmitter, 535

Direct-acting valve body, 2068

Discharge coefficient, 1630

Disconnect, electrical power system, 1909

Discrete, 649, 685, 743, 1276

Discrete control valve, 2067

Disk valve, 687, 2080

Dispersive chemical analyzer, 1804

Displacement, 191

Displacer, 1442

Displacer level instrument, 1441

Displayed chemical formula, 233

Dissociation, 1744

Dissolved oxygen measurement, 630

Distech model EC-RTU-L DDC controller, 2354

Distech model ECP-410 DDC controller, 2353, 2482

Distillation, 133, 1786

Distortion, valve performance, 2184

Distributed control system (DCS), 2362

Distribution power line, 1886

Diverting valve, 2076, 2149

Division, hazardous area, 2585

DIX Ethernet, 1056

DMM, 515, 991, 1285, 2789

DNR, 1079

DNS, 1079

Documenting calibrator, 1259

Domain Name Resolver, 1079

Domain Name Server, 1079

Domain Name System, 1079

Domain, frequency, 1868

Domain, time, 1868

Doping optical glass, 633

Doppler effect, 1679

Doppler ultrasonic flowmeter, 1679

Dot product, 94

Double-helical gear, 107

Double-ported globe valve, 2072

Double-precision floating-point number, 978

DP cell, 949

DP transmitter calibration fitting, 1363

DPDT switch contacts, 746

Drain hole, orifice plate, 1606

Drift, 1256

Drift test, analyzer, 1843

Drive, electric motor, 84, 2224

Droop, 2274, 2398

Drop, 1057, 1137

Dropper, computer virus, 2702

Drum instruction, Koyo PLC programming, 837

Drum sequencer, 834

Dry calibration, 1445

Dry contact, 659

Dry leg, 1424

Dry-block temperature calibrator, 1290

Dryer, compressed air, 2629

Dryer, instrument air, 694

Dryseal pipe threads, 568

DSO, 991

DTE, 1046, 1058

Dual-column chromatograph, 1793

Dump solenoid, 2669

Duplex, 1035

Dynamic braking, AC motor, 2238, 2240

Dynamic compensation, 2525, 2529
Dynamic friction, 2192
Dynamic Host Configuration Protocol, 1071
Dynamic luminescence quenching, 630
Dynode, 1831

EBCDIC, 984
Eccentric disk valve, 2080
Eccentric orifice plate, 1604
Echo Request, ICMP, 1069
Eductor, 216
Effective Isotropic Radiated Power (EIRP), 1218
Effective Radiated Power (ERP), 1218
EIA/TIA-232 serial communication, 1045, 1061, 1069, 1084, 1101, 2723
EIA/TIA-422 serial communication, 1049, 2723
EIA/TIA-485 serial communication, 1049, 1061, 1101, 2723
Einstein, Albert, 73, 250, 1797
EIRP, 1218
EIV, 2668
Ejector, 216
Electric circuit, 299
Electric motor valve actuator, 2110
Electrical conductor, 298
Electrical heat tracing, 1385
Electrical insulator, 298
Electrical interlock, reversing motor starter, 738
Electrodeless conductivity cell, 1751
Electrolysis, 273
Electromagnetic force, 230
Electromagnetic induction, 1667
Electromagnetic wave, 485
Electromagnetism, 340
Electron, 228
Electron capture detector, GC, 1781
Electron flow, 660
Electron orbital, 241, 242
Electron shell, 242
Electron subshell filling order, 247
Electronic manometer, 1295
Element, 228
Elevated voltage signal, 1006
Emergency isolation valve (EIV), 2668
Emergency Shutdown (ESD) system, 2655
Emerson AMS software, 971, 2059, 2144

Emerson DeltaV distributed control system (DCS), 971, 1179, 2059, 2364
Emerson model 1420 WirelessHART gateway, 1236
Emerson model 375 HART communicator, 1090
Emerson Ovation distributed control system (DCS), 2364
Emerson Smart Wireless Gateway, 1109, 1236
Emerson THUM WirelessHART adapter, 1234
Emerson, Ralph Waldo, 2730
Emission spectroscopy, 253
Emissivity, thermal, 1562
Emittance, thermal, 1562
Emulsion, 229
Encryption, 2722
Endothermic, 273, 278, 2161
Endress+Hauser magnetic flowmeter, 1675
Energy, 75
Energy balance, 73, 2402, 2518
Energy control procedure, 86
Energy in chemical reactions, 273, 278
Energy loss, flowmeter, 1624, 1734
Energy, in chemical bonds, 248
Energy, kinetic, 88
Energy, potential, 80, 278
Engine, internal combustion, 53
Engineering units, 1192
Enriched uranium, 2693
Enthalpy, 143, 144, 148
Enthalpy of formation, 278
Entropy, 75, 226
Epistemology, 1278
ePiX C++ library, 415
Equal percentage valve characterization, 2186
Equipotential, 317, 326, 613
Equivalent circuits, series and parallel AC, 361
Erosion, 2212
ERP, 1218
Error, controller, 2266, 2279
ESD, 2655
Ethernet, 1023, 1056, 1101, 1107, 1136, 1152, 1154, 1185, 1236, 2347, 2361, 2363, 2488, 2715, 2723
Euler's relation, 371, 388
Euler, Leonhard, 208, 371
European alpha value, RTD, 1511

Evaporative cooling tower, 168
Examine if closed (XIC), 810
Examine if open (XIO), 810
Exchanger, heat, 130
Excitation source, for bridge circuit, 327
Excitation wires, 4-wire RTD circuit, 1513
Exergy, 75
Exothermic, 273, 278, 2161
Expanded structural formula, 233
Explosion-proof enclosure, 2591
Extended floating-point number, 978
Extension grade thermocouple wire, 1545
External fault, power system protection zone, 1997
External reset, integral control, 2326, 2555
Externally-loaded pressure regulator, 2108

Fabry-Perot interferometry, 629
Fade margin, RF, 1223
Fade, RF, 1223
Fahrenheit, 123
Fail closed, 2118
Fail locked, 2118
Fail open, 2118
Fail-safe mode for a control valve, 2115
Fail-safe mode for split-ranged control valves, 2160
Failure rate, λ, 2614
Failures In Time (FIT), 2614
False positive, 2639
False state, PLC programming, 800
Farad, 346
Faraday's Law of Electromagnetic Induction, 362, 677, 719, 1962
Fast Fourier Transform (FFT), 1868
Fault tolerance, 2657
FCC – Federal Communications Commission, 1232
Federal Communications Commission, 1232
Feedback control system, 2258, 2509
Feedforward control strategy, 2511
Feedforward with trim, 2516
Ferrule, electrical connection, 589, 595
FF (FOUNDATION Fieldbus), 1129, 2367
FFT algorithm, 1868
Fiber optic cable, 623

Fiducial pulse, radar, 1474
Fieldbus, 505, 622, 970, 1091, 2367
Fieldbus coupling device, 2368
Fieldbus Foundation, 2367
FIFO shift register, 1263, 2538
Fill fluid, 572, 1320, 1322, 1324, 1325, 1331, 1359, 1366, 1505
Fillage, 1402, 1460
Filled bulb, 174, 1505
Filled impulse line, 1377
Filtering, negative (spectroscopy), 1821
Filtering, positive (spectroscopy), 1821
Final Control Element, 497
Fire signals, 982
Fire triangle, 2587
Firewall, computer, 2721
First Law of Motion, 74
First-order differential equation, 2419
First-order lag, 2419
Fiscal measurement, 1630
Fisher "Level-Trol" displacer instrument, 1443, 1451, 1458
Fisher "Whisper" low-noise trim, 2211
Fisher AC^2 analog electronic controller, 2337
Fisher E-body control valve, 3153
Fisher E-plug control valve, 2082
Fisher Micro-Flat Cavitation trim, 2202
Fisher model 1098EGR pilot-operated pressure regulator, 2117
Fisher model 2625 volume boosting relay, 2129
Fisher model 3582 valve positioner, 2131
Fisher model 546 I/P transducer, 957, 2098
Fisher model 760 pressure relief valve, 2654
Fisher model 846 I/P transducer, 2098
Fisher model DVC6000 valve positioner, 2132, 2142
Fisher model DVC6200 valve positioner, 2132
Fisher MultiTrol pneumatic controller, 2317
Fisher Provox distributed control system (DCS), 2364
Fisher ROC digital controllers, 2358
Fisher ValveLink software, 2144
Fisher-Rosemount model 846 I/P transducer, 962
Fission chamber, 1493
Fission, nuclear, 2409, 2684, 2693
FIT, 2614

Five-point calibration, 1274

Five-valve manifold, 1358

FIX/Intellution HMI software, 850

Fixed Programming Language (FPL), 783

Fixed-point variable, 976, 981

Flame ionization detector, GC, 1782

Flame photometric detector, GC, 1781

Flame safety system, 836, 2673

Flange taps (orifice plate), 1610

Flange, pipe, 560

Flapper, 917

Flare gas flow measurement, 1690

Flashing, 2197

Flashover, power line insulator, 1929

Flexure, 1254

Float level measurement, 1402

Floating control action, 2279, 2286

Floating voltage signal, 1005

Floating-point number, 977

Floating-point variable, 2377

Flooded displacer, 1454

Flow conditioner, 1626

Flow control (serial data communication), 1033, 1046

Flow integrator, 42

Flow programming, chromatograph, 1792

Flow prover, 1296, 1658

Flow switch, 682

Flow totalizer, 42

Flow tube, 1621

Flow-straightening vanes, 1626

Fluid, 170

Fluidized sand, 1289

Fluke brand multimeters, 2789

Fluke Documenting Process Calibrator (DPC) instruments, 1260

Fluke DPCTrack2 calibration management software, 1260

Fluke model 525A temperature calibrator, 1285

Fluke model 744 calibrator, 1551

Fluke model 744 Documenting Process Calibrator (DPC), 1259

Fluke model 771 clamp-on milliammeter, 894

Fluke model 801 differential voltmeter, 339

Fluke SV225 stray voltage adapter, 2792

Flume, 1649, 2051

Fluorescence, 630, 1826

Fluorescence quenching, 630

Flywheel, 98

Foam, 229

Follower, 2138

Force balance system, 924, 943, 945, 949, 959, 1335, 2307

Force-balance valve positioner, 2135

Fork terminal, 596

Form-A switch contact, 650, 746

Form-B switch contact, 650, 746

Form-C switch contact, 656, 657, 746

Formula weight, 263, 2504

Fortescue, Charles Legeyt, 456

FOUNDATION Fieldbus, 1091, 1129

FOUNDATION Fieldbus (FF), 2367

FOUNDATION Fieldbus H1, 1023, 1136

FOUNDATION Fieldbus H2, 1136

FOUNDATION Fieldbus HSE, 1136

Four-wire RTD circuit, 1513

Fourier series, 456, 1865, 2232, 2238

Fourier transform, 1212

Fourier, Jean Baptiste Joseph, 456, 1865, 2232, 2238, 2246

Foxboro (Invensys) I/A distributed control system (DCS), 2364

Foxboro FOXNET process data network, 2364

Foxboro INTERSPEC process data network, 2364

Foxboro magnetic flowtube, 1677

Foxboro model 13 differential pressure transmitter, 949

Foxboro model 130 pneumatic controller, 921, 2322

Foxboro model 14 flow totalizer, 43

Foxboro model 43AP pneumatic controller, 2320

Foxboro model 557 pneumatic square root extractor, 1594

Foxboro model 62H analog electronic controller, 2337

Foxboro model E69 I/P transducer, 952

Foxboro model E69F I/P transducer, 2098

Foxboro model IDP10 differential pressure transmitter, 1321, 1339, 1426

Foxboro SPEC 200 analog electronic control system, 2340, 2364, 2375, 2627

Foxboro SPECTRUM distributed control system (DCS), 2364
Fractionation, 133, 1786
Fragment, IP packet, 1067
Frame check sequence, 1033
Fraunhofer lines, 1806
Fraunhofer, Joseph von, 1806
Freezing point of water, 1287
Frequency domain, 1868
Frequency Shift Keying, 1024
Frequency shift keying, 1092
Fresnel zone, 1226
Fribance, Austin E., 2
Friction, static versus dynamic, 2192
FSK, 1024, 1092
Fuel cell oxygen sensor, 1848
Fugitive emissions, 2086, 2649
Fulcrum, 99
Fulcrum, torque tube, 1449
Full Variability Language (FVL), 783
Full-active bridge circuit, 335
Full-duplex, 1035
Full-flow taps (orifice plates), 1611
Fully developed turbulent flow, 203
FUN, Fieldbus addressing, 1153
Function block programming, 2345, 2368
Function, inverse, 25, 2043, 2749
Function, piecewise, 2056
Functional diagram, 536
Functions, transfer, 409
Fundamental frequency, 484, 488, 1205, 1865
Fundamental Theorem of Calculus, 26
Fusion splicing, optical fiber, 638

G DIN rail, 599
G, unit of acceleration, 1863
Gain and bias function block, 2523, 2534
Gain margin, 2428
Gain, amplifier, 1206
Gain, controller, 2267
Galilei, Galileo, 74
Galvanic corrosion, valve stem packing, 2092
Galvanic isolation, 732
Galvanometer, 328, 338
Gamma ray radiation, 1489
Gas, 170

Gas centrifuge, 2694
Gas centrifuge cascade, 2696
Gas centrifuge stage, 2696
Gas expansion factor, 1631
Gas Filter Correlation spectroscopy, 1821
Gas Laws, 198
Gas phase effect, radar level instrument, 1468, 1473
Gas, calibration, 264, 1258, 1299, 1833, 1847
Gas, span, 1299, 1833, 1847
Gas, zero, 1833
Gate valve, 687, 2068
Gate, logic, 743
Gauge line, 575, 1345
Gauge pressure, 186, 1308
Gauge tube, 575, 1345
Gauge, tube fitting, 577
Gay-Lussac's Law, 198
GE Multilin model 369 protective relay, 733
GE Multilin model 745 protective relay, 1995
GE Series One PLC, 770
Gear ratio, 96
Gear set, 106
Gear, bevel, 111
Gear, double-helical, 107
Gear, herringbone, 107
Gear, hypoid, 112
Gear, miter, 111
Gear, planetary, 109
Gear, single-helical, 107
Gear, spur, 106, 2112
Gear, worm, 113, 2112
Geiger counter, 1492
Geiger-Muller tube, 1492
General Electric "Magneblast" circuit breaker, 1915
General Electric model 121AC time-overcurrent protective relay, 1976
Generator, 300
Generator synchronization, 1901
Gentile flow tube, 1621
Gerlach scale, 181
GFC spectroscopy, 1821
Gibbs free energy, 226
Gibbs' phase rule, 159
Gilbert, 340

Global Positioning System (GPS), 383

Globe valve, 2068

Google, Internet search engine, 1081

GPS satellite system, 383

Graded index optical fiber, 636

Graphic User Interface, 1071

Graphite rupture disk, 2644

Gravitational potential energy, 80

Groth model 1208 pressure/vacuum safety valve, 2648

Ground, 322

Ground fault, 1992

Ground loop, 618, 1010, 1196, 1951

Ground-referenced voltage signal, 1004

Grounding, magnetic flowmeters, 1674

Group, hazardous area, 2586

GUI, 1071

Guide ring, pressure safety relief valve, 2652

Guided wave radar, 481, 1465

Hagen-Poiseuille equation, 207, 1640, 3150

Half-duplex, 1035

Half-life, of a radioactive substance, 1490

Half-wave dipole antenna, 488, 1204

Hall Effect sensor, 1333, 2132

Hand controller, 2526

Hand switch, 655

Hand valve actuator, 2114

Handshaking (serial data communication), 1033, 1046

Handwheel, control valve, 2115

Handwheel, valve, 2099

Hard alarm, 2374

Hard constraint, 2570

Hard determinism, network, 1152

Hard override, 2568

Hardware compensation, thermocouple, 1539

Harmonic frequency, 484, 488, 1205, 1865, 2246, 2794

Harmonic restraint, 87 relay, 2006

Harrier jet airplane, 2408

HART, 1024

HART analog-digital hybrid, 623, 970, 1086, 1265

HART communicator, 1090

HART multidrop mode, 1099

Hazardous location, 2584

Head (fluid), 208

Heat, 1502

Heat exchanger, 130, 2254

Heat of formation, 278

Heat of reaction, 274

Heat pump, 126

Heat tape, 1385

Heat tracing, 1383

Heat transfer by conduction, 128

Heat transfer by convection, 130

Heat transfer by radiation, 127

Heater, overload, 727

Heisenberg, Werner, 252

Helical bourdon tube, 1295, 1311

Hello World program, 853

Henry, 350

Herringbone gear, 107

Herschel, Clemens, 1621

Hertz, 274, 1287

Hess's Law, 279

HEU, 2693

Heuristic PID tuning example, 2468

High-limit function, 2551

High-performance butterfly valve, 2080

High-select function, 2549

High-side transistor switch, 661

High-speed pulse test, 2637

High-voltage circuit breaker, 1921

HMI panel, 845, 979

Hold-off distance, radar, 1475

Home position, CNC machine, 657

Honed meter run, 1633

Honeywell Experion PKS distributed control system (DCS), 2364

Honeywell model UDC3000 controller, 2346

Honeywell Radiamatic, 1556

Honeywell TDC2000 distributed control system (DCS), 2363

Hooke's Law, 91, 2121, 2130, 2145

Hot standby, 2631

Hot terminal, 452

Hot-swappable PLC I/O, 771

Hot-tapping, 1730

Hot-wire anemometer, 1718

HTTP, 1085
HTTPS, 1085
Huddling chamber, pressure safety valve, 2651
Human-Machine Interface panel, 845, 979
HVAC, 1512, 2083
Hydration, pH electrode, 1762
Hydraulic, 173
Hydraulic lift, 172
Hydraulic load cell, 1486
Hydraulic valve actuator, 2105
Hydrogen ion, 282, 284, 1744
Hydrometer, 193
Hydronium ion, 282, 284, 1744
Hydrostatic pressure, 178
Hydroxyl ion, 282, 284, 1744
Hyperterminal, 1028
Hypoid gear, 112
Hysteresis, 1274, 2435
Hysteresis error, 1254

I.S. system, 2592
I/O, 762, 771
I/O, analog (PLC), 780
I/O, discrete (PLC), 773
I/O, hot-swappable, 771
I/O, network (PLC), 782
I/O, remote, 772
I/P transducer, 898, 916, 2098
IANA, 1071
ICANN, 1071, 1079
Ice cube relay, 747
Ice point, thermocouple, 1533, 1539
ICMP, 1069
ICS-Triplex TMR safety control system, 2668
Ideal Gas Law, 197, 1506, 1697, 2757
Ideal PID equation, 2304, 2335, 2382, 2415
Identifier, Fieldbus device, 1154
IEC standard 61131-3 (PLC programming languages), 783
IEC standard 62591 (WirelessHART field instrument communications protocol), 1203, 1228
IEEE codes for power systems, 1979
IEEE standard C57.12.00-2010 for transformers, 2009
IFC spectroscopy, 1821

Ifconfig, utility program, 1080
Imaginary number, 369, 371
Immunity, noise, 1017
Impedance, 360, 390, 404, 405, 409
Impedance, characteristic, 474
Impedance, surge, 474
Impeller-turbine mass flowmeter, 1702
Impulse line, 575, 1345
Impulse tube, 575, 1345, 1359
In situ measurement, 1839
Inches of mercury, 178
Inches of water column, 178
Inclined manometer, 183, 1306
Inclined plane, 104
Increment (counter), 818
Index of refraction, 632
Indicator, 509
Inductance, 350
Induction cup protective relay mechanism, 2028
Induction disk protective relay mechanism, 1976, 1985
Induction motor, 718, 2233, 2799
Induction, mutual, 364
Induction, self, 362
Inductor, 350
Inferred variable, 1172, 1303, 1433, 1632, 2044, 2046, 2519, 2534
Information Technology (IT), 2691
Infrared thermocouple, 1557
Inherent valve characteristic, 2178, 2184
Inrush current, 719
Installed valve characteristic, 2178, 2184, 2495
Instantaneous overcurrent protection, 1983
Instrument air systems, 2629
Instrument Protective Function (IPF), 2655
Instrument tray cable (ITC), 1142
Instrument tube bundle, 1383
Instrumentation amplifier, 1001
Insulation, R-value, 129
Insulator, electrical, 298
Integer number, 974
Integer variable, 2377
Integral control action, 2279, 2286
Integral deadband, 2282, 2438
Integral orifice plate, 1614, 1636
Integral windup, 2282, 2373, 2399

Integral windup, limit controls, 2554
Integral windup, override controls, 2570
Integral, calculus, 8, 20
Integrating process, 2400
Integration, applied to RMS waveform value, 357
Integrator, 42
Interacting PID equation, 2305, 2336
Interactive zero and span adjustments, 1250, 1295
INTERBUS-S, 1053
Interface level measurement, 1395
Interference Filter Correlation spectroscopy, 1821
Interlock, reversing motor starter, 738
Internal combustion engine, 53
Internal fault, power system protection zone, 1997
International Morse code, 983
International Practical Temperature Scale (ITS-90), 1288, 1530, 1548, 1551
Internet Control Message Protocol, 1069
Internet Protocol, 1066
Interoperability, Fieldbus devices, 1177
Interposing, 757
Intrinsic safety, 1337
Intrinsic safety barrier circuit, 2593
Intrinsic standard, 1278, 1281
Intrinsically safe system, 2592
Inverse function, 25, 2043, 2749
Inverse square law of radiation, 1223, 1557, 1559
Inverted classroom, 3164, 3168
Inviscid flow, 201
Ion, 228
Ion-selective membrane, 2062
Ionic substance, 282
Ionization, 1744
Ionization chamber for detecting nuclear radiation, 1492
IP, 1066
Ipconfig, utility program, 1079
IPF, 2655
IPv4, 1068
IPv6, 1077
ISA 84, 783
ISA PID equation, 2304, 2382, 2415
ISA100.11a, 1038
ISEL Fieldbus function block, 2559

ISM 2.4 GHz radio band, 1232
Isolating diaphragm, 572, 1320, 1322, 1325, 1367
Isolation, galvanic, 732
Isomer, 228
Isopotential point, pH, 1775
Isothermal terminal block, 594, 1527
Isotope, 228, 2693
Isotopes, 237
Isotropic antenna, 1216
ITC, 1142
ITS-90, 1288, 1530, 1548, 1551
ITT safety valve, 2675

Jabber, 1040
Jacket, reactor vessel, 505
Jam packing, valve, 2088
Join Key, WirelessHART, 1242
Josephson junction array voltage standard, 1279
Joule, 293
Joules' Law, 312
Jumper wire, 2797

K-Tek brand magnetic float level indicator, 1407
Kallen, Howard P., 2
KCL, 322, 1991
Kelvin, 123
Kelvin resistance measurement, 1513, 1748
Kermit, 1028
Key, cryptographic, 2722
Keyphasor, 1875
Kinematic viscosity, 199
Kinetic energy, 88
Kirchhoff's Current Law, 322, 1991
Kirchhoff's Voltage Law, 320, 412
Knife-edge bearing, 1449
Knockout drum, 1443
Koyo CLICK PLC, 768, 799, 851
Koyo DL06 PLC, 769
KVL, 320, 412

L2F optical flowmeter, 1687
L_Type, FOUNDATION Fieldbus AI block parameter, 1192
Ladder Diagram, 761
Ladder Diagram programming, 753, 798
Lag time, 2422, 2532

Lag time function, 2536
Lambert-Beer Law, 1802
Laminar flow, 204, 207, 3150
Laminar flowmeter, 1640
LAN, computer, 2721
Lantern ring, 2088
Laplace transform, 415, 1212
Lapping valve plugs and seats, 2073
LAS, 1151
Laser-two-focus optical flowmeter, 1687
Latent heat, 144
Latent heat of fusion, 145
Latent heat of vaporization, 145
Law of Continuity (fluids), 205, 1581, 1653, 2197
Law of Intermediate Metals, thermocouple circuits, 1536
LD, 753, 798
Le Châtelier, Henry Louis, 2732
Lead function used for dynamic compensation, 2536
Lead time function, 2536
Lead/lag function, analog circuit, 2539
Lead/lag function, digital implementation, 2547
LEL, 1849, 2588, 2682
Lenz's Law, 620, 1985, 2239
LEU, 2693
Level gauge, 1396
Level switch, 669
Lever, 99
Lift pressure, safety valve, 2644, 2651
Limit switch, 657
Limited Variability Language (LVL), 783
Limiting case, 198, 206
Line current, 446
Line of action, 93
Line of Sight (RF), 1226
Line pressure effect, pressure transmitter, 1341
Line reactor, 2232, 2248
Line voltage, 446
Linear time-invariant (LTI) system, 435
Linear valve characterization, 2186
Linearity error, 1253
Linearization, 2050
Link Active Scheduler, 1151
Link Active Scheduler, FOUNDATION Fieldbus, 2368

Lipták, Béla, 2, 1420, 1653, 2176, 3148
Liquid, 170
Liquid interface detection with radar, 1470
Liquid valve sizing equation, 2167
Liquid-tight flexible conduit, 603
Live List, Fieldbus, 1155
Live zero, 858, 861, 878, 1249
Live-load packing, valve, 2089
Lo-Loss flow tube, 1621
Load, 300, 2259, 2272
Load cell, 332, 1482
Load cell, hydraulic, 1486
Load encroachment blocking, protective relay, 2035
Load line, 2181
Load versus source, 306
Load, process, 2510
Loading pressure, 2108
Lock-out, tag-out, 85, 1491, 1912
Lockout auxiliary relay, 2036
Logarithm, common, 1206
Logic gate, 743
Logic solver, 2666
Logix5000 PLC software, Rockwell, 833
LonWorks, 2353
Loop calibrator, 902
Loop diagram, 533
Loop sheet, 533
Loop, computer programming, 2376
Loop-powered transmitter, 311, 887
Loopback address, 1071
Louvre, 2083
Low flow cutoff, vortex flow transmitter, 1665
Low-limit function, 2551
Low-select function, 2549
Low-side transistor switch, 661
Lower explosive limit (LEL), 1849, 2588, 2682
Lower range value, 497, 1249, 1266, 1295
LRV, 497, 1249, 1266, 1295
LTI system, 435
Lubricator, valve packing, 2091, 2195
Luft detector, 1814
Lug, electrical connection, 589, 595
Luminiferous ether, 1056
Lusser's Law, 2604, 2672
LVDT, 1333, 2113, 2134

MAC address, Ethernet, 1057, 1068, 1154
Machine, simple, 99
MacNeil, Blair, 2483
Macrocycle, 1153
Madelung rule, 247
Magnetic flowmeter, 1668
Magnetic remanence, 1964
Magnetic shielding, 619
Magnetostriction, 1408, 1478
Magnetrol liquid level switch, 670
MagTech level gauge, 1398
Make-before-break switch contacts, 1954
Man-in-the-middle attack, 2709
Manchester encoding, 1023
Manifold, pressure transmitter, 1355, 1358
Manipulated variable, 497, 2256
Manipulating algebraic equations, 2747
Manometer, 182, 1294, 1304
Manometer, cistern, 1306
Manometer, inclined, 183, 1306
Manometer, nonlinear, 2048
Manometer, raised well, 1306
Manometer, slack tube, 1295
Manometer, U-tube, 1306
Manometer, well, 1306
Mantissa, floating-point number, 978
Manual loading station, 2526, 2553
Manual mode, 498, 2371
Manual valve actuator, 2114
Mapping, Modbus, 1109
Mark, 1022, 1024
Mask, Allen-Bradley SQO sequencer output instruction, 841
Mask, subnet, 1072
Masoneilan model 21000 control valve, 2069
Mass balance, 73, 2402, 2518
Mass density, 57
Master Terminal Unit (MTU), 972, 2356
Master-slave channel arbitration, 1036
Maximum experimental safe gap (MESG), 2586, 2591
Maximum working pressure, 1344
Maxon safety valve, 2674
Maxwell's electromagnetic equations, 486
Maxwell, James Clerk, 486
MCC, 729

MCC room, 607, 1911
Mean life (of a component or system), 2620
Mean Time Between Failures (MTBF), 2617
Mean Time To Failure (MTTF), 2617
Measurement electrode, 1757
Measurement junction, thermocouple, 1523
Mechanical advantage, 100
Mechanical interlock, reversing motor starter, 738
Median signal select, 2557
Medium-voltage circuit breaker, 1914, 1972
Memory map, 793
MEMS, 1331
Meniscus, 1305
Mercoid pressure switch, 665
Mercury, 1392
Mercury barometer, 1308
Mercury cell, 1549
Mercury tilt switch, 665, 671
Merrick weighfeeder, 1724
Metal fatigue, 1318
Metal rupture disk, 2643
Metallic liquid-tight conduit, 604
Meter run, orifice (honed), 1633
Metering pump, 2250
Metrology, 1278
Mho element, protective relay, 2029
Micro fuel cell oxygen sensor, 1848
Micro Motion Coriolis mass flowmeter, 1711
Micro Motion ELITE model Coriolis mass flowmeter, 1277
Micromanometer, 184
Micron, 633, 635
MIE, 2588
Mil, 1863
Miller, Richard W., 1630
Milton-Roy metering pump, 2250
Minicom, 1028
Minimum bend radius for optical fiber, 638
Minimum ignition current ratio (MICR), 2586
Minimum Ignition Energy, 2588
Minimum linear flow rate, turbine flowmeter, 1662
Miter gear, 111
Mixing valve, 2076, 2149
Mixture, 229

Mobile phase, 1776
Modal dispersion, 635
Modbus, 782, 1101
Modbus 984 addressing, 1108
Modbus ASCII, 1106
Modbus mapping, 1109
Modbus Plus, 1101
Modbus RTU, 1106, 1237
Modbus TCP, 1085, 1107, 1237
Modulus, bulk, 1459, 1461, 1479, 1636, 1681, 1682
Molarity, 263, 1297
Mole, 262
Molecular chemical formula, 233
Molecular substance, 282
Molecular weight, 263
Molecule, 228
Moment arm, 93
Moment balance system, 943
MooN redundancy notation, 2657
Moore Industries model IPT I/P transducer, 2098
Moore Industries model SPA alarm module, 518, 681
Moore Products "Nullmatic" temperature transmitter, 1508
Moore Products model 353 digital controller, 2344, 2347
Moore Products model 65 pneumatic square root extractor, 1594
Moore Syncro analog electronic controller, 2338
Morse code, 983, 1019
Motion balance system, 944, 945, 953, 2309
Motion control system, 2496
Motion-balance valve positioner, 2139
Motional EMF, 1668
Motor Control Center, 729
Motor control center (MCC), 607, 1911
Motor drive, 84, 2224
Motor overload protection, 726
Motor valve actuator, 2110
MOV, 2110
MTBF, 2617
MTS M-Series magnetostrictive float level transmitter, 1479
MTTF, 2617

MTU, 972, 2356
Mu metal, 619
Multi-column chromatograph, 1793
Multi-mode optical fiber, 635
Multi-segment characterizer, 2056
Multi-variable transmitter, 1100, 1470, 1635, 1716, 1786
Multidrop, HART, 1099
Multilin model 369 protective relay, 733
Multipath ultrasonic flowmeter, 1684
Multiplexer, 1002
Multiplication factor, nuclear fission, 2409
Multiplying relay, 2502
Mutual induction, 364
Mux, 1002
MV, 497
MWP, 1344

NAMUR recommendation NE-43, 2639
NaN, 977
Nassau model 8060 loop calibrator, 907
National Bureau of Standards, 1278
National Electrical Code (NEC), 1142, 2585
National Fire Protection Association (NFPA), 2585
National Institute of Standards and Technology, 1278
Natural convection, 135, 165
NBS, 1278
NDE, 713
NDIR spectroscopy, 1807
NDUV spectroscopy, 1807
NDVIS spectroscopy, 1807
NE, 712
NEC, 1142, 2585
Needle valve, 1364, 2070
Negative feedback, 37, 922, 936, 1383, 2120, 2331, 2405, 2409, 2413, 2427, 2429, 2431, 2478, 2490
Negative filtering (spectroscopy), 1821
Negative lag, 2408
Negative pressure, 189
Negative self-regulation, 2408
NEMA 7 enclosure, 2591
NEMA 8 enclosure, 2591
Nernst equation, 1755, 1767, 2062

Netstat, utility program, 1085
Network ID, WirelessHART, 1242
Network Manager, WirelessHART, 1231, 1233
Neutral pH, pure water, 284
Neutral terminal, 452
Neutralization, pH, 288
Neutron, 228
Neutron backscatter, 1489
Neutron radiation, 1489
Newton's Law of Cooling, 2419
Newton, Isaac, 74
Nichols, N.B., 2445
NIST, 1278
NIST traceability, 1280
Nitrogen-phosphorus detector, GC, 1781
Noise floor, 645, 1220–1222, 1225
Noise immunity, 1017
Noise, control valve, 2167, 2174, 2176, 2210
Non-bleeding pneumatic relay, 933
Non-contact radar, 1465
Non-dispersive chemical analyzer, 1807
Non-inertial reference frame, 1704
Non-metallic liquid-tight conduit, 604
Non-Newtonian fluid, 200
Non-retentive instruction, PLC program, 814, 824
Non-Return-to-Zero, 1022
Nonincendive circuit, 2592
Nonlinear manometer, 2048
NooM redundancy notation, 2657
Normal energization state of a solenoid, 712
Normal state of a circuit breaker auxiliary contact, 1974, 1982
Normal state of a PLC program contact, 755, 784, 800, 804, 810
Normal state of a relay contact, 745, 751
Normal state of a switch, 650, 652, 655, 657, 659, 664, 669, 678, 682
Normal state of a thermal overload switch, 726
Normal state of a valve, 699, 702, 708
Normal state of a vibration switch, 1879
Normal state of an 86 lockout relay, 2037
Normally de-energized (NDE), 713
Normally energized (NE), 712
Not a Number (NaN), 977
NOx emissions, 2062

Nozzle, 917
Nozzle, process vessel, 1399, 1454, 2634
NPN output switch, 661
NPT pipe threads, 568
NRC, Nuclear Regulatory Commission, 1493
NRZ, 1022
NSS Labs, 2715
Nuclear fission reactor, 2409, 2684
Nuclear radiation, 1489
Nuclear Regulatory Commission, 1493
Nuclear versus chemical reaction, 265
Null modem, 1048, 1061
Null zone, radar, 1475
Numeration systems, 974
NUN, Fieldbus addressing, 1153
Nupro pressure relief valve, miniature, 2646
Nyquist Sampling Theorem, 991

Octal base relay, 747
OFC, 640
Off-delay timer, PLC programming, 823
Ohm, 313
Ohm's Law, 313
Ohm's Law analogy for turbulent fluids, 2166
Ohm, Georg Simon, 313
Oil bath temperature calibrator, 1289
Oleo, degrees, 181
Omega model CL-351A dry-block temperature calibrator, 1290
Omega model OS-36 infrared thermocouples, 1557
On-delay timer, PLC programming, 823
On-off control, 2263
OOS mode, Fieldbus, 1192
Open Fiber Control safety system, 640
Open, electrical fault, 324
Open-loop test, 2396, 2397, 2400, 2407, 2412, 2430
Operating coil, protective relay, 1993, 2025
Optical fiber cable, 623
Opto 22 Optomux network, 1053
Orbital, electron, 241, 242
Order of magnitude, 938
Order of operations, 2750
Orifice meter run, 1633
Orifice plate, 1601, 2045

Orifice plate, concentric, 1602
Orifice plate, conical entrance, 1609
Orifice plate, eccentric, 1604
Orifice plate, integral, 1614, 1636
Orifice plate, quadrant edge, 1608
Orifice plate, segmental, 1605
Orifice plate, square-edged, 1602
Oscilloscope probe, ×10, 2543
Oscilloscope, differential measurement mode, 1199
Out coil, PLC programming, 814
Out Of Service (OOS) mode, Fieldbus, 1192
OUT_Scale, FOUNDATION Fieldbus AI block parameter, 1192
Output limit, PID controller, 2374
Output tracking, 2372
Over-damping, 428
Over-lean condition, 2587
Over-rich condition, 2587
Overload "heater", 727
Overload protective device, 726
Overreaching, protective relay, 2014
Override, hard, 2568
Override, hard versus soft, 2570
Override, soft, 2569
Overtone frequency, 1865
Oxygen control, burner, 2062

P&ID, 531
Packet, IP data, 1066
Packing lubricator, 2091, 2195
Packing, bellows seal, 2093
Packing, jam, 2088
Packing, live-loaded, 2089
Packing, valve, 2086
Paperless chart recorder, 513
Parallel damper, 2083
Parallel digital data, 987
Parallel PID equation, 2303
Parallel pipe threads, 570
Parallel versus serial digital data, 1018
Parity bit, 1029
Parshall flume, 2051
Partial stroke valve testing, 2637, 2670
Particle, 228
Parts per billion (ppb), 1840

Parts per million (ppm), 283, 1299, 1833, 1840, 1846, 1850, 1852
Pascal's principle, 175, 2208
Pascal, pressure unit, 171
Pass Token (PT), Fieldbus, 1155
Passivation layer, metals, 268, 2203, 2217
Password lockout, 2719
Password timeout, 2719
Passwords, 2709, 2711, 2713, 2715, 2718
Path loss, RF, 1223
Pauli Exclusion Principle, 241
Payload, computer virus, 2702
PC-ControLab software, 2463
PCOF, 634
Per unit, 865, 879
Percent, 865
Periodic Table of the Elements, 236
Periodic waveform, 1865
Permanent pressure drop, 1734, 2199
Permanent pressure loss, 219
Permittivity, 346
Permittivity, relative, 1467
PFD, 529, 2601, 2625, 2671
pH, 284, 1297
pH neutralization, 288
Phase change, 1287
Phase current, 446
Phase margin, 2428
Phase reference signal, vibration monitoring, 1875
Phase rotation, 439, 456, 465
Phase sequence, 439, 456, 465, 719
Phase shift, process dynamic, 2427, 2431
Phase voltage, 446
Phase-shift oscillator circuit, 2428
Phasometer, 380
Phasor, 373
Phasor diagram, 440
Photoelectric effect, 1831
Photoionization detector, GC, 1781
Photomultiplier tube, 1830, 1836
Photon, 250
Pickoff coil, 1656
Pickup coil, 1656
Pickup current, protective relay, 1984
Pickup, protective relay, 1977, 1983

Piecewise function, 2056

Pieruschka, Erich, 2604

Piezometer, 1580

Pigtail siphon, 1386

Pilot burner, 2676

Pilot valve, 926

Pilot, protective relay, 1999

Pilot-loaded pressure regulator, 2108

Pilot-operated control valve, 2106

Pilot-operated valve, for overpressure protection, 2654

Ping, utility program, 1069, 1073

Ping6, utility program, 1077

Pipe elbow flow element, 1623

Pipe flange, 560

Pipe hanger, 1485

Pipe taps (orifice plate), 1611

Piping and Instrument Diagram (P&ID), 531

PIS, 2655

Pitch, thread, 568

Pitot tube, 1617

Planck's constant, 250, 1797

Planck, Max, 250, 1797

Planetary gear set, 109

Plato, degrees, 181

PLC, 744, 754, 757, 761, 874, 1102

Plug, tube, 583

Plugging, AC motor, 2238, 2244

PMV model 1500 valve positioner, 2137

Pneumatic, 173

Pneumatic "resistor", 1641

Pneumatic control system, 501

Pneumatic deadweight tester, 1293

Pneumatic diaphragm valve actuator, 2097

Pneumatic piston valve actuator, 2102

Pneumatic relay, 928

Pneumatic valve actuator, 2097

PNP output switch, 661

Poise, 199

Polar form, complex number, 373

Polarity, 296

Pole, transfer function, 409, 411, 417

Pole-zero plot, 413

Polling, 1036

Polybius, Greek historian, 982

Polynomial expression, 427

Porpoising of process variable, 2462

Port number (OSI layer 4), 1085

Port, control valve trim, 2172

Port-guided globe valve, 2070

Position algorithm, defined, 2383

Positive displacement pump, 691

Positive feedback, 2408, 2429, 2431

Positive filtering (spectroscopy), 1821

Postel, Jon, 1071

Potential energy, 80, 278, 292

Potential transformer (PT), 732, 1887, 1932

Potential transformer (PT) safety, 1951

Pound per square inch, pressure unit, 171

Poundal, 1701

Power, 75

Power Line Carrier (PLC) communications, 1890

Power line carrier telemetry, 972

Power line charging current, 1999

Power reflection factor, 1470

Powers and roots, 2053

ppb, 1840

ppm, 283, 1299, 1833, 1840, 1846, 1850, 1852

Pre-act control action, 2283, 2287

Preamplifier, pH probe, 1770

Precipitate, 229

Precision potentiometer, 1285, 1549

Predictive maintenance, 1256

Pressure, 170, 171, 1391

Pressure gauge mechanism, typical, 1312

Pressure recovery, 219

Pressure recovery factor, 2198

Pressure Relief Valve (PRV), 2109, 2644

Pressure Safety Valve (PSV), 2109, 2644

Pressure snubber, 1364

Pressure switch, 664

Pressure, absolute, 186

Pressure, differential, 186

Pressure, gauge, 186

Pressure, hydrostatic, 178

Pressure, negative, 189

Pressure, within solids, 189

Pressure-based flowmeters, 1575

Pressurized Water Reactor (PWR), 163, 2684

Preventive maintenance, 2628

Primary Coated Optical Fiber, 634

Primary injection test, current transformer, 1965

Primary sensing element, 497

Prism, 1804

Probability, 2598

Probability and Boolean values, 2600

Probability of Failure on Demand (PFD), 2601, 2625, 2671

Probe Node (PN) token, Fieldbus, 1155

Problem-solving technique: applying numerical values, 2194

Problem-solving technique: divide and conquer, 1070

Problem-solving technique: limiting cases, 410, 956, 2612, 2772, 2774

Problem-solving technique: thought experiment, 506, 945, 1412, 1437, 1438, 1440, 1441, 1455, 1468, 1523, 1626, 1727, 1811, 1815, 2192, 2194, 2273, 2315, 2386, 2401, 2460, 2528, 2558, 2562, 2572, 2664, 2741, 2773, 2774, 2776

Procedure, energy control, 86

Process, 496, 2254

Process alarm, 2374

Process and Instrument Diagram (P&ID), 531

Process Flow Diagram (PFD), 529

Process load, 2510

Process switch, 516, 650

Process variable, 496, 2255

Product, chemical reaction, 265

Profibus, 782, 1091

Profibus PA, 1023, 1131

Profile factor, multipath ultrasonic flowmeter, 1685

Programmable Logic Controller, 744, 754, 757, 761, 874, 1102

Programming, chromatograph, 1792

Projectile physics, 89

Proof of closure switch (safety valve), 2676

Proof testing, 2636

Proportional band, 2270, 2285

Proportional control, 2264

Proportional control action, 2285

Proportional weir, 1647

Proportional-only offset, 2274, 2280, 2398

Protection zone, power system, 1997, 2014

Protective Instrument System (PIS), 2655

Protective relay, 462, 731, 1891, 1914, 1972, 2656

Protective relay codes, 1979

Proton, 228

Prover, flow, 1296, 1658

Proximitor, Bently-Nevada, 1872

Proximity switch, 659

Prussian blue, 2073

PRV, 2109, 2644

Pseudocode, 36, 46, 2376

PSI, 171

PSV, 2109, 2644

PT, 732

PT, potential transformer, 1887, 1932

PTFE (Teflon) valve stem packing, 2092

Puget Sound Energy, 1897

Pull string, 602

Pulley, 101

Pulse test, 2637

Pulse width modulation, 2228

Pump curve, 2184

Purge cycle, 2676

Purge flow rate, 1380, 1416

Purge flow regulator, 1382

Purged impulse line, 1380

PWM, 2228

PWR, 163, 2684

Quadrant-edge orifice plate, 1608

Quadrature pulse, 821

Quality factor, RLC filter circuit, 434

Quantization error, 988

Quantum theory of light, 251

Quarter-active bridge circuit, 334

Quarter-wave antenna, 489, 1204

Quarter-wave damping, 2447

QUB, Fieldbus, 1157

Quick-opening valve characterization, 2187

QUU, Fieldbus, 1157

R-value, 129

R-X diagram, 2019

Racking out, circuit breaker, 1914

Radar detection of liquid interfaces, 1470

Radar level instrument, 481, 1465

Radial damper, 2084

Radiation, heat, 1502

Radiation, heat transfer, 127

Radiation, nuclear, 1489

Radio Frequency (RF), 1142, 1218

Radio frequency interference from motor drive circuits, 2232, 2238

Radio wave, 485

Radioactivity, 237

Radiotelegraph, 1019

Raised well manometer, 1306

Raised-Face (RF) flange, 563

Ramp-and-soak setpoints, 2488

Range wheel, 949

Rangeability, 1652, 1768, 2153

Rangeability, optical flowmeter, 1690

Rangeability, pressure-based flowmeter, 1634

Rangedown, 1277

Ranging, 1247

Rankine, 123

Ransomware, 2704

Rate control, 2283

Rate control action, 2283, 2287

Rate limit function, 2551

Ratio control strategy, 2498

Ratio station, 2502

Ratio, gear, 106

RC phase-shift oscillator circuit, 2428

Reactance, 360

Reactance versus resistance, 384

Reactant, chemical reaction, 265

Reaction rate, 2451

Reaction, enthalpy of, 274

Reaction, heat of, 274

Reactor, power line, 2248

Reading, active, 2730

Real Gas Law, 198

Real number, 369

Real number (floating-point), 977

Receiver gauge, 912, 918, 923, 1597, 1599, 2047

Reclosing relay, 1973

Recorder, 512, 2444

Rectangular form, complex number, 373

Rectangular weir, 1645

Rectifier, SCR controlled, 2229

Recursion, computer programming, 2381

Red Lion Controls panel-mounted indicator, 511

Red-line editing, 750

Reduced-port control valve trim, 2172

Reducing union, tube, 581

Redundancy, 2361, 2631

Redundant sensors, 2662

Redundant transmitters, 2557

Reference electrode, 1759

Reference junction compensation, 1532

Reference junction, thermocouple, 1523

Reference probe, radar level instrument, 1473

Reference pulse, radar, 1474

Reflection factor, 1470

Reflection grating, 1805

Refractive index, 632

Regenerative braking, AC motor, 2238, 2242

Regulator, purge gas flow, 1382

Relation control strategy, 2506

Relative flow capacity, 2175

Relative gas density, 180

Relative permittivity, 1467

Relative permittivity, influence on radar level measurement accuracy, 1472

Relay, 497

Relay Ladder Logic programming, 753, 798

Relay, ice cube, 747

Reliability, 2596, 2609, 2622

Relief valve, 2109, 2644

Remanence, CT, 1964

Remote backup, protective relay, 2033

Remote PLC I/O, 772

Remote seal, 1368

Remote setpoint, 2371, 2488, 2490

Remote telemetry system, 972, 2360

Remote Terminal Unit (RTU), 972, 2356

Repose, angle of, 1464, 2060

Request timed out, error message, 1074

Required Safety Availability (RSA), 2671

Reseat pressure, safety valve, 2644, 2651

Reset coil, PLC programming, 814

Reset control action, 2279, 2286

Reset deadband, 2282, 2438

Reset windup, 2282, 2373, 2399

Reset windup, limit controls, 2554

Reset windup, override controls, 2570

Residence time, see Retention time, 2515

Resistance, 313, 360

Resistance versus reactance, 384

Resonance, 421, 482

Resonance, mechanical, 2699

Resonant wire pressure sensor, 1331

Resource block, Fieldbus, 1172

Response factor, chromatograph detector, 1785

Restrained differential relay, 1994

Restraint coil, protective relay, 1994

Restricted-capacity control valve trim, 2172

Retention time, 1776, 1779, 1795, 2514, 2515, 2680

Retentive instruction, PLC programming, 814, 824

Reverse indication, controller output, 884, 2121

Reverse valve actuator, 2116

Reverse valve body, 2117

Reverse-acting controller, 506, 2266, 2571

Reverse-acting pneumatic relay, 929

Reverse-acting transmitter, 535

Reverse-acting valve body, 2068

Reynolds number, 201

Reynolds number, for laminar versus turbulent flow regimes, 204

RF, 1142, 1218

RF flange, 563

RFI, 2232, 2238

Richter scale, 1768

Riemann sum, 19

Riemann, Bernhard, 19

Right-hand rule, 93, 304, 343

Ring terminal, 596

Ring-Type Joint (RTJ) flange, 563

Rising stem valve actuator, 2114

RLL, 753, 798

RMS quantities, 356

Robertshaw Vibraswitch, 1879

Rockwell ControlLogix 5000 PLC, 767

Rockwell PLC-5, 764

Rockwell SLC 500 PLC, 765

Root privilege, computer, 2714

Root, polynomial, 427

Root-mean-square (RMS) quantities, 356

Roots and powers, 2053

Rosemount Analytical X-STREAM X2 gas analyzer, 1820

Rosemount field-mounted indicator, 511

Rosemount Micro-Motion Coriolis mass flowmeter, 1711

Rosemount model 1151 differential pressure transmitter, 962, 1324, 1338, 1411, 1613

Rosemount model 1151 gauge pressure transmitter, 1373

Rosemount model 268 HART communicator, 1090

Rosemount model 3051 differential pressure transmitter, 900, 1327, 1338, 1418, 1634

Rosemount model 3051S differential pressure transmitter, 1329

Rosemount model 3095MV multi-variable transmitter, 1172, 1635

Rosemount model 3301 guided-wave radar transmitter, 2059

Rosemount model 3301 level transmitter, 1474

Rosemount model 5708 3D Solids Scanner, 2060

Rosemount model 648 WirelessHART temperature transmitter, 1241

Rosemount model 8700 magnetic flowmeter, 1675

Rosemount model 8800C vortex flow transmitter, 1666

Rotabolt self-indicating bolt, 564

Rotameter, 1380, 1416, 1642, 2096

Rotating magnetic field, 715, 2233, 2799

Rotating paddle level switch, 673

Rotork electric valve actuator, 2112

Router, 1068

RS-232 serial communication, 1045, 1061, 1069, 1084, 1101, 2723

RS-422 serial communication, 1049, 2723

RS-485 serial communication, 1049, 1061, 1101, 2723

RSA, 2671

RSView HMI software, Rockwell, 850

RTD, 1285, 1509

RTD table, 1511

RTJ flange, 563

RTU, 972, 2356

RTU Modbus frames, 1106

Run tee fitting, 582

Runaway process, 2407

Rung, PLC programming, 800

Rupture disk, 2643

Rupture disk, graphite, 2644

Rupture disk, metal, 2643

RVDT, 2113, 2134

SAE straight thread pipe fittings, 570
Safety barrier circuit, intrinsic, 2593
Safety Instrumented Function (SIF), 2655
Safety Instrumented System (SIS), 2655
Safety PLC, 2666
Safety relief valve, 2651
Safety valve, 2109, 2644
Safety, instrument transformer, 1951
Sage "Prime" model thermal mass flowmeter, 1719
Salt, 288
SAMA diagram, 536
Sample rate, 991
Sample time, 991
Sample-and-hold PID algorithm, 2434
Sand bath temperature calibrator, 1289
Saturated steam, 165
SCADA, 972, 1890, 1908, 2356
SCFM, 1696
Scheduled communication, Fieldbus, 1151, 1155
Schrödinger, Erwin, 252
Schweitzer "Best-Choice Ground Directional Element" protective relay logic, 2012
Schweitzer Engineering Laboratories model SEL-387L differential current relay, 1999
Schweitzer Engineering Laboratories model SEL-551 overcurrent relay, 1978, 1990
Scintillation optical flowmeter, 1689
SCOF, 634
Scram, 2686
Screwless terminal block, 592
Seal pot, used on wet leg of level measurement system, 1427
Seal-in contact, 735, 790, 1986
Seat load, control valve, 2126, 2133
Seating profile, electronic valve positioner diagnostic, 2145
Seattle City Light, 1897
Secant line, 34
Second derivative, calculus, 12
Second Law of Motion, 74, 97, 192
Second-order lag, 2424
Secondary Coated Optical Fiber, 634
Secondary emission, electrons, 1831
Security, 2596, 2597
Security, versus dependability, 2656

Segmental orifice plate, 1605
Segmental wedge, 1623
Segmented ball valve, 2081
Segway personal transport, 2408
Selective permeability, 1756
Selectivity, protective relay, 1997
Self-balancing bridge, 329
Self-balancing system, 922, 1334
Self-diagnostics, 2639
Self-induction, 362
Self-powered transmitter, 885
Self-regulating process, 2397
Sense wires, 4-wire RTD circuit, 1513
Sensible heat, 144
Sensing line, 575, 1345
Sensing tube, 575, 1345
Sensitivity, radio receiver, 1222
Sequenced control valves, 2147
Sequencer Compare instruction, Allen-Bradley PLC programming, 843
Sequencer Load instruction, Allen-Bradley PLC programming, 843
Sequencer Output instruction, Allen-Bradley PLC programming, 839
Serial digital data, 987
Serial versus parallel digital data, 1018
Series PID equation, 2305, 2336
Servo motor control, 2496
Set coil, PLC programming, 814
Setpoint, 497, 2257
Setpoint tracking, 539, 2346, 2373
Setpoint, remote, 2371, 2488, 2490
Shaded-pole AC induction motor, 722
Shading coil, 722
Sheave, 117
Shelf life, pH electrode, 1762
Shelf life, span gas, 1847
Shell, electron, 242
Shell-and-tube heat exchanger, 132
Shielded cables, 616
Shielding, magnetic, 619
Shift register, FIFO, 1263
Shift register, used to implement dead time, 2538
Shinskey, Francis Greg, 3
Short, electrical fault, 324
Shunt resistor, 897

Shutter, radioactive source lock-out, 1491

Siemens 505 PLC, 763

Siemens APACS control system, 2667

Siemens model 353 digital controller, 2344, 2347

Siemens Procidia controller GUI software, 2348

Siemens Quadlog safety PLC, 2666

Siemens S7-300 PLC, 767, 2715

SIF, 2655

Sightfeed bubbler, 1416

Sightglass, 1396

Signal, elevated, 1006

Signal, floating, 1005

Signal, ground-referenced, 1004

Signature, control valve diagnostics, 2144

Signed integer, 975

Silicon resonator pressure sensor, 1331

Silicon strain gauge element, 1318

Simple apparatus (intrinsically safe), 2594

Simple machine, 99

Simplex, 1035

Simultaneous systems of linear equations, 268

Single-ended analog input, 1008

Single-ended signaling, 1045

Single-helical gear, 107

Single-mode optical fiber, 635

Single-phase AC induction motor, 719

Single-phasing a three-phase AC induction motor, 723

Single-precision floating-point number, 978, 1121

Sinking current, 304, 774, 777

Sinking output switch, 660

SIP, 571, 1367

SIS, 2655

Skydrol hydraulic fluid, 694

Slack diaphragm, 1311

Slack-tube manometer, 1295

Slide rule, 1616, 2168

Slip speed, 719, 2233, 2234

Slip-stick cycle, 2437

Slope, pH instrument, 1772

Slurry, 2212

SMA style connector, 637

Smart instrument, 1265

Smart transmitter, 901

SMTP, 1085

Snell's Law, 632

Snubber, pressure, 1364

Society of Automotive Engineers (SAE), 570

Socrates, 3171

Sodium error, pH measurement, 1758

Soft alarm, 2374

Soft constraint, 2570

Soft override, 2569

Software compensation, thermocouple, 1540

Sol, 229

Solenoid, 341

Solenoid valve, 342, 698

Solid, 170

Solute, 229

Solution, 229

Solvent, 229

Solver, logic, 2666

Solving for variables in an equation, 2747

Sonic flow, 2208

Sonic level instrument, 1460

Source versus load, 306

Sourcing current, 304, 774, 777

Sourcing output switch, 660

Soxhlet, degrees, 181

Space, 1022, 1024

Span, 497

Span adjustment, 1250

Span gas, 1299, 1833, 1843, 1846

Span shift, 1252

SPDT switch contact, 746

SPEC 200 analog electronic control system, 2340, 2364, 2375, 2627

Species, chemical, 1776

Species, chemical composition, 284, 1758, 1780, 1808, 1810

Specific gravity, 180, 192, 1376, 1410

Specific heat, 1721

Specific volume, 180

Spectacle blind, pipe, 565

Spectroscope, 252

Spectroscopic notation, 244

Spectroscopy, absorption, 256

Spectroscopy, emission, 253

Speed of light, 632

Speed of sound measurement, transit-time ultrasonic flowmeter, 1683

Speed of sound through a fluid, 1681

Spiral bourdon tube, 1311

Split-range control valves, 2147

SPLT Fieldbus function block, 2159

Spool valve illustration, 2-way, 693

Spool valve illustration, 4-way, 707–709

Spool, pipe, 562

Spread-spectrum radio, 1232, 1234, 2722

Spring adjuster, valve actuator, 2123

Spring-loaded pressure regulator, 2107

Sprocket, 120

SPST switch contacts, 745, 1953

Spur, 1057, 1137

Spur gear, 106, 2112

Spurious trip, 2656

SQC instruction, Allen-Bradley PLC programming, 843

SQL instruction, Allen-Bradley PLC programming, 843

SQO instruction, Allen-Bradley PLC programming, 839

Square root characterizer, 1593, 2050, 3145, 3147, 3150

Square root extractor, 1594

Square root scale, 1597, 2047

Square-D Motor Logic Plus control, 740

Square-edged concentric orifice plate, 1602

Squirrel-cage AC induction motor, 718

SSH, 1085, 2715

ST style connector, 625, 634, 637

Stagnation pressure, 1578

Standard cell, 1283, 1549

Standard conditions, chemistry, 278

Standard cubic feet per minute, 1696

Standard enthalpy of formation, 278

Standard heat of formation, 278

Standard temperature and pressure (STP), 278

Starter, motor, 727

Static contact, PLC programming, 816

Static friction, 2192

Static pressure effect, pressure transmitter, 1341

Stationary phase, 1776

Stator, 715, 2233

Stator winding, 437

Status propagation, Fieldbus, 1174

Steady-state gain, 2413

Steam cut, valve trim, 2213

Steam jacket, 505

Steam table, 148, 163, 165

Steam tracing, 1383

Steam trap, 1383

Steam, industrial uses of, 164

Steam-hydrocarbon reforming process, 2502

Steam-In-Place, 571, 1367

Stefan-Boltzmann equation, 2061

Stefan-Boltzmann Law, 127, 1553

Stem connector, control valve, 2123

Stem packing lubricator, 2091, 2195

Stem valve, 932

Stem-guided globe valve, 2069

Step 7 PLC software, Siemens, 2702

Step-index, 636

Stiction, 2193

Stilling well, 1494, 1651

Stinger, pressure test accessory, 1363

Stoichiometry, 265

Stokes, 200

STP, Standard Temperature and Pressure, 278

Strain gauge, 331, 1317

Strapping table, 2058

Stress, 171

Strip chart recorder, 513

Strobe light, 380, 992

Stroboscope, 380, 992

Strong acid, 286

Strong base, 287

Strong nuclear force, 230

Strouhal number, 1663

Strouhal, Vincenc, 1663

Structural chemical formula, 233

Stub, 1057, 1137

Stuxnet computer virus, 2692

Sublimation, phase change, 154, 2699

Subnet mask, 1072

Subshell, electron, 242

Substitution, algebraic, 2760

Subtractive transformer windings, 1946

Superconductivity, 313

Superfluidity, 313

Superheat, 166, 2507

Superheated steam, 166, 2507

Superheated vapor, 149

Superheater, 2507

Supernatant, 229

Supervisory Control And Data Acquisition, 972, 1890, 1908, 2356

Surge impedance, 474

Susceptance, 360

Suspension, 229

Sutro weir, 1647

Swagelok instrument tube fittings, 577

Swamping, 177, 337, 1051, 1512, 1638, 2427, 3150

Switch, 649

Switch, dry contact, 659

Switch, process, 516, 650

Switch, wet contact, 659

Switching hub, Ethernet, 1064

Symbol, 798

Symbolic addressing, 798

Symmetrical components, three-phase power system, 456, 2010

Synchronization lamp, 1901

Synchronization, generator, 1901

Synchronous data transfer, 1021, 1027

Synchronous motor, 2233

Synchronous speed, 719, 2233

Synchrophasor, 383

Synergy automobile hybrid electric drivetrain (Toyota), 110

Système International, 72, 255

Systems of linear equations, 268

Table, RTD values, 1511

Table, strapping, 2058

Table, thermocouple values, 1530

Tag name, 798, 847

Tag name, naming conventions for, 848

Tangent line, 34

Tank circuit, 421, 482

Tank expert system, 1431

Tap hole finish, orifice plate, 1611

Tape-and-float level measurement, 1405

Tapered pipe threads, 567

Tare weight, 1482

Target flow element, 1620

Taylor analog electronic controller, 2338

TCP, 1067, 1083

TDMA, 1038

TDR, 623

Tee tube fitting, branch, 582

Tee tube fitting, run, 582

Tee tube fitting, union, 582

Teflon (PTFE) valve stem packing, 2092

Telegraph, 1019

Telemetry system, 972, 2360

TELNET, 1085, 2715

Temperature calibrator, dry-block, 1290

Temperature calibrator, oil bath, 1289

Temperature calibrator, sand bath, 1289

Temperature coefficient of resistance, RTD wire, 1509, 1511

Temperature programming, chromatograph, 1792

Temperature switch, 678

Temperature, defined for a gas, 1501

Terminal block, 590

Terminal strip, 590

Terminal, fork versus ring, 596

Termination resistor, 479, 623

Test diode, 895

Test switch, power instrumentation, 1952

Test Uncertainty Ratio, 1281

Test, computer programming, 2378

Texas Instruments 505 PLC, 763

Thayer ball mill controller, 45

Thermal conductivity detector, GC, 1782

Thermal energy, 1501

Thermal image, circuit breaker, 1565

Thermal image, electric motor, 1563

Thermal image, motor starter, 1564

Thermal imager, 1562

Thermal mass flowmeter, 1718

Thermal overload heater, 726

Thermal siphon, 135, 165

Thermistor, 1509

Thermocouple, 1285, 1522

Thermocouple burnout detection, 1552

Thermocouple compensation, hardware, 1539

Thermocouple compensation, software, 1540

Thermocouple table, 1530

Thermosiphon, 135, 165

Thermowell, 1566

Thin-layer chromatography, 1777

Third Law of Motion, 74

Thought experiment, 506, 945, 1412, 1437, 1438, 1440, 1441, 1455, 1468, 1523, 1626, 1727, 1811, 1815, 2192, 2194, 2273, 2315, 2386, 2401, 2460, 2528, 2558, 2562, 2572, 2664, 2741, 2773, 2774, 2776

Thread pitch, 568

Three-element boiler feedwater control, 2517

Three-valve manifold, 1355

Three-way globe valve, 2076, 2149

Three-wire RTD circuit, 1514, 1515

Throttling control valve, 2067

Tieback variable, Allen-Bradley Logix5000 PLC programming, 2556

Tilt switch, mercury, 665, 671

Time constant, 2420, 2532

Time constant, differentiator circuit, 2331

Time constant, integrator circuit, 2332

Time Distribution (TD) message, Fieldbus, 1155

Time Division Multiple Access (TDMA), 1038

Time domain, 1868

Time domain reflectometry, 1458

Time overcurrent protection, 1983

Time overcurrent relay, 1973

Time-Domain Reflectometer, 623

Timer circuit, watchdog, 2639

Timer, PLC programming, 823

Tofino industrial network firewall, 2721

Token Ring, 1037

Token-passing channel arbitration, 1037

Top Hat DIN rail, 599

Toroidal conductivity cell, 1751

Torque, 82

Torque tube, 1447

Torr, 188

Torricelli, Evangelista, 218

Toshiba magnetic flowmeter, 1675

Total heat of steam, 148

Total internal reflection, 632

Totalizer, 42

Toyota Prius hybrid electric automobile, 110

Traceability, NIST, 1280

Transducer, 497

Transducer block, Fieldbus, 1172

Transfer function, 435

Transfer function, pole, 409, 411, 417

Transfer function, zero, 409, 411, 417, 422

Transfer functions, 409

Transform function, 1212

Transformer, 364

Transformer, AC, 361

Transformer, additive windings, 1946

Transformer, subtractive windings, 1946

Transient fault, power line, 1929

Transit-time ultrasonic flowmeter, 1682

Transition zone, radar, 1476

Transition-sensing contact, PLC programming, 816

Transmation model 1040 loop calibrator, 907

Transmission Control Protocol, 1067, 1083

Transmission line, 474, 484, 622

Transmission power line, 1886

Transmitter, 497

Transport delay, 514, 1792, 2431, 2528

Trap, 1386

Trap, steam, 1383

Trend recorder, 512, 2444

Triaxial vibration probe array, 1874

Triconex TMR safety control system, 2668

Trim, 2067

Trim, in a feedforward control system, 2516

Trip curve, thermal overload, 729

Trip solenoid, 2669

Triple Modular Redundant (TMR) control system, 2668

Triple point, water, 152, 1287

Triple-cascaded loops, 2496

True state, PLC programming, 800

Tube bulkhead fittings, 581

Tube cap, 583

Tube connector, 581

Tube fitting gauge, 577

Tube plug, 583

Tube union, 581, 584

Tuning fork level switch, 672

TUR, 1281

Turbine flow element, 1654

Turbulent flow, 204

Turck Fieldbus coupling device, 1141

Turndown, 1277

Turndown ratio, flume flowmeter, 1653

Turndown ratio, optical flowmeter, 1690

Turndown ratio, pressure-based flowmeter, 1600, 1634

Turndown ratio, turbine flowmeter, 1657

Turndown ratio, weir flowmeter, 1653

Turns ratio, transformer, 96

Twaddell, degrees, 180

Twin-turbine mass flowmeter, 1702

Twisted, shielded pair cables, 621

Two's complement, 975

Two-wire RTD circuit, 1512

U-tube manometer, 1306

UDP, 1083

UEL, 2588

Ullage, 481, 1402, 1460, 1478

Ultimate gain, 2446, 2447

Ultimate period, 2446

Ultimate sensitivity, 2446

Ultrasonic flowmeter, 1679

Ultrasonic level instrument, 1460

Ultrasonic level switch, 674

Unbalanced signaling, 1045

Undependability, 2597, 2671

Under-damping, 428

Underreaching, protective relay, 2015

Uninterruptible power supply, 1975

Union tee fitting, 582

Union, tube, 581, 584

Unit conversions, 61

Unit reaction rate, 2452

Unity fraction, 61, 271

Universal Serial Bus, 1045, 1050

Universal solvent, 229

Unreliability, 2609

Unrestrained differential relay, 1994

Unscheduled communication, Fieldbus, 1151, 1155

Unsecurity, 2597

Unshielded, twisted pair (UTP) cable, 1060

Unsigned integer, 975

Up-down calibration test, 1274

Upper explosive limit, 2588

Upper range value, 497, 1266, 1295

UPS, 1975

URV, 497, 1266, 1295

USB, 1045, 1050

Useful life of a component or system, 2620

User Datagram Protocol, 1083

UTP cable, 1060

V-cone flow element, 1622

V-notch weir, 1645

V/F ratio, 2236

V/Hz ratio, 2236

V1 missile development, 2604

Vacuum circuit breaker, 1917

Vacuum tube voltmeter, 337

Vacuum, produced by a venturi, 216

Valence electrons, 246, 248, 298

Valve actuator, 2067

Valve body, 2067

Valve characterization, equal percentage, 2186

Valve characterization, linear, 2186

Valve characterization, quick-opening, 2187

Valve manifold, 1355

Valve packing, 2086

Valve seal overtravel interlock switch, 2676

Valve signature, electronic positioner diagnostics, 2144, 2196

Valve sizing equation, liquid, 2167

Valve stem packing lubricator, 2091

Valve trim, 2067

Valve, solenoid, 342

Valves, 2067

Varec pressure relief valve, 2645

Variable displacement pump, 692

Variable-area flowmeter, 1642

Variable-frequency drive, 84, 601, 883, 1036, 1102, 1133, 1231, 1552, 2224, 2234, 2415, 2567, 2630, 2785

Variable-speed drive, 2228

VCR, Fieldbus, 1157

Vector cross-product, 93, 304, 1705

Vector dot-product, 94

Velocity algorithm, defined, 2383

Velocity factor, multipath ultrasonic flowmeter, 1685

Velocity factor, transmission line, 481, 632

Velocity of approach factor, 1638

Vena contracta, 1601, 2201, 3144

Vent hole, orifice plate, 1606

Vent valve fitting, 1361

Venturi tube, 219, 1580

VFD, 84, 601, 883, 1036, 1102, 1133, 1231, 1552, 2224, 2234, 2415, 2567, 2630, 2785

Vibrating fork level switch, 672

Vibration loop, tubing, 585, 2099

Virtual Communication Relationship (VCR), Fieldbus, 1157

Virtual Private Network, 2722

Viscosity, 199

Viscosity, absolute, 199

Viscosity, kinematic, 199

Viscosity, temperature dependence, 200

Viscous flow, 201

VisSim Realtime computer software, 2483

Volt, 293, 295

Volta, Alessandro, 293

Voltage, 292

Voltage follower circuit, 936

Voltage signal, elevated, 1006

Voltage signal, floating, 1005

Voltage signal, ground-referenced, 1004

Voltage transformer (VT), 1887, 1932

Voltage-to-frequency ratio, 2236

Volume booster, 2128, 2158

Volume, defined, 57

von Kármán, Theodore, 1663

Vortex flowmeter, 1664

Vortex street, 1663

Voting system, 2560

VPN, 2722

VSD, 2228

VT, voltage transformer, 1887, 1932

VTVM, 337

Wade Associates, Inc., 2463

Wallace & Tiernan, 1295

Wally box, 1295

Wastewater disinfection, 503, 2260

Watchdog timer circuit, 2639

Waveform, periodic, 1865

Wavenumber, 1801

Weak acid, 286

Weak base, 287

Wear-out period, 2620

Wedge, 104

Weighfeeder, 1722

Weight density, 57

Weight-based level instrument, 1482

Weir, 1645, 2051

Well manometer, 1306

Weschler panel-mounted bargraph indicator, 510

Westinghouse model CO-11 overcurrent relay, 1962

Weston cell, 1283

Wet calibration, 1445

Wet contact, 659

Wet leg, 1424

Whip antenna, 489, 1204

Whitelisting, application, 2714

Wild flow, 2501, 2561

Wild variable, 2259, 2501

Winch, 83

Wind-up, controller, 2282, 2373, 2399

Wire duct, 608

Wire loom, 608

Wire pulling, 602

Wire, jumper, 2797

Wireless transmitter, 2434

WirelessHART, 1038, 1229, 2722

WLAN, 1040

Wonderware HMI software, 850

Work, 75, 293

Worm gear, 113, 2110, 2112

X-windows, 1071

XD_Scale, FOUNDATION Fieldbus AI block parameter, 1192

XIC (examine if closed), 810

XIO (examine if open), 810

Yagi antenna, 490, 1204

Yokogawa CENTUM distributed control system (DCS), 2363, 2364

Yokogawa DPharp pressure transmitter, 1331

Yokogawa model DYF vortex flowmeter, 1178

Yokogawa model EJA110 differential pressure transmitter, 1331, 1339

Zero, 497

Zero adjustment, 1250

Zero energy state, 1361

Zero gas, 1833

Zero sequence, 456, 463, 733, 2010

Zero shift, 1251

Zero, transfer function, 409, 411, 417, 422

Ziegler, J.G., 2445

Ziegler-Nichols "closed-loop" (Ultimate) PID
        tuning example, 2466

Ziegler-Nichols "open-loop" (Reaction rate) PID
        tuning example, 2464

Zirconium oxide, 2063

Zone of protection, power system, 1997, 2014

Zone, hazardous area, 2585

CPSIA information can be obtained
at www.ICGtesting.com
Printed in the USA
BVHW011645170220
572580BV00002B/30